The WORLD of TREES

The WORLD of TREES

HUGH JOHNSON

Consultant editor John Grimshaw | Preface by Thomas Pakenham

UNIVERSITY OF CALIFORNIA PRESS *Berkeley Los Angeles*

Editorial for the first edition

Sir George Taylor

Kenneth A Beckett

Roy Lancaster

Alan F Mitchell

JRP van Hoey Smith

Oscar Traczewitz

University of California Press, one of the most distinguished university presses in the United States, enriches lives around the world by advancing scholarship in the humanities, social sciences, and natural sciences. Its activities are supported by the UC Press Foundation and by philanthropic contributions from individuals and institutions. For more information, visit www.ucpress.edu.

University of California Press
Berkeley and Los Angeles, California

The World of Trees

First published in Great Britain in 1973 by Mitchell Beazley, an imprint of Octopus Publishing Group Limited.
www.octopusbooks.co.uk

An Hachette Livre UK Company
www.hachettelivre.co.uk

Revised editions 1984, 1993

This edition copyright © Octopus Publishing Group 2010
Text copyright © Hugh Johnson 2010

Library of Congress Control Number: 2010924311
ISBN: 978-0-520-24756-7 (cloth : alk. paper)

19 18 17 16 15 14 13 12 11 10
10 9 8 7 6 5 4 3 2 1

Commissioning Editor: Helen Griffin
Senior Art Editor: Juliette Norsworthy
Designers: Carole Ash, Jaz Bahra, Mark Kan, Yasia WIlliams-Leedham
Illustrator: Jaz Bahra
Project Editor: Jo Wilson
Picture Manager: Giulia Hetherington
Copy Editor: Joanna Chisholm
Indexer: Mike Hunt
Production: Peter Hunt
Design Coordinator: Gary Almond
Cover Design: Lia Tjandra
Printed and bound in China

Contents

Foreword

It's nearly 40 years since I dug a hole in the sticky clay of my lawn and planted my first exotic tree. It was a manna ash from Greece or Asia Minor. And the choice of that tree was inspired by a passage in the first edition of Hugh Johnson's wonderful book. Hugh described the manna ash flaunting its blossoms "like the velvet-framed bosoms of Nell Gwyn and her contemporaries on ancestral dining-room walls". I found the idea irresistible. If only the tree had found the Irish climate irresistible, too. It grew slowly and primly (it resented our cool summers) but is now a healthy 20 foot tall.

Since then I have successfully peopled my garden with trees first encountered in Hugh's book. I have planted giant sequoias and pygmy pines – enough scholar (or pagoda) trees to write a hundred theses, and enough handkerchief trees to blow a thousand noses. Hugh is a true gourmet who savours an amusing tree as he savours an amusing wine. But his knowledge of trees is encyclopaedic. And behind his sparky prose is a passionate plea. Get your spade and plant a tree, he tells us, to restore peace and elegance to our planet.

For this new edition Hugh has pruned out a few of his purple passages (but not the bit about Nell Gwyn, I'm glad to say). He has added the fruits of many more years' experience, many new species, and many crisp new illustrations. And he has brought into focus many a small treasure hunt that has changed the botanical world. Who would have predicted that a "fossil tree" – the dawn redwood (*Metasequoia glyptostroboides*) only discovered in China in 1941 – should now be a favourite tree in most parts of temperate Europe and America?

The book has also become more personal. You could call it the story of a love affair. We read about the trees that he has loved and lost (the southern beech called the rauli that died in his garden during a cold winter) and others that have lavished their affection on him, like Dombey's beech or the Chinese lace-bark pine.

Who speaks for trees, someone once asked me. Well, now I know the answer. It's Hugh Johnson.

Thomas Pakenham, Author of **Meetings with Remarkable Trees**
Co. Meath Ireland, summer 2010

Introduction

A traveller should be a botanist, for in all views plants form the chief embellishment.
Charles Darwin

Darwin could have been more specific. Not of all landscapes, but of most, he could have said: "trees form the chief embellishment". Of all plants they are the most prominent and the most permanent, the ones that set the scene and dictate the atmosphere. Trees define the character of a landscape, proclaim its climate, divulge the properties of its soil – even affirm the preoccupations of its people.

I wrote the first edition of this book 37 years ago, in the excitement of discovering for the first time the beauty and diversity there is in trees, and dismay at seeing the principal and biggest trees around me suddenly dying. In 1973 Dutch elm disease had just arrived. Within five years it had destroyed the landscape where I had come to live, in the east of England.

I remember a before-and-after sequence of two photographs in *Time* magazine. Before: main street in some New England town; all harmony; comely cream clapboard dappled with light from the crowns of an avenue of tremendous elms. After: the elms gone; comeliness has become nakedness. What was all proportion and peace is desolate.

It was partly in impotent protest that I started this book. What, I asked, can we do and what should we plant to restore the serenity we had lost? Over time the question has answered itself: in East Anglia oaks, ashes, and willows now fill many of the spaces and delineate the devastated skyline. But as soon as I started to investigate, to visit botanical gardens and get lost in forests, above all simply to look around me with an inquiring mind, I discovered a variety of beauty and meaning I had never suspected.

This is the aspect of trees that I wrote my book about. I didn't write about the essential part trees play as creators of the air we breathe, or as our most versatile and universal raw material. Is anyone in any doubt today about the necessity of trees to keep our planet in being?

My aim was to bring trees into focus for everybody like myself, who was aware of trees, loved trees, but in a way so vague that it now seems to me quite shameful. It was the personal account of a writer who found in trees a new point of contact with creation, a source of wonder and satisfaction which has the inestimable advantage of growing almost everywhere.

Nearly 40 years later I am astounded by my juvenile pretension; immensely grateful, though, that by plunging in I found myself a hobby, a passion, an obsession for life. It has led me to travel, to read, to talk with people similarly afflicted in a lifetime voyage of discovery. I have become a gardener, a tree collector, and a forester. I have begun to read trees with a degree of understanding, to develop preferences and prejudices, to see how beautiful the commonplace can be and how banal some of the novelties created by nurserymen.

Forty years on trees have become fashionable. A society that was grudgingly grateful for their shade now ventilates about their preservation. A tree can be a celebrity. Doubts about climate change throw into relief questions that have no answers – always a healthy situation. Tree science and technology have moved briskly. What was less predictable was that we should discover hundreds more kinds of trees. The introduction of new species accelerated through the three centuries of exploration that opened the world to the acquisitive West. The early 20th century marked a climax of new discoveries. Since the 1970s, though, we have been in a new era of exploration. Central America in particular is being scoured for plants as never before, but above all it is the opening up of China and eastern Asia, much of it for the first time, that is enriching our knowledge and our collections.

The standard work of scholarly reference for the Western tree world for the 20th century was W J Bean's *Trees and Shrubs Hardy in the British Isles*, eventually, in 1988, expanded to five volumes. In 2004 The International Dendrology Society (Dendrology is the study of trees) commissioned two botanists, John Grimshaw and Ross Bayton, to review all the new trees that had been found in the wild, and many that have been created, since the 1970s. Their *New Trees*, published in 2009, recorded and described more than 800 species. In some genera numbers have doubled. If not all these trees are yet available to be grown, or may be difficult or tender or just dull or ugly, or even potentially invasive weeds, they open endless vistas of possibilities. They multiply the genetic potential of many trees many times. Hybridization in trees is a slow business compared with most plants: in some genera it has hardly begun. The future has never been more exciting.

Not, I hasten to say, because difference for its own sake is necessarily a good thing. Indeed in most situations where tree planting is on the agenda the best plan is more of the same. The beauty of great forests, of most woods, and all avenues lies in homogeneity. Yet no one can deny that a theme with countless variations is a never-ending source of pleasure. It is true of wine, my other passion: it is even more true of trees. The theme is simply a plant with a trunk, branches, twigs, leaves. Yet think of a giant elm, and think of a cherry, stooping with soft flowers – and think of a resiny old fir crusted with lichen. Can anything make a more fascinating collection? For once possession is not the point. Trees are wonderfully public property. Wherever you go you are enjoying somebody's trees, and not a penny changes hands.

After 40 years of revelling in all this beauty and variety, enjoying the changing seasons, leaf-fall, bud and flower, planting and felling, trying to put sublimity into words, subconsciously collecting, I couldn't resist the urge to go back and revise my own words in the light of experience. More than half of this book is new, or at least completely rewritten; I hope the rest has stood the test of time.

What I hope this book does is to make vivid through words and pictures the essential differences between the great groups of trees, to tell their story, and then go on to enjoy the pleasures of their subtly, elaborately, almost endlessly varying designs – the species and varieties of familiar families from all over the temperate world. What a tree (or indeed a plant) family consists of, what its traits and likenesses are, what its needs or preferences, and how they relate to other tree families form the basic structure of the book.

I have tried not to be pedantic about what is or isn't a tree. There is a classic definition. It calls for a woody plant (capable of) growing 20 feet high and tending to a single stem (though it may have more). What do you do, though, if the thing is single-stemmed as can be, palpably tree-like and permanent, but has never grown more than 10 feet? Or what do you do if it is 30 feet high but hopelessly and incurably bushy?

My answer has tended to be inclusive rather than exclusive. It is one of the wonders of the subject that the first cousin of a 200-foot mammoth can be a two-foot pygmy. And the relationship adds enormously to the pleasure of growing the pygmy – which is all most people have room for.

A hundred years ago Lionel de Rothschild, one of the world's greatest growers of rhododendrons, was lecturing a city gardening club. "Gentlemen", he began, "no garden, however small, should be without its two acres of rough woodland." For most of the human race those must be two acres of the mind. But what pleasure there is, and how much to be learned about our place in God's universe, in going about them observantly, equipped with even a little knowledge.

For the first edition of this book I had the indispensable help of many generous people. Sir George Taylor, then Director of Kew, I must thank first for giving me the confidence to get started. Alan Mitchell of the Forestry Commission did not spare himself nor his

incredible store of information. Ken Beckett, former Technical Editor of the *Gardener's Chronicle*, Roy Lancaster, the curator of Hillier's arboretum in Hampshire, and Oscar Traczewitz, head forester of the International Paper Company, answered innumerable questions and put me right countless times. Dame Sylvia Crowe, past President of the Institute of Landscape Architects, was good enough to read my chapters on the use of trees, the garden, and landscape and make invaluable suggestions.

For this edition, inspired in part by the publication of *New Trees*, I owe my greatest debt to John Grimshaw, not only for making his own knowledge available to me but also for gently nudging mine down the paths of rigour and accuracy. He is a generous scholar, a formidable botanist, and a kind and entertaining friend. His first love is bulbs, but he showed no reluctance to elevate his gaze to trees. Andrew Lawson, whom I am tempted to call our national garden photographer, so definitive are many of his images, has done beautiful work for this book. If a large number of the pictures are my own it is because I was there and I know what I want to show you.

The publishing business of 2010 is a very different environment from that of 30 or 40 years ago. Gone are such primitive things as galley proofs, transparencies, colour separations, flowcharts, indeed whole art departments. Wonderful new software allows (indeed obliges) the author to work online. My immensely skilled, patient, and understanding editor Joanna Chisholm is 400 miles away in Edinburgh.

Mitchell Beazley, my publishers since 1970, have undergone major changes and moved house (in the middle of doing this book). The more credit, then, to my editorial friends, led by David Lamb and Helen Griffin, who have seen it safely to press, to Juliette Norsworthy who has overseen the design, and to Giulia Hetherington who has supervised the pictures. To them, with the customary claim that all the mistakes are mine, I offer an author's grateful thanks.

Hugh Johnson

How a Tree Grows

WHAT DISTINGUISHES A TREE from all other plants is the wooden structure it raises above the ground. By annual additions it builds a tall scaffold on which to hang its leaves, its flowers, and its fruit. Each new year's growth is no more than a crop of twigs. Yet those twigs contain in essence the identity of the tree; in time they become branches; eventually perhaps huge limbs. In the winter tower of tracery you can still detect the way the buds formed along the first frail shoots.

Each year's shoot springs from a bud, and ends by forming new buds. The buds are the resting stage common to most trees of the temperate world, where day length and temperature sharply differentiate the seasons. Each contains in miniature a whole new shoot – either of twig and leaves or of flowers. Even before the shoot of the current year has reached its final size the next year's buds are brewing. By early or mid-summer they are complete – even though the embryo growth they hold will not appear until the following early or mid-spring. So it was last year's spring that determined this year's growth, if not its extent. The tree's vigour and success in a given year are largely down to the moisture it receives.

One bud on each shoot is different from the rest. It is the bud farthest from the tree's roots – the last on the limb: the leader. The nearer this bud is to a vertical position over the centre of the tree the more marked its dominance will be. The force of gravity, you might say, is its authority. It is able to assert its dominance through the agency of hormones. It forms the hormone auxin in its growing tip; gravity distributes it to the buds lying behind. Auxin has the effect of moderating vigour – leaving more of the available sap for the leader to use in increasing its lead, and in forming the biggest of the buds, the one that will be dominant next year.

This is the plan for a tree with the simplest construction: for example a spruce. In a broadleaf tree the same chemistry applies – not to a single topmost shoot but to an increasing number of competing leaders as the tree's crown widens out.

Meanwhile what of the other buds, the ones that are being dominated and deprived? The shoots they produce are shorter, with a different function. They are the leaf and flower bearers. Their role is to fill out the canopy of the tree with leaves to take advantage of all the light, or to

It is not too fanciful (right) to see in this, the first shoot of an oak tree from its acorn, the pattern and even the likeness of the whole tree to come. Its first leaves are photosynthesizing and its hormones directing its growth.

Some 30 years later (left) the oak has settled its basic future structure. In the open, without competition for light, its hormone for dominance is shared between several leaders at random. They account for spreading branches – which may or may not survive external forces.

carry the flowers and fruit. Take away the leader, however, and the auxin-making function is inherited by the next shoot down: one that would otherwise have remained in subjection, perhaps a spur with a few leaves. A new leader is born.

The buds are many, but they are not formed at random. They stick to the characteristic pattern of the species, which is the same as the arrangement of its leaves. Each bud forms in the angle which a leaf makes with the twig.

The tree's ultimate aim is to form a head which will give it the maximum exposure to the light. In the forest all the light is overhead: side branches are shaded out and die – nothing but a narrow crown of up-reared branches can reach the light amid the press of other trees. But in the open field a tree builds up its head in its own characteristic way. There is no mistaking the feathering of a willow, tall fan of the elm, or the oak's zigzag rhythm.

Logically you would expect the final tree to consist of all its annual increments: to be able to count all the annual shoots since it was a seedling. Up to a certain age most conifers (which make a single whorl or tier of branches every year) allow you to do this. Some poplars and alders do the same. But with most broadleaf trees the scent quickly gets cold; there are far too many accretions – and accretions upon accretions – to be counted. One thing that simplifies it is the death rate among

All the life of a tree goes on around the circumference of the trunk and branches between the new wood and the bark. As it expands, last year's wood is added to the load-bearing centre as an annual ring.

twigs. Most of the older sideshoots are disposed of in this way. In 10 years, if a branch produced and kept only two sideshoots a year, the total number would be 19,683. In fact the count on a 10-year-old birch was 238. As the tree moves on beyond them, casting them into shade, they simply drop off. (Birches are notorious twig droppers.)

What complicates it are the external influences forcing or persuading the tree to take up a certain attitude. Of these gravity, light, and wind are the most important three.

WHAT IS A TREE MADE OF?

Between 80 and 90 percent of the bulk of a living tree is water, drawn from the ground by the roots. Of the remaining 10 to 20 percent, no less than 91 percent is derived from the atmosphere by the leaves, which are thus the tree's main feeding organs, collecting all the carbon and oxygen the tree needs. Apart from water, with its all-important supply of hydrogen, the roots provide nitrogen, potassium, calcium, phosphorus, sulphur, iron, magnesium, and other trace elements. The central bulk of the tree is effectively dead, no longer accumulating or releasing elements but storing carbon.

All the blue part of this tree is water. For comparison about 60 percent of the human body is water. The green part is composed of the elements listed below.

THE COMPONENTS THAT MAKE UP A TREE

Hydrogen 6.2%

Nitrogen 1.5%

Potassium 0.9%

Water 87.9% — Carbon 43.5%

Oxygen 44.4%

Other 3.5%

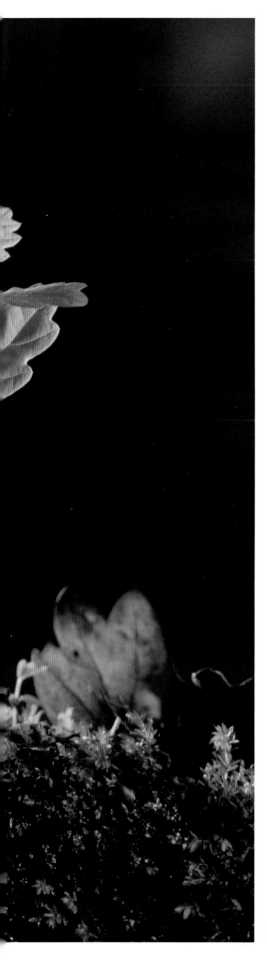

Buds into branches

Everything necessary for the next year's development is contained inside the buds, which follow a pattern characteristic of the tree in question. There are three possible budding and branching patterns: alternate (one bud at a time on alternating sides of the shoot), opposite (two at a time opposite each other), and whorled (three or more in a circle around the twig). The horse chestnut twig (bottom) follows three years of growth of an opposite-budding species. On the right are the eventual results of years of increments in three species with different budding and branching habits.

Alternate one bud at a time   Opposite buds two at a time   Whorled three or more buds at a time

OLD BUDS HELD IN RESERVE

A section of a tree showing how even suppressed buds, which are held in reserve in case the leader is lost, grow outward each year just enough to keep pace with the annual growth ring. The lower left-hand bud on the trunk is totally suppressed. The one above it has formed a short shoot. On the right a mature branch has been lost but a suppressed bud from when it was much younger has kept up with the thickening of the trunk around it. Given light it will probably break to form a new shoot.

Short shoot

Lower
bud

Suppressed
bud

THE RESULT OF A BROKEN MAIN SHOOT

Where a leading shoot with its power of dominating and suppressing the lower branches is broken its function is inherited by the stronger sideshoot just behind it. The process is easiest to watch in conifers like the spruce (right). As soon as a leader breaks and a side branch takes over to grow upright it changes its own budding and branching pattern. Instead of producing only sideshoots in one plane it starts branching all round in whorls (far right). Without this facility trees would soon become flat-topped and quickly be shaded out by their competing neighbours.

THREE-YEAR-OLD TWIG OF A HORSE CHESTNUT TREE

A three-year-old twig of horse chestnut shows three annual growth rings. Its centre is still pith which disappears with age.

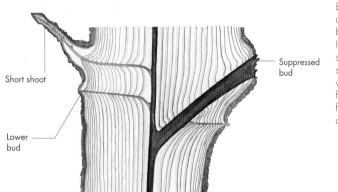

Year 1
growth

Two years ago the big sticky bud was here – scales from the scars are still visible. The leader's hormone deprived the side buds of vigour. They have grown only slightly in two years; in time they will probably be shaded and drop off.

Some of the buds formed two years ago in the angles of the leaves are still suppressed. The horse-shoe scar under the bud, where the leaf stood, still shows rings where its veins were attached.

THE TWIG IS FATHER OF THE TREE

The tree's mature shape epitomizes the way it has branched all its life, subject to its habit of shedding branches and to the effects of the elements. An oak starts with a main stem but by the age of 20 it is hard to spot a leading shoot.

A pine at 12 years is a pyramid of almost regular whorls of branches, while a willow is a prolific producer and shedder of ambitious twigs at acute angles. It is is easy to see the big tree as an extended version of a little one.

Oak – all random angles

Pine – abandons regularity

Willow – successive acute angles

SPRUCE SIDE BUDDING

INSIDE THE BUD

A cross-section of a bud shows the shoot and all its leaves ready formed. They are complete by mid-summer of the year before they appear.

The side buds of the spruce (left) always develop, in contrast to the horse chestnut where the side buds are generally suppressed by the terminal bud. The spruce buds elongate very rapidly.

The flower bud of the same plant, the common lilac. Here the whole flower structure – with scores of little flowers also ready formed – is packed within scales that are modified leaves.

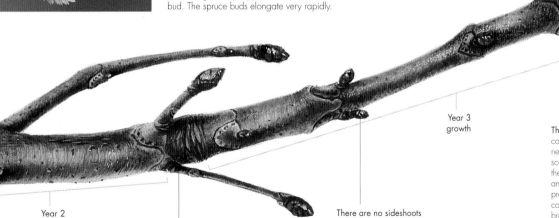

Year 3 growth

Year 2 growth

The sticky bud, which contains the coming year's new shoot, is enclosed in scales. These scales are themselves a form of leaf and are covered with protective resin (more common in conifers than broadleaves). Unless the main shoot is destroyed its side buds will probably get no farther than the little shoots on the left.

There are no sideshoots yet on the one-year-old twig, but buds are ready just above where the old leaves fell. One or both of these buds will shoot and bear leaves this year. The surface of the twig is marked with breathing pores ("lenticels") to supplement the activity of the leaves.

The two-year-old section of the twig lies between two rings of old bud scars. One of the buds has developed into a one-year-old sideshoot.

Last year's growth started at the bud scars. Each succeeding pair of leaves and side buds is at right angles to the one before.

How a Tree Works

IF YOU SCRATCH THROUGH the bark of any living twig you will find a thin green layer. Immediately within this lies the cambium (from the Latin for "exchange"). Only the cambium has the power of making new wood. If you destroy a band of it round the trunk the tree dies.

It is the remarkable power of the cambium layer to make three different kinds of new cells at once. Every year it deposits new wood cells on its inner side, thickening the trunk of the tree. So it has to grow itself, to surround the increased bulk. It also has to keep adding phloem, or inner bark, to keep up with the expanding circumference.

Within this narrow zone of bark, cambium, and the new outer ring of wood, the tree's whole circulation system functions. The sap goes up in the new wood and comes down in the phloem. So much has been known for centuries through the simple experiment of girdling. If you strip a ring of bark off a tree (leaving the cambium intact) the top continues to get its water supply: what suffers is the roots. The water coming down from the leaves with its store of carbohydrates, the

product of photosynthesis, fails to reach them. They stop growing and feeding: eventually the whole tree lacks its ration of minerals and nitrogen from the soil, and dies.

It is still a matter for wonder how frail leaves, transpiring as much as 300 feet above the ground, can have the power to pull a column of water up that distance to supply themselves. It used to be thought that the roots pushed it up – which is, of course, equally mysterious. Now it is known that the roots have some pushing power in spring, before there are leaves to pull, but that leaves are perfectly capable of this feat of suction. What makes it possible is the cellular structure of the wood, which takes the fullest advantage of the principle of surface tension. Water particles are highly mobile, but they are reluctant to part company with each other. It is almost as hard to tear them apart as it is to compress them. So if you pull one end of a long thin stream, the rest has the strength to hang on and follow.

A wood structure with longer cells that made the passage of sap more efficient was one of the evolutionary improvements on the ancient conifers that the newer

broadleaves made. There is a simple experiment you can do to compare the two. If you strip a small vertical piece of bark off a tree and stab the bare wood across the fibres with a penknife, a white mark above and below the cut shows where the air has filled the cells. On a broadleaf tree the mark is much longer than it is on a conifer.

The price for having longer cells, though, is the danger of air bubbles in the sap stream breaking the flow. Ice creates the problem. When water freezes (as sap obviously does in trees in a very cold winter) air dissolved in it forms bubbles. In small cells (as in conifers) they simply disappear when the sap thaws. But what happens in bigger ones? The water columns, once ruptured, might never be restored.

What has only recently been discovered is that most of the sap in broadleaves rises in the ring of wood made that very year. The sap need not rely even on the previous year's wood to give it a passage upward, so efficient is the cambium at making new wood in the spring before the leaves have started drawing the sap up. "This fabulous efficiency" – I quote Zimmermann and Brown's *Trees: Structure and Function* – "takes place at the expense of safety." It is the downfall of the elm: the fungi that kill it block the current year's wood cells, and with them the main stream of sap.

THE TREE'S INTERNAL PLUMBING SYSTEM

The circulation of sap (that is, water containing food) is shown in the diagram, right. The dark blue lines and arrows indicate water (drawn from the ground by the roots) moving upward in the outermost layers of wood to the leaves. Most of this water is lost to the air by evaporation from the leaves. The pale blue lines and arrows are the sap on its downward journey in the inner layer of the bark, outside the cambium. This pattern of circulation was discoverd by removing the bark or "girdling" and finding that it was the roots that were thus deprived of nourishment not the tree above. To the right, a magnified section of an annual ring or wood shows the long vertical cells in which the sap moves up the trunk, crossed at right angles in places by "rays" of shorter cells. Rays are used for conveying reserve supplies across the tree to where they are needed. Where the vertical cells appear to be interrupted they are simply hidden by ray cells. The only breaks in the tubes are the perforated cells ends, which help to support unbroken a column of water sometimes

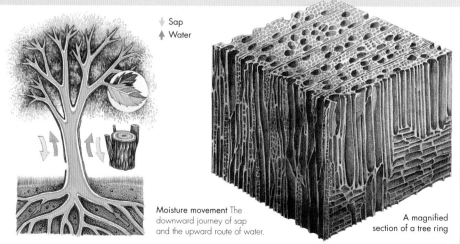

Sap

Water

Moisture movement The downward journey of sap and the upward route of water.

A magnified section of a tree ring

as much as 300 feet high. A comparison between "early wood", which the tree makes in spring, and "late wood", which it makes in summer, would show that early wood tube cells are much bigger,

having much more water to carry to the developing leaves, shoots, and flowers. The difference in cell size gives the characteristic lighter and darker shading of annual rings.

This cross-section of a tree trunk breaks it down into five principal concentric layers: bark, inner bark or phloem, cambium, sapwood, and heartwood.

The heartwood is dead. It is a receptacle for the tree's waste matter, which makes it toxic to most organisms that might feed on it (and also darkens it). Yet if air reaches the dead heart of many species of tree it soon decays and leaves the tree hollow. The only function of heartwood is to give the tree strength and rigidity. How a tree can bend its dead heartwood in order to reach out to the light is something that I have never seen explained.

The bark is a protective layer, sometimes (as on birches) a mere skin but on some trees (redwoods, for example) as much as a foot thick. This outer bark is continually being produced by its own specialized cambium. The position of the cambium determines whether the bark peels or cracks.

Sapwood is all the wood that still functions; it carries sap up the tree, stores nutrients, or transports them from one part of the tree to another. Most sap flows in the new ring of wood made in the current year. If this is blocked, as it is by such fungal infections as Dutch elm disease, the tree may die; the older rings cannot keep it alive on their own.

A branch collar: The tissues of a branch are formed before those of the trunk around it. Each year's new growth is followed by trunk growth overlapping the rings of the branch and forming a collar both integrated with it and surrounding it. The branch collar is vital to the strength and health of the tree and should never be cut off when a branch is removed.

If the cambium layer could be detached from the bark and wood it would be practically invisible, being only one cell thick. Yet the power of the tree to live and grow is contained in this infinitesimal film. The cambium continually produces xylem cells on its inner side (forming the wood) and phloem cells from its outer surface.

The texture of the bark is largely decided by the bark or cork cambium. In many trees it grows in shallow overlapping arcs; the bark eventually cracks at their edges as the trunk swells. In others (such as birches) it forms a thin peeling sheath near the surface which is constantly renewed.

The inner bark or phloem is a spongy layer providing an easy downward passage for the sap. The sap carries sugar from the leaves to feed new wood cells and supply the roots with energy.

The structure and mechanism that connect a branch to the trunk with what can be enormous strength are still not clearly understood. An American researcher, Alex Shigo, posits the solution above.

Growth rings, showing as (lighter) springwood, made when the tree is growing fast, and (darker) summerwood, when growth slows down.

The Leaves

THE PERMANENT WOODEN structure is peculiar to trees – and in a less rigid form to shrubs. The temporary organs that clothe it, the leaves, the flowers, and fruit, are not. Their functions are the same as in every other plant.

The leaf has the task of collecting food. Ninety percent of the solid matter that makes up the trunk, the branches, and the roots themselves is carbohydrates plucked, as it were, out of the sky by the feeding leaves. A leaf becomes self-supporting remarkably quickly. Before it is even half-grown it has started exporting nourishment to the rest of the tree.

Leaves get nourishment out of both the air and the ground. They do the first by photosynthesis: they hold up to the sunlight delicate vessels impregnated with chlorophyll and filled with water. The chlorophyll causes an exchange of hydrogen (from the water) with carbon and oxygen (from the carbon dioxide in the atmosphere). Carbon is the tree's chief food: from it a tree can make the sugars and starches it needs.

The leaves are extremely effective evaporating units, very nearly as efficient as an open water surface like a pond. Their turnover of water is far greater than photosynthesis demands. It is responsible for the whole circulation of sap around the tree. As the leaves lose water to the air they draw it up through the twigs, the branches, and the trunk out of the roots, and ultimately out of the soil, below them. With the water, which can be as much as two or three hundred gallons a day in summer, they draw dissolved minerals from the soil. This is the tree's secondary source of solid supplies – and also of course its whole system of getting the supplies to where they are needed.

Leaves have a short life – perhaps six months on the average. The longest any leaf hangs on a tree is seven or eight years (on certain firs). But most leaves, even of "evergreen" trees, fall and are replaced every year. Evergreen leaves can simply be defined as leaves that are adapted to survive the winter. The adaptation consists of a more miserly evaporation system: a thicker skin, generally a simpler shape (since the longer the margin of any surface in relation to its area the more water vapour will "flow off" it), sometimes a coat of hairs or felty material, and often a coating of wax, giving the leaves a bluish cast, into the bargain.

The only thing that can be said about leaf shape is that in itself it can't be very important: so many variants arise from tree to tree of the same species, and even from branch to branch and twig to twig. If the shape of a leaf were of fundamental evolutionary importance it would surely be more consistent. But look at the maples, to take an obvious example of diversity.

THE BIRTH OF A LEAF

Leaves tightly packed in the bud begin to expand with the arrival of spring. Once expanded, the leaves draw the sap up the trunk of the tree. The force which makes the sap rise to swell the leaves remains a mystery. The swelling of the leaves (of a horse chestnut, right) forces the enfolding bud scales (themselves a modified form of leaf) to hinge back. Once clear of the bud scales the leaves, already fully formed, unfold.

To say they have a family resemblance is sometimes as far as you can go. Are family resemblances determinative? Certainly good looks can run in families.

Most broadleaf trees have produced versions with "cut" leaves – leaves with jagged, deeply lobed, or embayed margins. Many conifers (see pages 76-163) have congested or otherwise eccentric varieties. They seem to make no difference to the tree's functioning.

Leaf colours can vary within species almost as much. A red pigment overlaying the green of the chlorophyll gives a "copper" effect. Sometimes a low level of chlorophyll makes the leaves yellow or golden. But lack of chlorophyll means less photosynthesis and hence less food: yellow-leafed and strongly variegated trees are slower growing.

The last act for many leaves is the most colourful. It is their function, before they die, to convert starch into sugar which the tree can store as reserve supplies. But cold nights prevent them passing on to the tree the sugar they have made, so it builds up in the leaf tissue. In many leaves the result is a red pigment. As the chlorophyll is decomposing at the same time, whatever pigment the leaf contains (in such cases red, but usually yellow) is no longer masked by green: its chrysanthemum colours emerge.

The return of all the nutrients in leaves to the earth. Fallen leaves are almost the entire make-up of a forest floor – rather less important on a lawn. The day the staff downed tools at Versailles for the Revolution, Marie-Antoinette is supposed to have complimented her gardener for the pretty idea of scattering leaves on the grass.

DECIDUOUS OR EVERGREEN?

The leaf is the tree's feeding and breathing organ. Below are the deciduous (i.e. one-season only) leaf of an oak and the evergreen leaves or needles of a pine.

Oak leaf (Deciduous tree) Deciduous leaves fall when cells break down in the area known as the "abscission layer" between the leafstalk and twig. Later a layer of cork forms over the wound producing a leaf scar. Some leaves (e.g. of young beech) fail to form such a layer. They die but cannot break off, so they hang dead on the tree until new leaves push them off in spring.

Pine leaf or needle (Evergreen tree) Evergreen conifer needles (those of a five-needled pine, left) have a harder cuticle and often a thicker layer of wax to slow down their transpiration – essential in winter in northern zones because of the lack of water. Their breathing pores are concentrated in lines, often white with accumulated wax, usually on the underside (or inner side where needles are bunched together).

A young oak leaf is coloured green by the chlorophyll which is its main active component and is tinted yellow by the pigment carotenoid. In autumn the tree stops making chorophyll; it fades first from the sap cells between the veins, leaving the veins green. With chlorophyll gone it is the yellow (carotenoid) and red (anthocyanin) pigments you see, until they fade and decomposed carbohydrates and oxidized tannins colour the leaf bark brown.

WORDS FOR THE SHAPES OF LEAVES

The essential shapes of leaves, perhaps surprisingly, are finite in number and have been quite simply defined by the terms shown below. Four aspects define the shape of a leaf: its overall outline, its margin, its tip, and its base. Follow these closely and all tree species can be pinpointed accurately. In addition, leaves are often compound, or made up of several parts, shown below. Length of petiole, or leafstalk, is another constant measurable factor.

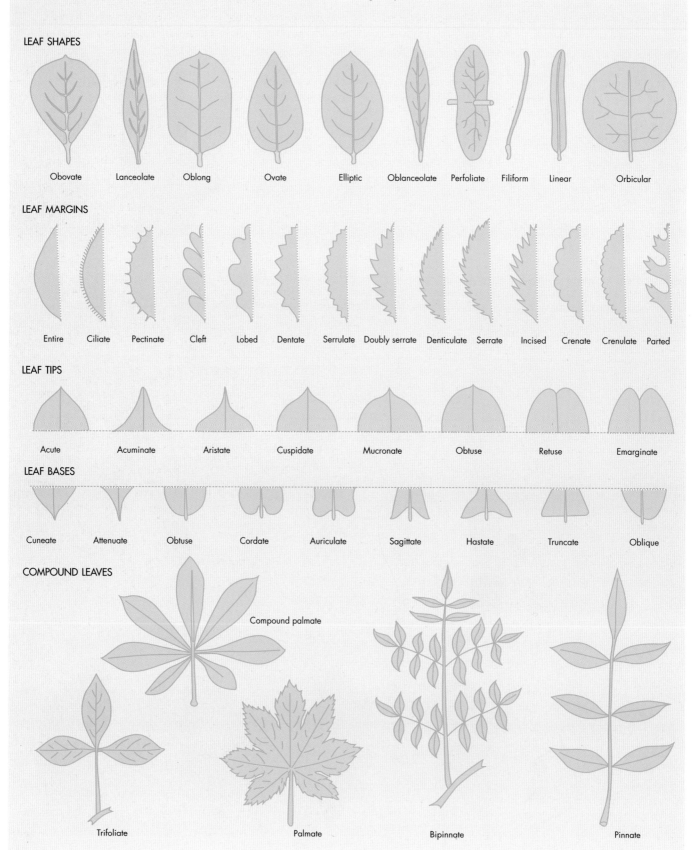

LEAF SHAPES

Obovate Lanceolate Oblong Ovate Elliptic Oblanceolate Perfoliate Filiform Linear Orbicular

LEAF MARGINS

Entire Ciliate Pectinate Cleft Lobed Dentate Serrulate Doubly serrate Denticulate Serrate Incised Crenate Crenulate Parted

LEAF TIPS

Acute Acuminate Aristate Cuspidate Mucronate Obtuse Retuse Emarginate

LEAF BASES

Cuneate Attenuate Obtuse Cordate Auriculate Sagittate Hastate Truncate Oblique

COMPOUND LEAVES

Compound palmate

Trifoliate Palmate Bipinnate Pinnate

How leaves provide food for the tree

The leaf is the tree's feeding and breathing organ. In deciduous trees they fall and are renewed each year, avoiding the risk of damage and the expense of their maintenance during adverse seasons. Few northern broadleaved trees are evergreen – these become much more common in southern latitudes, where their harder, often leathery or waxy leaves enable them to conserve water during dry periods. In conifers the reduced surface area and hard waxiness of the needles enable them to survive the winter without losing too much water.

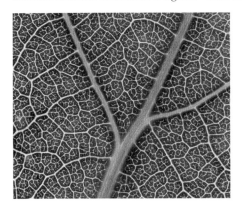

A network of veins spreads through a leaf (here of aspen), efficiently bringing water to all areas and transporting sugars from photosynthesis back to the tree for use in growth, or for storage as starch.

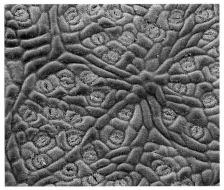

The flow of water is maintained through transpiration, the loss of water vapour through pores on the leaf surface called stomata, the rounded bodies visible here. Gases needed for and produced by photosynthesis and respiration also pass through stomata.

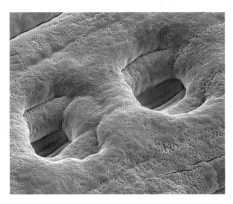

In a spruce needle the stomata are deeply set in a waxy groove, minimizing water loss. This is important for an evergreen tree, which often experiences dry conditions, whether in winter, when water is locked-up as ice, or in summer drought.

HOW DIFFERENT TYPES OF TREE FIND SUNLIGHT

Most trees, including conifers, bear leaves only at the tips of their shoots, thus forming a very shallow layer of greenery over the scaffold of their branches. It takes infinite subtle adjustments of shoot and stalk length and angle (and also of leaf size) to make the best use of the light (and thereby ensure the maximum photosynthesis). Beech (as the illustration right shows) is particularly effective at making an almost unbroken mosaic of leaves which casts deep shade. Some maple leaves have to grow stalks as long as themselves to manoeuvre into the light.

THE FOOD CYCLE

Plants convert the sun's energy into carbohydrate by photosynthesis, expressed in the formula

$$6CO_2 \text{ (Carbon dioxide)} + 6H_2O \text{ (Water)} \xrightarrow{\text{light}} C_6H_{12}O_6 \text{ (Sugar)} + 6O_2 \text{ (Oxygen)}$$

It is the basis for almost all life on earth (excepting only a few bacteria), providing both food in the form of carbohydrates and oxygen for respiration. The water is taken up by the roots, carbon dioxide enters via the stomata, and the reaction occurs in the chloroplasts through the mediation of the chlorophyll molecules. The sugars produced are transported in the plant's network of veins to where they are needed for growth or storage, and the excess oxygen escapes through the stomata.

TREES, WOOD, AND THE CARBON CYCLE

Photosynthesis captures energy from the sun, converting the building blocks of water and carbon dioxide into soluble sugars. These can be stored as starch or converted into other compounds, especially cellulose, for building the plant's structure. Cellulose is the primary building block of all plants; strengthened with another complex molecule called lignin it forms wood. Laid down in annual rings as the tree grows, wood has remarkable properties of strength and durability.

The timbers in the magnificent roof of Westminster Hall, constructed in the 1390s from oak, can not only be dated precisely by the pattern of their rings but also demonstrate the importance of wood as a long-lasting carbon sink. Massive quantities of carbon dioxide are removed by trees and bound up in wood, from which this gas is released only by decay or combustion. In the current situation of climate change caused by rising atmospheric carbon dioxide levels, trees provide one of the best options for sequestering carbon, hence the tree-planting options offered to offset emissions fom air travel. The use of wood for fuel, with its long-term cycle balancing input of carbon dioxide into tree growth and its output through combustion, is seen as a sustainable carbon neutral resource, but the logical conclusion is that for long-term carbon sequestration more timber should be used in the construction industry.

The roof of Westminster Hall contains 6,000 tons of oak from Surrey.

The Flowers

IT IS ALWAYS HARD not to fall into the trap of anthropomorphism – in other words, of attributing to inanimate objects our own feelings and motives. It is specially hard when you come to talk about sex. Why, we are bound to ask, do plants have to go through the whole risky business of sexual reproduction, entrusting their seed to the elements, or to insects, when it can give them no pleasure and must enormously reduce the chance of sucessfully reproducing at all?

Most plants, after all, can reproduce themselves by other means: either by sending out suckers from the root, or forming new roots on a side branch in contact with the ground, or just detaching a piece to start a new life on its own. A few depend on such means: their sexual system has broken down and their seeds are never fertile. Yet all plants have some arrangements for mixing their own characteristics with others of their kind. It is a prerequisite of evolution: a genetic lottery where a new combination may make a superior product.

And strange to say (anthropomorphism creeping in again) it is when they feel their life in danger, either through old age or because nutrition is lacking, that plants make the most efforts at reproduction and produce the most flowers. Richly manured orchards bear little fruit. Big flower (and fruit) years follow sunny and dry summers, when nutrients (except carbohydrates from the sun) are in short supply.

Flowers are the plant's sex organs. They exist to exchange genes (and hence characteristics) with neighbours of the same species. Chancy as it obviously is to try to establish physical contact between a speck of dust on one tree and a minute egg on another, this is what they have to achieve.

There are as many forms of flowers as there are plants. Indeed our system of classifying and naming plants is largely based on flower designs. There is such a thing as a "perfect" flower. In it both sexes are present, which should make things much easier. However (in the words of C C Sprengel, who in 1793 discovered the relationship between flowers and insects): "Nature wishes crossing to occur." All sorts of checks and barriers are built in to see that a flower does not fertilize itself. The pollen ripens at a different time from the

A bee visits the flower of an apple tree. On its visit to collect the pollen from the anthers for its hive it leaves pollen grains collected from other flowers on the stigma.

POLLEN UP CLOSE

Pollen is produced in the anthers of flowering plants and by the male cones of conifers, and acts as the carrier of the male gamete from plant to plant. Each pollen grain is protected by a hard coat, often with distinctive architecture, as seen here: wind-pollinated species have rounded pollen, while the irregular shape of the horse chestnut grain indicates that it is adapted to dispersal by insects. On arrival on a stigma a pollen tube develops, forming a route to the ovule for the sperm cells and enabling its successful fertilization. Wind-carried pollen is produced in huge quantities, as the chance of it reaching a stigma is slim, and this abundance can cause allergies (hay fever) in some people. On the other hand, its toughness means that it easily fossilizes, allowing the development of a permanent pollen record that can track climatic and ecological changes over time.

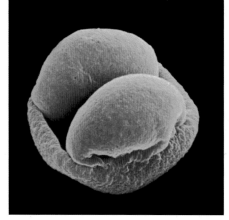

Scots pine pollen grain magnified 1,000 times

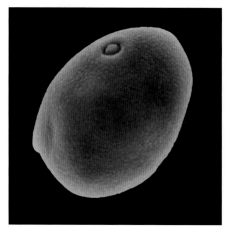

Alder pollen grain

HYBRID ODDS

The diagram below shows how the characters of the parents of a hybrid, though blended in the first generation of the cross (e.g. red and white flowers producing a pink offspring) are still retained as possibilities in the hybrid's genetic makeup. The first cross (F1) gets half its chromosomes from each parent. But it can make red- or white-flower pollen and ovules. These have four possibilities for recombination (F2): two of them pink, two like the original parents. The offspring of the second cross (F3) will be white, pink, and red. The white and red have reverted to the parental colour; only the pink shows its hybrid parentage.

FIRST GENERATION
Parent flowers

Sporophytic cells

SECOND GENERATION
First cross

Possible sexual cells of the hybrid

THIRD GENERATION
Second cross

FOURTH GENERATION
Third cross

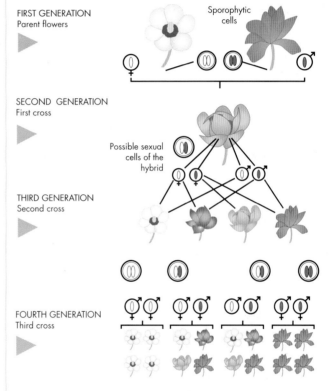

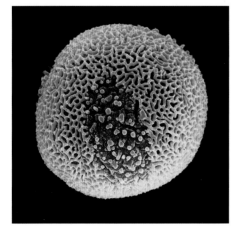

Plane tree pollen grain

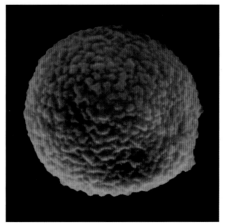

Poplar pollen grain

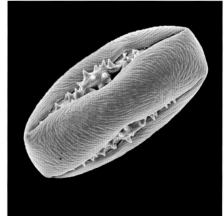

Horse chestnut pollen grain

ovule. Often the pollen is incompatible with its own tree's ovules. Or if by any chance they do couple the self-fertilized flower often falls off the tree before it can become a fruit.

What the flower hopes for (so to speak) is a fine breezy day for the pollen to be carried away in a golden shower to the next tree. And for the tree to windward to oblige with a similar gilding.

Wind-pollination tends to be the rule for forest trees. Their flowers, in consequence, are nothing much to look at and have no smell. The flowers of most of the big heart-of-the-forest trees (the beech family for example) are catkins that waste very little energy on display. Trees that need to attract insects to carry their pollen (which are often the trees of the open ground at the forest edge) go in for more elaborate designs. They are the cherries, the magnolias, the willows, and most of the chief ornaments of our gardens.

Most forest trees, in fact, have separate flowers for each sex (some for pollen, others for ovules) but have both types of flower on every tree. This is true, too, of all the conifers except the yew and its allies, *Araucaria* and juniper which have separate all-male and all-female trees. But most ornamental trees have both sexes in each flower.

To speak of conifers in terms of flowers is not strictly correct. Every time I make the slip (not difficult to do) I receive a rebuke from my botanical mentor. Write 100 lines: "Conifers do not have flowers; they have strobili. Their common name is cones." Thank you, sir.

Female conifer cones tend to be on the upper branches, while their pollen organs are on the lower: a simple precaution against the pollen falling too easily onto the ovules of the same tree.

Flowers (and cones of both genders) enable the forester or the nurseryman to combine characteristics he likes in two different trees by making a hybrid. The trees have to be genetically very close. He can no more cross a willow and an oak than a dog and a cat. But if he sees, say, two beautiful pines, one taller and the other straighter than all the rest, he can ensure the pollination of one by the other. He puts a bag over the branch end carrying the female strobili before they are mature, and injects into the bag pollen from the other tree. With luck the pine trees from the resulting seeds will be both tall and straight.

In the nursery business crossing has given us many of our most useful garden trees and extra vigour is a common characteristic of hybrids. Hybrids between two genera (as opposed to species) occasionally occur. The notoriously quick-growing Leyland cypress is an example. There remain, I should perhaps add, gardeners who believe no hybrid can have the grace and perfection of a natural species. It is too late to introduce them to CC Sprengel.

THE STRUCTURE AND FUNCTION OF FLOWERS

Flower size, shape, colour, and complexity vary greatly. Some are tiny, effectively reduced to the reproductive organs only; more complex ones, such as the crab-apple seen here, are built of whorls of structures surrounding the reproductive parts. The sepals (calyx) protect the bud, opening to allow the petals to unfurl their display. Next are the stamens, bearing pollen in their anthers, with female parts in the centre. Pollen is received on the stigma and the genetic material travels to the ovule, usually well protected in the ovary at the heart of the flower.

Petal

Sepal

Anther
Filament
Stigma
Stamen

Pedicel

Bud

Receptacle contains ovary and ovules

Leaf

Reproductive parts

Flowers exist for the purpose of reproduction through the fertilization of an ovule by pollen. The simplest mechanism is seen in conifers; their reproductive apparatus is minimal, relying on the wind for success, with no display function. Flowering plants, which evolved alongside insects, have a much more complex set of floral structures to protect the reproductive parts and attract pollinators – though some still depend on the wind. Flower shape, colouring, and timing reflect their pollination mechanism rather precisely. Showy flowers are effectively advertisements, competing with other species for the attention of a passing pollinator.

Conifers strictly speaking don't have flowers, their primitive reproductive parts being held in strobili or cones. At the shoot tip of this pine are the male strobili producing pollen; the familiar female cones are below

Catkins bearing tiny flowers are common in trees, as in this oak. Long male catkins produce masses of wind-borne pollen, to be captured by female flowers held in small clusters, usually on the same tree.

Flowers of the maple are "perfect", having both male and female parts. Although individually small the group is conspicuous and attracts pollinating insects such as honey bees and flies seeking nectar in early spring.

Members of the pea family, such as *Laburnum*, have characteristically shaped flowers. Showy and fragrant, they reward visiting insects with nectar. Successful pollination results in pea-pod like fruits full of seeds.

Paulownia **attracts pollinators** such as bumble bees that crawl into the tubular flower, guided by numerous spots and ridges inside. Pollen is deposited on the insect's back to be taken to another flower.

An example of a cultivated camellia in which the sexual parts have been converted into more petals. Cuttings are the only way of propagating sterile forms like this.

The Fruit

THE FRUIT OF A BROADLEAF TREE (or any flowering plant) is the female part of the flower – the ovary, with the ovule inside – fertilized and grown to ripeness. The ovule becomes the seed; the ovary the seed's covering. Taking the peach as an example, the ovule is the kernel of the stone, the ovary is its shell. Round it is a fleshy covering which is a development not of the flower itself, but of the stem just below it.

Conifers are technically quite different. Their ovules never have ovaries; the kernel has no shell. They may, however, (as juniper berries do) have a fleshy covering: again derived from their former stalks.

Fruit always has one simple purpose: to put as much distance as possible between itself and its parent tree. Some of nature's most inventive adaptations come into play to persuade the birds, the beasts, and the elements to cooperate.

When a tree clothes its seed in a substantial parcel of sweet flesh, as plums for example do, it is sacrificing a great deal of hard-won starch to make sure of interesting the birds. The flowers alone use a fair quantity to attract insects. Someone calculated that the 200,000 flowers on one cherry tree used 25 pounds of starch.

Almost all fruit serves simply to feed animals. Yet so big and so long-lived are the forest trees only 0.01 percent of acorns need to germinate for the oak population to remain the same.

Most seed ripens in the autumn – conifer seed in the second autumn after its fertilization. It is designed to spend the winter dormant either on the tree or on the ground, then to germinate the following spring. Winter cold is actually necessary to activate it and break its dormancy: a precaution against it germinating as soon as it falls only to be killed by the ensuing cold weather. When the gardener exposes his or her seed to cold and damp it is called "stratifying". Seeds have been known to stay dormant for 1,000 years.

In the ripe seed there are the beginnings of a little root and a tiny shoot, with one or more seed leaves which act as storage organs for food and usually as the first operating leaves of the new plant. These are the cotyledons, whose rather obscure influence pervades the whole world of plant classification. They remain within the seed supplying the initial first food that is needed for germination. Trees with two or more cotyledons produce wood in concentric rings: the classic tree pattern. Trees with only one grow as a cluster of fibrous bundles that gets longer but not fatter. The great "monocot" is the palm.

THE STAGES OF GERMINATION

The various stages in germination are shown from the cracking of the fruit wall (1), through to the fully grown seedling. The progress from seed to seedling begins first when, after the seed has been ejected (2) the root tip splits the outer covering of the seed (3). The cotyledons are fully developed at this stage, but remain in the seed. There is enough fuel stored in the seed to power the rootlet until, guided by gravity, it turns downward and buries its tip in the earth (4). From that moment it can supplement the seed's supplies and provide water for the seed leaves to swell up and emerge. The root grows deeper and the old seed shrivels. The final stages in germination are reached when the stem of the seedling straightens and the seed leaves fan out and start the process of photosynthesis (5). At this point, the tiny bud in their midst will swell to produce the stem and the first true leaves of a new tree (6).

1 The fruit wall cracks open (horse chestnuts germinate in autumn)

2 The seed is ejected

HOW SEEDS ARE DISPERSED

Animals In hoarding acorns for winter feed, squirrels inevitably lose some in the ground so that they are effectively planted.

Birds Thousands of species have their seeds distributed by birds who feed on their berries.

Wind Willows and poplars produce light seeds with a sail of cottony fluff which carries them miles. They need bare, preferably damp ground to survive.

Catapult Witch hazel is one of a number of plants that have contrived a mechanism to catapult their seeds into open ground, as much as 40 feet from the parent.

Air pockets (bladders) full of air enable the seeds of alders and other waterside trees to float downstream to a damp place to germinate.

Maple keys The dry, winged seeds are split in two at an angle of 60 degrees, making them aerodynamic enough to float to fertile ground. These shown here are red maple.

Each flower in the generous cluster (the terminal corymb) of a mountain ash (left) has followed the rose family pattern and covered its seed with flesh derived from the base of the flower and the adjacent stalk. Presumably they taste as good to birds as they look.

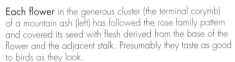

3 The root and shoot emerge   4 Side roots appear   5 The shoot reaches for the light   6 Root and shoot are both functioning

History in a Tree

EVERY YEAR THE GROWING TREE buries its past in another ring of functioning wood. The oldest part of the tree, the middle, having died, grows harder in death and provides the tree with its backbone. So long as it is sealed off from the air by living tissue on the outside it is virtually incorruptible. Each year's new ring remains just as it was when it was added: a faithful record of one year of the tree's history. Count them, and you know the tree's age (and a lot more besides).

The size of the new shoot the tree makes each year is decided by the weather of the previous summer when the bud was formed. Not so the annual ring. What affects its width seems to be (subject of course to such outside influences as fertilizers) the weather while it is actually forming. The rainfall of the previous winter is important, since it provides the ground water of the spring – the sap which will flood up the new ring as it forms. Even more decisive, because more variable, are the sun and rain of the current spring and summer. Growth rings, therefore, are an immediate record of the weather of the growing season.

But rings are rarely perfect. Trees are not so often round as oval, or fluted, or just plain lop-sided in cross-section.

A forest tree, sheltered on all sides and striving upward for the light, is likely to be the nearest to a perfect cylinder. The more open-grown and wind-buffeted a tree is the more it will taper, broadening at the base where the leverage of the top applies the greatest strain.

Conifers and broadleaves react in different ways to the strains of wind: conifers putting on extra growth on the leeward side, broadleaves on the windward. The same applies to the strain of gravity on horizontal branches: a conifer builds up the branch on its underside, a broadleaf on its upper. Both result in an oval cross-section, longer in the axis of gravity or prevailing wind.

Shipping a gigantic slab of a 3,000-year-old redwood over to England in 1851 was a piece of pure showmanship. Europe has no prehistoric trees: everyone was awed that anything should live so long.

Since the 1950s there has been much more constructive activity in the far west, based on the discovery of pines far older (though far smaller) than the redwoods. It is not their age alone, however, that makes them so exceptionally interesting. It is the strange conditions under which they grow, 10,000 feet up in the White Mountains of California, in a state of chronic drought. Where redwoods grow it rains without fail: every year they add a ring of wood about the same size. But where the bristlecone pine grows there is so little rain that the tree is a super-sensitive rain gauge. Every annual ring is different: and the story they tell goes back (so far) no less than 8,200 years.

There is no tree that old still alive.

This bristlecone pine (left) in the White Mountains of California may be 3,000 years old. Most of the tree is dead: the live part grows infinitesimally slowly. Dead wood does not rot in the desert atmosphere.

Botanists (right) at the University of Arizona prefer whole sections of trees for dating, but can take a "core" sample of a living tree (here a 250-foot giant redwood) without harming it. The core is extracted by hand with a fine auger.

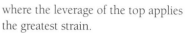

A TREE'S EVENTFUL LIFE

The annual growth rings of a tree record more than just the fat and lean years when it grew quickly or scarcely at all. The sketches on the right are a complete record of typical events in the life of a tree that started life hemmed around by other trees, was knocked over, burnt, saw its neighbours felled, and survived a drought – all circumstances legible in its distorted pattern of growth rings (below).

The tree was felled when 47 years old. A rough guide to the age of most normal trees is to allow one year for each inch of girth at shoulder height: e.g. a tree 10 feet around is about 120 years old.

At five years old the slender tree was knocked sideways. It reacted by growing twice as strongly on the lower side. Five years later it was upright again.

At 14 years old a ground fire destroyed the bark and cambium on the windward side. They grew over the wound by degrees: it took six years to close entirely.

Adjacent trees gradually deprived the tree of light and moisture. When it was 27 years old "thinning" of the woods brought it into the open. There was a great leap in growth.

After six years of rapid growth came a period of drought; its effects on the tree are visible for six rings. Serious drought can slow down growth for years after normal rainfall returns.

The oldest living specimen is 4,900. But the Laboratory of Tree-Ring Research at the University of Arizona has developed a technique for matching samples of wood from living and dead trees even from broken-off bits lying on the desert floor to build up a continuous series of rings.

The rings are microscopically narrow. On one sample there are 1,100 rings in the space of five inches. But the sequence of relative widths never repeats for more than a year or two. Any substantial sample has enough rings either to be unique or to overlap with another. A computer soon finds out which and puts it in its place.

The university expects eventually to push the records back to 10,000 years of weather – back in fact to the centuries when the last ice age was in retreat. The value of the work to weathermen is obvious. But in 1969 another startling issue came to light.

Wood whose age was exactly known was tested by the Carbon-14 method, the accepted means of dating prehistoric sites. But the answer came out wrong. The pines revealed that the basic assumption of C-14 dating, that the carbon in the atmosphere is at a constant level, holds good only for the last 3,500 years. Before that time the errors mount up rapidly: there was a 700-year error within the millennium before 1500 BC.

As a result the presumed dates of some of the most important early structures have had to be changed. Stonehenge has been backdated 1,000 years – upsetting the notion that technology was inherited by western Europe from the East. The implications of all this have yet to be gone into fully. But the fact remains that a Californian desert tree is able to bear witness to events long before history began.

Roots and Soils

THERE IS A DISCIPLINE about the top-hamper of a tree. It obeys its own specific rules of growth. In contrast the tree's roots are opportunists: they go wherever they find the feeding best. E H Wilson told the story of a particularly happy-looking ivy on the walls of Magdalen College, Oxford, which turned out to have got into the cellar and drunk a whole barrel of port. How many glasses he had drunk that evening can only be a matter of speculation.

We know much less about the roots of trees than about their branches, simply because they are out of sight. Very rarely has a tree been excavated *in toto*; more often we see the roots more or less complete when a tree is blown over by a gale.

The roots have three jobs to do. They anchor the tree in the ground; they are the tree's chief source of water; and they have to forage for the elements from the soil that are a small but vital part of every cell.

The first root every tree (and every plant) puts out from its germinating seed is a taproot. The taproot is a sort of emergency action to plug into the supplies as quickly as possible. Thereafter most trees soon concentrate on exploiting the topsoil, normally the richest in organic matter. Most feeding roots of most trees are found within six inches of the surface.

Spruces, beeches, and poplars are examples of trees which rarely root deep,

A **fallen forest tree** reveals where its roots have been growing. Often they are limited to the first foot or so of the forest floor with no deeper roots to anchor them at all. The topsoil is constantly replenished by leaf-fall.

The shape of a tree's root system (above) depends less on the type of tree than the type of soil. Certain trees predominate on certain soil types, but where their roots go is a question of tapping the supply of nutrients, water, and oxygen with the least physical obstruction in the way. Above are some of the roots of an oak in "brown earth", the typical soil of deciduous woodlands in areas of relatively low rainfall, slightly acid but rich in bacteria which quickly break down into humus –

the surface layer of vegetable litter. Flowers and fungi typical of such woodlands (known as "dry oakwood") are:
1 Wood meadow grass
2 Bilberry
3 *Lactarius quietus*
4 Pignut
5 *Boletus pulverulentus*
6 Inkcap
7 Stinkhorn
8 Wood sorrel

The soil is about five feet deep, free draining, loose-textured, and fertile. Most of the tree's feeding roots are in the top 12 inches, the topsoil layer with the most humus. There are no definite "horizons", or layers of different material. Topsoil fades off into subsoil; "sinker" roots find matter to exploit right down to the bedrock. The upper subsoil gets its rich rust colour from iron which oxidizes readily in well-aerated conditions and is the most visible of all minerals in the soil. In more acid soils the iron

oxides dissolve and are "leached" into lower layers, as in podsol. Here they remain evenly distributed on all levels, as do the other essential minerals (phosphorus, potassium, nitrogen, sulphur, calcium, and magnesium).

even when young. Firs, most oaks, and many pines on the other hand go on with a taproot for several years, which makes them much harder trees to transplant. What happens then depends on the soil. In most temperate trees, roots do not penetrate very deeply, one of the deepest recorded being 21 feet from a pine on sand – the most permeable of soils. Far more commonly the taproot withers away and side-roots take over.

Roots go where it is easy to go. If they can they will follow old worm tunnels or spaces where previous roots have died and rotted. They have to be wary only of the water table: the level below which the soil is permanently wet. Their policy here is brinkmanship: they need the water, but they need oxygen too. They like to keep a toe in, but if they are submerged for long the roots of most trees drown. On the whole a fluctuating water table is not such a safe bet as the rain permeating from above. Beeches are normally shallow-rooting trees. After a summer-long drought in England it was widely remarked how little they seemed to have suffered. But that autumn it rained prodigiously. Next summer beeches died – of drowning. They had sent down emergency roots into the water table.

As the crown of the tree spreads, the roots extend to keep up with it. They are at their most active under the drip line, where the rain falling on the crown runs off around the circumference.

Root growth is almost constant, stopping only in freezing weather. It is only new and growing roots that function actively in collecting supplies. In the area where the action is – just behind the root tip – root, root hairs, their fungal associates, and soil are intimately knit together.

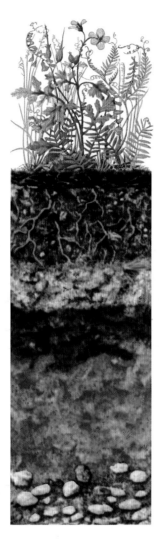

Woodland known as "damp oakwood" has a "gley" soil. Gley occurs where drainage is bad, the soil has little oxygen, and iron cannot oxidize. Often an impermeable layer of clay keeps the water level high. Airless, waterlogged soil below is seen as a dark grey "horizon". Oak, ash, and hazel grow well but root shallow. Herbs include wild angelica, bramble, woundwort, herb robert, tufted, vetch, vernal grass.

"Podsol" is a common soil type where rainfall is high, evaporation is low, and the net water movement in the soil is downward. The calcium in the surface is dissolved and "leached" to lower levels, leaving the topsoil very acid. Bacteria cannot flourish; there are no worms; there is little oxygen; leaves are broken down into humus very slowly. Surface plants are male fern, mushrooms, twin flower, winter green, bilberry, bell heather.

Typical chalkland soil ("rendzina") is a very shallow layer of fertile topsoil with good humus breakdown and efficient drainage. Below is nothing but chalk or limestone. The surface can (through "leaching") be acidic even a few inches above pure chalk. Beech and ash are typical rendzina trees; most conifers are unhappy. Herbs include dog's mercury, ramsons, fescue, bugle, avens, wood sanicle.

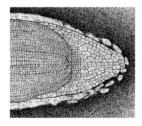

A root elongates in the area behind the tip. The cap produces new cells to be sloughed off as the tip advances. These discarded cells lubricate the root's way through the soil.

Root hairs help the root absorb water and mineral salts. This is the start of the transpiration stream which ends with evaporation from the leaves.

Not only are cells from the root tip constantly being sloughed off to lubricate its passage, but drops of acids (chiefly carbonic) are exuded from the growing cells to dissolve the elements they need, which the root hairs absorb into the tree's plumbing to be transpired by the leaves.

Most roots have allies: fungal friends or mycorrhizae with which they form a close association and which are characteristic of the tree (as truffles are of certain oaks). The fungi act as go-betweens. The roots supply them with sugars; they supply the roots with minerals from the soil. The reason why alkaline soil is fatal to many trees is that it kills these essential partners.

When a German botanist, Bernard Frank, first described mycorrhizae in 1883 botanists were sceptical. How could a plant and a fungus be mutually beneficial? Now we know that 90 percent of plants depend on some sort of fungal association and are learning to use the fact to the advantage of what we plant. There are various ways of giving a new tree a boost by adding mycorrhizae to the soil, from scattering granules impregnated with the relevant mycorrhizae to the planting hole to watering on a solution containing them after planting. In public tree collections, Kew Gardens for example, it is now standard practice.

How Trees are Classified

THE BASIC UNIT of classification is the species by which we mean a group of similar organisms, which grow together in nature, breed together, and produce offspring like themselves.

When we talk about a genus we mean a group of such species, consistently different in detail, not normally interbreeding but usually linked in a fairly obvious way. The English settlers arriving in New England recognized the oaks as oaks, even though the species they knew back home was not among them. From ancient times these two broad categories of natural objects have been instinctively acknowledged.

But a more searching scrutiny of the natural world has also suggested (to Aristotle in the first place) that there are broader groups and deeper relationships to be found. It is easy to think up ways of classifying anything. You could make lists of trees with yellow flowers, or peeling bark, and have a system of a sort. But it would prove nothing because it would be limited to the characteristic you happened to choose. You could neither deduce nor predict anything from it. Aristotle guessed that there is a natural order of relationships where everything has a place. There was no clue, though, as to where to look for it.

Even Linnaeus, who is known as the father of modern taxonomy (the science of classifying natural objects), was in the dark about the "natural order". He classified plants by their sexual characteristics: number of stamens, ovaries, and so forth. His instinct was right, yet he admitted his system was artificial and would be superseded when the key came to light.

What natural links he did see he incorporated in a "fragment of a natural system". So far as it goes it still holds good. Taxonomists who followed him built on it until by the middle of the 19th century most of the genera of flowering plants had been assigned to "families".

Yet still the key, the link (if any) between genus and genus, between family and family, was unguessed at. It was left to Charles Darwin to supply it: plants are alike because they have common ancestors. Darwin said: "All true classification is genealogical … community of descent is the hidden bond which naturalists have unconsciously been seeking."

Since Darwin's day taxonomy and phylogeny (the science of the ancestry of things) have marched side by side. Evolutionary history has been likened to a tree. Only the most recent shoots are visible, but breaks in the canopy show that two or three of the shoots come from one branchlet, or maybe that two or three branchlets come from one branch. The main bulk of the tree – the trunk and limbs which represent now-extinct ancestors in the early stages of evolution – is invisible. The taxonomist's job is to reconstruct it from what he can see: the plants of today. Proof of ancestral links 100 million years ago are not easy to find. Fossil evidence has hardly ever helped, largely because the taxonomically significant bits – flower parts – rarely make legible fossils. Even tools

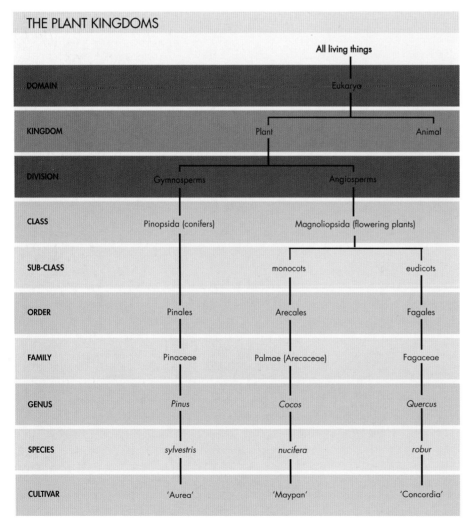

THE PLANT KINGDOMS

		All living things	
DOMAIN		Eukarya	
KINGDOM		Plant	Animal
DIVISION	Gymnosperms	Angiosperms	
CLASS	Pinopsida (conifers)	Magnoliopsida (flowering plants)	
SUB-CLASS		monocots	eudicots
ORDER	Pinales	Arecales	Fagales
FAMILY	Pinaceae	Palmae (Arecaceae)	Fagaceae
GENUS	*Pinus*	*Cocos*	*Quercus*
SPECIES	*sylvestris*	*nucifera*	*robur*
CULTIVAR	'Aurea'	'Maypan'	'Concordia'

The classification of a garden plant starts (at the bottom) with the smallest division: cultivar. It works up in bigger and bigger groupings. *Cocos* is a member of the palm family (Arecaceae), the Arecales order, the monocot sub-class, Magnoliopsida class, Angiosperm division of the plant kingdom.

NATURAL PHILOSOPHERS

Categories of objects, of experiences, of character, of plants and animals are the building blocks of understanding the world. Until the Renaissance, plants were only closely studied and categorized for their usefulness to man, not their relationships in a natural order. Nor was the nature of that order grasped until Darwin wrote *On the Origin of Species*.

Theophrastus (died 287 BC), a pupil of Aristotle, was the first man to classify plants as "woody" or "herbaceous", though he is better known for his classification of "characters" of men.

Charles de L'Ecluse (Clusius) (1526–1609), was from Flanders. He made the first real botanical (as opposed to medicine-herb) garden in Leiden in Holland. He bridged the gap between herbalism and modern gardening and botany.

Carl Linné (Linnaeus) (1707–87) This great Swedish naturalist is considered to be the founder of modern botany. The binomial system for naming plants was established by his work.

members is the flowerhead, a group of tiny florets making a composite flower.

Flower structure can vary quite a lot within a family, or even a genus, so it is not always a reliable guide to relationships. The maples (*Acer*) could, on the strength of their flowers alone, be chopped up into several different genera. Yet maples all have such similar fruit, all have the opposite branching pattern, and most of them have such obviously related leaves that to split them up would be pedantry and lose the link of common ancestry. More surprising is to find that horse chestnuts (*Aesculus*) are now regarded as being closely related to maples (shared characters visible to the naked eye include opposite leaves and a tendency to palmate lobing of the foliage), and that both belong to the largely tropical family Sapindaceae, which also contains exotica such as lychees.

such as microscopes, allowing us to peer into aspects of plant structure such as wood anatomy and the shape of pollen grains, have only taken us so far.

Modern technology has come to the rescue, with DNA-sequencing techniques now making it possible to look deep into a plant's past. Where taxonomists used to look for differences to pull plants apart, they now look for similarities to bring them together and shared gene sequences demonstrate relationships with little room left for speculation, with the result that there has been a wholesale re-examination of the phylogeny of the flowering plants, to ensure that Darwin's "community of descent" from a shared ancestor is accurately reflected in plant classfication. To the layman, some of the reassignments seem strange, separating plants long-thought to be closely related, or vice versa, but the demonstration of true lineages is intellectually very satisfying.

Members of some families are obviously related. The daisy family, Asteraceae (or Compositae), is an example of how far apart ecological adaptation can take related plants. It includes trees, shrubs, vines, herbs, even succulents – in every climate, on every soil, using every method of pollination and seed dispersal. The common link between all the family's

HOW TREES ARE NAMED

How simple it would be if a genus and a species name were enough to pinpoint any plant precisely. Unvexed by horticulture, or at least by the human desire to select and improve, there would not be too many deviations from the basic binomial rule. Instead a straightforward unimproved species in a garden is now an exception. The very ancestry of most garden plants is lost in the long process of breeding and selection, in many cases repeated every year to have new things to put in catalogues.

If there is an exception it is among trees. They operate (or most of them do) at a gentler pace. There is still infinite room for natural genetic variation. Some species (Lawson cypress, for example) come up with all sorts of wild ideas each time they sow their seed. Even the common beech gives a planter a range of colour options. On the other hand both botanists and nurserymen have sharp eyes for any deviation from the norm. There is fame, and possibly money too, in spotting the first birch with variegated leaves or maple with round ones. (You can try to breed a blue rose, but these would be beyond any breeder.)

The rules of taxonomy try to be clear about the ranking of plants that don't conform, and where exactly they fit in to what you might call the unnatural order. I suspect few people, beyond horticultural editors, follow (or are greatly concerned with) the range of possibilities. But they can help you understand how the taxonomic mind works.

The species is the basic category of plant, but it is not the smallest. Within the species there may be several varieties – often local forms with an ecological reason (extreme exposure, for example) for being the way they are. It is a matter of

botanical judgment what constitutes a species and what a variety. As Arthur Cronquist of the *New York Botanic Garden* said: "Custom, in such matters, is merely the sum of a series of individual opinions, plus inertia."

Below the basic species the first rank is the sub-species, or a natural plant that is not quite different enough for botanists to give it the dignity of specific rank, but forms a recognizable population in the wild. The name of a sub-species (or subsp.) follows the specific name (example: *Magnolia campbellii* subsp. *mollicomata*). A variety (or var.) is similar, but differs less markedly from the norm (*Betula utilis* var. *jacquemontii* is an example). A forma (or f.) is a botanical rank used to denote exceptional plants that do not usually form populations (for example a white-flowered individual in a generally pink-flowered species), but is seldom used for trees.

Horticulturists add to the taxonomic ranks to describe selections or hybrids made in gardens. Most important of these is the cultivar (or cv), which has been chosen for some quality that will only remain in existence if the plant is reproduced vegetatively – that is by cuttings, and thus represents a single genetic entity, or clone. A culivar name is indicated by single quotation marks, eg 'Quercifolium'. A "group" indicates the existence of a number of very similar clones which have something so clearly in common that they seem to be a cluster within the range of possibilities: copper beech seedlings, for example, gathered together as *Fagus sylvatica* Atropurpurea Group. The intensive breeding of rhododendrons has led to a number of such groups.

Plant names follow a pattern set down by international convention. In the early days of botany, names were long and complex, describing a plant's characters. *Quercus foliis deciduis oblongis superne latioribus; sinubus acutioribus; angulis obtusis* denoted the common European oak. In 1753 the Swedish naturalist, Linnaeus, cut through all this, instituting a system that separated name and description, still in use today. He said that a name should be in two parts only, the first word stating the genus, in this case *Quercus*, the second representing the species. For the oak he chose the Latin word *robur*, meaning strong and so the binomial name *Quercus robur* L. was coined.

To this day, when a new plant is found it must be described in print (not online), in Latin (at least briefly), and given a binomial name. To give credit where it is due, the author's name is tagged onto the binomial, usually in shortened form (e.g. L. for Linnaeus in the example above) and becomes part of the full name of the plant in botanical works (but not in this). Only in exceptional cases can the first name validly given to a plant be overturned, so a botanical author gains a certain immortality. About a quarter of a million flowering plants now have such binomial names. On pages 368-73 is a list of tree names and their exact meanings.

So a pretty tree is standing before you, somehow familiar, but you don't recognize it. It is a situation all tree-lovers face, often. Is there a fail-safe way to deduce its identity? As my wine-tasting friends say: "A glance at the label is worth 20 years' experience." In this case there's no label.

What are the clues: how do you analyze the twigs and leaves, the flowers and fruit, the trunk and branches in front of you to produce an answer? To eliminate the possibilities that don't apply is the only way to begin. That is the method of the botanical "key". It is not for the impatient, but a key is a systematic analysis, working from the basic (are the leaves opposite or alternate?) to the minute (are there hairs on the leaf veins?).

There are people who seize the essence of a plant almost at first glance, others who have to work at it – and doubtless others who just don't get it. Certainly one of the best ways of training your observation – of anything – is to draw it. A little sketch can sharpen your eyesight remarkably. A key, though, is the only way to be certain. You are working on the extreme outer fringe, the outermost leaves, of the vast evolutionary tree in all its ramifications going back to creation.

PLANT CLASSIFICATION AND RELATIONSHIPS

PRIMITIVE NON VASCULAR PLANTS

Systems of plant classification have evolved through the centuries. DNA analysis has brought new ideas and insights, resulting in a realignment of traditionally recognized groups. The diagram shows how tree families are currently grouped, in an order that reflects their evolutionary sequence from primitive to advanced. Ferns appeared about 300 million years ago (mya);

GYMNOSPERMS

FERNS   CYCADS   GINKGO   CONIFERS

FERNS
(tree ferns): reproduction like that of all ferns in two stages: the spores under the leaves have no sex. They drop off and grow into little flat plants (above). These have sexual spores which create another tree fern.

FERNS
(cycas family): primitive palm-like gymnosperms with short thick trunks. Female plants bear clusters of large naked ovules, fertilized by swimming sperm, resulting in large seeds.

GINKGOACEAE
(ginkgo family): has the oldest genus of tree still living, alone in its family, unaltered for 100 million years. The ginkgo's naked seeds relate it to the conifers: it has catkins for male flowers.

TAXACEAE
(yew family): typical conifers in leaves and habit, unusual in being either male or female. Single-ovule female "flowers" develop into open-ended berry-like "cones".

CUPRESSACEAE
(cypress family): tiny scale-like leaves pressed close to the twigs. Small roundish cones start fleshy and ripen woody. Male "flowers" are at branch tips.

PINACEAE
(pine family): narrow leaves in spirals, though sometimes in rows or tufts. Woody cones have two seeds per scale; bracts of "flowers" often remain visible in the cone.

BASAL ANGIOSPERMS
(*Illicium*: star anise illustrated): basal angiosperms are the most primitive extant flowering plants. Only a few hundred species survive, including water lilies and the star anise family. Insects pollinate the usually rather simple flowers.

MONOCOTS
(palms: coconut illustrated): unbranched stems of uniform thickness with terminal spiral of divided frond-like leaves. Flowers small, usually unisexual. Three ovules; one matures into seed with massive endosperm.

THE INFLUENCE OF DNA ON TREE CLASSIFICATION

Deoxyribonucleic acid (DNA) is the material of inheritance. Its elegant molecular structure, a double helix, was elucidated in 1953. Attached to the "backbone" of each strand of the molecule are chemical groups called bases, which always pair in the same way: adenosine (A) with thymine (T), cystosine (C) with guanine (G). The resulting series A–T, C–G, in infinite combinations, is the genetic code; through slight mutations and recombinations of the sequence the code is both distinctive for each species and unique to each individual organism, which means it can be used in applications as diverse as paternity or forensic tests, or the elucidation of plant evolution.

Sophisticated modern techniques allow scientists to study the genetic code relatively easily. By observing the pattern of recognizable mutations, passed faithfully from ancestors to descendants, it is possible to work out relationships between species, genera, or families in a way that is not always possible through the study of morphological characters. In consequence former classifications have had to be revised to reflect this.

  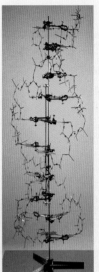

James Watson (b.1928) (above) is an American geneticist. Building on the work of others he and the British Francis Crick (1916–2004) (above centre) demonstrated that the DNA molecule (above right) is a double helix that can replicate itself, probably the most important scientific breakthrough of the 20th century. For their discovery they shared the Nobel Prize for Medicine with Maurice Wilkins in 1962.

recognizable *Ginkgo* fossils date back 270 mya. Conifers organize their reproductive structures into cones, but recognizable true flowers did not appear until about 140 mya, when the angiosperms (flowering plants) emerged. From the earliest flowering plants with the most primitive flowers, now known as basal angiosperms, the rest rapidly evolved, with angiosperms succeeding conifers as the dominant trees 100–60 mya. Traditionally divided into monocotyledons and dicotyledons from the number of seed leaves produced, angiosperms are now split into four groups: basal angiosperms (including the star anise family, Illiciaceae), monocots (including palms), magnoliids (including *Magnolia*), and eudicots (the vast majority). Within these groups are families combined by their broad ancestral relationships.

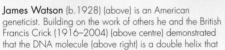

ANGIOSPERMS

EUDICOTS

MAGNOLIIDS (magnolia family): bisexual flowers with stamens and petals attached below protruding female organs which develop into a cone-like mass of seeds with a substantial food store or endosperm.

MYRTACEAE (myrtle family: eucalyptus illustrated): bisexual flowers in clusters; the many stamens often have coloured stems. Fruit a berry, drupe, or capsule. Leaves leathery and opposite.

SALICACEAE (willow family): sexes on different trees, male and female flowers both catkins. Fruit a capsule of many hairy seeds without endosperm. Leaves alternate and simple.

LEGUMINOSAE (pea family: pea illustrated): most flowers irregular with five petals. Fruit a pod with two or more seeds, no endosperm. Leaves usually compound.

ROSACEAE (rose family: rose illustrated): bisexual flowers usually with parts in fives. Fruit (*Rosa*, here) depends on development of "receptacle" and number of ovules. Leaves have tiny leaflets at base.

FAGACEAE (beech family): male and female flowers on same tree, male flowers tassel-like. Fruit surrounded by a woody cup. Seed has no endosperm.

ACERACEAE (maple family): opposite leaves and "keys" for fruit consisting of two winged nutlets joined at their bases. Flowers have five petals, five sepals, but eight stamens.

OLEACEAE (olive family): opposite leaves and bisexual flowers, usually with four petals and two stamens. Fruits vary: olive has drupes; ash has winged nuts.

Trees and the Weather

Seaside trees (left) often show the effects of constant pummelling by the prevailing wind. It suppresses normal new growth on the windward side while shoots expand in the comparative shelter of the lee side until the whole tree is lop-sided. This is a hawthorn by Morecambe Bay, Lancashire.

Nothing more graphically illustrates climatic change in a small space than the slopes of Mount Kilimanjaro (right). Dry plains give way to damp forest, to thinner woods, then scrub; then rock, and finally ice.

CLIMATE IS THE CRUCIAL FACTOR in deciding what trees grow where. Its broadest movements in geological time have governed the evolution of tree species. Relatively recent climate changes have settled the present natural distribution of these species around the globe.

The ancestors of all our trees were tropical plants. In the tropics the seasons are only vaguely marked by temperature changes: what alters most from one time of year to another is rainfall. Most tropical plants are evergreen. They are programmed to grow either constantly or intermittently whenever there is enough moisture, without fear of damage by cold.

Trees that adapted to life in the temperate zones, with marked winters, did so by learning to live with the seasons. A temperate-zone tree in its own habitat is precisely adapted to the exact weather pattern its parents and grandparents have undergone. It is what we call "hardy", because it has learnt to do the right thing at the right time and not risk exposing its vulnerable stages of growth to weather that can harm them. It has developed a clock that tells it when, judging by past experience, it is safe to get started in the spring, and when it had better pull in its horns for the winter.

The most graphic illustration of how trees find (literally) their own climatic

level is seen as you climb a mountain. The vegetation changes from full-fledged forest at the base through successive stages of what amounts to hardiness until soil runs out and conditions impossible for plants set in. On the higher slopes the length of time the snow lies and the number of days above 50°F enabling growth become determinant to what species you will find.

As soon as a tree is moved (or its seed is planted) out of its accustomed zone it is in potential danger. It seems odd that a larch from Siberia (for example) would be anything but delighted to be moved to a softer climate. You would expect it to luxuriate in the longer growing season, while still being totally "hardy", whereas what happens in practice is that it is lured out of its safe dormancy by higher temperatures than it expects too early in the spring. It starts growth, only to be cut back by late spring frosts. This happens repeatedly, and it dies.

The converse happens when a southern tree is moved north. It is relatively safe in the spring: it waits for warmer weather. But it keeps growing too late in the summer and its new wood is still soft and green when the first autumn frosts strike. Again, lack of hardiness really means being programmed for the wrong climate.

Perhaps more surprising is the difficulty trees experience in moving from the west

Untimely frost destroys new growth, leaves, and flowers on a magnolia. Timing is as important as low temperature.

if even a week can be added to the growing season by getting seed from 100 miles farther south without the trees suffering, the trees have added (say) a whole year's growth in 20 years, or five percent. In practice much better results, even up to 30 per cent, are sometimes achieved.

Where winter damage – as opposed to damage by late or early frost – is concerned, trees from different zones are on a more equal footing. Many, if not most, temperate-zone trees can stand being frozen solid while they are dormant (although some are more used to it than others, and none like it to happen suddenly). What more often hurts, and can kill, is winter drought.

While the ground is frozen in winter there is no water available to the roots. But high winds and often low humidity keep on evaporating water from the branches. Even the twigs of leafless trees transpire to some extent in these conditions. So the tree begins to dry out. On evergreens it shows in the browning of the leaves by the end of even a normal winter. A longer period of intense cold with clear skies and a high wind can easily prove fatal.

The Arnold Arboretum at Boston has perhaps (of all scholarly institutions) the most experience of testing temperate trees in hard winters. It argues that if one statistic is to be used to try to map plant hardiness zones it should be minimum temperature. Rainfall is important; soil has its influence; microclimates are vital. But on a broad front it is the lowest regular winter temperatures it experiences that decide whether a tree will survive or not.

On pages 36-7 the hardiness zone map originally developed by Dr Alfred Rehder at the Arnold Arboretum for the whole of North America is matched by a similar map on the same principles done for Europe. On pages 38-9 maps show the natural forest vegetation of the same areas, so that in theory it is possible to follow the apparent effects of minimum temperatures (and other conditions) on the selection of species to be found in the wild.

A **blanket of snow** protects against excessive transpiration and deep frost at high altitudes – here in the Austrian Tyrol.

coast to the east of North America or from the East to Europe. The difference between a continental climate and an oceanic one can be more upsetting to the rhythm than a simple move from north to south. Western conifers are as unhappy in New England as oaks from Ohio are in Britain or France. Often it is lack of summer heat to ripen the new wood ready for a hard winter that condemns a tree.

The most extreme example is a cold-climate tree that is moved to the sub-tropics. What can happen is that its buds may fail to open at all. Built into its schedule is the need for a cold spell (the winter) to break their dormancy. If there is perpetual warmth it is stuck: in all probability it will die.

In forestry the question of provenance (i.e. exactly where the seed comes from) is clearly of the greatest importance. The forester's object is to extend the growth period of trees as far as possible without putting them in danger of frost damage. There is little room for manoeuvre, but

Zones of Tree Hardiness in North America and Europe

WHETHER A TREE will live and grow in any region, given suitable soil and adequate moisture, depends on the lowest temperatures it will encounter (see pages 34-5). On this page we reproduce the map of the hardiness zones of North America compiled by the United States Department of Agriculture on the basis of average annual minimum temperatures – a map

that has long been taken as a guide by the horticultural industry. Opposite is a map of Europe on the same basis.

As an example of how they work, the phrase "zone 9" in the text identifies a tree that will survive in temperature zone 9 – as marked on the maps – or south of it, but only in exceptional circumstances in the next zone northwards, zone 8. Such broad

generalizations of course have many exceptions: altitude lowers the temperature locally; big lakes have the opposite effect.

The North American map is dominated by the influence of the great land- (and ice-) mass to the north. You can picture the situation as cold radiating from somewhere in Manitoba. Only the mountain ranges (and to some extent the warm Gulf of

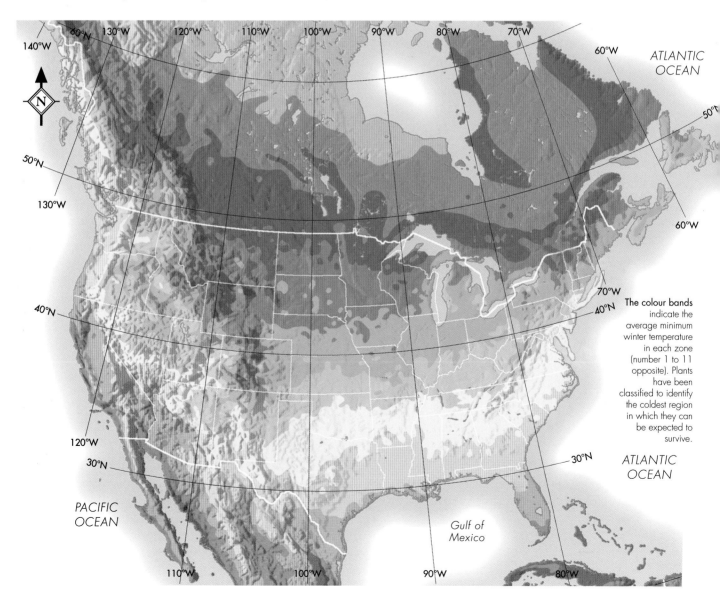

The colour bands indicate the average minimum winter temperature in each zone (number 1 to 11 opposite). Plants have been classified to identify the coldest region in which they can be expected to survive.

Mexico) distort this continental influence. Otherwise minimum temperatures rise in concentric bands to the sub-tropical fringes of southern California and Florida.

The rainfall map (right) shows that most of the major centres of population in North America fall in zones of adequate average rainfall for most trees (more than 30 inches a year) – though this does not mean that trees will not benefit from deep watering in dry spells. The annual figures do not show, for example, that California's rain comes mainly in winter, making watering very necessary in summer. In most of the United States more rain falls between May and October than between November and April.

The overwhelming influences on Europe are the Atlantic Ocean, the cold land mass of Asia, and the warm air mass over the Sahara. The ocean with its warm Gulf Stream is so effective in keeping winter temperatures up that, mountains apart, the hardiness zones of Britain and most of northern Europe run west-east rather than north-south. Where the Atlantic and Saharan influences coincide in southern Spain, winter temperatures are as high as in Florida, ten degrees of latitude or 700 miles farther south.

Rainfall also follows a mainly west-east pattern, but as in America annual levels are adequate in most populated regions.

HOW MUCH IT RAINS

Rainfall has as much bearing on the growth of trees as temperature. In temperate regions, for most trees, the more it rains the faster and bigger trees will grow. The Hoh rainforest on the Olympic Peninsula in Washington has 14 feet of rain and the biggest trees of several species. Europe's mountains and its west coasts have the most rain.

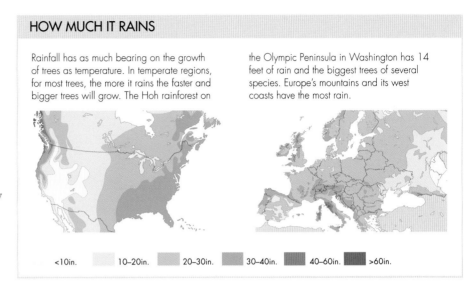

| | <10in. | | 10–20in. | | 20–30in. | | 30–40in. | | 40–60in. | | >60in. |

HARDINESS ZONES: MINIMUM WINTER TEMPERATURES

Hardiness, as understood in British horticulture, is the ability to withstand winter temperatures down to zone 8.

Zone	
1	Below –50°F (–46°C)
2	–50 to –40°F (–46 to –40°C)
3	–40 to –30°F (–40 to –34.5°C)
4	–30 to –20°F (–34 to –29°C)
5	–20 to –10°F (–29 to –23°C)
6	–10 to 0°F (–23 to –18°C)
7	0 to 10°F (–18 to –12°C)
8	10 to 20°F (–12 to –7°C)
9	20 to 30°F (–7 to –1°C)
10	30 to 40°F (–1 to 4°C)
11	Above 40°F (4°C)

Europe has a much narrower range of minimum temperatures, and thus hardiness zones, than North America, despite being as far north as the northern states. Only mountains and sheltered inland areas see any extremes of cold.

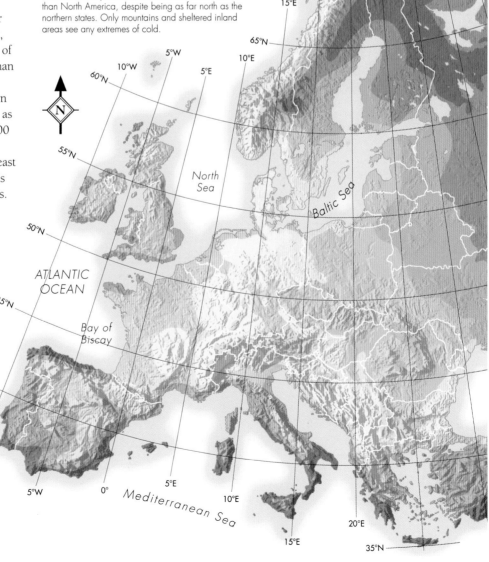

Natural Forest Vegetation

THE CLIMATE (BOTH prehistoric and modern), the soil, the mountains, the rainfall, the length of nights, and angle of the sun are factors that shape natural forest vegetation. On these pages are maps of its outlines in North America and Europe.

The European pattern is relatively simple. Starting in the north, after tundra (where trees, if any, are stunted) comes the vast northern coniferous forest of pine and spruce, sprinkled with birch and, farther

east, fir, larch, and spruce. South of a line on about the latitude of northern Scotland deciduous trees come into the picture in greater numbers. The northern limit of the oak is roughly along this line. All northern Europe comes into the mixed coniferous/deciduous belt with oak, beech, fir, pine etc as the principal trees, except where mountains make islands of coniferous forest or where heath (Scotland, north Germany) or grassland (southwest

Russia) occupies very poor soil. The Mediterranean zone is approximately marked by the northern limit of the olive. Its forest is largely evergreen (pine, oak, cork oak, carob) alternating with areas of aromatic shrubs and herbs (myrtle, thyme) dotted with olive, juniper, and cypress.

North America is much more complex. The forest zones of the east, compared with the hardiness zones on pages 36-7, show the same strong continental influence,

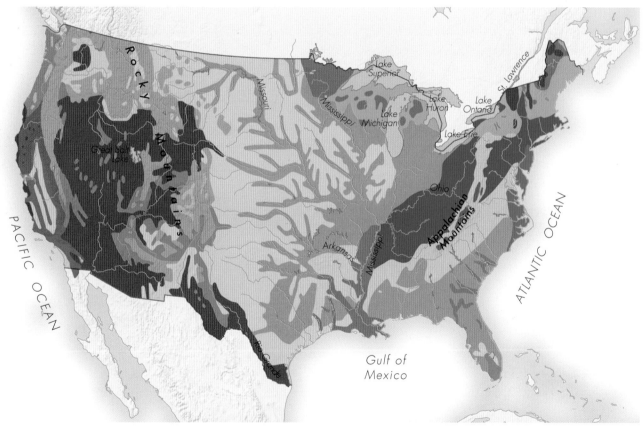

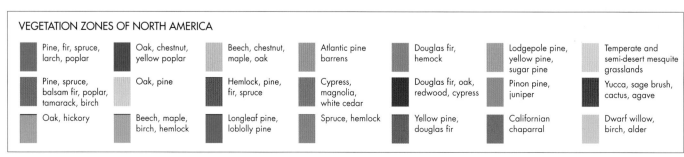

VEGETATION ZONES OF NORTH AMERICA

- Pine, fir, spruce, larch, poplar
- Pine, spruce, balsam fir, poplar, tamarack, birch
- Oak, hickory
- Oak, chestnut, yellow poplar
- Oak, pine
- Beech, maple, birch, hemlock
- Beech, chestnut, maple, oak
- Hemlock, pine, fir, spruce
- Longleaf pine, loblolly pine
- Atlantic pine barrens
- Cypress, magnolia, white cedar
- Spruce, hemlock
- Douglas fir, hemlock
- Douglas fir, oak, redwood, cypress
- Yellow pine, douglas fir
- Lodgepole pine, yellow pine, sugar pine
- Pinon pine, juniper
- Californian chaparral
- Temperate and semi-desert mesquite grasslands
- Yucca, sage brush, cactus, agave
- Dwarf willow, birch, alder

modified by river valleys and mountains. In the west the oceans and the mountains are dominant largely through their effect on rainfall (see page 37). The overriding factor in Europe is less one of temperature zones (the vegetation bears little relation to the hardiness zones on page 37) than of day length. Northern Scotland is on the same latitude as southern Alaska. At these latitudes conifers with their relatively small but permanent leaf area begin to have marked advantages.

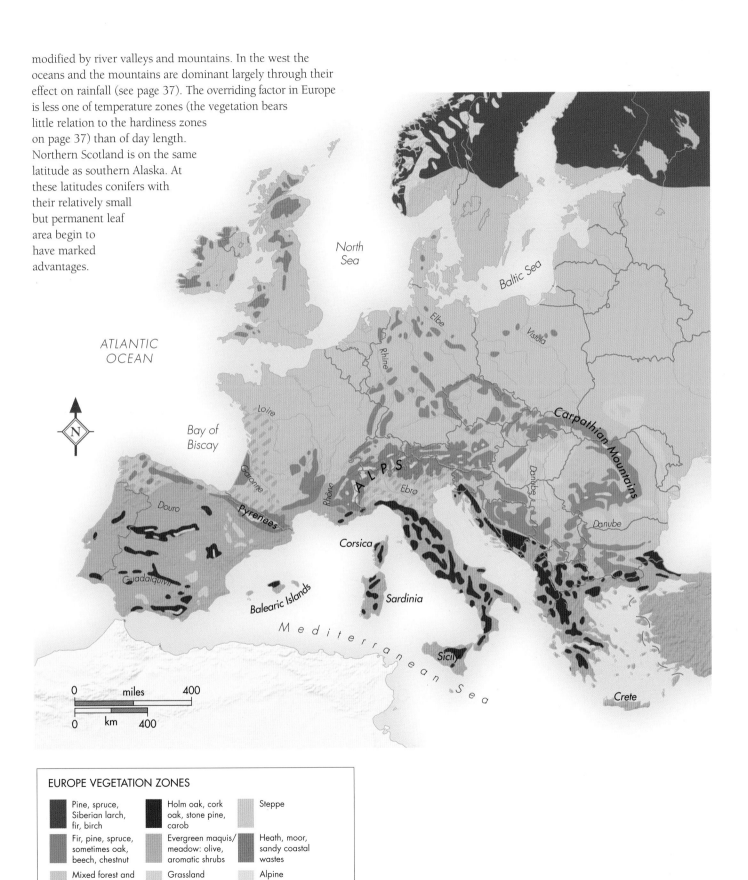

North Sea

Baltic Sea

Elbe

Vistula

Rhine

ATLANTIC OCEAN

Carpathian Mountains

Loire

N

Bay of Biscay

Garonne

ALPS

Ebro

Danube

Danube

Rhône

Douro

Pyrenees

Corsica

Guadalquivir

Sardinia

Balearic Islands

Sicily

Mediterranean Sea

Crete

0   miles   400

0   km   400

EUROPE VEGETATION ZONES

Pine, spruce, Siberian larch, fir, birch

Holm oak, cork oak, stone pine, carob

Steppe

Fir, pine, spruce, sometimes oak, beech, chestnut

Evergreen maquis/ meadow: olive, aromatic shrubs

Heath, moor, sandy coastal wastes

Mixed forest and meadow: oak, beech, fir

Grassland

Alpine

The Advance of Spring

The hours of daylight affecting the timing of flowering and leafing of the six different tree species in the key opposite are given for the five American cities indicated on this map. Day length interacts with temperature to give northern plants a growing advantage in late spring.

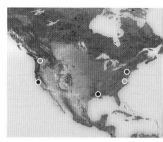

THE RIPPLE OF BURSTING buds, of woods growing amber with catkins, white with dogwood, tender green with the opening leaf, moves inexorably northward in the spring. Even in the dead of the year there are stirrings of swelling buds, early catkins on willows, eccentric bushes with winter flowers. But it is when the last frosts of spring pass that the full flush transforms the woods. In Florida and Naples this is as early as the first week of March. In New England not until about eight weeks later, about the first of May. At Edinburgh it is the last week of April.

The rising temperature is the prime mover in releasing leaves and flowers from their protecting buds. Spring moves north at a predictable pace in flat and open country. But local factors can cause wide variations on the same latitude. Latitude clearly bears little relation to the relative timing of spring in Europe and eastern North America.

Altitude is important. The temperature drops 1°F for every 300 feet you climb. The practical effect of this on tree growth depends on other local conditions, but 300 feet extra altitude has been observed to shorten the total growing season in the eastern States by five days, in the Alps by six, and in Scotland by 12.

Cities have the reverse effect. With the warmth and shelter they provide they can greatly advance the spring. London has pear blossom three weeks before a country garden only 50 miles away. Even a sheltering house wall can bring a tree into blossom a week or two ahead of its neighbour standing unprotected in the open.

Day length is the other deciding factor. However unseasonably warm it may be at the end of winter, or however miserably cold in the spring, trees are governed in the cycle by the photoperiod: the number of hours of darkness.

In the tropics day length varies no more than do the seasons: day and night are always about equal; trees grow year-round. As far north as Oslo, however, there is a bare seven hours of daylight in January, compared with

NORTH AMERICA						
MNTH	LATITUDE	LOCATION	EARLY SPRING	MID-SPRING	LATE SPRING	LIGHT
JAN	47°N	Seattle				8h 53m
	42°N	Cambridge				9h 25m
	39°N	Washington DC				9h 42m
	37°N	San Francisco		🌿	🌸	9h 52m
	30°N	Baton Rouge				10h 25m
FEB	47°N	Seattle		🌸		10h 21m
	42°N	Cambridge				10h 36m
	39°N	Washington DC				10h 47m
	37°N	San Francisco	🌸	🌷		10h 53m
	30°N	Baton Rouge		🌷	🌸🌿	11h 10m
MARCH	47°N	Seattle			🌷	11h 53m
	42°N	Cambridge			🌿	11h 55m
	39°N	Washington DC			🌷🌿	11h 55m
	37°N	San Francisco	🌸	🌿🌿	🌿🌿	11h 56m
	30°N	Baton Rouge	🌸🌿🌿🌸			11h 58m
APRIL	47°N	Seattle		🌸	🌸	13h 32m
	42°N	Cambridge		🌸	🌷	13h 17m
	39°N	Washington DC		🌸	🌸	13h 6m
	37°N	San Francisco	🌿			13h 5m
	30°N	Baton Rouge				12h 51m
MAY	47°N	Seattle	🌾			15h 4m
	42°N	Cambridge	🌸	🌾		14h 34m
	39°N	Washington DC	🌾🌿	🌿🌿🌿		14h 16m
	37°N	San Francisco			🌿	14h 11m
	30°N	Baton Rouge				13h 39m
JUNE	47°N	Seattle				15h 53m
	42°N	Cambridge				15h 14m
	39°N	Washington DC				14h 55m
	37°N	San Francisco				14h 42m
	30°N	Baton Rouge				14h 5m

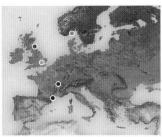

Latitude and hours of daylight can be less decisive in European gardens than the influence of the Gulf Stream. Its benign influence gives Britain in particular a uniquely temperate climate for its latitude. The coast of Norway shares some of the benefit.

17 in June. At the March equinox the north catches up with and overtakes the south in daylight hours. Thereafter its plants can begin to make up for their shorter growing season by using the extra light.

The interaction of temperature, day length, and local factors gives some surprising results. For example a magnolia which blooms in London in mid-March waits until mid-May in Fribourg in Switzerland, five degrees of latitude to the south – but in the northern foothills of the Alps. The same magnolia flowers in late March in both Seattle and Washington D C (eight degrees to the south) – but here it is the continental climate of the east which holds it back. It is interesting to speculate on the relevant influences using these tables in conjunction with the hardiness zone maps on pages 36-7.

Seasonal variation – the phenomenon of an "early" or a "late" spring – is totally unpredictable. When a cold spell holds up flowering for an unnaturally long time it can result in trees that normally flower one after the other coming out together.

In most years, however, the one thing that is predictable is the order of flowering and coming into leaf. This is the really essential information for any planter. With it he or she can dispose their trees either to give one short season of glory when they will be the envy of the neighbourhood with every tree aglow, or a calculated series of detonations so that for months on end there is always one tree flowering somewhere in the picture. Or a sequence of set pieces, of groups of complementary colour and form, which must be the object of every gardener who has the facilities and the space. The year-long sequence of special features (flowers, foliage, and fruit) of 100 species is given on pages 358-9.

MNTH	LATITUDE	LOCATION	EARLY SPRING	MID-SPRING	LATE SPRING	LIGHT
JAN	60°N	Oslo				6h 40m
	56°N	Edinburgh				7h 33m
	52°N	London				8h 14m
	47°N	Fribourg				8h 53m
	44°N	Montpellier				9h 14m
FEB	60°N	Oslo				9h 17m
	56°N	Edinburgh				9h 42m
	52°N	London				10h 1m
	47°N	Fribourg				10h 21m
	44°N	Montpellier				10h 33m
MARCH	60°N	Oslo			✿ ✿	11h 44m
	56°N	Edinburgh	✿		✿	11h 47m
	52°N	London	✿	✿ ✿		11h 49m
	47°N	Fribourg	✿			11h 52m
	44°N	Montpellier				11h 53m
APRIL	60°N	Oslo	✿ ✿	✿ ✿ ✿	✿	14h 28m
	56°N	Edinburgh			✿	14h 6m
	52°N	London			✿ ✿	13h 50m
	47°N	Fribourg	✿		✿ ✿	13h 32m
	44°N	Montpellier	✿	✿		13h 22m
MAY	60°N	Oslo		✿		17h 4m
	56°N	Edinburgh	✿	✿	✿	16h 15m
	52°N	London	✿ ✿	✿ ✿	✿	15h 37m
	47°N	Fribourg	✿	✿		14h 55m
	44°N	Montpellier	✿			14h 44m
JUNE	60°N	Oslo		✿		18h 50m
	56°N	Edinburgh				17h 35m
	52°N	London				16h 43m
	47°N	Fribourg				15h 55m
	44°N	Montpellier				15h 29m

EUROPE

The tables on these pages show the hours of daylight at different latitudes in Europe and America from January to June, together with the actual blooming and leafing dates of six species of trees. The dates are averages worked out over some 20 years.

A Word on Forestry

GREEDIER FOR SUNLIGHT than most plants and far better equipped to reach it, trees are what is dramatically termed the climax vegetation of almost any land with enough soil to support them. Left to themselves (which means not eaten by mammals) they inexorably overtake the ground-hugging plants, the herbs, and eventually the bushes and plants. Then they fight it out among themselves.

Over time it is fascinating to watch one species starting to dominate, then another with more reach and staying power. Fence off a piece of arable land and watch grasses take over for a year or two. Depending on

the nature of soil, pioneer plants such as foxgloves make their appearance in year two, flower in year three, then give way to the inevitable brambles. Little whippy birches are scarcely noticeable at first. Are there conifers nearby? Their evergreen shoots soon show up, with little hollies and very likely the weedy *Rhododendron ponticum* showing ambitious shoots from the start. Less conspicuous at first are the pale stems of rowan, or ash, or hornbeam. It is easy to miss their few first leaves among the brambles.

Come back, though, four or five years later and their still-slender sapling trunks are taller than the other plants. Another five years and they are jostling for the light, forming a canopy to rob lesser plants and compete with their peers. The management of this competition is the essence of forestry.

I have played the forester in two very different environments. One is in the centre of France, the Allier, where the eventual aim is to grow oaks to perfect maturity: timber to make fine furniture and the best wine barrels. The cycle from planting to harvesting averages 180 years.

The other is near the coast of north Wales, in the Snowdonia National Park, where the commercial return (there are other less tangible ones) comes from rapidly grown softwood. The rent-paying crop is Sitka spruce, destined for building or paper, and the average cycle 40 years or so.

In the Allier the soil is poor and lean, the rainfall low, and late frosts are relatively common. Oaks grow extremely slowly, but that is the point. With narrow growth rings they make fine-grained timber, almost impermeable by air (or wine). It fetches a fortune. In the Forêt de Tronçais (my woods were on the edge of it) trees ready for felling are auctioned individually where they stand, the buyers gathering round. One tree can fetch €20,000.

In north Wales the soil is also poor, gritty rendzina over granite or slate, but the rainfall high – up to 70 inches a year – and the climate generally mild. Late frosts are

The Forêt de Tronçais is France's noblest oak forest, with its most valuable timber. It takes 200 years to grow one of its perfect trees.

rare. The native woodland is oak, here, too: branchy wide-spreading trees mulched all over with mosses and lichens, springing from rocky banks and easily dominating their accompaniment of birch, rowan, and holly. Ash, beech, and a few sycamores are the other big trees, sometimes growing in sheltered gullies to 100 feet or more. "The forestry", as the locals call the interloping crop of foreign conifers, was planted, mainly in the 1960s, on sheep pastures. Their splendid drystone walls, bright green with moss, run like neolithic ghosts through the darkness of the firs.

The two systems of forestry are different in almost every way, the difference aggravated by the separate traditions of Britain and the Continent. What they have in common (and what differentiates forestry from other cultures – especially horticulture) is that the aim is a long,

Conifers in the mountains of north Wales. The bare ground, left centre, is a recent clearfell just replanted. Areas of trees of different heights show previous cuts and replants.

larch to Scotland) little thought was given to the open landscapes they supplanted. In the 1960s, when it was Sitka spruce that every hill-country forester in Britain planted, the trend was to plant brutal blocks of them, dark rectangles that cut across the sweeps of moorland. Not surprisingly they were resented.

The land was (and is) prepared for planting by ploughing the peaty upland turf. In wet places deeper drains have to be cut. The Sitka is the most tolerant tree: it will grow on rock or peat (or in good soil). It will even grow in a marsh – though slowly. Even for Sitka some drainage is vital.

There are foresters who see monoculture as the most efficient way to make money, others who see the land and the market as the two ends of a more complex equation. What tree will create the most valuable timber on the different soils of the property? What timber will be in demand in 40 years' time? Bet-hedging, intelligent reading of the land, a knowledge of tree species, and simple curiosity lead to a more complicated system of plantings.

A disadvantage of the standard spruce is inflexibility. The sawmill pays the best price for trees of a certain size. At the rate these trees grow the window for cutting them is only a few years, before their butts (the lower end of the trunk) grow too stout for the sawmill and have to be wasted. Larch is more flexible; it makes better timber (notably long-lasting, hard, and weather-proof with almost oak-like qualities) over a longer period. You can cut it young and sell it or leave it standing to gradually build up value.

Douglas fir can have high value but is less tolerant than spruce. It shuns marshland and benefits from shelter when it is young. Pines (Corsican or Scots: Californian lodgepole was a disastrous experiment of the 1960s) are reserved for rocky outcrops or shallow drier soil. There are silver firs (grand fir, noble fir) that grow fast and straight, and certainly add tone to the forest, but they suffer the disadvantage of minorities. The customer (the sawmill)

straight ideally branchless pole.

The best-quality oak in France is grown tall and straight. The best, tallest, and straightest trees are gradually isolated over the course of two centuries, their less perfect neighbours felled generation after generation until they stand wide spaced in broad clearings. They are felled in the winter after a rich harvest of acorns. The forest floor is cleared of bushes and weeds, the acorns fall and germinate as thick as grass. The huge trees fall on this carpet of seedlings. Within five years they have occupied the ground to the exclusion of everything except, often, brambles. Already they have started their 200-year race to become the next champion seed trees. Some 25 years later foresters will drive rides through the saplings to start the process of elimination once again.

English oak growing since the 17th century has been different; scarcely forestry at all, in fact. Oaks were planted with warships in mind. It was the characteristically angled or curving branches that were needed for shipbuilding. They were grown far apart to encourage their branching, either in pasture or woods where the competition was regularly coppiced – cut down to a stool to give the oaks space. Medieval man, moreover, was not so foolish as to let trees grow bigger than he needed and then have to saw them up. Houses were built with young trees when they had reached the right dimensions. Coppicing (of oak, among others) produced a steady supply of straight poles every 30 or 40 years.

A forest of exotic conifers is by definition an artificial creation. When the idea was first introduced (in the 18th century, by such as the Duke of Atholl bringing the

tends to look at them with narrowed eyes. Western hemlock is often shunned, despite making good timber very quickly, because it tends to flare out to a cumbersome butt. Norway spruce is simply not so fast or so tolerant as Sitka. So it tends to be a one-horse race.

The process of making the forester's dream, the fast and flawless tree, is a recent one. Since the 18th century landowners have realized that it makes sense to leave one or two of the biggest trees to reseed and restock land they have cut over. But juggling with chromosomes is very much a 20th-century idea.

In some cases it means running seed orchards, where the fertilizing of trees can be done under supervision and the results controlled. A typical forestry seed orchard would be a few acres of young pine trees, too young to bear seed themselves. On them are grafted the upper, seed-bearing branches of the best of the older trees.

In the orchard the flowers are within

A recent clearfell of 40-year-old Sitka spruce with the 3,000-foot Cader Idris in the background. The felling tractor has left the trees already measured, sorted, and piled by sizes ready for collection by the "forwarder".

reach. It is easy to protect them from random pollination and fertilize them only with pollen from another chosen tree. Then in due course it is easy to collect the cones and know just what seed you are sowing.

The trees are planted out as tiny plants in their second year of life. They are carried up the hill in a sack and planted in a little notch cut with a spade, two yards apart in straight lines. For a year or two you can scarcely find them in the grass, the willow-herb, the bracken and ferns, foxgloves and brambles. Then your eye picks up the line of green tufts. So, often, does the nose of a weevil that feasts on the bark of the baby trees. It often means two visits with a knapsack sprayer to solve the problem. Sheep are a menace. Seedling birch, sometimes rowan, can seem to outpace the planted trees at first. It is often worth a weeding operation with a strimmer in the third or fourth year to show who's boss. Remarkably soon the lower branches of the conifers smother the competition. By year eight or nine Sitka spruce has generally asserted its formidable power.

There is still plenty to do in the forest. Roads and tracks are vital. A forest, it is

often said, is as good as its access. Accidents happen: the main threat is wind, when a strong gust brings down lines of half-grown trees. Once there is a gap other trees become vulnerable. The "wind-firm edge" is an important concept: windward lines of trees that have become accustomed to buffeting and grow strong roots to resist it. Break that bulwark and acres of timber can collapse.

Fire is the other insurable danger. The standard practice is to start "thinning" the trees after 15 years or so. Sitka spruce by this time is 30 or 40 feet high, dense, and fiercely prickly: no one can enter the tight press of trees. The first thinning consists of cutting rides or "racks" by felling a single line every 15 yards. The "thinnings" are mostly leggy trees of little value except as chipwood or biofuel. Now the forester can see from within how the crop is doing. More important, light can penetrate and the inside of the forest start to benefit.

Some foresters believe in thinning again relatively soon. Others may wait another 15 years before cutting more swathes, this time taking out tree trunks usable as posts or for small construction. The remaining trees meanwhile have only sharp snags of old

lower branches killed by the darkness, but trunks – arrow-straight, of substantial girth, and above all 20 feet or so – produce lusty green upper branches and a yard-long leader.

Foresters talk about "continuous cover'" and "uneven age", a promised land in which a forest, even of exotic conifers, achieves a perpetual equilibrium, with trees reaching maturity among their children and grandchildren of many generations. Compared to an even-age plantation it looks unchanging and serene: an ecologist's dream. The way to get there is by assiduous thinning, leaving mature trees widely spread, and encouraging their natural regeneration by seeding. Seedlings, of course, will include many things beside the main crop. Birches will soon be abundant. After a few years there will be rowan, oak, larch, pines, perhaps beech. Bracken can be a menace. It will look beautiful and be musical with bird song. The problems come when you need to remove trees (income must come from somewhere). Felling and extracting big trees inevitably damages the surrounding plants, scrapes the bark of others nearby, and ruts the ground.

"Extraction" is always a messy business. Men with chain-saws are the exception in 21st-century forestry. Increasingly ingenious tractors are now designed to roll over steep, rocky, marshy, obstructed terrain, climb over piles of timber, and fell trees at high speed with surgical precision.

A hydraulic arm reaches out with a head that clamps around the base of each tree and measures its girth (recorded by the computer in the cab – length is measured in feet, girth in centimetres!). A concealed chain-saw then automatically cuts it and, as it falls, the hydraulic hand and arm lift it to the horizontal. It is grabbed by ratchets that pull it through the head, stripping off the branches, until the calipers measure the required girth for another cut. A big tree is measured and cut into perhaps two 16-foot "logs" (ideal for joists or planks), two 12-foot "bars" (used for palettes and battens), and two or three "poles" (thinner spars for fence posts, chipwood, or biofuel). The machine deposits each category in (relatively) tidy piles as it goes. The "brash" of stripped branches and foliage, forms a carpet over rocks and gulleys on which the

tractor rides to its next prey. A hundred trees would be a morning's work for one man with such a "harvester".Next on the scene is the "forwarder", an astonishingly flexible tractor and trailer equipped with a grab crane. The forwarder picks up the

already sorted logs, bars, or poles and bumps and rolls its way to a roadside stacking point to unload. From there timber lorries carrying 25 tons ferry it down forest roads, negotiate winding lanes, and use the fastest route to the sawmill.

A harvesting tractor has an hydraulic arm carrying a computerized felling head that grips, measures, cuts, and piles the logs.

The graded logs are stacked at the roadside ready for loading on 25-ton lorries that have to negotiate the narrow forest roads.

The haul from the steep, rutted, stump-strewn felling area to the road where a lorry can reach to carry away the harvest is undertaken by a "forwarder", a heavy tractor and trailer flexible enough to cross chaotic, rocky, and often boggy ground. The front wheels with immense tyres are suspended separately; the rear ones are tracked. The cabs of forest vehicles are heavily reinforced to protect the driver.

The Ice Ages

WE SHOULD GLANCE back a moment at the origins of the forest. When the club mosses reared up from the swamp and ferns formed the first trees – perhaps some 400 million years ago – they existed in a blanket of carbon dioxide. There was no living creature; there was no breathable air. But there was photosynthesis.

The coal in the ground is the carbon from that unbreathable sky. Tens of millions of years of forest, building trees with carbon and leaving oxygen as the by-product, was the origin of our world. Life on earth was made possible by trees.

We must leap unknowable millennia of evolution to reach the relatively recent past. The period known as the Tertiary, from (perhaps) 38 million to (perhaps) 12 million years ago, was furnished with trees shaped just as we know them. What was different was their numbers. The forest was boundless; it reached from pole to pole. And it was homogeneous; the same species, from a range far wider than we know today,

Fossil evidence that hickories, historically confined to North America and China, grew in Spain in the Miocene period (5–23 million years ago).

might occur anywhere in that great sea of trees.

In early Tertiary times the climate was distinctly warmer than today: the vegetation of Europe was like that of modern southeast Asia. By the Pliocene (i.e. late Tertiary) era gradual cooling had killed off the sub-tropical components of the northern temperate forest or caused them to become mere herbs, leaving a rich mixture of all the families found around the world today at these latitudes. Asia, America, and Europe alike had swamp cypress, magnolia, sequoia, sweetgum, ginkgo, incense cedar, umbrella pine, as well as oak, beech, and the other common trees we know.

This was the situation about a million years ago when the climate began to change for the worse, and went on changing till the sea froze at the poles, and trees died, and their offspring too, unless they could find a place unoccupied by other trees where the ice had not yet reached.

There was not one ice age, but four: Gunz, Mindel, Riss, and Würm (for such are the comic-opera names science has given them). They were leisurely affairs. Between Gunz and Mindel (which was the coldest, and saw the ice reach its maximum dimensions) there was a temperate period of something like 60,000 years, when conditions may have been similar to our time. Come to that, we are quite probably merely enjoying another interglacial period ourselves.

Vast numbers of tree species perished for ever as the arctic cold came south. Geography was the decisive factor. Where there was a line of retreat, a continuous land mass, species could fall back in tolerably good order. America and eastern Asia provided these conditions. Where there was a high east-west mountain range, or a sea with no way around, vegetation had its back to the wall. Between them the Alps, the Pyrenees, and the Mediterranean cut off the last hope of most of the species of Europe (which up to that time had been as rich in trees as the rest of the world). Only about 36 genera, and very few species of these, survive as European "natives"; whereas in China there are only six of the whole temperate world's genera missing.

The southern hemisphere, particularly Africa, suffered earlier than the north: a great glacier over South Africa advanced into the tropics some 250 million years ago; retreat to the north was prevented by an ocean where the Sahara now lies. Tropical Africa today has nothing like the richness of flora of South America or southeast Asia.

There were special conditions which sometimes allowed survival from the ice. One was extreme drought: in a place where rain (or snow) rarely fell, in the lee of a great mountain range, the ice may never have built up. Some botanists believe that such refuges were numerous, particularly at high altitudes, and account for many of today's local species. On the other hand there were conditions which extinguished

GONDWANALAND

The theory of continental drift holds that one huge southern land mass broke up and has been drifting apart for some 100 million years, taking plants with it. It was based on a plant fossil found in Gondwana in northern India and is said to account for the presence in (only) the southern continents of such plant families as the Proteacae. There is continuing debate about whether such an ancient split could explain, for example, the distribution of such genera as *Nothofagus*, found only in Australasia and South America.

375 million years ago

255 million years ago

The last ice age (left) reached its fullest extent of glaciation 20,000 years ago, covering most of North America, all of Scandinavia and northwestern Asia, some of northern Europe, and all the northern oceans. The ice

sheet built up chiefly where westerly winds brought rain. Siberia largely escaped glaciation, as did China, which became a refuge for such trees as could stand the cold. In America trees retreated south; in Europe most perished.

a race even where the ice was miles away: there is no harder barrier for a plant to cross than a closed community of other plants, taking up all the available space. This may well have been the case in, say, the south of France, which was never frozen but which failed to provide a refuge for many plants from the north.

The total effect of the ice ages was an enormous acceleration of evolution. Without them the world would still be in the age of reptiles. The effect of refugee

plants streaming across the continents was to stimulate genetic crossing, to favour what would previously have been unsuitable mutations. Nor is the process by any means at a standstill today. We can watch the pines for example still in full cry, reconquering the hills.

Fourteen thousand years ago the ice retreated for the last time (so far). The variations in climate since that time have been relatively slight, but enough to favour different trees at different times.

What flourished when is known to us primarily by the analysis of the pollen grains found in recent lake sediments in the strata of peat. In the immediately post-glacial "sub-arctic" period, the areas where the ice had been were still raw mineral soil and an unchecked wind allowed only the toughest trees – dwarf birches and willows – to make the first colonizing moves.

Since then, starting some 11,000 years ago, there was a build-up through a cool, then a more continental and extreme climate to a climax of relative warmth and wetness – the "Atlantic" period of between 7,500 and 5,000 years ago, the era when some of the bristlecone pines still alive today were seedlings.

Their lifetime has seen a reversal of the previous trends; another dry, continental, and extreme period followed by about 1,500 years of lower temperatures and more rain, the era (in human terms) of the classical civilizations and the Dark Ages. About AD 1000 another trend becomes discernible – now with plenty of still-living trees to give evidence: the relatively warm and dry period in which we live.

To speak in terms of broad trends is the best anyone can do, reviewing such a span of time. What is worth remembering, however, is that the answer to any particular question need not lie in the great averages of history. Cataclysms and catastrophes are as much a part of evolutionary history as the gradual attrition of the ages. Whether or where a species exists can be decided by one great frost; one plague of locusts; one unaccountable wave of natural competition. The movements of nature can take millennia – or they can happen in a flash.

Before the ice ages, some 34 million years ago, in the forests of Colorado, new studies have found the fossil evidence of 100 species of plants, including *Sequoia*, palms, *Ailanthus*, *Koelreuteria*, roses, *Mahonia*, *Amelanchier*, and many oaks and pines, all indicating a warm temperate climate.

One million years ago, just before the first ice age, typical European forest vegetation included the present oaks, birch, spruce etc. but also scores of trees we think of as American or Asiatic: magnolias, redwoods, hickories, white pines, incense cedars, umbrella pines, plum yews, and ginkgos.

185 million years ago

100 million years ago

15 million years ago

The Old World

TREE DISTRIBUTION SINCE the last ice age has been the slow setting to rights of what the ice disjointed. At first the process was infinitely gradual, through colonization and evolution. But later, since the dawn of civilization when man began moving plants about, redistribution has happened at a faster and faster pace – so that today, in countries which the ice overran and which have since seen a succession of civilizations, the question of what tree is native has little meaning.

Native when? When the ice advanced? Or retreated? Before man set about destroying the forest? In the earliest written account?

We know little of the early years. We know that the peach tree and the mulberry came from the Orient, the walnut from the Caucasus, the fig from Persia. But what trees the Romans brought, or what plants they moved from one part of their empire to another, are questions we can no longer answer.

Plants of usefulness are cultivated and transported before plants of ornament. If the Romans took the sweet chestnut and the walnut with them to France and Britain it was for their fruit. But the Romans were rose-lovers as well as gourmands and fanatical ablutionists.

Diggings at Chichester (their port of

Regnum) have uncovered a palace garden made exactly to Pliny's instructions to the very size and depth and spacing of the flowerbeds. It would be surprising if the nurserymen of Rome did not export to officers abroad the plants they loved at home.

The systematic study of plants had been initiated by the Greeks. It was continued by the Romans. Then in the Middle Ages it lapsed into a mixture of carpenter's know-how, herbalist's mumbo-jumbo,

and poetical symbolism. Chaucer made a catalogue of trees, which is a fair sample:

The Renaissance brought the classical

> The bilder ook, and eek the hardy asshe;
> The piler elm, the coffre unto careyne;
> The boxtree piper; holm to whippes lasshe;
> The sayling firr; the cipres, deth to pleyne;
> The shooter ew, the asp for shaftes pleyne;
> The olive of pees, and eek the drunken vyne.
> The victor palm, the laurer to devyne.

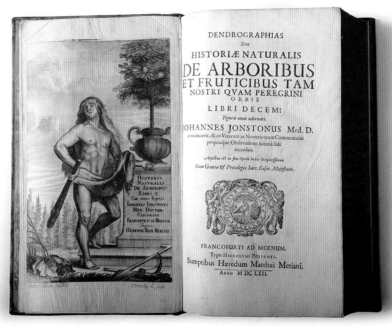

The first fully illustrated tree encyclopedia I know was written by my namesake, John Jonston, presumably a Scot, and published in Frankfurt by Matthew Merian in 1662, the same year as Evelyn's *Sylva*. It would be better known if it weren't in Latin.

DRAMATIS PERSONAE

Philosophers, herbalists, gardeners, apothecaries, explorers, and landscapers have all made contributions to our advancing knowledge of plant life – of which trees are only a highly specialist corner. There were of course parallel scholars, in China, India, and Arabia, but until early modern times almost all their interest was in the medicinal and pharmacological uses of plants. The study of plants' family relationships is a relatively recent idea.

The school of Aristotle and his pupil Theophrastus laid the foundation of Western biology in Athens in the 4th century BC.

Albertus Magnus (1193–1280) was the greatest exponent of Aristotle's botanical work. He taught at Cologne and Paris.

John Gerard (1545–1612) was author of the famous *Herball or General Historie of Plantes*, published in 1597.

Jean Bauhin (c.1541–1613). His *Historia Plantarum Universalis* set new standards in accurate and concise plant description.

systems to light again. Learned apothecaries in the north of Europe found that the plants of ancient Greece failed to tally with the local flora. Modern botany began with their "herbals".

The first herbal in English was Turner's, published in 1546. Gerard's popular and readable *Herball*, largely plagiarized from the work of the Dutchman Dodoens, appeared in 1597; John Parkinson's *Paradisus in Sole* in 1629. Parkinson was an ardent collector and may himself have introduced new species into cultivation.

At the same time, what had seemed an inexhaustible supply of timber began to look disturbingly thin. By the beginning of the 16th-century royal statutes were enjoining replanting. The Great Wood of Scotland (the Caledonian Forest) is described as "utterly destroyed". Books of husbandry (the first, Fitzherbert's, in 1523) began to give instructions for planting trees.

The first botanical expeditions (in the same period) covered the ancient world, most of it then in the Turkish empire. One might say that consolidation and realization of the trees of the Old World, including the world of the Bible, took up to the middle of the 17th century. In about 1600 Europe received the horse chestnut from Asia Minor; by mid-century she had the cedar of Lebanon.

There is a great stocktaking in John Evelyn's *Sylva* of 1662. The monarchy had been restored to an England worse off than ever for timber – particularly for oak, without which there could be no navy.

As at Vaux-le-Vicomte (above), the vast formal gardens of France were conceived as settings for self-satisfied people in huge numbers. Trees were mere building materials.

National security literally depended on oak trees, as France realized in the following century. Evelyn was commissioned by the Royal Society to discourse on growing trees. He did it with such gusto (and apparent knowledge) that his discourse was still being printed a century later.

While England was worrying about warships, France was getting on with the serious business of planting avenues. The same year as *Sylva* saw the culmination of the French style of landscaping: Le Nôtre's design for the park of Vaux-le-Vicomte, soon to be surpassed at Versailles.

Landscape design had advanced hand in hand with baroque building. Concern was entirely with form; plants were treated as far as possible as lifeless absolutes – an avenue and a colonnade came to the same thing; so did a hedge and a wall. Handled by masters, trees did what they were told.

As Kip's engravings (below) of English country houses of the period show, however, there were fewer masters than gardens. A banal formality was the general rule. The first botanic gardens were not much advanced in design from the knot gardens of the Tudors.

John Tradescant the Younger (d. 1662) and his father John the Elder (d. 1638) introduced many new trees to England, including some from Virginia.

André Le Nôtre (1613–1700). The chief practitioner of the majestic French style of landscaping; he designed the parks of Versailles and Vaux-le-Vicomte.

John Evelyn (1620–1706) was a courtier of King Charles II, passionate gardener, and encyclopedist of, among other things, trees.

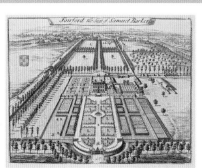

English landscape design was under heavy French influence when Johannes Kip did his famous engravings of country houses. This is Fairford in Kent c.1750.

The New World

WHETHER IT WAS actually the younger Tradescant – who went three times to Virginia – or correspondents of his father, or other returning colonists who brought them, we shall never know. But by 1656 the plant list of the "museum" the Tradescants kept in their Lambeth garden included many of the biggest and most striking of the trees of the American east coast: the black locust or false acacia, named *Robinia* after Tradescant's friend Jean Robin, curator of the Paris Jardin des Plantes; the tulip tree; the swamp cypress; the eastern red cedar. The Tradescants not only collected plants. Their "curiosities" of all kinds became the basis of Oxford's Ashmolean Museum.

John Evelyn mentioned the tulip tree (which he called the Virginia popular) and *Thuja*. He spoke of "great opportunities … we have of every day improving our stores with so many useful trees from the American plantations".

The man who seized the opportunities was the Bishop of London, Henry Compton, a nobleman and former mercenary. From his palace at Fulham he directed the spiritual affairs of the colonies, sending missionaries among the Native Americans, and choosing for his men such as could spot an unknown plant and, spotting it, get it safely home.

His Fulham palace garden became an

important centre for the plants of America. Among the trees introduced there (by his missionary John Bannister) was the first magnolia grown in Europe, the swamp bay (*Magnolia virginiana*). Black walnut, box elder, balsam poplar, balsam fir, scarlet oak, flowering dogwood were others.

The first home-grown American botanist was John Bartram, the son of a Quaker settler at Philadelphia, who farmed the banks of the Schuylkill river. By a happy chance he was put in touch with a London linen-draper named Peter Collinson, another Quaker, who had a passion for plants, particularly for trees, and (not being in a position to send out missionaries) was

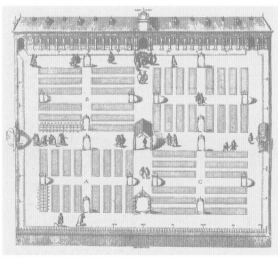

The Botanic Garden of Padua, founded 1545, was Europe's first. Leiden (left) began in 1590 with the Flemish botanist Clusius. He encouraged East India Company ships to collect plants and started the Dutch craze for tulips.

The Royal Botanic Garden at Kew (right), originally a private royal collection, now the world's greatest botanical institution, is dominated by its pagoda, built in 1761.

1600	1625	1650	1675	1700	1725
John Tradescant visited Algeria, Russia, and the Levant collecting plants. The Virginia Colony founded. Horse chestnut introduced.	**New trees** start to arrive in northern Europe.	**Harvard College** founded in Massachusetts. Cedar of Lebanon introduced to England. Carolina founded. Anglo-Dutch war. Tulip tree introduced from America.	**London plane introduced.** Chelsea Physic Garden founded.	**Kaempfer in Japan.** Queen Anne on the throne. William Kent at school. Cunningham in China. Act of Union with Scotland. Bank of England founded. Isaac Newton. J S Bach. Battle of Blenheim. *Robinson Crusoe* published.	

Civil War in England. Mulberry comes from China, hopes of a silk industry. Larch reaches England. *Robinia*, western plane, swamp cypress arrive from America.

William and Mary on the throne. Pennsylvania founded. *Liquidambar* and black walnut introduced to England.

Weeping willow introduced from China. Start of Georgian England. *Magnolia grandiflora* introduced from America. Linnaeus develops his system. Walpole is first prime minister. Handel's *Messiah*. Slavery in Georgia. First larch forestry in Scotland

looking for an American to collect plants and seeds for him. Their correspondence began in 1732 and continued until Collinson died a third of a century later. By then Linnaeus was ready to describe Bartram as "the greatest living botanist in the world".

His discoveries numbered some 200 species. He found them by wandering about alone in Native American country, from the Great Lakes down to Georgia. Courage has always been as important as eyesight to people who seek new plants.

The list of Bartram's trees is awe-inspiring. It includes the sugar and silver maples; the American ash, elm, and lime; the black, red, and white oaks; *Magnolia grandiflora*; the river birch; the longleaf and the shortleaf pines. Nor was it entirely a one-way traffic. The cedar of Lebanon (in 1746) and the horse chestnut (in 1753)

were two of the trees Collinson sent Bartram for his own collection, America's first botanic garden.

What started as a private arrangement became almost a public institution as the results began to be seen. Collinson shared the costs with subscribers: at first the young Lord Petre, whose estate was at Ingatestone in Essex; then the dukes of Richmond, of Bedford, of Norfolk; then Frederick, Prince of Wales, father of King George III, who with his wife, Princess Augusta, lived at Kew House by the Thames, just west of London.

The story of Kew, the greatest of the world's botanic gardens, begins here. The Prince and Princess provided a nucleus for the mushrooming interest in new plants. They employed the great William Kent (and later Capability Brown) to landscape their grounds. They asked the Earl of Bute, a Scot and (one might almost say therefore) no

mean botanist, to supervise their planting. The Prince died in 1751, but Princess Augusta pressed on. In 1759 she appointed William Aiton the first Curator of the Royal Botanic Garden. He was another Scot, trained at the already famous Chelsea Physic Garden by its celebrated curator, Philip Miller, the author of *The Gardener's Dictionary*. Between them Aiton and his son were at Kew for no less than 82 years.

What perhaps has more to do with trees is the interest some of the great landowners began to take in the new introductions. The Duchess of Beaufort at Badminton, the Duke of Argyll at Hounslow, and the Duke of Atholl in Perthshire became passionate tree planters. Lord Weymouth at Longleat planted so many of the new white pines from New England that they became known by his name. Records begin here for the growth of many trees that are still alive today, with every biographical detail ever recorded.

What was strange and disappointing, however, was the ignorance of all this by the great landscapers. England was at that very moment going through the greatest redecoration she has ever had. The old parks and gardens were being rooted up wholesale, to be replaced by the placid quasi-natural landscape favoured first by Bridgeman and Kent, then by the all-powerful Brown.

As Humphrey Repton, their more sophisticated successor, remarked: "Their trees are of one general kind, while the variety of nature's productions is endless, and ought to be duly studied."

The 18th century saw the making of England's great "natural" landscape gardens. Capability Brown's Stowe, shown here, is typical. But the great influx of new trees was virtually ignored.

1750	1775	1800	1850	1875	1900

Foundation of Kew Gardens. Capability Brown starts work. France loses America. Benjamin Franklin flourishing. Bartram's plants go to Collinson in London. *Ginkgo* and *Sophora* introduced. The British Empire expands in India.

Declaration of Independence. Banks visits Australia. Yulan magnolia arrives. First *Eucalyptus*. The Scottish Enlightenment. Adam Smith's *Wealth of Nations*. Invention of the toilet. Metric system introduced in France, Humphrey Repton.

Napoleon. Foundation of the Royal Horticultural Society by Josiah Wedgwood. Lewis and Clark cross America.

Queen Victoria. Florida joins the States. Darwin's *Origin of Species*. First Japanese plants, maples. Larch. Gold rush. First Himalayan rhododendrons. The Great Exhibition. First transatlantic cable.

Olmsted designs Central Park. Chinese and Japanese plants start to flow in numbers. Sargent at the Arnold Arboretum. *Magnolia campbellii*. *Davidia* from China. Colorado spruce introduced.

Freud publishes. Picasso paints. Queen Victoria dies. Wilson in China; a flood of plants. Many new rhododendrons. First *Nothofagus*. The Wright Brothers fly. Traffic lights invented.

1825, Douglas in the west. Western conifers introduced. First railways.

The World Expands

CHINA IS THE MOTHER OF GARDENS. The Chinese emperors loved and collected plants for almost 5,000 years. Large parts of China have been cultivated for so long that the notion of a wild plant is laughable: everything desirable is cultivated; everything else extinct. Species of the most prized plants – paeonies, chrysanthemums, camellias – have been improved and treasured for century after century. One can imagine the head gardener of the Emperor of China puffing up his paeonies to appropriate dimensions.

But China had no intention of letting the barbarians in. After Marco Polo's 13th-century visitation there was a pause of more than 300 years before another contact was allowed, under the pressure of a Western world curious and hungry for trade. It was the 18th century before any Westerner got even a toehold to look for plants.

The toehold was one of the East India Company's "factories" on the coast; the Westerner was James Cunningham, a surgeon with the company, who visited Amoy in the Formosa strait in 1698 and Chusan, an island south of Shanghai, in 1700. He was forbidden to ramble outside the port, but he procured paintings of Chinese plants and specimens of what the nurseries had to sell.

The French were more subtle – and more successful. They had no trading rights but gained entry to the country by sending missionaries trained in skills the Chinese lacked: glass-blowing; clock-making; engineering. By this means Father Pierre d'Incarville reached Peking in 1742 and stayed for 15 years. He had the commission of France's greatest botanist, Bernard de Jussieu, to collect for the royal gardens and the Jardin des Plantes in Paris. He also had London correspondents – among them the insatiable Collinson. The fruits of his labours included the tree of heaven, the silk tree (*Albizia julibrissin*), *Toona*, *Sophora*, the Chinese *Thuja*, the Chinese juniper, and the so-called pride of India (*Koelreuteria paniculata*).

All connections with China, however, were subject to the imperial whim. D'Incarville had been lucky; after him the doors closed again. But a glimpse had been enough: chinoiserie was all the rage in Europe – in furnishing, in decoration, even, for a while, in gardening (witness the pagoda built in 1761 at Kew).

With Japan the story was much the same: a tantalizing glimpse, followed by drawn blinds.

The earliest confrontation between the Japanese and the Europeans had not been a happy one. The Europeans were kept effectively at arm's length; only the Dutch

An expedition that never was. This painting of 1771 by John Hamilton Mortimer was recently identified in the National Library of Australia as the commemoration of Sir Joseph Bank's proposed second voyage to Australia with Captain Cook. The figures (left to right) are the naturalist Daniel

The first *Camellia* in the west was *C. japonica*, probably introduced from Japan to Portugal in the 16th century. Linnaeus named it after George Kamel.

A "strange oak" from Japan, *Castanopsis cuspidata*, was introduced by Philipp Franz von Siebold in 1830.

Not all plant introductions were successes in Western cultivation. The Chinese horsetail pine (*Pinus massoniana*) is one that failed.

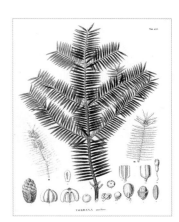

The Japanese nutmeg yew (*Torreya nucifera*) was in cultivation in England by the 1760s; nobody knows how it arrived.

Solander, Sir Joseph, Captain Cook, John
Hawkesworth, who was to record the expedition,
and Lord Sandwich of the Admiralty, the sponsor. They
never sailed because Banks insisted on taking such a
large staff that their accommodation was deemed a

danger to navigation. It is the earliest known portrait of
Captain Cook. Mortimer was mortified; in the embarrassment
of the cancelled expedition his painting was never exhibited.

Wasamaki (*Padocarpus macrophyllus*) reached
England early in the 19th century as an essential
ingredient of Japanese gardens and an omen of
good luck. It never caught on.

By the time his successor, Carl
Thunberg, a pupil of Linnaeus, reached
Yedo in 1776 there was a new spirit
abroad in Europe. Captain Cook had
visited Australia; Francis Masson (the first
professional plant collector) had returned
from the Cape; the botanist William
Roxburgh had left for Calcutta. The
French were heading for Peru.

The flood of new kinds of plants from
South Africa was to have the most impact
on European gardening. Francis Masson,
sent out by Kew in 1772, had a sensational
haul. *Pelargonium*, many bulbs, heathers,
and countless colourful daisies make their
appearance at this time. No country has
offered us more garden colour than the
Cape. Its forests, though, gave us little,
if anything.

Nor for a long time did those of
Australia (the first *Eucalyptus* was
introduced in 1790 but proved tender).
The monkey puzzle, arriving from Chile
at the same time, drew collectors' eyes
toward South America, but southern
Europe, the Caucasus, and increasingly
the west of India were seen as the most
promising source of new trees to try.

The new spirit was summed up in the
person of Sir Joseph Banks, Lord Bute's
successor at Kew and one of the most
knowledgeable and enthusiastic (and
richest) patrons that exploration has ever
had. He himself had sailed with Cook.
Now he saw to it that every expedition
was equipped with a competent young
botanist. And, incidentally, that the loot
should come to swell the great collection
he was making at Kew.

The chart of tree introductions on pages
358-59 illustrates perfectly the situation at
the end of the 18th century. Introductions
from the Mediterranean had come in a
steady flow for centuries. A colossal flood
from eastern North America had steadied
to a stream, largely from the work in the
southeast of another emissary of Kew,
John Fraser. But the rest – the Far East,
the American west, and the whole of the
southern hemisphere – were about to be
ransacked for the undreamed-of wealth
of trees that was to treble the number
known to the Western world within
the next century.

were allowed any trade at all, and they were
confined to an island in Nagasaki harbour,
except for annual envoys permitted to bring
presents to the emperor at Yedo – as Tokyo
was then called.

Engelbert Kaempfer, a German doctor and
botanist in the employ of the Dutch East India
Company, made the journey to Yedo twice, in
1690 and 1691, and gave the first report of
Japanese trees. The ginkgo (a Chinese tree, but
imported to Japan, like the art of gardening
itself, at a very early period) was his most
important discovery; he also sent back to
Holland maples and flowering cherries. He
was the last European botanist to visit Japan
for more than 80 years.

Giants and Gentility

The journeys of David Douglas through the wilderness of the American west for the Horticultural Society in London brought in a spectacular harvest of many of the world's biggest trees and the most valuable for European foresters. First he explored the Columbia river basin for two years before crossing the continent. His second trip was to Monterey and the central coast and his third back to the Columbia.

THE FIRST NEWS THAT THE Pacific coast of America held something quite exceptional for the botanist came from Captain Vancouver's voyage of 1792. His botanist – I almost said his Scotsman – was Archibald Menzies. With justifiable excitement Menzies reported the Douglas fir, the Alaska cedar (or Nootka cypress), the Sitka spruce, and the coast redwood: all trees twice as high as anything he had seen before.

John Bartram had dreamed of an expedition to the west many years before. His scheme eventually took place, 37 years after his death, when Thomas Jefferson organized the 40-strong Lewis and Clark expedition. But it lacked Bartram, or even a lesser botanist. The expedition returned from the coast (having spent two-and-a-half years on the way) with a very modest collection of seeds, without apparently having even seen Menzies's great trees.

When in 1824 London got round to sending a collector Sir Joseph Banks was dead, and Kew (temporarily, for political reasons) out of favour and short of funds. It was the Horticultural Society (founded in the year Lewis and Clark set out from St Louis) which commissioned the trip. At the recommendation of the great William Jackson Hooker, then at Glasgow but later to be the most illustrious director Kew has ever had, the man chosen was David Douglas, son of a stonemason from Scone in Perthshire, a self-taught botanist who had impressed Hooker as having unusual qualities.

Courage is common currency among plant collectors. The stories of their exploits have made many a good book. But by any standards Douglas was special. He was a Highlander with an obsession. He addressed himself to the northwestern wilderness with the energy of a fanatic. The Native Americans learnt to treat this strange "grass man" (as they called him) with utmost respect.

In a Stewart tartan coat, and with tea (of which he drank gallons) as his only comfort, he travelled thousands of miles of trackless country, very often quite alone, carrying on his back (and above his head through icy torrents) a prodigious load of specimens, seeds, cones – even, at one time, two living eagles. To be wet through, to be wounded, to be starving apparently meant nothing to him if he scented a new tree further on. His first expedition to the Columbia river (not counting the eight-and-a-half month voyage out, of which no less than six weeks were spent waiting to cross the dangerous sand bar at the river mouth) lasted from April 1825 to March 1827, when he left for home overland by the Hudson's Bay Company's "express".

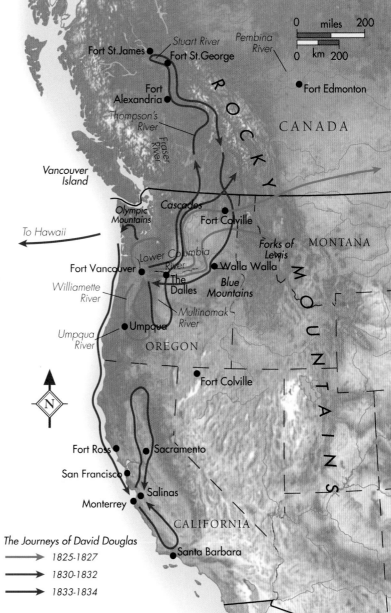

The Journeys of David Douglas

⟶ 1825-1827

⟶ 1830-1832

⟶ 1833-1834

His haul from this journey in trees alone (and there was much else besides) included the Douglas fir, the Sitka spruce, the noble and lovely firs, the bigleaf maple, the ponderosa, the bigcone, the sugar and the western white pines. He wrote to Hooker: "You will begin to think that I manufacture pines at my pleasure".

He reached the Columbia river on his second expedition in 1830. This time he also went to California and made Monterey his base. All the time he was sending home not packets but chests of seeds. His job was not just to discover but to introduce, which he

Loudon's *Suburban Gardener and Villa Companion* (1838) set the scene to which the discoveries of the dauntless botanists were introduced. Gardening reached a low ebb at the very moment when it received its greatest opportunity. The word used for such a menagerie of plants, grown in the most artificial ways, was "gardenesque". Loudon classified gardens from first-rate to fourth-rate, like ships of the line.

The first crossing of North America to the Pacific was made in 1804–06 by captains Lewis and Clark. For botany it had more symbolic than practical significance.

did so effectively that his finds were able to go immediately into circulation to scores of collections. His trees were becoming well known while he was camping in the wilds.

St Petersburg's famous botanical garden was among his clients. He was invited by the czar to make the appalling journey homewards through Alaska and Siberia. In 1832, full of enthusiasm, he set off from Fort Vancouver. But this expedition came to grief in the wilds of British Columbia. He had to return, canoe wrecked and dispirited, now blind in one eye.

Douglas met his end in Hawaii 18 months later. He fell into a bull pit and was killed by a wild bull. He was only 35.

For 20 years after Douglas's death the west was alive with botanists and the world agog to see what they found. Thomas Nuttall arrived by the Hudson Bay route the following year (and found the dogwood of the west, *Cornus nuttallii*, which Douglas must have been very unlucky to miss). Theodor Hartweg, sponsored by the Horticultural Society, sent home seed of the redwood, the Monterey cypress, and the chinkapin. A group of Scots landowners formed the Oregon Association and sent John Jeffrey (who found the Jeffrey pine and the western hemlock and introduced Douglas's grand fir). Veitch's famous nursery at Exeter sent William Lobb, who had pioneered exploration in Chile. He brought back, among new trees, the giant sequoia, the western red cedar (*Thuja plicata*), the white fir, and the lovely Santa Lucia fir. That was in 1853: the gold rush was on; there were hotels in the west.

If there is a contemporary of Douglas's to compare with him in prodigious output, though in a different field, it is John Claudius Loudon. Douglas delved; Loudon (you might say) spun. Loudon was the author of the seven volumes of the *Arboretum et Fructicetum Britannicum*, the first attempt to describe, from both the botanist's and the gardener's point of view, the vast range of trees then available, most of them new. He was also the editor of his own magazine and author of (among other thick books) *The Suburban Gardener and Villa Companion*. This gives all too clear a picture of the world to which these trees were being introduced. Loudon on plant introduction: "It is the beautiful work of civilization, of patriotism, and of adventure, first to collect these all into our country, and, next, to distribute them to others."

Loudon divided gardens into four classes – from first-rate to fourth. Most modern gardens would make him add an eighth- and ninth-rate. He is full of good practical information and advice. But whereas the school of "natural" landscapers swept all garden fussiness aside, ignored new trees, and dealt only in "prospects", Loudon preached a more dangerous creed: "When it is once properly understood that no residence in the modern style can have a claim to be considered as laid out in good taste, in which all the trees and shrubs employed are not either foreign ones, or improved varieties of indigenous ones, the grounds of every country seat, from cottage to mansion, will become an arboretum".

CONTEMPORARY GIANTS

Archibald Menzies (1754–1842), naval surgeon and botanist, sailed on Vancouver's *Discovery* on its voyage of exploration to the northwest in the 1790s. He was the first botanist to see the colossal conifers of the Pacific coast.

Thomas Nuttall (1786–1859), from Liverpool, tried several times to cross North America and finally succeeded at the age of 50 after a gruelling journey over the mountains. From 1822 to 1834 he was Curator of the Harvard Botanic Garden.

John Claudius Loudon (1783–1843), journalist and encyclopedist, made the first complete record of hardy trees and their implications for horticulture in his *Arboretum et Fructicetum Britannicum*. He also "conducted" *The Gardener's Magazine* from 1826 to 1843.

Frederick Law Olmsted (1822–1903) was inspired by Sir Joseph Paxton's design of Birkenhead Park, Liverpool, to compete in designing Central Park, New York (1858). Having won the competition he went on to be a professional landscape architect.

Explorers in the East

James Cunningham (d.1709?), the first botanist to introduce Chinese plants (in 1698), is commemorated by the Chinese "fir", *Cunninghamia*.

Pierre d'Incarville (1707–1757), a Jesuit priest, whose introductions, from Peking, included the tree of heaven, pagoda tree, and pride of India.

BOTANISTS WERE RUNNING hither and thither like ants as the 19th century opened. The only places where they still met more than physical difficulties were in the Far East. Work there was by maddeningly remote control: East India Company factories on the coast of China were the only point of contact. Plant-hungry collectors (of whom there were more and more) resorted to commissioning the masters of East Indiamen to visit the nurseries of Macao and Canton on their behalf. They got beautiful cultivated flowers, but still no idea of the trees and shrubs of the unknown interior.

The enterprise of a Bavarian eye doctor, Philipp von Siebold, was richly rewarded when he managed (on the strength of his ability to cure cataracts) to steal into Japan and take a first proper survey of her flora. That was in 1823. But the frustrations continued until China was finally bullied into opening her frontiers in 1844, and until Commodore Perry and his warships made it plain to the Japanese (in 1853) that what America was not given willingly she was quite prepared to take.

Part of the story of who went where in that amazing hunting ground, and what they found, is told in the accompanying map and portraits. There was (and is) a lot to absorb in a single sustained draught – which is more or less the way it happened. We have still hardly taken stock of the superb flora of western China – particularly of the south-eastern reaches of the Himalayas – and more will continue to be discovered for years to come. One thing which has clearly emerged, though, is the close relations between the species of China and North America. Their similar ice-age history continually crops up in parallel survivals; closely related trees differing only in details.

Once the oyster of Japan was opened the first to profit was a young nurseryman, J G Veitch of the great Exeter family firm, who in the 1860s sent home the Japanese larch, *Magnolia stellata*, and much more. Carl Maximowicz from St Petersburg found, among many important plants, the splendid birch that bears his name; George Hall from Harvard sent *M. kobus*, Thomas Hogg from New York katsura and *Cornus kousa*; and Ernest Wilson produced another rich haul.

The same period saw the exploration of Sikkim and the hugely fertile middle heights of the central Himalayas, initially by Joseph Hooker, son of the director of Kew (who was later to take his turn in that august chair). The eastern Himalayas were a dangerous hunting ground ("moist and crumpled" was Forrest's understatement). Nepal and Bhutan still have unexplored regions and no complete account of their plants. The great British craze for rhododendrons starts here.

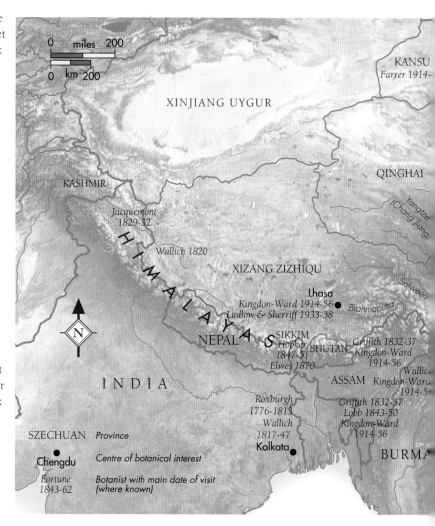

0   miles   200
0   km   200

KANSU
Farrer 1914-

XINJIANG UYGUR

QINGHAI

KASHMIR

Jacquemont 1829-32

Wallich 1820

XIZANG ZIZHIQU

Yangtze (Chang Jiang)

Lhasa
Kingdon-Ward 1914-56   *Brahmaputra*
Ludlow & Sherriff 1933-38

Salween

NEPAL   SIKKIM
Hooker 1847-51   BHUTAN   *Griffith 1832-37*
Elwes 1870   *Kingdon-Ward 1914-56*

Wallic
ASSAM   *Kingdon-Ward 1914-5*

INDIA

Roxburgh 1776-1813   *Griffith 1832-37*
Wallich 1817-47   *Lobb 1843-50*
Kingdon-Ward 1914-56

SZECHUAN   *Province*
Kolkata
BURMA
Chengdu
Centre of botanical interest
Fortune 1843-62   *Botanist with main date of visit (where known)*

Philipp Franz von Siebold (1791–1866), held the same physician's post as Kaempfer and Thunberg. He introduced Japanese crab, hydrangeas, *Paulownia*, camellias, azaleas.

Frank Kingdon Ward (1885–1958), from Manchester, explored and collected for more than 40 years between Tibet and Burma. Plants named *wardii* include a rhododendron, camellia, holly, and cherry.

Father Armand David (1826–1900) was a distinguished French zoologist and botanist. Apart from *Davidia*, he discovered numerous species called *davidii* and *armandii*.

Dr Clark Abel
(1780–1826) visited
Peking in a diplomatic
mission, reported the
Chinese elm, and
introduced the
ornamental apricot
and the shrub *Abelia*.

Victor Jacquemont
(1801–1832) reported
the deodar cedar.

Alexander von Bunge
(1803–1890) collected
in north China for St
Petersburg; discovered
Pinus bungeana.

Thomas Lobb
(1809–1863) collected
for the London nursery firm
of Veitch (who also sent
William Lobb to Chile and
California). A *Cryptomeria*
bears his name.

William Kerr (d.1814)
was sent from Kew to
China by Sir Joseph
Banks. The Chinese
juniper is his most
important introduction.

Grigori Nicolaevich
Potanin (1835–1920)
explored "Tebbuland"
(plants from here are
called "tangutica").
The Chinese larch
bears his name.

Father Jean Marie
Delavay (1838–1895)
was the first European
botanist in western China
and discovered scores of
good garden plants in
Yunnan. A magnolia and
silver fir bear his name.

Paul Guillaume Farges
(1844–1912) sent seeds
to Vilmorin, including
the first *Davidia*. He
discovered many
rhododendrons and
the splendid silver fir that
bears his name.

Augustine Henry
(1857–1930), an Irish
doctor, began to botanize
out of boredom at Ichang.
His haul from central
China included 10 new
maples, cherries, elms,
Chinese honey lucust, *Tilia
henryana*, and *T. oliveri*.

Frank Meyer
(1875–1918) was sent
to China by the US
Department of Agriculture
in 1905. He drowned in
the Yangtze river. He
imported millet, rice, and
Acer tataricum subsp.
ginnala, and *Juniperus
squamata* 'Meyeri'.

Reginald Farrer
(1880–1920) and Euan
Cox (1893–1977)
botanical writers, visited
Yunnan in 1919 and
introduced the coffin tree
juniper (*Juniperus recurva*
var. *coxii*).

Joseph Rock
(1884–1962), born in
Vienna, collected for
the Arnold Arboretum in
"Tebbuland", which he
described as "a garden
of Eden". A beautiful
mountain ash bears his
name.

Frank Ludlow
(1885–1972) and
George Sherriff
(1898–1967) made the
most famous of recent
Tibetan expeditions,
finding mainly primulas
and rhododendrons.

George Forrest
(1873–1932), "gentle and
brave", introduced
31,000 plants, including
309 new rhododendrons,
a beautiful silver fir, and a
snakebark maple, both
called *forrestii*.

Robert Fortune (1812–1880)
was the first collector in
China to have relative
freedom. He was sent by the
Horticultural Society in 1843.
His most famous exploit was
smuggling the forbidden tea.

Sir Joseph Hooker
(1817–1911), later Director of
Kew Gardens, returned from
the Himalayas in 1850 with
magnificent rhododendrons.
He introduced the Himalayan
birch, amongst other trees.

Ernest Henry "Chinese"
Wilson (1876–1930)
made the greatest individual
collection from China and
Japan, including paperbark
maple, giant dogwood, and
Magnolia wilsonii.

Collectors and Creators

WHAT HAPPENED TO THE FRUITS of all these expeditions? Who were the collectors and do their collections still survive? Did they take Loudon's advice and ban everything not imported or improved from their grounds? Did a school of gardening emerge that could keep up with such a wealth of wonderful new material?

The answer to the last question is yes – eventually. To the second last it is no – thank goodness. The collections made in the 19th century do still exist, many of them: but they have reached full maturity; often overmaturity.

Perhaps the most important tree collection to be founded in the 19th century was the Arnold Arboretum, attached to Harvard University. It occupies part of the Boston park system, which was magnificently landscaped by Frederick Olmsted in the 1860s. It depended, as Kew had done, on a giant to get it going. The giant was Charles Sprague Sargent, a patrician Bostonian whose name crops up everywhere in this or any book on trees. Sargent himself

introduced a number of species of flowering cherries and crab-apples from Japan. But most important he created the study of native American trees, particularly eastern ones, and their suitability for ordinary gardens. Today, despite Boston's hard climate, the Arboretum is one of the world's greatest collections, particularly strong in flowering trees: crabs and cherries and hawthorns.

After Sargent's death Ernest Wilson, two of whose Chinese journeys had been for the Arboretum, took charge. But he was killed in a car accident in 1930. Among the important books to emerge from the Arnold Arboretum are Sargent's own great 14-volume *Silva of North America*; Alfred Rehder's *Manual of Cultivated Trees and Shrubs*; several very enjoyable and rather lushly written books by Wilson; and Donald Wyman's *Trees for American Gardens*.

What is now the National Arboretum of France was founded about the same time. The name Vilmorin is another that constantly arises in tree talk. Maurice de

Charles Sprague Sargent (1841–1927) was America's greatest dendrologist. He was the first Director of the Arnold Arboretum, travelled in the East, employed "Chinese" Wilson, and perhaps most important set America to studying her rich ornamental tree resources.

Vilmorin was one of a family that had patronized botanical collection (and run a nursery) since the 18th century. At Les Barres, near Montargis in the Loiret, he received seeds from all over the world and started the great collection that is now run by the Office National des Fôrets as the national arboretum. When de Vilmorin died in 1918 it was given to the state and used to train foresters.

In England the great private achievement of the past 150 years was the Holford family's arboretum at Westonbirt in Gloucestershire. Two generations of Holfords concentrated for 96 years on their superb collection, which is now the British national arboretum, run by the Forestry Commission. The Royal Botanic Gardens at Kew has one of the world's greatest tree collections, shared between the original site on the Thames west of London and Wakehurst Place in Sussex, a 180-acre garden created by the Loder family and

Bedgebury National Pinetum in Kent has been planted by the Forestry Commission since 1925 as the nation's reference collection of all the conifers that can be grown in England's exceptionally benign climate.

THE PUBLISHED RECORD

The literature of trees is the best way of following the successes of collectors and the advance of dendrological knowledge. (Dendrology is the study of trees; its root the Greek for a tree, *dendros*.) Encyclopedic volumes, illustrated with more or less accurate engravings, began in the 1660s with John Jonston in Frankfurt (see page 48) and Evelyn's celebrated *Silva*. Subsequent landmark publications are illustrated here. Engravings gave way (in part) to photography with the work of Elwes and Henry before the First World War.

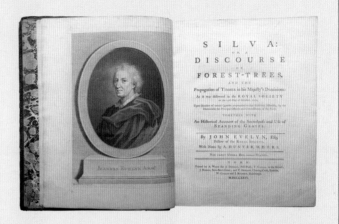

John Evelyn's *Silva – A Discourse on Forest Trees* was written in 1662 as the first paper given to the newly formed Royal Society, the "natural philosophers" of Britain, and remained the standard work for more than a century. My copy is the new edition edited by Dr Hunter and published in 1776. There are masterly engravings of the principal trees, and a great deal of anecdote.

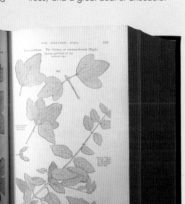

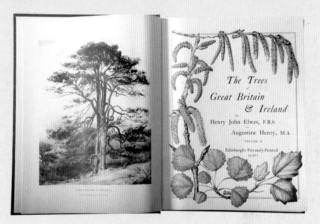

Arboretum et Fruticetum Britannicum is the ponderous title of an exhaustive catalogue by John Claudius Loudon, the greatest horticultural encyclopedist of the 19th century. I have the abridged edition of 1875; only 1,162 pages, with engravings of all the species which are described "with the exception of half a dozen".

The seven volumes of Elwes and Henry's book, privately published in 1906–1913, publicly for the first time only in 1969, are one of the greatest compilations of tree data and descriptions. This epic work celebrates a former golden age of tree scholarship and arboretum-making.

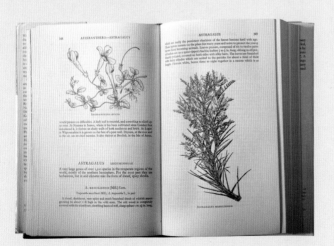

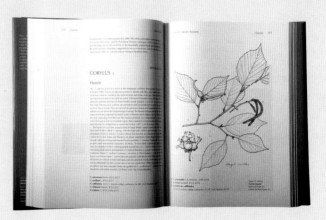

W J Bean's still standard reference book, *Trees and Shrubs Hardy in the British Isles*, swelled from two volumes to five in 70 years as knowledge and experience has been amassed. In 1914, when his book was first published, many of the Chinese species mentioned were so recently discovered that they had never flowered in cultivation.

The latest in this great line of tree catalogues is *New Trees*, in 2009, by my collaborator John Grimshaw and his collaborator Ross Bayton, listing all the species added to cultivation since 1970. The line drawings by Hazel Wilks follow a great tradition of precise delineation of plants.

William Robinson
(right), Irish gardener and
journalist, is credited with
having put paid to the
formality and artificiality of
high Victorian gardening.
His books *The Wild
Garden* and *The English
Flower Garden* were
highly influential. As editor
of *The Garden* magazine
he became close friends
with Gertrude Jekyll. The
two between them gave
gardening its Arts and
Crafts movement.

Gertrude Jekyll (left)
brought a painter's eye to
plants and gardens. Her
Surrey garden, those she
designed with the architect
Edwin Lutyens, and her
many books brought a
lyricism to gardening
which has been emulated
ever since.

bequeathed to the National Trust. In the
1920s the National Pinetum at Bedgebury
in Kent was added to the country's virtual
reference library of living trees of the
temperate world.

Britain has never been in danger of
losing the lead as the greatest tree-collecting
nation. Perhaps because of the poverty of
her natural flora, certainly because of her
mild and gently rainy climate, the number
of considerable collections of exotic trees
is immense. In Scotland it is staggering.
Perhaps the same combination of national
characteristics and climatic conditions have
made the British collect trees and lay down
their famous cellars of wine.

The literature of trees kept pace with
the collections. The first decade of the
20th century saw the publication of the two
greatest reference books on arboriculture.
The first, which was privately published
for subscribers, was the seven volumes of
Elwes and Henry's *Trees of Great Britain
and Ireland*. As a piece of scholarship it
is unlikely ever to be repeated. Both the
authors were immensely travelled (Henry
was an early collector in China) and they
knew between them where every tree grew

and to what size. They had no difficulty in
citing specimens in Ecuador or Edinburgh,
illustrating many of them with those
inimitable full-plate platinum photographs
of the period.

The second book was W J Bean's *Trees
and Shrubs Hardy in the British Isles*, which
is still the standard reference work on the
subject (and not just in Britain).

That turn-of-the-20th-century period
was vitally important in questions of
taste as well as scholarship. The gruff,
opinionated William Robinson joined forces
with the shrewd and sensitive Gertrude
Jekyll. Both wrote extraordinarily influential
books and both made gardens along quite
new lines. Perhaps we should throw into
the pot the influence of Josiah Conder,
whose book on the landscape gardening
of Japan showed the West for the first
time what the great oriental tradition of
gardening was really like – in photographs.

Robinson's approach is easy to illustrate
with a quotation: "The idea that every
choice tree in our pleasure grounds should
be set out by itself like an electric lamppost
is deeply impressed on the gardening mind".

What he and Gertrude Jekyll advocated

was a form of gardening exactly opposite
to the laborious formalities of the Victorian
age. It needed much more knowledge of
plants and far greater sensitivity and
observation. It looked easy, but in fact was
far harder than bedding out in rows. What
they wanted was to use the thrilling new
exotics in a context of natural (or natural-
looking) woodland so that both they and
the natives came to life in a new way.

It is hard to know where to begin or
where to end in a list of the great gardens
that benefited from their ideas. Certainly
British wealth was never more effectively
used to create beauty than by men such as
Leonard Messel at Nymans in Sussex, Lord
Aberconway at Bodnant in Wales, or Sir
Eric Savill in Windsor Great Park. The
influence of these practical examples of
constructing the earthly paradise by really
knowing the plants of other countries and
blending them with one's own has been
felt all over the world.

The serious study of trees up to this
time had remained the field of institutions
and a small minority of landowners. By
the mid-20th century there was a wider
audience. In 1952 two Belgian brothers,
Robert and Georges de Belder, combined
with German and Dutch friends, all
passionate horticulturists, to found the
International Dendrology Society, open to
all, amateurs or academics, who are serious
about the study of trees. Their activities,
tours, meetings, and publications
have formed a worldwide network of
information and experience some
2,000 strong in some 60 countries.

Plant hunters of the last 40 years

Can there still be new trees waiting to be discovered? A resounding answer came in 1994 when a ranger found the Wollemi pine a few hours' drive from Sydney. It is true the great majority of new and improved kinds appearing in nurseries are either varieties of those we know or nursery-bred developments from them, but no one can doubt that there are thousands of genotypes out there in the wild which are waiting to be discovered. Many well-known exotics have only been introduced as fresh seed or plants from the wild once or twice; it is difficult to claim they are fully representative of their wild populations. They may not be "new" any longer, but they have no less significance if they swell the gene pool of their species.

Since the 1970s the pace of plant exploration has again accelerated as planes have succeeded ships; communications of every kind become faster, cheaper, and easier; and countries have opened their formerly sealed borders.

The reopening of China after the Mao years was the great landmark. In 1980 a Sino–American botanical expedition to Hubei, led by Ted Dudley of the Arnold Arboretum, was the first Western plant-hunting journey into China since 1949. Perhaps the greatest value of such trips was to make friends and establish lines of communication. China is a prime focus of exploration. So are the mountains of

Like stout Cortez … A Sino–American expedition in 1980 gave foreign botanists their first view of Chinese mountains and their flora in 30 years. Ted Dudley and James Luteyn of the Arnold Arboretum survey the forests of Hubei. Part of their findings was that collections should be intense in any area to get a fair sample of genotypes.

southeast Asia. If there is an equivalent on the other side of the globe it is Mexico and central America – another region where plants successfully retreated during the ice ages. Both regions have given us scores of new garden plants, many of them trees. Trees, however, take a long time to establish and assess. Most of them have no track record in cultivation and have yet to be distributed.

Modern plant hunters fall into two categories; the official from botanic gardens and universities, and the private or amateur. The age of ventures such as Veitch's by major commercial nurseries, egged on by their competitive customers, seems to be over. It is through individuals, possibly less inhibited by red tape, that most new plants come into circulation. Examples are Carl Ferris Miller, who established his own arboretum in Korea, Dan Hinkley of Heronswood Nursery in Washington, Charles Howick of Northumberland, J C Raulston from North Carolina, Martin Gardner from Edinburgh, Maurice Foster from Kent, and Tom Hudson from Cornwall. Happily there are more and more enthusiasts eager to plant and test what they introduce. The internet carries countless records, ranging from the coldly scientific to the simply breathless, documenting first-hand experiences of new plants. The dendrological chat-room is something new to history.

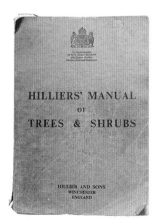

This battered paperback, largely compiled by Roy Lancaster, has been my companion since it was published in 1971. It is the closest thing to a complete portable guide to trees in cultivation.

Roy Lancaster worked for Hillier Nurseries before starting a career as a plant hunter, writer, broadcaster, and the leading popularizer of rare trees in Britain. His books include *Plantsman's Paradise*.

Dan Hinkley is a horticulturist, traveller, and writer who established Heronswood Nursery in Washington as a leading source of new plants in the USA. His books include *The Explorer's Garden*.

Bleddyn and Sue Wynne-Jones have made some 60 expeditions, mainly to east Asia, to collect plants for their nursery in north Wales. At Crûg Farm plants still only have numbers rather than names.

Choosing the Species

TREES ARE THE MOST telling plants in any landscape. In a garden they are decisive. They control more than its character: its sense of scale; its sense of privacy. If trees are used boldly they can create powerful effects with an almost incredible economy of means.

The various functional uses to which trees are put in gardens and streets and parks are discussed on the following pages. But the first essential is to choose a tree; to know what forms, colours, textures have what other characteristics; to be able to weigh up the various siren voices (such as speed of growth, colour, flowers).

Consider the position of the person with room for one tree outside their house – my dilemma when I lived in London. I wanted it to do everything a tree can. I thought bright gold or silver foliage would be nice, so long as it had brilliant autumn colour. But then of course I wanted it to be evergreen, to get full value from it in winter. I knew there were trees with lovely pink and peeling bark, so I wanted that. And the flowers were to be amazingly double, persistent, and fill the house with scent – and of course lead with hardly a pause to large scarlet fruit which the birds found bitter (but I, of course, found sweet).

(I planted a plane.)

Every tree-lover with a garden goes

Humphrey Repton (1752–1818) (left) succeeded "Capability" Brown as England's principal landscape gardener in an age becoming more interested in the new exotic trees arriving from America and the Far East. Unlike Brown he wrote cogently about his theories and practice. Portrait on ivory by John Downman c.1790.

through the stage of picking the most different trees: the ice-blue spires; the bright yellow turbans; the leaves like dinner plates. There is a special bed in my vegetable garden for them while I make up my mind where they will go. The answer very often should be nowhere.

For there is a link, even in the suburbs, between the garden and its environment. And the clue is to find it, and exploit it. You can enjoy exploiting contrasts so long as they maintain the link. The great mistake is to confront the environment by planting the most different trees you can find.

Almost everywhere there is a dominant

tree, or trees; even in the suburbs. America's lovely style of having undifferentiated private areas (the very opposite of Europe's tight fencing and hedging) in a common sward, often with big trees in common, or apparently so, sets the scene perfectly.

Where suburbs have gone over – as so often in England – to nowhere trees like the dreary purple-leafed plum, they need bringing back to the dominance of something real. It will be to everybody's relief when a big forest tree starts asserting itself. It could even (to be in keeping) flower, if it – let us say – were a tulip tree or a big wild cherry.

DISTINCTIVE SILHOUETTES

Before deciding on the species, or even the genus, of tree to plant it is a good idea to consider what overall shape best fits the space you have in mind. Perhaps take a photograph of the setting and sketch over it the silhouette of the ideal tree to fill it and complement the surroundings. For most of the year you will be more aware of the overall shape of a tree than its colour and detail. These are some of the most familiar and distinctive of a vast variety of shapes. With long practice a dendrologist can distinguish closely related species or even varieties by the peculiarities of their outline, even when they are bare in winter. You may have seen them running across gardens to confirm their suspicions.

Vase-like *Zelkova*

Multi-trunked *Amelanchier*

Picturesque pine

Let us consider the feeling that a tree is only really there if it is covered with flowers; a common feeling among gardeners, especially with small gardens. First, it is true that the frothy mass of blossom gives the sense of spring in a way nothing else can. But even if you pretend that the two weeks of blossom is really two months – the whole spring – that leaves 10. Not many flowering trees earn their keep in a small space on this basis. Nearest to it are the few cherries that colour well in autumn, or perhaps the crabs that bear fruit long after their leaves have gone.

The same frantic feeling of wanting

Forming prospects (above), shifting hills, planting clumps, damming and diverting streams all have an 18th-century flavour today. Yet today far more landscaping is being done than in the 18th century. Earth-moving is now much cheaper and easier. And we have a wider knowledge and choice of trees than did Humphrey Repton (at work, above).

Repton became famous for his *Red Books* (above) in which a painted flap revealed what he proposed for a landscape. This before-and-after view showed his proposal for felling the trees on an island in the lake at West Wycombe, Buckinghamshire. It is included in his *Observations on the Theory and Practice of Landscape Gardening*, published in 1803.

more than just a tree is expressed in the need for coloured leaves. A single red-leaved, silver-grey or gold-variegated tree, among green ones, can have the effect of enhancing and setting off both itself and its surroundings. By the streetful, on the other hand, red or yellow trees have the very opposite effect, like a whole meal of soup or sweet.

Does the same thing apply to odd-shaped trees? Should we prefer the standard big head in standard green?

The answer lies in the surroundings. Does your neighbour 20 yards away have a weeping willow? One is enough. If there are

Columnar cypress

Tower-like elm

Cloud-like ash

Conic spruce

Weeping willow

none in the street, and you have 50 feet each way to spare, there are few more rewarding trees to grow: among the earliest in spring with their golden young leaves; among the latest to fall, still fresh and green in mid-autumn.

Is an evergreen really better value? Function aside (if you need it to hide the power station, there is no argument) the seasons emphatically expressed by a deciduous tree are much more satisfying. It didn't take me long to love the winter sketch of my plane tree against the sky as much as its fat buds bursting, its big leaves forming, its new shoots lengthening, and all the rest of its miraculous performance.

As far as the army is concerned (if I remember my cadet days accurately) a tree is either a Christmas tree or a bushy-topped tree. But as soon as a civilian starts analyzing the range of tree characters it is amazing how many clear categories appear. To take form (in the sense of overall shape) alone: I can think of trees that are tower-like, cloud-like, vase-like, tiered, weeping, multi-trunked (like huge shrubs), columnar, globular, pyramidal (or more strictly speaking conic), mound-like, and just plain irregular and picturesque, like

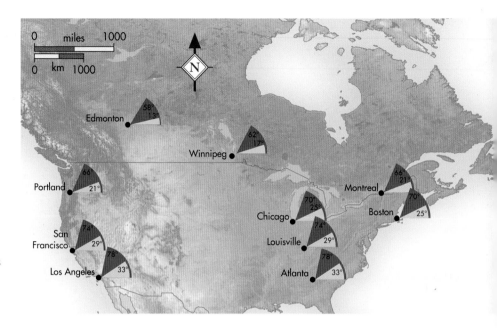

an old Scots pine. Humphry Repton (whose late 18th-century writings on landscape design are still the great textbook) pointed out how – to go no further than army distinctions – the jagged lines of Christmas trees produce a sense of drama (he calls it harshness), whereas the cloud-like outlines of bushy-topped trees give an air of serenity. He went on to urge the association of each with the most unlike architecture: bushy-topped trees with Gothic (which is

vertical in emphasis) and Christmas trees with Grecian (which is horizontal).

The more open and characterful forms belong in the foreground of the picture. It is easy to see that a silver birch, an olive, or a Lebanon cedar (or a tight column such as an Italian cypress) provides a good frame for what is beyond, whereas a solid billowing shape such as a sycamore or oak or a broad dark pyramid such as a fir gives the impression of blocking off the view even if it really hides no more of it.

The strongest effects are undoubtedly obtained by letting one form dominate a scene, using another only to emphasize how very cloudy, moundy, towery, or whatever your principal theme is. To have one of everything is bound to produce too diffuse and uncertain an effect, a problem Repton called "flutter" – he distinguished "flutter" from "intricacy".

I also rather fancy you would arrive at a state of "flutter" by planting two contrasting forms in equal numbers – say a homogeneous mixture of oak and larch. Both would be more appreciated if there were twice as much of one as of the other. Of course this is not a rule which only applies to trees: equal areas of grass and paving in a garden produce a similarly uneasy effect.

Pattern and texture are only slightly less important than form – though perhaps harder to pin down into categories. The most obvious example of pattern is the bare deciduous tree, where the balance,

The shade of trees is not only a necessity in such gardens as this, in southern Italy, but also offers opportunities for pattern-making and the creation of atmosphere. The garden is La Landriana, near Anzio. The English designer Russell Page set out mopheaded orange trees and balls of box to make chequers of light and shade that shift all day.

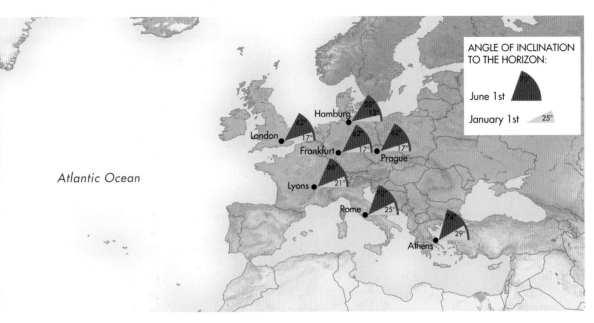

ANGLE OF INCLINATION
TO THE HORIZON:

June 1st

January 1st   25°

Atlantic Ocean

Hamburg   58° 13°
London   62° 17°
Frankfurt   62° 17°   Prague   62° 17°
Lyons   66° 21°
Rome   70° 25°
Athens   74° 29°

The angle of inclination of the sun to the horizon varies much more from summer to winter in the higher latitudes. In London it rises to 62 degrees in June and falls to 17 degrees in January while in Los Angeles its mid-summer height is 78 degrees but its January height is still 33 degrees.

THE PATH OF THE SUN

The sun's path through the sky is very much shorter in winter. The diagrams (below) show the course of the sun in January and June in two places, Boston (or Provence) and Miami. In summer it travels around about two-thirds of the compass, rising in the northeast, climbing high at noon (in Miami almost to the zenith), and setting in the northwest; in winter it travels only about one-third of the compass, rising in the southeast and setting in the southwest. The smaller diagrams show the total daily shadow of a tree in the same places.

Boston 42° North

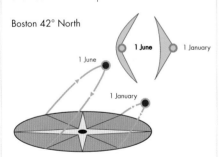

1 June   1 January
1 June   1 January
1 January

The trajectory of the sun and its shadows in Boston in June and January is the same as in Provence.

Miami 26° North

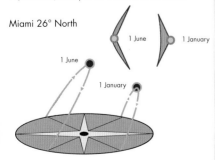

1 June   1 January
1 June   1 January
1 January

The trajectory of the sun in Miami (and the Canary Islands) creates shorter shadows.

the boldness or delicacy, the thrust or swoop of its twigs and branches are dominant in the winter scene. But another is the positive modelling of some trees: beech with its graceful layers of foliage; oak with its firm hummocks; or palm with its arching fans.

Huge leaves (the tree of heaven's for example) create patterns; small ones (of elm or many of the Japanese maples) are more a matter of texture. The glitter of English holly or the marvellous mattness of the fern-leaf beech, absorbing all light, are other examples of distinct tree textures.

Here light is the crucial factor. Modelling and texture (and also colour) are obliterated if the light is behind them; bare branch patterns are lost if the light is behind you, the viewer. This is a question that has exercised great minds. Leonardo da Vinci was firm: "The trees which are between you and the sun are far more beautiful than those which have you between the sun and themselves". Repton held the same opinion; illustrating it with two views of the Thames at Purley, one in the morning with the light full on the banks, the houses, trees, and boats, the second in the evening when all these objects are silhouetted against the low sun. "All natural objects" (he went so far as to say) "look best with the sun behind them; all artificial objects with the sun on them".

But then both Leonardo and Repton were thinking big: of great landscapes where detail was lost in distance and where

there was more colour from the shafts of the sun or the lilac haze of distance than from jolly variegated shrubs and Colorado blue spruces in the foreground.

Photographers are crucially aware that light from behind them shining full on a view (or a tree) flattens the picture, filling in the shaded contours that give it form and interest. Light from behind the subject (the picture taken *contre-jour*) can be wonderfully revealing for details in close-up but reduces distant objects to mere silhouettes.

In garden terms the message is that views to the south (in the northern hemisphere) should always be treated as having the light behind in winter, and from both sides and behind in summer. And views to the north as having the light in front in winter, and from both sides and the front in summer. Bearing in mind of course that the sun in mid-winter is about a quarter as far above the horizon as in mid-summer, which makes an enormous difference to what it reaches.

So the place for deciduous trees is to the south of the house and for evergreens to the north (where they also, happily, give shelter). But for trees of any kind whose colour or texture (or flowers) you particularly enjoy – assuming that the house is your viewpoint – the south is the angle to avoid. This is only to touch on the question of light in the landscape, which can be elaborated on for gardening, as it can for painting, *ad infinitum*.

Planning for Planting

"IT IS DIFFICULT TO LAY DOWN RULES for planting. Time, neglect, and accident will often produce unexpected beauties" – Humphrey Repton again. The trick is to know a "beauty" when you see one, and to be able to make the most of it.

Planning and planting are subjects in the round. There is no one right approach to them. One can consider the available material; the objectives; the techniques. Yet there remains a body of lore gleaned from long experience that is worth fitting to the circumstances.

A Dutch landscape designer (a change from Repton, at least) produced an epigram that bears a lot of thinking about: "The axe is my pencil." If Repton had thought of it he'd have said it, because three-quarters of the grand new effects he created for his clients, and demonstrated to them in advance with before-and-after "slides" (as he called them) in his famous *Red Books*, he contrived simply by selectively cutting down what was there already. This is cold comfort for the owner of an empty new garden. But it is surprising how often it does apply once you start to look at a place with open eyes, to sum up what Lancelot Brown referred to as its "Capabilities".

There is besides a wicked satisfaction in cutting down trees and clearing bushes which anyone who only plants, and would defend any growing thing with his or her life, should experience and realize. The most memorable moments to me in taking over a new garden have been when a tree came down, to almost feel the new view rushing in to make its contribution.

Heaven forbid that anyone should think I am advocating indiscriminate clearing. The modern world is full of unlovely things crying out to be hidden by trees. Taking stock should be a lengthy business. One must contrive to "see" the place as it will be with some of the existing trees removed. One must wait four seasons, to find out how leaf-fall changes the total effect. One must go through the site with great care, even if it is only "scrub": there may be

saplings in it which will make a marvellous new start if they are left in place. And finally one must do the job by degrees, and be ready to call a halt if an unexpected effect is achieved. "Accident will often produce unexpected beauties."

What one can hope for by clearing is to "borrow" some of the surroundings for one's own enjoyment. In the 18th century nobody thought it particularly wilful if a landlord shifted the village a mile or two to clear a new prospect from his library windows. It is surprising how much one can achieve even while leaving the village exactly where it is.

"A plain space near the eye gives it a kind of liberty it loves." This time a quote from William Shenstone, an 18th-century English poet whose garden (the famous Leasowes, near Birmingham) was rather more distinguished than his verses.

It sounds rather like another excuse for clearing, or at any rate for not planting. It can't apply, in any case, where the whole garden is "near the eye" and a plant must either be there or nowhere. It was a basic

tenet of the English landscape school, though, that the house should be set in plain lawns and the eye encouraged to go out to the limits of the ground (and beyond). Balustraded terraces around the house came later as part of the general Victorian fussiness. It is lovely to look out from under trees, but not to gaze slap into the side of a big one at short range. William Robinson went (characteristically) too far when he said: "No forest tree should ever be planted near a house."

Close into the wall is often an excellent place for a tree, if the foundations are sound and you are prepared to take a little trouble directing the young branches. I will always remember a house in Connecticut that had been built round a stand of pines, with an upstairs gallery right among the trees, so you could touch them. It wouldn't have been up Shenstone's street, though.

"Plant close together" – expensive advice from almost every landscape designer (and not just from the nursery trade). Trees like the company of other trees. To "dot a few starveling saplings on an open lawn" is a recipe for slow-starting, discouraged specimens. Capability Brown always planted in thick clumps – which his clients usually forgot to thin out later as they were meant to. The technique in those days was not to pay a nursery for a few very expensive standard trees (which in any case take years to shake off the dishmop look and develop much real character) but to plant lots of small ones, mixed in with shrubs. In the open country five or six hollies or hawthorns went in for every one big tree.

The great argument for planting big trees today is that they are harder for vandals to kill. In public places this overrides everything else. But it is doubtful whether they get away faster in anything except the short term – up to say five years. In 10-15 years it is very likely that a tree planted at half the height will have raced ahead.

It is not necessarily a natural look you are after in planting a garden. But the late Nan Fairbrother (whose book *New Lives, New Landscapes* puts her among the most influential – and enjoyable – of modern writers on the subject) had a few words to say about the Fitted-Carpet Complex. She pointed out that the normal situation in nature is a tree layer above a shrub layer above a herb layer. Anything that separates these is a garden style. Fairbrother gives the example of road sides where the householders have to watch the traffic across a wide strip of mown grass between tree trunks, whereas the natural regeneration of the area, with a nudge

The art of planting to look natural was keenly studied by the 18th-century landscapers. These two illustrations from Humphrey Repton's *Observations* show how, by planting some trees very close together, by planting trees and bushes of different heights at the same time, even by planting some trees at an angle to each other, you can create the effect of natural open woodland (above left). Below he shows how irregular planting can easily look artificial.

A landscape philosopher studies human anatomy as well as scenery. According to Humphry Repton (this is he) the human eye takes in a vertical picture 80 degrees deep, of which only 27 degrees are above the horizontal and under your brows. You therefore do not see the sky above this angle. At 200 feet distance you take in the whole of a tree 107 feet high.

in the right direction and a spot of fertilizer, could have given them a belt of beauty, with a rich mixture of self-sown native trees and shrubs punctuated by the row of trees (if, indeed, a row as such was needed at all).

How much you can use perspective and other tricks to enlarge the apparent size of your garden – or for that matter reduce it – is a question to ponder. Trees here are the crucial factor. They play the essential role of visually linking land and sky. Without them there is a sense of bleakness: man's puny six feet is too open to comparison with the globe, not to mention the firmament. You have only to stand in the middle of the vast and treeless parterres at Versailles to know the moon-walking feeling.

Happily, however, the human eye, which scans the surroundings for 90 degrees from left to right, peeps out through a comparative slit at the scene above the horizontal. Your eyebrows limit you to seeing 27 degrees of the scenery above eye-level at, say, five feet from the ground.

A simple calculation based on this shows that to fill your horizon and give you a sense of enclosure (if that's what you want) a tree 100 feet from your eye need only be 60 feet high, at 50 feet only 30 feet high, and at 25 feet only 17½ feet – as many little urban gardens attest.

Various ways suggest themselves of using this information. Essentially, you have here the clue to the future effect of new tree planting, and of establishing a vertical scale as well as a ground plan.

A surveyor's pole, 10 feet high with the feet marked off in alternate red and white stripes, was Repton's instrument for gauging the interaction of height and distance. He is guilty of the truism "objects appear great or small by comparison only". Your Japanese with his "distancing pine" would of course agree. You plant the big trees at this end of the garden and the small ones at the far end, and (without a 10-foot pole to give the game away) your boundary slinks off into the middle distance.

It has always been a planter's axiom to use uneven numbers of one kind of tree. Three trees, or five, or seven, but never two, or four, or six – unless in an avenue or as a definite pair flanking a gate. The point that matters is that you should decide whether an effect is supposed to be regular or not – and in either case go the whole hog. Irregularity means trees of different sizes, spaced widely in one place, crowded together in another. Regularity should be exact, or it looks tatty. In the long run, as any ancient avenue of trees will tell you, irregularity wins.

Trees for Shelter, Seclusion, and Structure

THE JAPANESE, WITH THEIR formalized style of garden design, would never decide (or should I say dare?) just to stick in a tree where they thought it would look nice. They would have to rationalize the situation – by giving the tree a title. That is the Tree of the Sun at Teatime or the Mobile-Phone-Mast-Hiding-Tree, they would say.

There is a lot to be said for their approach. In a small garden with no space to spare it is almost essential for each tree to have a *raison d'être*. Even in a big one the more controlled and purposeful the planting the greater will be the effect. Besides, there is no end to the functions of trees once you start to analyze them.

There is shelter (from the weather). Nearly always essential on at least one side of a garden, the side of the prevailing wind. And often essential on the side of the cold winds of early spring as well. Shelter belts should consist of native trees if possible, to blend in with the landscape. If they are evergreens they give the garden the same feeling of having a protecting shell all the year around. The great landscaper Sylvia Crowe said such wise words that I would be doing you a disservice not to quote her. She said, in essence: "Shelter belts should always be homogeneous: two species are plenty. Get an idea and make a thing of it: a belt can't be too tall and thin, for example: if that's the character you're after, make it a single line of Lombardy poplar. Never be indeterminate; to go all out for something is the only way to a rich effect with simple (or any) materials." That is what she said about shelter belts, but it is wisdom for any aspect of the garden when you think about it.

There is shade – the avowed purpose of all the street trees in America. In southern Europe pines and other conifers are widely used to great effect. Deciduous trees are much more popular in the United States, their leaves having a more cooling effect in very hot weather – and indeed casting more shade. A big shade tree very near a house

Looking southeast from the front of the house (left). A yew hedge defines the front yard and a simple screen of Lombardy poplars on the west, giving shelter from prevailing winds in summer, provides a strong structural base line to the view across a pond to a distant urn. The poplars have been topped lower at the far end to emphasize the perspective.

Trees give shelter, seclusion, shade in summer, and a frame to the view both of the house (right) and from it (top left).The reflection of the house in the pond is also framed by refections of evergreen trees: a deodar cedar on the left, pines on the right.

In the arboretum (left), north of the house, snow emphasizes the distinctive dome shapes of *Acer pictum, A. palmatum,* and *Cotoneaster franchettii* var. *sternianus,* with *Eucalyptus dalrympleana, Quercus agrifolia* from California, and a golden Scots pine behind.

can be almost like air-conditioning. (It can also, however, play havoc with its roots and branches if you are not careful.) In Britain few trees are planted primarily for shade – but we seek the shade of trees in grass to lie under every summer nonetheless.

There is screening, when the aim is to hide something extraneous. This can involve anything from a whole building to a bathroom window. In most cases trees are the only possible medium. The point to bear in mind here is that the screen planted near the eyesore has to be far bigger than the one planted near the eye. You can hide a whole house with one bush in the foreground; over against it you will need a whole row of trees.

There is seclusion, when the aim is to conceal your garden from those on the outside; trees give privacy.

There is structure: the role of hedges and walls where the scale is fairly modest, but in the landscape the job of trees is to define spaces. The old English landscape of wood, field, and meadow, all compartmented by tall lines of elms, or lower barriers of hawthorn, is the perfect example. In a garden in winter the structure tends to fall back on the evergreens, the thing to remember when planting them.

There is the sense of scale – a speciality of the Japanese gardener, who would persuade you that he or she has half the coastline of Japan in their backyard. Their Distancing Tree is a small pine on a small mound which with half-shut eyes you might take for a big pine on a mountain. Trees influence the whole sense of the size of a garden – and more. This aspect is discussed in the previous pages.

There is the question of framing of views, which we have touched on. It is worth noting how photographers instinctively scramble to the nearest tree to frame a picture of a new hotel or power station. I suspect they often have to take a branch with them to hold in front of the camera.

And finally there is the tree as sculpture: triumphant, central, and intended to be admired for its own sake. In Japanese gardens this is often the tonsured pine leaning negligently over a stone at the water's edge. In Western terms it is usually known as "the lawn tree". It is worth long pondering whether a magnolia or a cedar or a strawberry tree or a weeping ash deserves this place of honour.

In the first edition of this book in 1973 I showed pictures of my own garden in Essex. Below are comparable pictures taken over the past few years – the bones the same, but many new trees. *Plus ça change …*

In the walled garden (right) to the west of the house tree planting is rather more formal. Apple trees pruned to parasol shapes give shade on the lawns while two rows of clipped upright junipers line the central box-hedged path. Clipped pyramids of box mark the corners of each flower border. A row of clipped upright cypresses line each wall and a row of apple trees are pruned into a contrasting flat mushroom shape on the lawns. Being kept low they give shade on the lawn but not on the flower beds. An octagonal iron pergola in the centre supports an old pinot noir vine. Lombardy poplars beyond screen the house from a farmyard.

A little pool in a hollow, (right) approached by stone steps, has been thickly planted on the north and east sides with lush-foliaged trees to give a sense of seclusion. Larches, swamp cypress, and *Metasequoia* are the strong verticals. Around them cluster trees with good autumn colour: smokebush, *Parrotia persica*, *Liquidambar*, and a Caucasian maple whose red summer shoots contrast with the blue foliage of Scots pine behind.

Planting Trees

IF YOU ARE PREPARED TO TAKE enough trouble you can move a tree of almost any size without killing it. The catch is that the trouble increases sharply above a fairly small size, perhaps 15 feet, while the chance of success diminishes. Sir Henry Steuart would not agree with me, but for us ordinary folk it is one of life's trying little facts.

Sir Henry was a 19th-century Scottish landowner who shifted full-grown trees around his park like milk churns around a dairy. He boasted that he had never lost a tree, and never spent more than 12 shillings on moving one. When you consider that for this he employed 10 men for two weeks disinterring every scrap of root and reinterring it just down the road (so beautifully that his trees never needed stakes or guy ropes) you will see why America filled up so quickly with ingenious Scotsmen in the 19th century.

To have any hope of successfully moving an established big tree you need either a vast labour force or a huge machine. However huge the machine, though, it will probably kill the tree unless it has been laboriously prepared beforehand. This means digging a trench first round one side of the tree, then the other (with a year in between), and filling the trench with leaf mould. The tree will react by filling the leaf mould with fibrous feeding roots, making a compact-enough root system to move without losing half of it. Nurseries that grow trees with the intention of selling them extra-large undercut the roots each year with the same object. Or, more and more routinely, they grow the tree in a container from the start.

The demand for instant results has had a revolutionary effect on tree nurseries in the past decade. One of several developments that reduce transplanting risk and make the moving of substantial trees a routine (though still expensive) business is research into the behaviour of tree roots in pots.

All gardeners know that roots in a pot start to spiral when they reach the side.

The result is a congested mass of long twisting roots with no escape. A tree, which can spend years in a pot, is the worst case.

But root behaviour can be modified. One means, the Airpot, is shaped like a honeycomb with holes. Roots are directed to the exits, where they stop growing, sending out branches instead until the pot is a dense mass of short roots, easy to plant out. Another discovery is that a white pot discourages long spiralling roots; they head down to the darkness below, again making planting out easier.

But it is best to buy trees small. As a counterblast to the feeling of impatience which small trees seem to engender in most people I can do no better than to quote the inevitable Humphrey Repton – at his most dogmatic: "There is no error more common

THE NURSERY CHOICE

Broadleaf trees can be bought at different ages and sizes, trained in different ways. Traditional nurseries dig them up from where they are lined out in rows and either sell them with bare roots or wrap their roots to maintain the original soil round them.

Bare-root

Balled

Pot-grown

Trees are sold by a nursery with their roots either bare of soil, in which case they must be planted, either temporarily ("heeled in") or permanently, straight away; balled (i.e. lifted from the ground with a ball of earth around the roots and wrapped in burlap or sacking – these also need rapid planting); or grown in pots or cans – these can be moved at any time, but preferably not in summer.

Bush form (i.e. untrained) (above) is best for multi-stemmed trees, or any tree you want to train yourself.

Feathered (ie with all the lower branches intact) is normally the best way to buy specimen trees. They are cheaper and have all their natural character. It is up to you whether you leave the lower branches or not.

Semi-standard and standard are smaller and larger versions of the same thing: a tree with its lower branches removed and the top trained into a balanced head with a leader.

Semi-mature trees are standards grown on in the nursery to 12 feet or more. They tend to be disproportionately expensive and should only be used where the threat of damage to smaller trees is too great.

Moving full-grown trees (left) is no new notion: though modern machinery has made it much less laborious. But even under ideal conditions and at the right season (in autumn or spring) there is risk. Bigger trees need more continuing care.

Semi-standard and larger trees (right) must be securely tied to a stake for the first two years after planting. A practical form of tree tie consisting of a plastic belt with a buckle, going through a band which keeps the tree and stake apart. A tall guard is normally necessary only in public places. Damage from rabbits (or lawn-mowers) can be prevented with a plastic sleeve which covers the bottom of the trunk. In hot-summer areas the bark of young trees can suffer from scorching after planting out. Big trees should be planted with the same orientation to the sun as in the nursery.

trees (which are normally planted considerably bigger). Some of them are outlined below.

In places with a deadly winter all planting is best left to early to mid-spring, when the ground is thoroughly thawed and not too sodden. In more temperate places, deciduous trees are best planted about the time their leaves fall. Their roots can grow surprisingly in mild winter weather, giving them a better chance to put on leaves than if they were planted in the spring. Cold spring winds are also less likely to dehydrate them if they have had a winter in the ground to make roots. Evergreens in the same places can be planted either in early autumn or spring: if in the spring not until the ground is well warmed up.

Trees in pots or cans are in theory safe to plant at any time. Obviously the best planting season for them is the same as for any other trees. If you plant them in mid-summer you must water them much and often.

than to suppose that the planter may not live to see his woods unless they consist of (fast-growing) firs and larches … man outlives the beauty of his trees, where plantations do not consist of oak."

Lacking the facility Repton must have had of glancing the other way for 20 or 30 years, I have an alternative case to make: that the planter gets the greatest pleasure from intimate contact with a little tree and the chance to watch it grow bud by bud, perhaps even more than when it is bigger.

I have kept a letter I received from Alan Mitchell, one of England's greatest dendrologists, when I first started planting. "Very young trees", he wrote, "have singular advantages that seem to escape many people … There is great aesthetic pleasure in seeing trees at their formative and most vigorous growth … when flourishing young foliage is at eyelevel and leading shoots four feet long can be seen and measured." Which is my experience exactly.

Planting a tree seems simple. You dig a hole and stick it in. This, literally, is the way forestry trees in their tens of thousands are planted: there is no time for fancy business. There are touches of finesse, however, that make a great deal of difference to garden

PLANTING A TREE

Whoever said: "A one-dollar tree needs a ten-dollar hole" understood the matter right. It is well worth taking time and pains to give a tree a good start. It will need to be watered and weeded regularly for at least one year, and preferably two, if it is to fulfill its potential and grow fast.

Dig a square hole twice as wide as the root spread. Break up the bottom of the hole with a fork to ensure good drainage. Drive a strong wooden stake into the hole at one side; make it deep enough to keep the tree rigid.

Try the tree in the hole. Note that with bare-root trees the soil line on the trunk must be the same as before. If necessary refill the hole until the tree is the right height. Then spread the roots out into their natural position (mainly horizontal).

Start backfilling by hand, sifting soil carefully between all the roots and rootlets. Pack them in tight. Mix the backfill soil with compost or leaf mould to make it crumbly if necessary. Add mycorrhizae if you can.

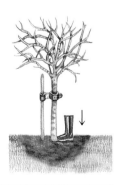

When the hole is full (all the soil should go back in) firm the ground round the trunk, putting all your weight on it. Pull the tree upwards by the trunk: it should not move at all. Water it generously and keep it free of grass and weeds.

Maintenance

I DON'T KNOW IF ANYONE has sat down cold-bloodedly to calculate how many of the trees planted for ornament survive to do their job. Forestry trees have a good record: 90 percent success is the average. With ornamental trees the failure rate is far higher, despite the fact that so much more trouble is taken, and despite the expensive nursery rearing. It is not only harder to get a four- or five-year-old tree off to a good start, but the risk of damage and the consequences of neglect are more serious. Trees need maintenance. There is a huge difference between a property where the trees are looked after and one where they are supposed to look after themselves.

The first need of trees in the year after planting is water. If the rainfall of the area is less than 30 inches a year, reasonably spaced out through the seasons with a good proportion in the spring, three or four heavy waterings in the first year, and for the next two or three years if possible, will make an enormous difference. An invention called the Treegator, though not pretty, is a practical help. It is a water bag with a porous base that fastens around the trunk and slowly leaks its contents into the roots.

The roots of a newly planted tree are so near the surface, and so limited in extent,

CUTTING OFF A BRANCH

A branch should never be cut off in one piece: the bark will tear. The first step is to cut off the greater part a foot or two from the main limb. Start all cuts from underneath. Make the final cut just outside the branch collar.

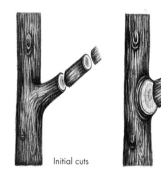

Initial cuts   Final cut

A tree surgeon has a dangerous job. Before he climbs he throws a rope over a top branch and rigs a harness. Essentially he hangs from his belt and keep himself in position with his hands and feet.

that grass growing over them competes for whatever rations are going. The tree's response is to shade out the grass with its branches – but nursery training removes the low branches that would give the shade. The tree must be helped. Botanical gardens and new commercial orchards always keep a three- to five-foot circle of earth hoed free of grass and weeds around every young tree. In a normal garden this is laborious and not very pretty. The alternative is to mulch the tree with something that will keep down grass and weeds and at the same time help to keep the soil moist. In short grass the perfect material is to hand: the turf you have cut out to plant the tree, simply

turned upside down and stamped down again round the stem. In long grass the answer is a thick pile of cut long grass (and weeds) to cover the ground for two or three feet all around the stem to a depth of four to six inches. Nothing can grow through the pile: it slowly rots down to add a layer of humus to the soil.

There is normally no need to interfere with the branches while the tree is small. It knows what is expected of it. The temptation is there, nonetheless, to see the twigs of today as the limbs of the day after tomorrow and make sure they form a harmonious, balanced pattern. Most of all to see that the trunk or a central leader has the dominance that will give the tree stature. If two leaders appear to have equal status, choose one and carefully cut off the other.

Some trees grown in the open without competition begin to be complacent, to stop growing upward and round out their crowns while they are still quite small. It is perfectly sound practice in such a case to shorten the side branches and give the tree a vertical emphasis again. Rather than just cut back any branch and leave a stump in mid-air choose a fork with a shorter angled branch you can leave, cutting off the longer arm of the branch flush at that point.

This sort of adjustment to older trees, particularly where it is concerned with repairing damage and fighting disease,

No tree likes to grow with short-mown grass covering its root zone. It is hard to find an attractive mulch; here long grass is piled round fruit trees. In Kew Gardens tired trees are rejuvenated by injecting air into the root zone and keeping the whole area bare of grass for several years.

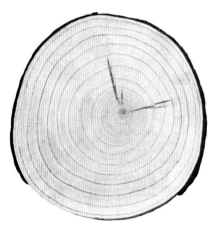

The effect of fertilizers. This Sikta spruce had a diameter of two inches after 20 years on poor land. Then 70 pounds of phosphate per acre was applied. The tree added half an inch diameter that year. Six years later its diameter had trebled.

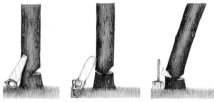

There is an art in safely cutting down a tree and making it fall exactly where planned (left). The first stage is the "undercut": a wedge of wood is removed with two cuts, one horizontal and the other at an angle downwards to meet it (above). The undercut must point exactly where the tree is intended to fall. The second stage, the felling cut, is made from the opposite side starting at a point just above the undercut and sloping slightly down to it. The tree will pivot on a "hinge" of wood and the direction of its fall will be controlled. Big trees with hard wood can be further controlled if the felling cut is stopped short and the tree pushed over by driving in a broad iron wedge with a sledgehammer.

is called tree surgery. It is not a job for amateurs. There are real tree-cowboys who can stand (no hands) on a branch 80 feet from the ground while they use a chain-saw. I have got as far as learning to use the ropes and harness to get up there, but I shudder at the thought of actually doing anything (except hang on) at that height.

Research into the inner structures of trees (see How a Tree Works, page 14) has had an influence on how surgeons remove branches. The advice used to be to cut off a branch as close to the trunk as possible in the belief that anything in the nature of a stump was an invitation to rot. Thanks largely to Alex Shigo, an American who spent years of his life dissecting trees with a chain-saw to study how buds become branches and what happens when branches die, the function of the collar at the base of each branch is now clear: it is vital to the healing of a wound and its future strength. Leaving a stump is bad; removing a branch flush with the trunk is worse.

Never take off a branch in one piece. There is nothing you can do to stop it falling by its own weight before the cut is finished, and taking with it a torn piece of the bark. Remember that the life of the tree lies just inside the bark.

The only reliable method is to cut the branch off a foot or two from the trunk (undercutting to prevent tearing) and then make a separate log of the short stump you have left. Saw from underneath until the blade jams in the cut. Then saw downward just outside the branch collar to meet the undercut. If there are ragged bits of wood where the two saw cuts met, saw or chisel them off. It used to be considered good husbandry to cover the cut or any wound with fungicidal tree paint; the modern view is that wounds are best left open to the air to heal.

Unfortunately the gardener quickly learns that many trees which to him or her are a feast to the eye are quite a different sort of feast to all manner of creatures. A tree stripped of leaves by caterpillars or aphids is a pathetic sight, and trunk borers which work out of sight are no better: they can rapidly kill a tree. In some areas they preclude the planting of a susceptible species.

Nature is redder in tooth and claw in this regard in North America than in Europe. The extremes of the continental climate, above all the wrecking power of ice storms, also help to make tree upkeep much more of a problem in the United States and Canada. The wide range of pests and diseases which threaten the appearance, and sometimes the life, of trees is outside the scope of this book. But you should keep an eye out for signs of trouble, even in a single branch, and try to find the explanation for the problem.

TOOLS OF THE TRADE

Some of the tools used in pruning and maintaining trees are shown on the left (not to scale). Secateurs are for pruning twigs and unwanted new growth. Lopping shears are good for branches up to one inch diameter – the long handles give good leverage. A pole saw combines long reach with a cord-operated lopper. A chain-saw with a petrol engine is the standard tree-felling tool today. A chain carrying alternate blades and "rakers" to clear the sawdust runs in a track around a steel plate. It is noisy, hot, and heavy, but much quicker and easier than an axe. A folding pruning-saw is a useful pocket implement for minor bits of surgery. The vertically positioned pole-handled chain-saw is used for cutting high branches.

Pruning and Other Arts

HEDGES ARE THE EVERYDAY example of trees bidden by man. We accept the idea of enough beech trees, say, or yews, to make a substantial wood, all being restrained by cutting every year to take the form of a mere wall. Yet other forms of tree discipline are frowned on, especially by romantic Anglo-Saxons, as unnatural.

The Japanese think differently. To them a natural pine, whether compact or straggling, is less appealing than a bidden pine engineered into dancing a sort of pine ballet. The garden is a stage, and the actors are called upon to do more than just stand around in their ordinary clothes.

Are we in the West perhaps too functionally minded? We cordon an apple tree to make it bear more fruit. We pollard a willow to get osiers for fencing and baskets. But if we have any fancy notions about shapes it is usually to carve quite unvegetable things in topiary – be it quasi-architecture, the Sermon on the Mount (in an ancient Warwickshire garden), or even a hunt in full cry, which is frozen in yew on the point of crossing the drive to a French château. Topiary has been considered so fundamental to gardening that a Roman gardener was a topiarius. Its modern revival is giving form and style to many otherwise unremarkable gardens.

Not that functional forms are without beauty. The methods for maintaining the maximum number of fruiting spurs on an apple or a pear tree give the symmetry and order of the espalier, the fan, and the cordon. Where there used to be a demand for regular supplies of long straight poles there are woods of coppiced hazel or chestnut: every six or seven years each tree was cut back to the ground, where it formed a "stool" and sent up a porcupine of new shoots. They were not very lovely perhaps when they were in production, but old, abandoned coppice is some of the prettiest woodland. In gardens the idea is taken up with trees distinguished for their foliage: regular cutting down giving a bush, usually with bigger leaves than those of the natural tree, forced by roots out of proportion to the top. I have seen golden poplar, *Eucalyptus*, and *Paulownia* hacked mercilessly to the ground, all bounding back in the same year with eight-foot shoots covered with mammoth leaves.

Pollarding is the term for the coppice principle applied six feet above the ground, where grazing beasts can't reach the new shoots. It is the coppice of the meadow, as opposed to the woodland. The word pollard comes from poll – meaning head.

Willows are the most commonly pollarded trees, but oak, ash, and hornbeam have often been cut for firewood in this way in the past. Sometimes in fact with old spreading trees it is hard to see whether they were once cut off six feet above the ground or not. Generally you can say that if all the branches start from the same point on the trunk it is a pollard.

Pollarding is little used for ornament. The happiest effect I have seen achieved with it has been that of instant fen-land: a group of willows with red-barked young shoots standing glowing in the winter sunlight above their reflections in a pond.

Bonsai is the end of the road.

The deliberate use of all a tree's forces and (as it were) instincts to make a toy is gardening of a very sophisticated kind. It has special value for a city-dweller, whose chance to observe the forces of nature is limited and distorted to the point, one fears, of psychological danger. One little tree in a pot can teach you more than two weeks in a forest about the processes of

A common form of pollard in fen country: the willow (or osier) cut back every year or two for its pliant new rods for weaving into baskets or fences. The trunks of such pollards often rot inside and become hollow sleeves, still supporting a flourishing crown. Varieties with coloured young bark treated in this way in spring become gay shockheads the next winter.

does; it restricts their long-term growth of shoots and roots, but at the cost of a horrible mutilated sight in the square all winter.

One underlying principle affects all pruning: every time you cut off a growing tip you transfer the energy due to it to the next bud or growing tip back down the line. There is no guesswork in what will happen when you cut off a twig. The last remaining bud will take up the struggle and go whichever way it is pointing. If you want a branch to grow to the right you must choose a bud pointing that way and cut off what lies beyond it, as close to the bud as possible, leaving no "snag" or stump to die.

Any branch you cut below its lowest bud, simply leaving a stump, will do one of two things, neither of which you want. Either it will die or it will produce epicormic shoots of no beauty or value: just a bunch of straight whiskers. With this in mind it is not difficult to pick on the branches you want to encourage to grow straight on and those you want to change direction; those you want to shorten or remove and those you want to be dense and bushy.

The trick is to see the tree's natural tendency and help it along. As in so many arts, the Japanese are the masters. But who in the West will go to the length of snipping every new shoot on a tall pine to reduce its length, and returning in the autumn to pluck off the surplus needles?

plant life, if you work with it, as the Japanese do, to bring out its essential character. Bonsai only works if you study, follow, and stress the natural growth pattern of the tree.

Big-tree modelling comes to much the same thing, with the difference that the roots have no restrictions: the size of the tree is controlled only by the amount

you stop each year's growth. In practice of course this does restrict how much the roots grow, too.

The French, with surprisingly little aesthetic sense, are addicted to what they term *élagage*: the periodical (often annual) lopping of branches with such savagery it looks as though they want to teach the trees a lesson. In a sense that is what it

Hornbeam hedges (left) at the Prieuré Notre-Dame d'Orsan in the centre of France are famous for their whimsy. Little birds and bigger beasts spring out from hedges and arches. Hornbeam is as fast and biddable as any topiary material.

The Yew Tree Avenue (right) at Clipsham, Rutland is a collection of 150 absurd smooth-shaven yews, not in a garden but lining a public path. It was the idea of the Head Forester of Clipsham estate in 1870. Like the beech hedge of Meickleour, Perthshire (see page 259) it is now kept in trim by the Forestry Commission.

The Conifers

CONIFERS BEAR CONES. That is their hallmark. Not all cones look like cones. Nobody would suspect a juniper berry, for instance, of being any such thing. But technically a plant whose female flower offers its eggs naked to the world for fertilization, and then develops into a "fruit", usually of wood, to enclose them while they ripen, is a conifer. And all conifers are trees – or at least woody plants. There are no herbaceous conifers.

The conifers include the world's biggest plants – by far. There is a startling gap between the record heights of broadleaf trees and conifers. In the American championship table there are no broadleaves more than 180 feet or so. Then a pause. Then a list

of conifer champions from 250 feet right up to 380 feet.

Again, conifers hold all the age records. Millenarians are commonplace. The arguments concern how many thousands of years the oldest of the pines has been growing.

In the natural course of evolution the heyday of the conifers is over. They are one of the oldest plant families. They flourished long before the broadleaves. But they have been on the decline, numerically and territorially, for millions of years.

Yet man has found them a more efficient plant for his or her purposes, for timber and paper, than the up-to-date broadleaves. Partly it is their willingness to grow under the unlikeliest

circumstances; on land too poor or wet or exposed – above all too far north – for farming. Partly it is their simple design with all the emphasis on a thick trunk – the branches being no more than something to hang the foliage on. But mainly it is their speed: the conifers are adapted to use the low sun of the higher latitudes to its utmost; they nearly all keep their leaves and thus have the means to make and store food year-round. In the cold north, therefore, they add wood to their bulk faster than any broadleaved tree.

There are about 650 species of conifers, in 50 genera in only eight families. They all grow in the temperate zones – most of them moreover in the colder parts; if they grow in the tropics at all it is only on mountains with a temperate climate. The vast majority – 33 out of the 50 genera – are confined to the northern hemisphere. Nearly all of them are evergreen trees with tough, dark green, needle-like leaves which make dense and deeply shady forests. In the gloom their lower branches rapidly die and drop off; their trunks develop into flawless pillars, often rising to an awesome height.

Yet from the same race, through odd freaks and sports, come small, precisely formal, and even miniature trees of exquisite detail and often brilliant colouring – man's other reason for taking the conifers under his protection.

The Ginkgo or Maidenhair Tree

ONE FEELS A CERTAIN RESPECT for a creature which has simply declined to evolve. I believe there are a few reptiles, lowly crabs, and some insects which have been much the same for 100 million years or so. But for a forest tree to survive its relations, its descendants, the conditions of its birth: to look unmoved on the drift of continents, the rise of mountain ranges, the coming and going of aeons of reptiles and ages of ice – to survive all this unaltered, for 200 million years, argues a degree of tenacity. Not to mention a sound design. There are gingko trees in Hiroshima that even survived the nuclear bomb.

The ginkgo's unmistakable leaf is as startling as Man Friday's footprint among the tangle of tree-ferns in fossil strata of the Paleozoic era. The tree ranged worldwide in those days. Before the modern conifers evolved, long before the rise of the broadleaves, the ginkgo flourished in America, in Asia, in Australia, and on (what is now) the Isle of Mull. It went into decline even before the Ice Ages, it seems. And yet it remained viable. There was still somehow a niche for this strangely sophisticated primitive.

By the time man came on the scene the ginkgo had retreated to the mountain forests of Zhejiang in easternmost China, and Sichuan in the far west. It may still exist there in a natural state. Nobody seems sure. But its re-emergence is due to the priestly planting of it in temple gardens, first in China, then in Japan. An eighth-century Chinese writer mentions it. A 16th-century one calls it the duck's-foot tree – referring to the shape of its leaves. Its modern name is the Japanese version of the Chinese *yin-kuo*, meaning "silver-fruit". (The fruit was another reason for planting it: the kernels, roasted, were eaten when liquor was flowing freely; the peanuts and cashews of the ancient Chinese.)

Life in the West

The West heard of the ginkgo when Kaempfer wrote of his visit to Japan in 1690. In 1730 the first plant arrived in Europe, in Utrecht, and in 1754 Kew Gardens bought the tree, which still grows there, from a nurseryman in east London. It was at this stage that it was christened maidenhair tree (*adiantifolia*) from the maidenhair fern, the only plant with leaves of similar shape. Thirty years later it arrived in Philadelphia … and its new lease of life began.

One of the advantages of living on after your era is that your enemies have all, as they say, gone before. Doubtless there were epidemics of ginkgo disease, and hordes of ginkgo-eating insects millions of years ago. But today, as oaks wilt and elms wither, the ginkgo can just look inscrutable. Even urban air-pollution leaves its odd duck's-foot leaves fresh and wholesome. It will grow in shallow concrete tubs by a street where buses snort foul fumes. It is the answer to the Parks Department's prayer – unless it turns out to

Ginkgo autumn colour is short-lived but lovely; a clear yellow which emphasizes the fineness of the foliage. The formal stance of young trees makes them ideal for avenues. This one is in Yamagata Prefecture north of Tokyo.

Sunlight behind these ginkgo leaves emphasizes their peculiar shape: like long-stemmed fans with more or less wavy edges and a notch in the middle. Leaves on young shoots are divided almost in two by the notch.

The fruit of the ginkgo is rarely seen in public places; female trees are best avoided because the fleshy covering of the edible nut rots with a powerful, unpleasant smell.

The design of the ginkgo has changed little in 150,000,000 years. This fossil leaf found in North Dakota is part of the evidence that the ginkgo grew worldwide. Eventually it retreated to the mountains of central China.

The stiff structure of a young ginkgo can continue into old age. Some are distinctly upright, even fastigiate. Others, like this veteran in Kew Gardens, planted in 1762, eventually relax and launch tumbling cascades of lush foliage.

be female. Ginkgo fruit is far from desirable, stinking as it rots; the ones you see planted in the streets are all males. Neither in Europe nor America, though, is it planted very often simply as the eccentric beauty it is. A young ginkgo's branches rise stiff and straight, rather like an Atlantic cedar's: in summer its shape is much softened by the pale green leaves drooping on long stalks; in winter, sculptural (if you like it) or stark (if you don't). With age the branches dip at the end and fan out. It may just be fancy, but the ginkgo's distant relationship to the modern conifers seems to be reflected in its stance, even in old age.

The tree's best moment is in the autumn, when the leaves turn from their well-sustained green to clear butter-yellow, with no hint of orange or brown. They fall soon after, but the fleeting effect on a big ginkgo, fuller and more pendulous in figure, is marvellous.

Some 70–80 feet is normally full size: one at Milan reached 125 feet, and they may reach half this height again in China. Hot summers certainly encourage them. They have a reputation for slow growth, but recently planted street trees in London are growing two feet a year. As to the spread, much depends on the clone: nurseries have distinguished at least 25 varieties for habit or the colour of their leaves. The ancient creature still tries new tricks. Among them is a very narrow upright one known as the 'Princeton Sentry' propagated in America for street planting, but the most graceful trees are the broad and droopy kind.

Extract of ginkgo? Millions take it in the faith that it will keep their minds active and memories keen. Whether it works or not, it seems appropriate for a creature so ancient.

The Pines of North America

PINES ARE TO THE CONIFERS what oaks are to the broadleaves: the most widespread, most varied, and most valuable trees of their order. The biggest family of conifers goes by their name, the Pinaceae. It includes the firs, spruces, cedars, larches – almost all the needle trees. But the actual genus *Pinus*, the pines proper, is limited to 110 or so species with certain clear and obvious characteristics, of which the easiest to see and remember is the relatively long evergreen needles in tight bundles, each bundle (of up to five needles, according to species) wrapped at its base in a papery sheath.

The yearly growth of each shoot of a pine takes the form of a "candle". Firs and spruces and the rest add to their branches every year more or less horizontally; pine candles tend to grow vertically, only settling down to their place as part of the branch later, as they grow heavier.

In contrast to its aspiring new shoots, however, a pine is much less forceful than its cousins in its defiance of gravity. A fir or a spruce is a spire as a young tree, and a spire it remains. Most pines, where they have room to expand, take a course in middle age which brings them nearer to the widespreading broadleaves. The result – an eccentric flourish of bold branches, often on a bare stem, usually with beautifully coloured and patterned bark – is one of the most triumphantly picturesque of all trees.

As old trees pines are the standard-bearers of the skyline. But the owner of a young pine has, if anything, even more to enjoy: a vigorous plant at eye level, fantastically rich in its detail, its thick and sappy shoots bristling with bright new needles, embossed with male and female parts of splendidly original and suggestive design.

Panamerican

The natural range of the pines is enormous. They grow from the Arctic Circle to the Equator. If anything they favour rugged conditions: drought and extreme exposure on mountains; on sandy seashores, where the subsoil is permanently frozen; or, like the famous Jeffrey pine, on a rocky dome high above Yosemite, where there seems to be no soil at all.

If pines have a headquarters in our era it is in Mexico, whose tropical highlands they seem to relish. With their evolutionary tactics of interbreeding they keep botanists busier there than anywhere else on earth. Not all, though, have such lively social lives; some species, isolated on disparate ridges for millennia, remain antique originals. Of the hundred-odd known pine species 30-odd are natives of North America. They divide readily into northeastern, southeastern, and western stables. Anyone planting one should certainly consider the trees of his or her own area first – not forgetting, of course, the imports from Europe, Asia, and Mexico.

The great pine of the northeast is the white or Weymouth pine (*P. strobus*). The old New England forest, which was relentlessly butchered over a period of 300 years, contained vast stands of it, with trees 200 feet high and eight or nine feet across. Its beautiful soft white wood supplied half Europe's needs, as well as building boats and panelling mansions throughout the eastern States.

It was introduced to Britain in the 18th century by Viscount Weymouth, who planted it at Longleat. In Britain it has never been a great success: but in the warmer climate of France and Germany there are good woods of it. Its thin, pale needles of a delicate blue-green make it a beautiful garden tree: given plenty of light it will keep long branches low-down, where you can see them, but it will make a giant eventually – far too big for the ordinary garden. In its place I grow *P. × hunnewellii*, its cross with the Japanese *P. parviflora*, a slighter tree but with charmingly dishevelled blue-green needles and an almost

Northern pitch pine (right) is the almost indestructible pine of dry barrens in the northeastern United States. Chalk soil is the only thing it won't stand. It makes a medium-sized tree with long branches and an open crown. Its habit – unique among pines – is to produce tufts of needles from the trunk. The needles are 3½–4½ inches long and grouped in threes.

The jelecote pine (*Pinus patula*) (right) is one of several from Mexico with limp, thin, drooping needles. Sadly this elegant tree is on the borderline of hardiness in Britain.

The Jeffrey pine (left) in Yosemite National Park. It differs from the very similar ponderosa pine in its narrower crown and big clawed cones. It seeds at higher altitudes than the similar ponderosa.

ridiculous number of delicate little cones – one of our favourite Christmas decorations.

Neither of the other two northeastern pines – the red pine (*P. resinosa*) and the pitch pine (*P. rigida*) – are nearly as big or as beautiful. The red is a sombre tree, dark green with dark red bark, its foliage heavy like that of the Corsican pine; the pitch pine is generally considered a last-resort tree for really rough conditions, where it can cut a rugged, romantic figure. Its peculiarity of sprouting tufts of green leaves from its trunk makes it easy to pick out. In Canada the jack pine (*P. banksiana*) and in the eastern States the scrub pine (*P. virginiana*) are small trees which play similar tough roles.

Southeastern pines are of enormous economic importance, and of great interest and vigour. They cover thousands of square miles of torpid, otherwise unprofitable country. Yet none of them is exactly a beauty; in areas with cooler summers there are better alternatives. The strange one is the longleaf pine (*P. palustris*). Long indeed are its

needles – as much as 18 inches. But its real quirk is to crouch, a mere grassy mound of potential pine tree, for as much as three or four years before it starts to put on height. In its "grass stage" it can survive forest fires and at the same time build up a strong root system to boost its growth. The southern forester uses four principal pine trees, and does a complicated equation to know which to plant where (the factors are speed, quality, and resistance to rust disease). Longleaf is slow-growing but fairly resistant. Shortleaf (*P. echinata*) is even slower, but more resistant. Loblolly (*P. taeda* – it deserves the soubriquet of *Pinus tedious* no more than several others) is the biggest, best, and fastest, at any rate on rich land, but highly susceptible to rust. Slash pine (*P. elliottii*) is the best and fastest on poor soil, but again liable to the disease. Slash has another advantage: a thicker bark, which makes it possible to burn the forest floor to suppress competition while the trees are young.

Farouche frontier trees

GREAT WESTERN PINES

"You will begin to think I manufacture pines at my pleasure" was the report home of David Douglas, exploring the seaward slopes of the Sierras for the first time in the 1820s. In size, vigour, and variety the western American pines are the world champions.

P. coulteri
Bigcone pine; claw cones weigh up to five pounds.

P. ponderosa
Most important western timber pine with huge range. To 240 feet.

P. jeffreyi
Cones are up to a foot long and its thick twigs smell of violets.

Beautiful fissured bark ranging in colour from ochre to pink is characteristic of ponderosa pine. So are the needles in stiff round bunches at the ends of shoots. It shares the same mountains as sugar pine and grows to almost the same size.

The western pines are in a rather different league. David Douglas, having canoed and portaged for weeks on end to find the sugar pine (*P. lambertiana*), called it "the most princely of the genus: perhaps even the grandest specimen of vegetation". Having met the grandfather of all sugar pines in a valley in the Siskiyou mountains of Oregon I can only say Amen. The monster stands 270 feet or so, not as a pole with a shaggy top, as western giant trees tend to be, but mightily branched. The upper branches of sugar pines are their distinguishing feature: straight, and often of immense length. Even a 200-foot tree tends to be a well-proportioned T, so each branch near the top must be 70 feet long, weighed down only slightly with its cylindrical two-foot cones at the tips. Nobody would call it a garden tree, or even a beautiful one. "Sugar" refers to the sweet gum it exudes – chewing gum to Native Americans. In European collections it has grown to about 100 feet and then perished, being, like all American five-needled pines, liable to the rust disease.

The only other western pine which reaches a comparable size is the ponderosa (*P. ponderosa*), which grows alongside the sugar pine on the same mountain slopes, and reaches its apogee (250 feet or so) in the same Siskiyou valley and nearby Grants Pass. The early settlers' name for the ponderosa was the bull pine, which well expresses the weight and vigour and solidity of this most impressive production. The bark forms into huge rectangular plates of warm pink, etched in grey. The needles are dense and stiff, up to almost a foot in length. In ideal conditions ponderosa has grown 120 feet in 50 years, perfectly straight, and only slightly tapering. Of all the big western pines this is the one to choose for a (big) garden in Europe. It transplants well to almost any soil, farouche frontier air included. It is rarefied reading, perhaps, but I love to pore over the classic *Forest Trees of the Pacific Slope* by a forester botanist, George B Sudworth, who worked for the American Bureau of Forestry when the West was wild. He trekked thousands of miles by mule to write precise descriptions of the native trees of a vast area from Alaska to Mexico and record on

exactly which watersheds, plains, or bluffs they grow. "Knowledge of trees", he wrote, "is gained through long study by a partly unconscious absorption of small, indescribable but really appreciable details." He could tell a sugar pine from a ponderosa across a canyon by the cut of its jib. The excitement of the newly opening West is palpable in every line.

Jeffrey pine is similar to ponderosa with more of a liking for high places. Western white pine (*P. monticola*), seen in the mountains, is not unlike sugar pine – though a far better garden tree, the western equivalent of the New England white pine, with beautiful soft foliage. It has the same rust problem, though. Of the other California mountain pines perhaps the most rewarding to grow is the bigcone (*P. coulteri*) – not a tall tree but a broad one. The reward here – and hence the name – is the cones, as big as a baby's head and armed all over with eagle's claws.

Of the bristlecone pine (*P. aristata*), one of the oldest living creatures on earth, I have written on pages 26-7. Anyone who wants to try this patriarch in his or her own garden will find it very slow. For the first few centuries the curious little dewdrops of resin on the leaves are probably the best part. The knobcone (*P. attenuata*) is another small and slow species. Its curving cones, as solid as hand grenades, stick close to the trunk and never fall or open until a forest fire comes their way.

The shore pine (*P. contorta*) seems to be the most adaptable. It grows from the sea coast right up to 11,000 feet (where the variety is called lodgepole). Not only is it adaptable and hardy, but it also has beautiful dense foliage, usually from head to foot, and striking red buds which make it look as though it is flowering. Its disadvantages, which British foresters have discovered (there was a craze for planting lodgepole in the 1960s) are crooked growth – at least in the form that reached Britain – and excessive fertility. Its seedlings appear in vast numbers as forest weeds and can grow six feet tall in their first three years.

Seaside and south

The remainder of the western pines belong on the coast, in milder conditions, with the onshore gale as their chief foe. The most important is the Monterey pine (*P. radiata*), perhaps the lushest and most lordly of the whole tribe. Nature had restricted this incredibly vigorous, but rather tender, plant to a few square miles around Monterey in southern California. Man has changed all that: it is now the chief forestry pine of the southern hemisphere. Results in New Zealand are sensational – with devastating results for the landscapes it rapidly dominates. In its fifth year one tree there put on 20 feet. No wonder kiwis these days see exotic trees (and animals – except of course sheep) as the enemy. Where it is happy the Monterey pine grows vast branches of prodigious weight and keeps them densely clad in bright bottle-green needles. Even in southern England it grows four feet a year.

The bishop pine (*P. muricata*) is along the same lines, but slightly duller in colour, often somewhat slower and hardier, an excellent front-line defence against sea winds. In the same seaside context the Torrey pine (*P. torreyana*) has a limited southern range. In the wild it is a short, open, spreading tree with its bundles of five needles as heavy as darts: the leaves of plants below are punctured by them as they fall. In better soil it grows fast to a large size.

In New Mexico the piñon pine (*P. cembroides*) is grown for its seeds rather than for its looks. They are the pine nuts which were an important food for the Native Americans. Mexico proper is so prolific in pines that many should be much better known. Perhaps the most glamorous pine of all is the Montezuma (*P. montezumae*). It spreads pale grey needles in wide whorls round deep, resinous, orange buds almost like some titanic Michaelmas daisy. An old tree is a vast low dome of these stiff, brush-like shoots, but such specimens are rare, and others given the name can be less dramatic.

The jelecote or spreading-leaf pine (*P. patula*) could hardly be more different. The most feminine tree of all the pines, it has pale, drooping needles, as thin as threads, hanging from the tips of gently curving limbs. For all its languid appearance it grows like the devil in a warm climate, and can become an invasive nuisance. The same warm-temperate forests shelter *P. pseudostrobus*, a better, stouter forestry tree with similar long pendulous needles in fives.

A fourth Mexican pine which occurs here and there in collections abroad is the Mexican white pine (*P. ayacahuite*). To the nonexpert its smooth grey bark and fine needles make it a recognizable relation of the white pines of the eastern and western States and indeed Asia. Its intercontinental cross with *P. wallichiana* of the Himalayas occurred at Westonbirt arboretum in Gloucestershire in 1904 and bears the name of its founder, as *P.* × *holfordiana*.

Monterey pine is the fastest. I planted this tree at home in 1973 to commemorate the first edition of this book. In 2010 it had a girth of eight feet.

The needles of the Montezuma pine (right, in Sussex) are pale greyish green, up to a foot long, grouped around bright red-brown buds. Unfortunately this beautiful tree is particularly susceptible to hard frost.

The Pines of Asia

LOOKING AT A PHOTOGRAPH OF JAPAN it is sometimes hard to know whether the scene is wild coastline or suburban garden. The lanterns and the carefully placed stones soon give the game away, but the main visual element which links the two, and carries at least a momentary conviction, is the pine trees. Wind-contorted, shank-shrunk dotards of pine trees are the soul of the Japanese landscape. A gardener's business is to bend a sapling to his whim, to give it premature grey hairs by constant bullying, by weighing down branches, by pinching out shoots, by a salon-full of cosmetic tricks. It is an art Japan learned from China.

The principal victim of all this flattering attention – and the one on the willow-pattern china – is the Japanese white pine (*Pinus parviflora*). Its natural form already suggests the more highly coloured coiffured version. It is short, wide-crowned, often with several stems; it holds its twigs and needles in flat plates, facing upward. The needles are short, dark, grey-blue, with a slight twist and a paler inner surface which gives them a certain vivacity.

Japanese trees came west with a rush in the 1860s. The Japanese white, red, and black pines all appeared in the West within 10 years. The red pine (*P. densiflora*), however, is so similar to the Scots pine that there is little need for it, except to a collector. This is the skyline pine of the Japanese garden: the ragged, red-barked head on the quasi-distant hill. Its semi-dwarf variety, 'Umbraculifera', is the most interesting for the Western gardener. It makes a broad dome, no more than 12 feet high but with age at least as much across; a tent full of snaking branches. There is also a variety called dragon's-eye ('Oculus-draconis') with variegated foliage.

There would be no difficulty in growing either of these species in any temperate garden or in a tub, or even on a plate, as bonsai. The red is very much faster than the white to start with; apparently it has reached 100 feet in California, but 50 feet is the most it has done in Britain. On the other hand the white, like most five-needled pines, transplants well when it is bigger.

The tough one
The Japanese black pine (*P. thunbergii*) can be bracketed with the European black pines – particularly the Corsican. It has proved itself the best of all windbreak trees for the seashore in New England. Poor sandy soil does not worry it: it slowly forms a head of long unpredictable branches, bearing tufts of dark foliage – another opportunity (which the Japanese do not miss) for crimping and curling and theatrical effects.

The repeating pattern of black, red, and white pines, whether it has a sound botanical basis or not, certainly helps one to classify, to recognize, and to remember certain types. There are two more

Pinus hwangshanensis must have been admired for centuries, but was only introduced to the West in the 1970s. So far it has not developed the almost cedar-like habit it shows here in its native Huangshan, the Yellow Mountains of Anhui in eastern China.

Asiatic trees with typical 'white pine' characteristics – silky
light green foliage, with the young bark smooth and grey. The
Himalayan white pine has been known by various names (*excelsa*,
griffithii, the Bhutan and the Nepalese pine, as well as the correct
P. wallichiana). I have mentioned its marriage at Westonbirt (see
page 83). In *Trees for American Gardens* Donald Wyman names
Seattle and Philadelphia as two places where it does particularly
well, whereas Boston is too cold for it. The best trees in the British
Isles (above 100 feet) are in Devon and Ireland, the mildest areas.
The Himalayan white pine is one of the few five-needled pines to
resist the rust disease. It grows quickly to start with, then settles
down at 50 feet or so as a broad, open tree of beautiful colour and
gleaming texture.

The Chinese white pine (*P. armandii*) is much rarer. I have seen
splendid young trees with long, grey-green needles growing at
the French National Arboretum at les Barres, south of Paris. In my
collection it soon died. At all events it becomes ragged in maturity.

Pine of privilege

The connoisseur's Chinese pine is the lacebark (*P. bungeana*), a tree
which often makes several stems – and the more the better, since
the bark is the main attraction. Areas of white, buff, grey, rust-red,
purple, and green are revealed where the bark comes off. The bole
is smooth and bulky, in fact not unlike a London plane in feeling
but with a silver sheen. This is the pine of Beijing gardens; in the
country you are more likely to see the dull dark *P. tabuliformis*.
From seed collected in Beijing, *P. bungeana* has grown 15 feet in
15 years for me and is just starting the peeling habit.

Chinese pines are oddly rare in the west and, although recent
expeditions have introduced several, most are rather dull. Perhaps
the most characterful is *P. hwangshanensis*, growing as billowing
masses among the rounded rocks of the tourist-trodden Huangshan
mountains of Anhui, for all the world like a prototype of Japanese
cloud-pruned trees in a Zen garden.

Korea makes an excellent contribution with a pine rather like
the Swiss cembran, only, to my mind, much more decorative.
The Korean pine (*P. koraiensis*) has the same way of keeping its
upholstery from top to bottom, but it is a fine shade of deep grass-
green, with what seems an abnormal amount of needles, worn with
a distinct flourish. It is a well-proportioned tree, and this extra
bushiness makes it quite vibrant with colour and life. 'Silveray' is
a very pretty bluish cultivar.

One more valuable tree from the Far East must find a place
here – though it is only the most distant relation of the pines. The
Sciadopitys is usually called the umbrella pine, because its long,
succulent, bright green needles radiate from the branches like the
spokes of a half-open umbrella. It grows slowly in a perfect full-
dress cone shape: a formal but very satisfying feature on a lawn
or among low shrubs. There are 100-foot umbrella pines in Japan
(and a 65-foot one in Kent), but 30 feet is a more probable
maximum in most gardens.

If I were asked to choose one from all these for a small garden
I would plant the Japanese white pine in one of its selected forms.
Its needles have such a stylish twist.

The shadows (above) of
Japanese white pines cross
raked gravel in a formal
Japanese garden. The
white pine is the small and
often contorted tree in the
famous willow pattern.

Lacebark pine (right)
is one of the most
ornamental of all pines.
In the wild it tends to have
an almost white bark. In
cultivation the bark is grey-
green, flaking to show a
patchwork of colours.

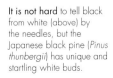

The male "flowers" (left) of a garden form of the Japanese white pine, *Pinus parviflora* 'Bonnie Bergman'.

It is not hard to tell black from white (above) by the needles, but the Japanese black pine (*Pinus thunbergii*) has unique and startling white buds.

The Bhutan pine (top), *Pinus wallichiana*, grows fast and straight. Here, in Bhutan, it is being forested for timber.

Japanese umbrella pine (above) bears no relation to the Italian version. *Sciadopitys verticullata* is an odd-ball fossil relic from Japan and the origin of Baltic amber.

The Pines of Europe

FEW WOULD DENY pride of place to the Scots pine (*Pinus sylvestris*) among the pines of Europe – even perhaps among the pines of the world. I find its beauty unrivalled. The richness of its colouring, its wild poise set it apart. No matter how grey the winter's day, the papery bark, flaking in butterfly wings of salmon and green, glows with the warmth of a fire in the sky. And if the bark is richly red, the leaves can be no less richly blue. To handle a heavy, resinous spray and see the unripe cones, little jade carvings in the deep sea lights and shadows of the needles, is an intoxication.

In its natural range, from Siberia to the Highlands to the Mediterranean, the Scots pine is unique. It was the sole north European pine to survive the Ice Ages. In the Alps the cembran pine (*P. cembra*) came through in some mountain refuges. Only along the shores of the Mediterranean and the Atlantic did half a dozen other species survive.

Of the great Caledonian forest where the Scots pine grew only a few remnants have come down to us, among them one of the most beautiful and romantic woodlands in the world: even its name, the Black Wood of Rannoch, has primeval vibrations. It lies at the head of a glen, on the high moorland round Loch Rannoch, in the centre of Scotland. Among the vast pines, not tall but broad and somehow darker than their descendants, widely spaced in the heather and bilberry, grow great fountainous gold birches, at the same time taller and more weeping than any you have seen. A little alder grows by the burn, which slides on long granite slabs to the loch. A rowan stoops under a hundredweight of scarlet berries.

Until recent times, hard as it is to believe, these few pines of Rannoch Wood were the only pines in Britain. Early man had massacred the rest. An untravelled contemporary of Shakespeare's would never have seen the tree the English have come to know, with cheerful ignorance, as "the fir".

The male "flowers" (left) of Scots pine cover weak shoots in late spring, while the needles are still emerging. Female cones are dark red, in bunches of two to five on the end of vigorous new shoots: they take two years to ripen seed, in the interim turning dark green.

Trees on Rannoch Moor in Argyll (right) – this is Loch Tulla – are reputed to be the surviving remnants of the Caledonian Forest, hence the last descendants of British native pines from the Ice Age.

The Arolla or cembran is the skier's pine of the Alps, often the highest tree on the slopes, tall, dark, and narrow, above the bare-branched larches. This is in the Dolomites, where the dwarf *Pinus mugo* comes up to its thighs.

The Corsican pine is southern Europe's biggest. In southern Italy it has grown to 180 feet, a long-branched tree with dense, black foliage. An excellent windbreak. The Austrian and Crimean pines are close relations.

PINE CONES AND TREE SHAPES COMPARED

What all pines have in common: their bunches of needles in little sheaths, two to five at a time, their upright "candles" of "flowers", and their hard and heavy cones (as well as their taste for tough conditions) still leaves room for wonderful variety of colour, form, and habit.

P. sylvestris
Often seen bushy-topped on a long trunk with flaking, glowing-red bark.

P. nigra subsp. *laricio*
One of the toughest, straight, evenly branched with an airy crown.

P. halepensis
Often makes a broad head of fine pale green needles.

The stone (or umbrella) pine (above) could be called the Roman pine. It reaches its perfect form in the city and the Campagna around. In combination with the dark verticals of cypresses it lies at the heart of Italian gardening.

Pinus pinaster (left) is the pine of the sandy Landes in southwest France, Europe's biggest man-made forest.

North Germany, the Baltic, and Scandinavia grow the same pine. In France it is simply known as "*pin sylvestre*"; no *écossais* about it. At higher latitudes and altitudes it narrows its crown, becoming almost a spire, but more important is the shade of green, or greeny-blue, which seems to be bluer the farther north you go. Scotland's can be almost as pale as steel. There is also a yellow-leaved version that turns bright gold in frost.

The Alps have their own pines: the slow, solid, columnar cembran or Arolla pine of Switzerland (*P. cembra*), which could almost be called the ski tree (its high woods have a vital role in preventing avalanches), and farther east the Austrian black pine (*P. nigra* subsp. *nigra*), closely related to the black pine of Corsica and the south of Italy, *P. nigra* subsp. *laricio*, syn. (the hort-speak for also known as) *P.n.* subsp. *maritima*. You can still inspect the specimens of these close cousins in the Cambridge Botanic Garden planted by Charles Darwin's tutor, John Stevens Henslow, its creator. It was Henslow who recommended Darwin for the Beagle voyage, and his inquiries into such related species or subspecies as these pines set in train – perhaps – Darwin's evolutionary enquiries.

The prize for endurance
Of all European pines the Austrian is probably the toughest: it will transplant well even at 10 feet tall; it will grow in clay (which most pines would shun); it doesn't mind lime in the soil. It has special value in growing on ground that has been churned up by contractors, where the topsoil has been buried or pilfered. You see

it in such unpromising places as JFK airport and the filling stations on Italian autostradas, where its big bushy crown makes a quick sight-screen and windbreak. It is a rough and rather dark and dull tree, but I have one to break the northeast wind behind the house, and I wouldn't be without its long strong branches for all the flowering cherries in Japan. With age its bark forms thick ragged plates of purply-grey that jut out angrily from the trunk. The Crimean variety of the same tree, *P.n.* subsp. *pallasiana*, syn. *P.n.* var. *caramanica*, is even more heavily branched and arguably more handsome. A pity the names of this noble group are such a dog's breakfast.

Both the common black pines of Europe are trees of use rather than beauty. The Corsican version, which is also the Calabrian, is a useful forestry tree for poor sandy ground, quicker growing than the Austrian, narrower and less branchy. At home in Calabria it has grown as tall as 180 feet, and in Corsica to 150 feet. The Italian navy used it for masts – but the French thought that too much resin made it "as hard and translucid as horn". The size for a Scots pine mast (which was the best, in the terms of the 18th-century mast-brokers of Riga on the Baltic) was more than 18 inches in diameter and about 80 feet long. Smaller trees were sold as spars or termed "Norway masts" – since Norway had no big trees to sell.

Unquestionably the most distinctive of European – perhaps of all – pines is the stone (or umbrella) pine of Italy (*P. pinea*). It was a godsend to the Roman and Renaissance gardeners, who were in the habit of confusing vegetation and architecture: it provided them with the perfect foil for the black pillars of their cypresses – a dense black canopy on a nice tidy pole. The pine tree's tendency to flatten out on top is epitomized in the stone pine: its twigs seem to multiply in the crown to leave no chink for light, while all the lower branches disappear without a trace. It is a reasonably hardy tree and could be grown in more northern gardens than it is – though the warmer it is the better it grows. Indeed it should be, for there are few big trees that send out such a strong cultural signal and which interfere less with what goes on underneath.

In the kind of country where the stone pine grows there is often another less shapely pine with grey to pale green needles and a crooked leaning trunk – the Aleppo pine (*P. halepensis*). The seaside part of Provence is typical Aleppo pine country (the hills between Marseilles and St Tropez are an example), and its forests are fresh green and elegant-looking as a result. Unfortunately Aleppo pines are also notoriously inflammable. Old trees are rare. Alas it needs Mediterranean warmth; northern Europe is too cold for it.

The turpentine pine

In some ways the most useful pine, though not a glamorous tree, has been the maritime pine or pinaster (*P. pinaster*). The pinaster's place is the seashore. It was the means by which the biggest man-made forest on earth was developed on shifting sand in southwest France during the late 18th and early 19th century. The Landes were virtually impassable, roamed by shepherds who lived on stilts. In 1789 M. Bremontier started to plant some 12,500 acres of sand-dunes, open to the Bay of Biscay, with pinaster and broom seed together. He laid brushwood over the top, and waited. The result was to fix what was thought unfixable. Then three million acres were added to cultivation in the wake of his experiment. Today the pines of the Landes are still tapped for turpentine and used for railway sleepers and pit props in coalmines – but it is the third and fourth generations of them, and in the meanwhile their roots and needles have made what was a barren wilderness so fertile that farms are taking it over.

One scrub pine of the Alps (*P. mugo*), not really a tree but a bush, is well known to rock gardeners in various forms. But there are two other European hill pines which are little known and which are even tougher and less choosy about soil than the cembran of Switzerland. The two-needled Bosnian pine (*P. heldreichii*, also known as *P. leucodermis* or indeed *P. heldreichii* var. *leucodermis*) is one we should see much more of, especially in small gardens. It holds its branches in tight to its sides in a manner that displays its pale, mottled, even silvery trunk (*leucodermis* translates as "white skin"), grows slightly quicker than the cembran, and is certainly more attractive in youth, and has bright blue young cones into the bargain. On Mount Olympus in Greece there are huge trees of it with mottled bark the Germans call *panzer* or *panther*. The other European hill pine, also from Yugoslavia, is the Macedonian pine (*P. peuce*). It makes a broad, dark, well-furnished spire under almost any conditions. At the Conifer Conference in London in 1972 – a defining moment in the assessment of conifers – it was described as "undoubtedly the most useful five-needled pine for landscape planting".

Fresh young pine cones (left) are a joy to handle and examine, like exquisite pieces of carving. This is the Macedonian pine (*Pinus peuce*), dense with its five-needle bunches, bright green, and blue-white on the underside.

The cheerful, graceful tree (right) *Pinus halepensis* is almost unknown in northern Europe outside the Mediterranean. Aleppo is the second city of Syria and the western end of the ancient Silk Road. It spreads through Turkey and Greece to Algeria and Spain.

The Silver Firs of North America

TO MOST OF US THE SILVER or true firs are probably the least familiar of the common members of the pine family. The word "fir" is loosely bandied about. It has been applied to every pointed tree; almost every conifer. But the true firs (*Abies*) are a conservative genus. They rarely dominate the scene; each species keeps to a relatively narrow range: many, indeed, are well on the way to extinction. Superficially they look very like spruces, and moreover all firs tend to dress rather alike.

Yet as a sensuous experience the fir tree is hard to beat. There is, first of all, its smell. Firs have been called balsams in North America for a long time. They seem to exude the scented gum from every pore. Their barks carry fat blisters of sweet resin; their shoots are full of it; their leaves, crushed, coat your hands with the delicious glue. Every species has its own peculiar scent: of lemon, or turpentine, or tangerine. The form of the tree expresses vigour under stern control. Within the strictest spire shape of any tree,

the liveliest branching pattern darts into space. The view up the side of a fir tree is of thousands of little aircraft zooming out from the centre: shoots and sideshoots create a design of crystalline energy and rigour against the sky.

A juicy rigidity is equally well expressed in the needles: plump, leathery, and firm, apparently adhering to the stout shoots with little round sucker pads. Dried branches of most firs keep their needles firmly on: spruce sprays disintegrate. There is rigidity in the way the branches thrust out horizontally in regular whorls, rarely drooping as spruce branches do. The cones stand up on the branch ends, and, when they break up and scatter their seeds, leave their centre spikes standing there; spruce cones hang down, and fall off, leaving nothing on the tree.

There is a simple test to tell whether a tree is a fir or a spruce. Pull a living needle off a twig. If it leaves a neat round mark, a slight dent, it is a fir. If a little piece of bark comes too, leaving

The female strobili (botanists get upset if you call strobili "flowers", even with inverted commas) of silver firs are often as pink and plump as raspberries. This is the grand fir (*Abies grandis*), with straight two-inch needles set on the sides of the shoot.

The nine-inch cones of the noble fir (*Abies procera*) are ornamented with long-tailed bracts. The blue-grey needles are one inch long, hiding the shoot with their bases, then curling around from the sides to cover the top of the shoot.

Red fir (right) in the Mount Shasta National Forest, northern California. The name comes from the colour of the trunk. Red fir often keeps a fine conical top, its lower branches swooping down in graceful curves.

a torn scar, it is a spruce. When spruce needles fall naturally, though, the minute pegs which form their bases stay on the twig, making it rough to the touch. Fir branches are pegless, and consequently smooth.

The cones of firs are the most attractive cones of all, but since they mostly grow high up on tall trees they are hard to collect. Some species have green or brown ones; most are dark and dusty purple, and so solid-looking that their habit of collapsing in your hands is quite disconcerting. Those of Asian species especially often start off a wonderful Prussian blue.

Botanists have named about 50 species of the firs. In a tribe as homogeneous as this there arise disagreements over what is a species and what a variety. Populations sometimes flow into each other with enough intermediate specimens to keep everyone guessing. All are limited to the cooler hills and mountain ranges of the northern hemisphere. On the whole they are relatively self-indulgent trees, liking deep moist soil and generally not attempting the cold dry hills of the far north.

North America has nine of the firs: two in the east and seven in the west. The eastern species, the balsam fir of Canada and the Lake States (*A. balsamea*), and Fraser's fir (*A. fraseri*), of the Great Smoky Mountains, are on an altogether smaller scale than the western ones, hence more often grown in gardens. Fraser's fir had a bad 20th century, blighted by a woolly aphid-like pest that in places killed 80 percent of the population. The balsam certainly puts up with the worst climate of any of them, growing slowly to a maximum height of about 60 feet. It keeps a good shape, a dark spire complete with lower branches, even when crowding has killed the foliage. The scented needles are used to give the "pine-woods" smell to soap and such – and in Quebec the resin from the trunk is collected as glue. "Canada balsam" is the finest cement for glass in optical instruments. The wood, strange to say, is the one part which has no smell of resin.

On the biggest scale

It was, inevitably, David Douglas who discovered the biggest of all the firs, the grand fir (*A. grandis*). This tree of prodigious vigour thrives where the rainfall is highest – Vancouver Island has grand firs nearly 300 feet high, towers ascending out of sight into the mist dripping thick with ferns and moss. But even more startling is its speed. Atlantic rain apparently suits it as well as Pacific, or better: trees in Scotland have grown 160 feet in 50 years. Even in drier areas it is a quick grower: 50 feet in 25 years in Essex.

There is a childish simplicity about its design. The leaves are straight, parallel-sided, of shiny mid-green, neither dark nor light. They are set each side of the twig in flat rows, as straightforward as the teeth of a comb. Their undersides have two pale blue lines – but these you rarely see; the branches hang flat and low. Where it stands above its fellows, exposed to the wind, the grand fir often loses its top. But an accident like this is only an excuse for another display of strength: three or four new tops will grow as tall as the first, creating a massive square head.

The red silver fir (*A. amabilis*) occupies ground only slightly higher; it too likes to have its roots in deep soil, to have a long

Red silver fir (*Abies amabilis*) (above) grows in dense stands from Alaska south to Oregon, identified by its dense glossy green foliage on drooping branches and its pale ash-grey bark, smooth on young trees. It has reached 200 feet, rather slowly, but proves temperamental in Europe.

The grand fir (*Abies grandis*) (right) is the one to grow for a healthy massive tree at high speed. This is one that has grown 50 feet in 25 years in my garden, lustrously leafy to the ground despite low rainfall.

growing season, and an equable climate. But its name is for beauty rather than size or speed. Its Latin name translates literally as "lovely fir". Thick lustrous leaves set off its pale and smooth grey trunk blotched with white stains of resin. Though a foothill tree, it bears a pointed crown which is the mark of snow-covered heights.

The red fir (*A. magnifica*) and the noble fir (*A. procera*) keep to the middle altitudes of the Cascades and the Sierras from Canada to southern California. Noble fir has the north of the territory and red the south. The related *A. durangensis* extends this group south through Mexico to Guatemala; it is one of several Central American firs, where they are confined to the higher mountains. There is no striking difference between these trees.

The noble fir has the biggest cones of any, and bears them most freely. Its needles are shorter, paler, more crowded, with grooved upper-sides (which red firs never have). But their general appearance is the same: of immense vigour somehow restrained within a narrow space, as if there were a glass funnel inverted over the tree. In western gardens both do equally well; in exceptional eastern gardens fairly well; in European gardens the noble fir, on the whole, rather better. Planted experimentally in our Welsh woods it starts slowly, looking for two or three years as though the weeds will choke it, then shoots skyward, its rigidity making it conspicuous. In nature it is the bigger tree, to nearly 250 feet.

Chosen for beauty

One of the mountain firs, oddly enough, is the most cultivated of these Americans in gardens. The Colorado white fir (*A. concolor*) seems to be the most adaptable to conditions in the east and in Europe. Its natural range is enormous; it grows all the way from New Mexico to Oregon, where its variety Low's silver fir (*A.c.* subsp. *lowiana*) merges with grand fir. The tendency of its long needles toward blue-green has produced garden cultivars in shades of piercing silver-blue: *A.c.* 'Candicans' and *A.c.* 'Violacea' are two of the best known, and there are many dwarf variants.

The alpine fir (*A. lasiocarpa*) has a similar, rather more northerly, range, yet its garden performance has not been good. Its variety, the cork fir (*A.l.* var. *arizonica*), from the southernmost part of its range, is healthier, and also prettier, with silvery-blue needles.

The Santa Lucia or bristlecone fir (*A. bracteata*) is the odd one out. It is a rare tree, not often cultivated and found naturally (but not easily) only in canyons in the Santa Lucia mountains, behind Monterey and Big Sur overlooking the Pacific. To me it is the most beautiful and memorable of all firs. Where the others have plump needles with blunt ends, crowding and curling along the shoots, Santa Lucia's needles are long, slim, straight, and hard, brilliant green with two white lines underneath ending in a sharp point. They fan out below the shoot: on top they surge forward to the tip where, in place of a fat, round, resinous bud, there is an elegant parcel of folded parchment. The "bristle" cone is equally eccentric: a fat little brown barrel decorated with what look like long green feelers. Best of all, though, is the Santa Lucia's shape: a sharp, broad-based spire, terraced with branches which dip in the middle under the weight of drooping, gleaming branchlets, and turn up at the end like a spaniel's tail.

The noble fir (*Abies procera*) (above) is perhaps the most formal and rigid of its tribe, its branches well spaced and straight, catching the light with silvery gleams. These are 25-year-old trees in north Wales; Japanese larch on the right.

The Colorado white fir (*Abies concolor*) (left) has long pale blue needles arranged in a simple open pattern, principally along the sides of its shoots.

The Silver Firs of Europe and Asia

DECORATIVE, STRONG-GROWING SPECIES of fir abound in Asia, but not all of them have been properly tried out in the Western world. Since most of them are too big for widespread garden use, and forestry prefers local kinds, they remain in a quite unwarranted obscurity. Even among Europe's three native firs only one, the common silver fir, has been planted widely. The Spanish and Greek firs are acknowledged handsome and hardy, yet we are only just beginning to put them to use.

The silver fir of Europe (*Abies alba*) was for many millennia the second tallest European tree. In Britain it was the tallest until its reign was brought to an end (at 180 feet) by a Douglas fir in 1955. From its natural home in the mountains in a band stretching from the Pyrenees through Burgundy and the Vosges as far as Poland the silver fir was introduced as a forestry tree throughout Europe. It reached Britain in 1603 and grew so big that it was forgiven its susceptibility to spring frosts. Until the northwest American conifers came on the scene more than 200 years later it had no competition for speed of growth and eventual height. Even then it was a relentless aphid which decided the question in favour of the newer exotics. None of the first trees survives, but there are some 17th-century specimens still left in Scotland: Dawyck Botanic Gardens in the Borders boasts a craggy old fir in addition to its famous beeches. Silver firs tend, perhaps because of frost damage, to make big-branched trees rather than spires: huge, open-headed, grey-barked, dark-foliaged constructions of no special elegance or beauty. They are at their best when, as younger trees, their bark is as silvery as a beech's.

Hard firs from hot climates

Two firs that are very much alike, perhaps survivors of the same race long decimated by the Ice Ages, have taken up their residence at the extreme corners of southern Europe – Greek fir (*A. cephalonica*) in Greece and Spanish fir (*A. pinsapo*) in the mountains behind the Costa del Sol in southern Spain (and in the Atlas mountains of Morocco). This Mediterranean lineage gives blue-grey trees with stiff, hard leaves emerging hedgehog-like from all round the shoot; uncomfortable to handle.

Abies numidica is another, from Algeria – a rather denser and less prickly tree. Where the Spanish and Greek firs are particularly worth planting is in chalky, well-drained districts where other firs have reservations about growing. They are not fast, but they have character from the start. There have been others; in Sicily there are only 30 trees left in the wild of the indigenous *A. nebrodensis*.

Spanish fir is native only to the Sierra de las Nieves, the dry mountains near Ronda, inland from Malaga. It grows splendidly in dry gardens, even in relatively cold northern climates.

SILVER FIR PROFILES COMPARED

The profile of a species, even one as deliberate as a silver fir's, depends to some extent on its growing conditions. Lack of nutrients, and especially light, can distort it. Other things being equal, though, these are the probable profiles of five. *Abies nordmanniana* is perhaps the most reliable.

A. alba   A. nordmanniana   A. koreana   A. pinsapo   A. cephalonica

Silver fir shares the forest with beech in the Vercors National Park, south of Grenoble in the alpine foothills of France. The park is a limestone plateau with deep gorges, tough going in the hill-climbing stages of the Tour de France.

Caucasian trees in general have a fine reputation. The Caucasus mountains lie on the same latitude as northern California (and Tasmania, another rich plant kingdom, in the southern hemisphere). With the Black Sea on one side and the Caspian on the other, with Russia to the north and Iran to the south, these mountains are one of the great meeting places of weather systems – and also of genetic contributions. Mount Ararat lies just to the south. One might easily fancy the garden of Eden was in these hills – in which case the Caucasian fir (*A. nordmanniana*) would have been one of its more distinguished decorations.

A young plant of the Caucasian fir is the most delectable of evergreens. You look down on dense tiers of long, shiny, deep green needles. Underneath they have two icy stripes marking the pores. The tree is always one of the happiest, best-fed looking of the firs, right into its gianthood, when its long branches, still dense and shining, cascade in rich tiers to the ground. It has a characteristic, too, that makes it a better choice of Christmas tree than northern Europe's old standard, *Picea abies*. Its needles don't drop off when it is brought indoors.

The supply of beautiful firs from farther east, from the Himalayas, China, Korea, and Japan is almost overwhelming. Probably the best known of them at the moment is the Korean fir (*A. koreana*), which is obliging enough to cone prolifically at eye level. The ripe cones remind me of those decadent displays of Fabergé imitating confectionery in ivory and rubies, or scrambled eggs in amber and gold: they are plump, and piercing purple, with brown contour lines and spangles of translucent resin – and like presents in dreams they crumble when you pick them up.

Two Japanese firs – the Nikko (*A. homolepis*) and the Veitch (*A. veitchii*) from Mount Fuji – are comparatively widely planted.

A COLLECTION OF CONES

Silver fir cones have this in common: they are cylindrical and stand upright on the branches. Beyond that their patterns show the mysterious originality of natural adaption to conditions where each species evolved. The beautiful Santa Lucia fir (*Abies bracteata*) is distinct for its thin protruding bracts.

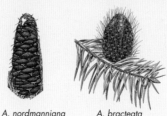

A. pinsapo   A. cephalonica   A. alba   A. nordmanniana   A. bracteata   A. forrestii   A. koreana   A. smithii

The Korean fir (left) is the one most often grown for its cones. Even as a young tree it can bear heavy crops. They start deep purple and turn brown before they split up.

The Nordmann fir from the Caucasus seems lustrous with health, like several trees from this region of great forests. It scores as a Christmas tree because its needles remain attached to the shoot, even when they are dry.

The undersides of the needles of many silver firs are marked with white or silver bands of stomata or breathing pores. The excellent *Abies fargesii* was discovered in China by the missionary Father Farges and introduced by E H Wilson in 1901.

The latter is one of the easiest to recognize, at least as a big tree, from its odd habit of forming an armpit in the trunk under each rather spindly branch. It is a gleaming tree, with the white bands under the leaves, which are very conspicuous on the new shoots. The Nikko fir has the same effect: each leaf strains upward, hiding its green side and showing its white. There are fine differences of leaf length or shape between these species. But often the most noticeable feature is the colour of the new shoot within the covering of leaves. The Veitch fir has brown shoots, the Nikko fir white; a third Japanese species, Maries' fir (*A. mariesii*), has almost red shoots and its winter buds are coated with resin.

Firs of the Fathers
A number of Chinese firs also bear the names of their European discoverers, including *A. delavayi*, which commemorates the Abbé Delavay, who virtually initiated botanizing in west-central China. A short list of the best Chinese firs would also certainly include George Forrest's variety of the fir named after the Abbé Delavay: *A. delavayi* var. *forrestii* (reclassified *A. forrestii*). This particular fir sums up for me the juicy feeling of well-being which makes the

firs such satisfying trees. My notes here get as far away from objectivity as they do at the end of a good wine-tasting: "thick squirrel-brown shoots: needles crowding all round, brilliant white beneath; buds are red globes, shiny and succulent with resin; very bunchy and rich-looking".

Another on the Chinese list would be Farges fir (*A. fargesii*), which was brought back by E H Wilson from the same area. Another would be the Manchurian version of the Nikko, *A. holophylla*, which Alan Mitchell at the 1972 Conifer Conference singled out as a "good doer". In mild and rainy regions any of these firs will soon become substantial trees, though like most conifers their true beauty is best seen when they are relatively young.

One couldn't close the list without at least one of the Himalayan firs. The east Himalayan fir (*A. spectabilis*) – a tree very like the *A. delavayi* I have described – grows on the highest Himalayan slopes and the west Himalayan fir (*A. pindrow*) on lower ground. The west Himalayan fir has distinct long thin needles dividing at the end into two tiny spikes. It has survived for 25 years in my garden without ever looking as though it is enjoying it.

The Douglas Fir

WITH THE GREATEST RESPECT to the giant sequoias, the Douglas fir (*Pseudotsuga menziesii*) can be fairly called the world's biggest *beautiful* tree. There is no suggestion of an amazing survival from another era about the Douglas; it is in the prime of life, with a natural range all the way from Canada to Mexico – as well as close relations in China and Japan.

The Douglas fir stands alone, aloof from others whose names it partly shares. It is not a silver fir. Its cones hang down, and remain in one piece. It is not a spruce either. Its leaves have tiny stalks and suckers like the fir's. It deserves better from botanists, however, than the Latin name they gave it. For *Pseudotsuga* means "false hemlock". The "pseudo", with its implication of moral turpitude, I (on its behalf) most categorically refute. It was never a hemlock, and it never pretended to be. Happily the rest of the Douglas's nomenclature celebrates the two deserving explorers who found it and introduced it to cultivation. Archibald Menzies, sailing with Captain George Vancouver's expedition in 1791, was the first. David Douglas, the tireless envoy of the London Horticultural Society, sent home the seeds in 1827 and its common name remains his most conspicuous memorial.

The forest that Douglas found is also still unplundered in a few spots in the northwest. Mount Rainier National Park is one. Penetrating even 100 yards into a stand of old growth Douglas fir makes it hard to believe that this is the country Douglas crossed and recrossed, sometimes with Native American guides but often on his own. Every few metres a fallen log the size of a truck stops

The waters around British Columbia and the state of Washington are a busy highway of log rafts, with Douglas fir one of their most valuable components. It makes ideal construction timber. Whole trees once made the masts of sailing ships.

Strange trident-shaped bracts like lizards scuttling for shelter are the clue to identify Douglas fir cones. Exact provenance of the seeds is important in deciding the trees' growing season.

Grown in the open, away from the forest (this one is in New Zealand) Douglas fir often makes a broad branchy tree, easy to recognize by its dark, deeply rutted, and weathered-looking bark. It can reach a fair size, even in low-rainfall areas.

your progress. Great rotting snags tower above the brilliant emerald of the mounting young trees; Douglas and silver fir, western red cedar, red alder. Rivers, of course, were the only way of covering long distances. But to portage through these woods …

The seeds from which those first cultivated Douglas firs grew crossed the continent in this way. The cones were entrusted to the "express" run by the Hudson's Bay Company – an annual expedition of a score of *voyageurs* who paddled and clambered 2,000 miles from Hudson Bay to the mouth of the Columbia River to collect pelts.

The Douglas fir is a relatively easy tree to recognize. The clearest indicators are the buds at the ends of the twigs, which are reddish-brown and pointed, very much like rather fat beech buds. If the tree has cones, they are unlike any others: egg-shaped, hanging, with three little paler points, an upside-down trident, extruding from beneath each dark brown scale. The bark of a big Douglas fir is unmistakable: immensely craggy and rutted with dark corky cracks. The hang of the whole tree is distinctive too: the foliage is dense and dark and soft, weighing down the sideshoots in swags below the branches.

In the hills of the Pacific Northwest occasional trees are so pendulous that they could, at a distance, be Brewer's spruce. Like all the western trees, though, they become stiffer, narrower, and paler as they go south and higher into the mountains. The Rocky

Mountain form of Douglas fir has the varietal name *glauca* – though it is by no means bright blue – and is hardy where the big vigorous green tree is not (in northern New England, for instance). Var. *glauca* is also happier on chalk or lime. Ordinary Douglas fir will grow in a limy soil but more slowly, and can look faintly chlorotic.

Given reasonably good soil the Douglas is one of the best timber trees on earth. It makes wood like good-quality pine at a prodigious rate. A tree planted 100 years ago is now, at more than 200 feet, one of the tallest trees in Britain. It stands by a woodland pool on the Duke of Atholl's estate in Dunkeld in Scotland. Climbing this tree and its rivals (one in Argyll, two in Wales, at Powis Castle and Lake Vyrnwy) to see which is currently Britain's tallest has become a competitive sport: 63.79m (210ft) (the Argyll tree) was the 2009 figure, which looks modest enough beside the tallest Douglas ever measured, in Vancouver in 1902, at 415 feet and 77 feet circumference. The tallest still standing (as far as anyone knows) is known as the Doerner fir, near Roseburg in Oregon, a promising 329 feet high and weighing, I read, as much as a loaded Boeing 747. But American loggers are not prone to sentiment, and the Douglas has kept them in profitable business for 150 years.

I have seen hillsides too steep to climb in the Pacific Northwest being planted with Douglas by aeroplane: they were seedlings from selected trees planted in tiny pots that fall like bombs and dig in on hitting soft ground. In Europe the Douglas is usually planted on good sites with some shelter and reasonable drainage. Given good conditions, though, Douglas will grow away so fast that its slender leading shoot wanders off-course and seems to tie itself in knots, only to recover later. Compared with a rigid, prickly spruce a Douglas fir is a friendly tree, soft-textured despite the horsepower under its bonnet.

Douglas firs (left) above Metlako Falls along Eagle Creek in Mount Hood National Forest, Oregon. In close forest they regularly achieve the forester's idea of perfect form – unbranched and tapering gradually for 100 feet or more.

The cones (right) hang like those of spruce, among the soft deep green needles. The end bud of the shoot is shiny brown, rather like that of a beech. A mountain form of Douglas from the Sierras has bluer, thicker needles and smells of turpentine.

The Spruces of North America

THE SPRUCES REMIND ONE in their primitive spikiness that they evolved in the age of reptiles. Those thin unwavering shoots groping at the sky are like tentacles. Few spruces, indeed, are graceful. They seem to lack the plump poise of the firs, their nearest relatives. They are uncomfortable trees; their needles hard and spiny, their twigs rough, the bark thin and scaly. With age they become gaunt and gappy, their branches skeletons of corroded old twigs. The beauty of their new shoots is out of reach as they grow old. You have to be a forester to appreciate the sturdy pole which holds the tattered top against the sky.

There are about 40 species of spruce, as there are of silver fir. The spruces, on the other hand, are more varied than the firs in their habit, their colour, and manner of growth. They contain in their number a handful of trees more strikingly ornamental than any other conifers – so ornamental, in fact, that it is hard to know whether to think of them as *haute couture* or Carnaby Street.

The spruces far outnumber the firs. From the mountains of the south, where the two share the same tastes and habits, the spruces fan out northward into much colder, wetter country. The tree map of the great land masses of the north, Siberia, Canada, and Alaska, shows the astonishing hardiness of the race.

Timber factories

Three spruces, the red (*Picea rubens*), the black (*P. mariana*), and the white (*P. glauca*), grow in the far north. Their range extends south to the Great Lakes; the red as far as the mountain tops of the Great Smokies. Their importance lies in their wood, which makes some of the best paper pulp – or has done in the past. Demand has exceeded the ability of the forest to regenerate.

As ornamental trees they are only planted where nothing else will grow. But the spruces have the peculiarity, much more than the firs, of producing odd branch "sports" of dwarf form or unusual colour – and the northern spruces are no exception. *Picea glauca* var. *albertiana* 'Conica', a bright green bushy form of the white spruce found in the Rockies in Alberta, is one of the best-known garden conifers – though bearing little resemblance to its parents. *Picea mariana* 'Doumetii' is a similar production of the black spruce.

The biggest member of the spruce family (*P. sitchensis*) comes from the land of giants – the northwest. Almost inevitably it was David Douglas who introduced it to cultivation. It takes its name from Sitka, the old Russian capital of Alaska on Baranof Island off the southeast Alaskan coast. Along the fog-bound straits and inlets

The perpetual mists and cold of the mountains of Princess Royal Island, British Columbia are home to the Sitka spruce. This immensely vigorous and hardy tree is a staple of forestry in far-northern Europe. It can grow to 300 feet and apparently even forms the nose cones of Trident missiles.

The shoot of the Sitka spruce (above), magnified five times, shows the woody pegs at the base of the flattened needles and the plaited effect of the ribbed shoot itself. The leaves are dark green above; this shot from below shows the breathing pores, white with a waxy coating.

Black spruce trees and willows (above) in fall colour along Igloo Creek at Denali National Park, 200 miles north of Anchorage, Alaska. The crushed foliage of the black spruce is strongly scented of balsam. Its range runs all the way to Newfoundland.

CONES OF NORTH AMERICAN SPRUCES COMPARED

P. pungens
Colorado blue spruce cones mature from red to brown with thin, wavy scales; about four inches long.

P. mariana
Black spruce is a small dense tree with purple young cones only an inch long, turning brown as they age.

P. rubens
Red spruce from the east has rigid little cones and short needles all around the shoot.

P. glauca
White spruce cones are 2½ inches long with rounded scales. The tree grows from Alaska to Novia Scotia.

there, where whales and log rafts loom up intermittently out of the gloom, the spruces grow trunk to trunk to the water's edge. Melancholy ramrods of trees almost 300 feet high are found, their moss-grown bases splaying out as though to get a purchase in the saturated soil. When they die, fall, and rot they provide perfect seedbeds for the seed of their kin, resulting in plumb-straight rows of seedlings, and eventually trees, across the forest floor. Starting their life on a log makes these trees put down buttress-roots.

Farther south, on the Oregon coast, the Sitka is the first tree to front the sea wind on crags and promontories above a beach buried deep in jetsam lumber. The grey water-worn snags below, the grey wind-torn branches above, epitomize the staying power of the spruce. The same qualities have made it one of the most-planted forest trees in northern Europe in this century. One of David Douglas's originals still stands at Curraghmore in Ireland, 180 feet high. During its lifetime some two million acres of swampy exposed ground in Ireland, Scotland, Wales, and England's wettest counties

have been made productive by its kin – the cause of endless debate between those who see them as money and those who resent their presence in sublimely wild landscapes. No other tree will tolerate peaty bogs with minimum drainage. There is no mistaking the Sitka: its green-grey needles are fiercely sharp to touch.

The beauty of the family

It is barely 100 miles from the coast where Sitka grows to the heights of the Siskiyou mountains. But the spruce of the Siskiyous belongs to a different world. Brewer's spruce, or the weeping spruce (*P. breweriana*), is certainly the most beautiful tree of its race – and certainly the most mournful of all conifers. Its mountain home, above the valley where the world's biggest pine trees grow, is a garden of rare and graceful plants: the Lawson cypress is wild here, growing with pendulous western hemlock and Douglas fir, tufted incense cedar, tangerine-scented white fir, brilliant-leaved dogwood and big-leaved maple, and

Two spruces, Engelmann and the Colorado blue (left), share vast expanses of the Rocky Mountains with Aspens. Engelmann takes the higher, drier ground, above 9,000 feet. It has a clean brown trunk and two-inch dark cones. The Colorado blue likes moisture, grows above 7,000 feet, has a twiggy trunk, and three-inch pale buff cones. Blueness of foliage is variable in both.

Colorado blue spruce (*Picea pungens*) is the Colorado state tree. Forms as blue as this are rarely seen in the Rockies but give joy to gardeners with their air of icy formality.

Engelmann spruce growing at Wakehurst Place, the country branch of Kew Gardens in Sussex. There is a blue cast to its needles but it makes a more natural-looking, softer tree than *Picea pungens*.

Unmistakably a spruce (right); the upper crown of the tree where the female "flowers" are concentrated is thick with hanging cones. This is the typical wild form of Colorado spruce, dense and dark blue-green. The species is best known in gardens by its pale blue variety, Colorado blue spruce.

neat, dark green bulwarks of Sadler oak. This weeping spruce, though, would stand out in any company. Its arching branches carrying long shawls of green-black lace hold arcs of sunlight or snow, receding tiers in different angles of profile; an exquisitely satisfying design.

But whereas Lawson cypress, equally limited in nature, has been prolifically fast and easy in cultivation, Brewer's spruce needs wooing. It is slow to start, demands moisture (in the air as well as in the ground), and takes several years before its weeping habit really shows. Branch cuttings grafted in a nursery weep sooner, but need staking to develop a leading shoot. In the eastern States it rarely does well; in Europe, on the other hand, there are superb examples, although it does not seem to live long. One in the British national arboretum at Westonbirt grew to 40 feet in 35 years. It is a tree to try, and to keep trying.

There could not be a greater contrast than the blue spruce of Colorado (*P. pungens*). High altitude and bright light give the

species its stiff habit of growth, the grey-blue cast in its leaves. But the cultivars a rejoicing nursery trade has found are surely bluer and stiffer than God ever intended. They include the most piercingly ice-blue of all plants. Various cultivars of its forma *glauca*, notably 'Koster' and 'Moerheimii', have a totally artificial air: they rise in stiffly perfect terraces of ice to a stiffly perfect pinnacle, making them perhaps the hardest of all plants to place successfully in a garden. It is worth remembering, too, that they are best in youth; gaps and dead twigs show up all too clearly on a pale ground.

The answer is to compromise, and plant the much less common, softer, and more graceful Engelmann spruce (*P. engelmannii*). The glaucous version of this (forma *glauca*) is also very pale blue. But in place of the stiff icy terraces it has long furry tails for shoots; the needles run along the shoots toward the tip. The species is an important timber tree of the Rockies, with a more northerly range than the Colorado blue spruce.

The Spruces of Europe and Asia

HALF OF ALL THE SPECIES OF SPRUCE are natives of China. We have two or three good ones from Japan and the Himalayas. As usual, Europe's contribution in numbers is minimal – only two species. On the other hand, they are probably the two most attractive of all the spruces to the gardener.

The Norway spruce (*Picea abies*) is Europe's tallest native tree: 200 feet is within its reach. It was the first of the pointed trees to re-enter Britain, whence the Ice Ages had expelled it. As long ago as 1500 it was planted and admired. As the source of much of Britain's light lumber (deal, as we call it) which was bought in the Baltic, it also brought the word "spruce" into the English language: it derives from Pruce, the old English name for the Kingdom of Prussia. Today in most of northern Europe it is, at any time of year, "the Christmas tree". It is certainly the continent's oldest, with an individual in Sweden reckoned as 9,550 years old.

From the forestry point of view Norway spruce has one advantage over the faster-growing Sitka (*P. sitchensis*): it is frost-hardy from the start. There are cold areas where sapling Sitka suffers. For the gardener it is a dull, unrewarding tree; its only attraction comes at Christmas: its importance lies in its profusion of offbeat cultivars. It comes in drooping, upright, golden, dwarf, bush, prostrate, tabular, globular, congested, and exploded forms.

The best of the bunch

Much more beautiful and useful in its normal form, however, is the exemplary Serbian spruce (*P. omorika*). There are no difficulties with limy soil, or indeed soil of any kind, with this adaptable tree. Moreover, it can manage in city air. It grows tall (but not too tall), stays slim (but is satisfyingly solid at the bottom), and forms a strong but graceful pagoda pattern with its short arching branches. It flatters buildings by introducing a harmony and rhythm which they often lack. Most important of all, it grows fast: two or three feet a year to about 50 feet – and on, in the right conditions, to 100. In my pigeon-pestered garden it loses its top (a too-tempting perch) frustratingly often, but has a wonderful way of putting matters right, far out of my reach, by promoting one of its top circle of branches to leader. With all these qualities it is hard to think of anything else you could stand on a 10-foot square base which would do the environment so much good.

The oriental spruce (*P. orientalis*) is not quite as exotic as it sounds: it comes from nearest Asia, from the Caucasus mountains – the home of a number of our best garden trees. One can't be dramatic about the oriental spruce: its strength is simply in its neatness. It has short needles which sit tidily on its twigs, a model of grooming. For all that, it makes a big tree, and a handsome one, very striking in spring when it appears to be covered in red flowers.

Dragon spruce (above) is China's equivalent of Europe's Norway spruce, collected by E H Wilson in Sichuan in southwest China on one of his astonishingly productive expeditions in 1905. It is rough to the touch even in comparison with the Sitka spruce.

The narrow, almost pagoda-style spires of Serbian spruce (above) are unique, reliable, indeed almost indispensable as elements in an ornamental landscape. It comes from limestone hills on the borders of Bosnia and Serbia but can grow happily on acid soil.

Spruces (left) are essentially gothic trees. Caspar David Friedrich, prince of German Romantic painters, might have added a Calvary to this scene in the Vicenza province of Italy.

Spruce needles (this is *Picea abies*) are nearly always square in section (*P. omorika* has flat needles) and have breathing stomata on all sides. Pull one off and it leaves a tear on the twig, but when it falls off dry it leaves a tiny stump.

The Morinda spruce (right) from the western Himalayas, is Asia's nearest rival to the weeping Brewer's spruce – and rival, here, to the 18th-century pagoda at Kew. It has reached 100 feet in Scotland but can suffer from late frosts in spring.

It also has a cultivar, *P.o.* 'Aurea', which greets the spring with startling creamy-white (instead of green) new shoots. A bit gimmicky, but effective nonetheless. I grow a very sombre cultivar with needles as close to black as green, which I find extremely elegant. I have labelled it *P.o.* 'Atrovirens'.

None of the far-eastern spruces can be reckoned essential garden material, like the above, and indeed are seldom grown. Encomiums have been lavished on one or two, particularly the Sargent spruce (*P. brachytyla*), which is from western China. The dragon spruce (*P. asperata*), which has been called China's version of the Norway spruce, is one of the best known, valuable for its densely needled, dark, and stolid appearance. Its Latin name celebrates the way it identifies itself to anyone who is rash enough to stroke its branches: a file is less asperate. The dragon's bark is its bite, but beware of the tiger-tail's needles.

The Japanese tiger tail (*P. torano*), with creamy-buff shoots and chestnut-red buds, has the sturdiest leaves of all spruces. Stout, stiff, and extremely sharp, its shoots bristle with arms like those of medieval Japanese knights.

Of the other Chinese spruces one very pretty one is well known

for its brightly coloured, almost flower-like, ornamental cones: indeed it is probably the only spruce which is planted primarily for that reason. (The corresponding blue-coned fir is *Abies koreana*.) The Likiang spruce (*P. likiangensis*), from Yunnan in southwest China, becomes spectacularly uxorious in the spring, loading its branches with crimson male and bright red female young cones, the females (which turn into the familiar woody cones) standing up boldly at the tips of the branches until pregnancy weighs them into the correct spruce position, dangling down. It is fair to add that this performance only starts when the tree is 25 or 30 years old. Nurseries sell grafted plants for this reason. Those pressed for either time or space can plant the dwarf *P. abies* 'Pusch'.

Of all the Asiatic spruces possibly the best known is the Himalayan weeping spruce (*P. smithiana*). Were it not for the existence of Brewer's this would be a respectable entry as the weeping spruce. As it is, this much bigger (to 128 feet in the south of England), broader, less densely furnished tree, tending indeed to become decidedly skeletal with age, must take second place. It may be the answer in parts of the US where Brewer's is unhappy. But, as a young tree, it has trouble with late spring frosts.

The oriental spruce (left) from the Caucasus has the shortest needles of any spruce, resulting in the tidiest, most elegant spire. *Picea orientalis* 'Aurea', here, puts out pale creamy new shoots, a brilliant sight in spring.

The jewel-like intimate detail (above) of conifers are often worth close examination. *Picea abies* 'Pusch' is a miniature cultivated in order to display the bright pink future cones when the plant is only one foot high.

SOME SPRUCE CONES AND NEEDLES

Even Linnaeus was confused about the names of spruces and firs. He called the Norway spruce *Pinus abies* ("pine fir") and the common silver fir *Pinus picea* ("pine spruce") – exactly the wrong way round. The Victorian John Loudon followed him. At one time the silver fir was called *Abies picea* – but enough of this. The needles of spruces vary more than their cones, which always hang down. Needles vary from painfully rigid to relatively soft.

P. abies
The cones of the common Christmas tree are shining brown when fresh, their scales jagged at the end as though bitten. The needles drop when they are dry.

P. smithiana
The needles of this Himalayan spruce stick out all around the twig, with prickly points. Its cones are matt brown, cylindrical, to seven inches long, with rounded scales.

P. orientalis
The oriental spruce has needles only half an inch long on its slender branches, giving it a neat, well-groomed look. Its slender cones are a rich purple before fading.

The Hemlocks

ONE OR OTHER OF THE HEMLOCKS (*Tsuga*) is a familiar sight from Alabama to Nova Scotia, and from San Francisco to Alaska; between them the four local species cover most of eastern America and much of the west coast. In Europe, on the contrary, it is a surprise to most people that there is such a tree at all. (If hemlock means anything to them it is the waterside weed that Socrates is supposed to have taken as poison. It seems that the first *Tsuga* to be brought to Europe, *T. canadensis*, has a hemlock-like smell.) *Tsuga* is one of the few important genera of conifer that occur in America and Asia but have no representative in Europe and is rarely seen in European gardens. Even in our forests, where it grows fast and straight, it is under-appreciated.

In appearance the hemlocks (all except one) come somewhere between a spruce and a yew. They are botanically close to the spruces; the yew likeness lies in their flat, blunt-ended, relatively broad needles. The hallmark of the hemlock, however, lies in its manner of growth. Of all the needle trees it is perhaps the most graceful in texture and detail. Without drooping or dangling it suggests elegant repose.

You need to look no further than its leading shoot to see the hemlock's philosophy. It nods in a sleepy curve. Instead of neat ranks, or jostling crowds, its needles look faintly dishevelled. The side ones are longer than the top ones, and no two exactly agree on the direction to go. But they are a lovely soft green on top, marked white underneath, and dense enough to look almost furry.

Three Americans

The important differences between the hemlocks are questions of habit and habitat rather than twigs and leaves. They all answer the description above except the mountain hemlock of the Pacific Northwest (*T. mertensiana*), whose needles surround each shoot, pointing onward and outward, more suggesting a cedar than a yew. Soft hemlock-green, in this case, is also overlaid with the bloomy

Hemlock foliage (left) is soft and fine, with the needles arranged in two ranks on the shoots. Unripe cones formed from the female flowers of eastern hemlock ripen dark brown but remain only one inch long. Hemlock seeds germinate freely even in the shade of big trees.

Shelter and moisture are essential for the lush growth of western hemlock, Crabtree Valley, Linn County, Oregon. The western hemlock, biggest and best of its race, has a characteristic drooping leading shoot.

The weeping form of *Tsuga canadensis* 'Pendula' (left) is small and slow-growing; one of the best conifers of this shape. The ordinary variety makes dense, hardy hedges, much used in New England.

Mountain hemlock (right) in the Cascade Mountains, Oregon. They have the short branches and narrow profile of trees used to long months of snow. Their leaves have a grey-blue sheen like most high-altitude conifers, but their nodding tops proclaim them hemlock.

grey cast typical of the mountain tree – a subtle and telling colour in the garden.

The familiar pattern repeats itself: the eastern hemlock is the tough one, usually shorter in stature. The western hemlock (*T. heterophylla*) is the giant, the spoilt child of moist ferny litter, tall shelter, and perpetual mist. Where it succeeds (and the moisture and shelter are both important) it can be breathtakingly graceful. Its long branches form loose layers, softer and more sloping than those of a cedar but almost as broad and architectural. It keeps a tall, narrow crown with a nodding shoot even as a tree 140 feet high. What clinches it for me is the foliage – a sort of subdued brilliant green. Why do European foresters neglect it? It is often rejected by sawmills because with age its trunk splays out into buttresses, and it is embarrassingly prolific with its seedlings.

The eastern hemlock (*T. canadensis*) very often grows up with several leading shoots, to make a spreading shape. Its foliage has an odd feature in a line of little blunt needles lying upside down, with their white undersides showing along the top of the shoot. Eastern American gardeners have found it so adaptable that they think of it more as raw material for shaping (particularly into a hedge) than as a tree with ideas of its own. Among its own ideas (it has almost as many as the Norway spruce) are gold, upright, and dwarf versions, and an extremely effective weeping one – perhaps the best low evergreen weeper for the lawn. The maximum height of this form (*T. canadensis* 'Pendula') is nine or 10 feet, but its spread might be 20 or 30.

The Carolina hemlock (*T. caroliniana*) is to all intents a local form of the eastern one. If there is any advantage in either of the two Japanese hemlocks (*T. diversifolia* and *T. sieboldii*) or the Chinese *T. chinensis* it is only a still-denser supply of needles. Formerly the sole discouragement from growing any of this graceful group is their distaste for limy soil, which is strong in the western, the Japanese, and the Carolina, and less so in the mountain hemlock. The obliging eastern one hardly seems to be bothered by lime at all. Since the 1980s, however, a woolly pest from China, looking like cotton buds under the branches, has devastated the population of both the eastern and Carolina hemlocks. Chinese hemlock appears immune and clearly has an important future where its cousins are under threat.

The True Cedars

THE LAYMAN USES THE NAME CEDAR for any tree with dark, spice-scented timber. Thus America has the incense, the western red, the eastern red, the Alaskan yellow, and the Port Orford cedar; Japan the plume cedar and the Hiba cedar; the giant *Juniperus procera* is the African pencil cedar. The botanist uses much stricter rules and recognizes only four, none of them from America (or the Far East) and none of them remotely like the American "cedars". For once I would say (as a committed layman) that the botanist is right.

There is nothing like the true cedars in the whole coniferous realm. Not even the pines produce a tree of such majestic architecture. Partly no doubt because of their biblical background – they are the trees most mentioned in the Bible, and always as an image for fruitfulness and strength – Europeans regard them with something approaching awe. Countless cedars in English churchyards and on the lawns of country houses are celebrated in village lore as having come back with the Crusaders. Villagers' geography, in this case, is better than their history. Few of the giant trees which cut such an apparently timeless figure are more than 150 years old. On the other hand crusading country is just where they do come from.

Close relations

Three of the four species of cedar come from the shores of the Mediterranean; the fourth comes from the western Himalayas. The cedar of Lebanon (*Cedrus libani*) is the most celebrated and by far the oldest in cultivation. The remaining grove of ancient trees on the slopes of Mount Lebanon (at 6,200 feet to 4,000 feet below the summit) is a tiny remnant of a great forest that has been exploited since the ancient Sumerian civilization of Mesopotamia. Deforestation, indeed, aided by such eager customers for cedar as the pharaohs and Solomon for his temple, destroyed the fertility of the legendary Fertile Crescent. The biggest grove has been a place of pilgrimage for centuries; from all accounts the cedars have remained exactly the same since travellers began writing about them. There are roughly 400 of them, as well as others in a dozen more sites: the biggest 48 feet in girth and calculated (by guesswork) to be about 2,500 years old. The biggest natural woods of Lebanon cedar surviving are in the Taurus mountains in southeastern Turkey, composed of a hardier subspecies, *C. libani* subsp. *stenocoma*, with shorter branches than the Lebanon trees but usefully growable as far north as Boston, Massachusetts.

The great Sir Joseph Hooker, incidentally, poured cold water on the idea that Solomon's temple was built of these trees. In his view builders' merchants of Solomon's day were no better botanists than they are nowadays: the wood they called cedar was probably a sort of juniper.

Despite such recollections of antiquity it is still remarkable that the first cedar of Lebanon to be planted in Britain, the forerunner of a great revolution in English gardens, is alive today. It is in the rectory garden at Childrey in the Thames valley, planted there by Dr Pococke, the rector, in 1646, soon after his retirement as the Embassy chaplain in Constantinople. Today its trunk is 26 feet in girth.

John Evelyn did his best to popularize the cedar of Lebanon in the 1670s, but it was not until the earliest trees started to bear cones that they could be planted on a massive scale. At first their hardiness was in doubt; a terrible winter in 1740 killed large numbers. But they soon became a craze; the Duke of Richmond

The Atlas cedar (above) originated in the Atlas mountains of Algeria and thrives on heat. This tree at the Château de Laàs in Béarn, in southwest France, enjoys warmth and high rainfall.

For decades it grows straight, sometimes with several trunks; then it loses its leader and assumes its familiar flat top. The cedar of Lebanon at Knightshayes (right) in Devon is typical of the cedar's long association with noble country-house architecture.

CEDARS COMPARED

The Atlas and Lebanon cedars are botanically very similar. Few can tell them apart by their details of needles and cones, though their habit and colours are often distinct.

If anything the Atlas cedar cones are more cylindrical and less tapering than those of the cedars of Lebanon. The needles of all cedars (like larches) grow singly on new shoots but in bunches on short spears from older ones.

The deodar or Himalayan cedar is a softer tree, with longer needles and drooping leading shoots. The cones form in rows along the upper side of the branches, light grey at first, like fat eggs, then ripening to yellow-brown.

The ancient cedars (left) on Mount Lebanon epitomize the strongly horizontal table-topped shape that makes such a perfect foil for classical building. The remaining cedar groves are the last remnants of a great forest, used for building timber since the Pharoahs.

A magnificent mature deodar (below), in the garden of Middlethorpe Hall in York, may have been planted by Lady Mary Wortley Montagu in the 18th century. It has the tabular format of old cedars but with the softness of its longer shoots and needles.

planted 1,500 at Goodwood in 1761. They were the only exotic trees that Capability Brown used in his landscapes (had he seen pictures of old ones, or was he expecting pointed trees?) and they continued to be planted in great numbers until the 1830s, when the new giants from Douglas's travels began to take their place. A survey done in Hertfordshire, a great cedar county, after the gale of Christmas 1990, showed how few were planted in the 20th century. There were hardly any young trees to succeed the scores that were blown down.

The cedars, in fact, are not nearly so slow-growing as their stateliness might suggest; they can reach 50 feet in 30 years in good conditions, and grow stout and impressive in an equivalently short time. Some make single monumental trunks, many burst out in a sort of organ-pipe formation of as many as 20 or even 30 competing stems. The tallest in Britain, at Eden Hall, Cumbria is 108 feet. Not until they are getting on in years, though, do they strike their splendid attitudes, forming dark plateaux low over the lawn, at rooftop height, and again, high above pediment and belfry, spreading a black table in the sky.

The demand for blue

The Atlas cedar (*C. atlantica*) does not have quite the same classical education as the cedar of Lebanon. Its arrival, from the Atlas mountains of Algeria, was not until 1839. A forest of it was planted in 1862 on Mont Ventoux in Provence – a magnificent sight today. It is a bold man, however, who claims to be able to tell the difference between the Lebanon and the Atlas. The only certain way is with a magnifying glass: the Atlas's needles have a minute translucent spine at the tip.

The ordinary green Atlas cedar is not planted nearly so much as the variety 'Glauca', the blue one which is said to be the most popular of all ornamental conifers today. I have not actually seen one in a window box, but it is remarkable where people do plant them. One sees their desperate pale blue gestures over every little garden wall in some parts of France – particularly the southwest. It is hard to think of any tree less suitable for a small space: colossal spread is its very essence. And heaven knows how you would fell it in a backyard.

Many think the deodar (*C. deodara*) is at its best as a young tree. With longer needles in a quieter colour, a lovely soft silver-green, it nods and droops in delicate attitudes. It tends to maintain its basic conical shape into maturity, rather than rearing all its upper branches together to a high plateau. In outline and texture it remains soft rather than rigid. In its native Himalayas it makes what must be some of the world's loveliest forests.

The fourth true cedar comes from Cyprus. Compared with the others it is unspectacular – and consequently a rarity in cultivation though much fancied, I believe, for bonsai. In its forests high on the Cyprus Mount Olympus, though, where its partner is the oriental plane, its near-black foliage makes a wonderful contrast to the planes' pale trunks. It grows slowly to about half the size of a Lebanon cedar, and is notable only for its dark, much shorter leaves – whence came its Latin name, *C. brevifolia*.

Cedrus brevifolia (above) is a species (or perhaps subspecies) limited to Cyprus. Its needles are shorter and generally darker green, the tree smaller, and its adaptation to gardens more problematical. The trees in the forests of Paphos are said to be suffering. This tree may be endangered.

The needles of cedars (right) are produced from mature twigs in dense tufts, 20 or so at a time. The Cyprus cedar is distinguished by the tiny size of its needles, often less than half an inch long. When cedar cones ripen and fall they leave a central spike standing on the branch.

European larch, showing its autumn gold here, is a timberline tree among dwarf and black pines in the wilderness of Fischleintal in the Dolomites, in northeast Italy, where the geological evidence of the Ice Ages is dramatically exposed in limestone pillars and cliffs.

The Larches

OF ALL THE CONIFERS which are widely planted as forest trees the larches make the prettiest woods. It must be because they celebrate the seasons that we find them so friendly and amenable. The advantages of evergreenery are obvious enough – yet we love the larches for being different: for flushing feathery green in mid-spring; for glowing gold in late autumn; even for spending the winter as a tangle of bare black rigging, when their colleagues are demonstrating the superiority of their leathery leaves.

Their leaf dropping lets in the light so that a rich undergrowth of shrub and herb and creature can flourish, and contributes year by year to the fertility of the forest. Yet it by no means detracts from their vigour, as you might think it would do. The larch is one of the fastest-growing of all trees; certainly the fastest to make strong and heavy wood with almost oak-like qualities.

Close to, the larch's summer rig is remarkably like the cedar's. They are clearly the closest kin, with their leaves in little rosettes like those of no other tree. The larch bears no resemblance at all to the two other deciduous conifers, the swamp cypress and the dawn redwood; the confusion comes in telling larch from larch.

North America has three larches, two of them excellent trees. Both Russia and China have two species – of which one forms the world's northernmost forest, on the latitude of the northernmost point of Alaska. The Himalayas, Japan, and Europe have one species each. Possibly more important than any of the species, however, is a hybrid, the fruit of a union romantically formed at a ducal seat in the Highlands of Scotland.

A Scots romance

The family of the dukes of Atholl has had trees in the blood for generations. It was James, the 2nd Duke, who first saw the possibilities of the larch for reafforesting his naked Perthshire hillsides. The European larch (*Larix decidua*) had been brought to England from the Alps, its native home, in 1620; John Evelyn had described it; but it had remained a rarity. In 1738 the duke planted his first larches beside Dunkeld cathedral on the banks of the Tay, in sight of some of the best salmon water in the world. One of these famous "mother" larches, as they are always called, can still be seen. It is a matronly figure, as broad as it is high (which is 105 feet) with massive curving limbs. The seeds from this tree and its companions were planted on a steep slope nearby called Kennel Bank. One of them now stands 131 feet high; its trunk, scarcely tapering at all, soars 90 feet without a single branch, then on to a distant, delicate head of pale, weeping branches.

The 4th Duke was so smitten with the larch that he planted 17,000,000 of them. His example was followed all over Britain,

The common European larch comes from high up in the drier parts of the Alps between France and Austria, often mixed with dark Arolla pines. It was introduced to Scotland as a forestry tree in the 17th century: a sensation because no one had seen a deciduous conifer before.

so that by the mid-19th century larch was the most important plantation tree. At that stage an aphid started to prey on it, with disastrous results.

In 1861 the nurseryman James Veitch introduced the Japanese larch (*L. kaempferi*) to cultivation. The 7th Duke of Atholl, as larch-minded as his forebears, was quick to plant it, alongside Kennel Bank. By 1904 the first hybrids had come into the world.

As so often happens, the hybrids were better than either parent. The Dunkeld larch proved faster, more resistant to insects and disease, and just as beautiful. Today all three are planted in America and Europe, but the hybrid tends to be first choice.

The two American larches are similar in all respects except size. The tamarack, or eastern larch (*L. laricina*), ranges from Alaska to Illinois. It is a tree of swampy ground, and rarely grows taller than 75 feet – much less where the ground never dries out. It is ideal for gardens in really cold areas; its little rose-like cones are very decorative. The western larch (*L. occidentalis*) is a native of Montana and Idaho, where it can reach 175 feet – scarcely a garden tree.

The Himalayan larch (*L. griffithii*) makes a graceful garden tree in milder regions. But the one that deserves to be far better known is a Chinese tree – the false larch (*Pseudolarix amabilis*) – which grows very wide and not so high (or fast). What is different is its texture. The perfect little open rosettes of needles are the softest, silkiest green; there's a porcelain quality about the depth and texture of its autumn apricot-gold. It would need a broad space in a garden – perhaps 30 feet across eventually. But it would be my first choice.

Larch swept all before it in the 18th century. For hardiness, speed of growth, and the quality of its almost oak-like timber it had no serious rival. So versatile is the timber that its uses ranged from ship-building to the panels on which Raphael painted his angels. Here in its adopted Scottish home it grows in fruitful association with mature Scots pines.

LOOKING AT LARCHES

The deciduous needles of larches arise from the twig: either singly or in tufts of up to 40 at a time growing from a woody spur. The female "flowers" are bright red cones, ravishingly pretty to see with the soft pale green of the new leaves, and become tough little cones about an inch long. The European and Japanese larches hybridized at Dunkeld in the Highlands in 1900 to produce *Larix x eurolepis*, more vigorous than either parent.

European larch
This has buff shoots. Its hybrid with Japanese larch is a similar colour with reflexed cone scales.

Japanese larch
Orange to red shoots give the tree in winter a warmer look. The scales of its cones curve back when ripe.

American larch or tamarack
This has yellow/brown twigs, reddish bark, and tiny cones.

The most beautiful of the larches, though not the most robust, is the golden or false larch (*Pseudolarix amabilis*), from eastern China, introduced in 1852 by Robert Fortune. It slowly makes a long-branched tree with long, soft pale green needles that turn clear golden yellow in autumn. Sadly it shuns alkaline soil.

The young cones of Japanese larch are like plump little eggs, opening as they ripen and curling back their thin, rounded scales. The tree is more particular about soil than its European equivalent but grows rapidly to 100 feet in acid soil with high rainfall.

Monkey Puzzles and Kauris

THE ORIGINALITY OF THE GINKGO is not to resemble any other conifer. It brings its own one-off design unaltered from the distant past with no kin for comparison. The race that looks most obviously Jurassic Park should appear, if I were to stick strictly to the supposed order of plant evolution, as a later development, between the larches and the yews. I put the monkey puzzles and their family here because they stand out anywhere as primitive creatures; almost scary.

I shall never forget going to see them in their home on a Chilean volcano. It was mid-autumn, and snow had fallen on the cinders of Volcan Llaima. A recent eruption had left huge mounds and trenches of ash and clinker black against the snow. The forest was pure *Araucaria araucana*; nothing but the gaunt trees, the biggest like scaly parasols, their progeny clustering round, gesticulating with spindly limbs. The smallest were comically child-like, some meek and tidy, some waving their fists, scampering, it seemed, through the drifts.

Araucaria araucana (the name comes from the native tribe of the region) had arrived in Europe by 1800. The busy botanist Archibald Menzies helped himself, the story goes, to the nuts from the Governor's dessert table, sowed them on board ship, and back in London gave a tree to Sir Joseph Banks. In 1844 William Lobb, collecting in Chile for Veitch's nursery, sent home enough seed to start a craze for what rapidly became known as the monkey puzzle (or Chile pines), and popular with children as an excuse to start pinching one another.

The vogue soon set up monkey puzzles as thick as television aerials in the suburbs of late-Victorian England. The highly flavoured taste of the time would combine these reptilian relics with bogus-rustic arbours and geometrical beds of scarlet salvias.

My ideas about these bold reptile-trees (there is no other way of describing their bright green scaliness) were abruptly altered when I saw them being tried out in what are known as "forest plots" on the west coast of Scotland. Where one monkey puzzle looks quite embarrassingly stiff and self-conscious – at least as a young tree – a whole clump of them takes on quite a different air. Seven or eight monkey puzzles, six feet apart and 20 feet high, would make the most magnificent garden sculpture: ferocious-looking, I'll admit.

The other araucarias – the Norfolk Island pine (*A. excelsa*), the candelabra tree (*A. angustifolia*), and the bunya bunya pine (*A. bidwillii*) – have been considered up to now too tender, alas, for northern Europe or most of the United States. Alas because they are really much better-looking trees than the monkey puzzle.

Monkey puzzles live alone, with no other trees, amid the clinker and ash of volcanoes on the Chilean side of the Andes. This is Volcan Lanin. Almost their only accompaniment is the bizarre bamboo *Chusquea culeou*.

A branchlet of a wollemi pine (right), a giant tree on which dinosaurs used to feed, lies on top of a fossil containing the same species. The wollemi pine was thought to be extinct until a park ranger discovered a living specimen in New South Wales in 1994.

The bunya bunya pine (below) is the least exotic-looking of the *Araucaria* with relatively soft leaves that could almost hide in an English wood.

The reptilian-scaly branchlets (left) of the monkey puzzle. Each leaf is hard and tough as leather, tipped with a sharp point.

The wollemi pine (right), so far. The sensational Australian prehistoric newcomer has reached this stage in cultivation: not outlandishly strange, bonny, and bushy, if not beautiful. Perhaps the foliage most resembles a *Cunninghamia*.

Monumental, certainly (above). Elegant? Perhaps. The regularity of the Norfolk Island pine's comb-like branches makes it the easiest tree in the world to identify at a distance. Not a tree for Britain, and only just possible in the south of France.

The Norfolk Island pine is the beauty of the family. It grows for many years with the hypnotic symmetry of a paper cutout, each perfect whorl of branches holding aloft four perfectly matched green combs of leaves. The bunya bunya is less formal and more luxuriant, with mighty swags of scaly foliage. Like the monkey puzzle, however (and indeed most conifers), the bunya bunyas grow ragged in old age and should be put out of their misery.

Celebrity

By the end of the 20th century you might have thought (and we all did) that the last Jurassic parkland tree had been discovered; that in the age of Google Earth there was no corner of the globe where a mastodon of a tree could hide. Botanists, we know, are still finding – or creating – new species in remote China and South America, but it is 60 years since something prehistoric has re-emerged.

Then in 1994 a ranger named David Noble from the National Parks Service of New South Wales spotted a strange tree in the forest just north of Sydney, not far west of the Hunter Valley. It was identified as a new species of *Araucaria* (but *Araucaria* leaves have no midrib, and their cone scales are fused together), dubbed the wollemi (with the accent on the "i") pine, and launched on a celebrity career with all the pizzazz of the Age of Conservation. Genus: *Wollemia*; species: *nobilis* – remembering its discoverer, but subtly, in bot-speak, meaning "illustrious"; the best of its kind.

No tree (or perhaps no tree since the wellingtonia in 1853) has been introduced with such a fanfare. The Botanic Gardens of Sydney and Kew had the notion – why had no one before? – of using its notoriety to raise funds for its conservation. Small plants quickly sold at premium prices.

Surprisingly, for a tree from the New South Wales bush, the

The wollemi pine (below) produces its cones in winter on the tips of its branches, unlike any other conifer. Its buds are covered with a white waxy substance.

The eventual cones (below) of the monkey puzzle are almost globes, six inches or more across, full of fertile nuts. The cone shatters before falling.

wollemi pine seems extremely robust in cold and wet conditions. Its home is beside a creek in a gorge of acidic sandstone. Winter is frosty – for which the tree seems prepared, with white resin-covered buds – nor are its leaves specially formidable; dull green and narrow, drooping sightly from their stalk, giving it a faintly hangdog look. Uniquely, its cones and their male counterparts hang from the tips of the branches. Whether in cultivation it will grow as a single-stemmed tree or with several trunks is not yet clear: it makes new buds up the stem, which suggest potential bushiness. It has, of course, been planted everywhere. I have planted one in Essex in neutral soil, where the rainfall is 20 inches, and one in Wales, where it is 60 and the soil is acid. So far the Welsh plant is faster and happier.

The kauri (above) of the Queensland rainforests, *Agathis robusta*, near Cairns, has the same immense trunk as the New Zealand species, *A. australis*, and reaches over 150 feet.

The primitive reproductive system of the (gymnosperm) conifers (right) looks pretty naked. This is the future cone of an *Agathis*, the ovule protected only by scales rather than the flowering parts that provide angiosperms with a protective ovary.

Maori canoes

New Zealand's mighty Kauris belong in the genus *Agathis*, but you could argue that they are araucarias in spirit – and indeed the kings of the family. The family resemblance lies in the huge cylindrical trunk and long leather leaves. *Agathis* is a genus restricted to the warmest parts of New Zealand. Australia, Melanesia, and tropical southeast Asia, with only tentative guest appearances elsewhere (notably in Tresco and southwest Ireland).

Kauri once dominated the rainforests north of Auckland, but the quality of the timber has been its downfall: barely 0.01 percent of its original forest has survived. Logs 80- or 90-foot long formed the

Maori battle canoes that held 180 warriors, and later settlers were relentless in exploiting it. Its valuable resin was mined from the forest, where it persists in the ground, as copal for varnish. Several Kiwi families now famous for their wine started as gum diggers; most of them for some reason from Dalmatia.

Maori woodcraft was so ingenious that they would prepare a tree to become a canoe years before felling it, clearing the forest around and stripping the bark off one side. The other side, continuing to grow, grew heavier while the stripped side began to rot, becoming easier to hollow out. When the tree was felled the canoe was already in embryo with its keel heavier and harder than its topside.

The Plum Yews and Podocarps

THE YEW FAMILY, TAXACEAE, is represented across much of the northern hemisphere by its principal genus *Taxus*, but a few others lurk in the forests of Asia. The only ones to be occasionally found in cultivation are the nutmeg yews (*Torreya*), whose distribution stretches from eastern Asia through California to Florida. All have spiky, hard-needled foliage with positive points and large, plum-like fruits. Although none is a tree that would justify itself in a small garden, the California nutmeg yew (*T. californica*, named after the same botanist as the rare Torrey pine) makes a respectable tree: the biggest in Britain has reached 72 feet. After 35 years in my collection it is a 25-foot bush with one short trunk and many wandering branches; no beauty, but useful shelter. The Japanese species (*T. nucifera*) has smaller, relatively broader leaves, but like most torreyas thrives best in areas with hot summers.

Yews for highbrows

Looking very similar, and differing only in microscopic arrangements of bracts around the fruit are the plum yews (*Cephalotaxus*), but this distinction is sufficient to place them in their own family, Cephalotaxaceae. (*Kephalos* is Greek for a head, which is little help.) A taxonomist in mild mood might be inclined to lump nutmeg and plum yews together, and a gardener certainly would. At first glance they could easily be confused with the yews (*Taxus*), but the leaves of both *Torreya* and *Cephalotaxus* are much longer, more spaced out, and lighter green. Like *Torreya*, the fruits of *Cephalotaxus* are nuts enclosed in flesh like a small plum, not open at one end as the *Taxus* fruit is. *Cephalotaxus* grows into a small, irregular, rather open tree whose main charm is in the strong pattern of its leafing.

The Chinese plum yew (*C. fortunei*) is probably the most-planted of the group. Its long, soft, light green leaves are ranked along the sides of its shoots so that each branch presents a flat upper surface like a young grand fir branch, but bolder in detail.

The Japanese plum yew (*C. harringtonia* var. *drupacea*) is similar, again, a bush rather than a tree, with the leaves, paler underneath, tilting up to give a shallow V-shape to the shoot. My specimen has remained, after many years, a committed bush of horizontal inclination, although an excellent form imitates the Irish yew almost to the life, with the same way of leafing radially all around the upward-pointing branches, the only obvious difference being that the leaves are bigger. The fruits of both (it is hard to think of them as cones) are light green ripening to purple, juicy, and inviting-looking. Squeezed hard they exude delicious-smelling milky resin. They are plants that grow well in the shade and tolerate chalky soil, and would certainly make a worthwhile contribution to many gardens.

A splendid willow-leaf podocarp (left) holds centre stage at the Savill Garden in Windsor Great Park, no doubt puzzling many visitors. It is endemic to southern Chile, with its high rainfall, and may find the London climate rather dry.

Japanese plum yew (above) (*Cephalotaxus harringtonii*) has particularly lively foliage, looking faintly like a bird flapping its wings. Its name commemorates a Regency dandy but its chemistry may contain a potential cure for leukaemia.

Podocarps are more abstruse

The podocarps are largely a submerged clan as far as European and American gardeners are concerned. They are (with the araucarias) the principal conifers of the southern hemisphere. In South Africa, for example, yellow-wood (*Podocarpus latifolius*) was formerly the chief source of timber. Podocarps, in fact, are almost the only native South African conifers. In Australia "brown pine" and "black pine" are both podocarps. In New Zealand *Dacrycarpus dacrydioides*, the biggest of the family, up to 200 feet, is called "white pine" and *Podocarpus totara* is just totara, the big-bellied tree the Maoris carve into fantastic figures. Few cross into the northern hemisphere, but the northernmost, from Japan, is an exceptional garden tree. The few excellent hardy species from New Zealand are still seldom seen in cultivation, but some high-altitude species form excellent low shrubs, often flushing into growth with bright new shoots.

The fruit of the podocarp (technically a cone) is a purple or red berry to which a nut is attached or embedded. Podocarp leaves vary from yew length to willow length, all narrow but the longest ones too tapering at the ends to be described as needles. Their best quality is the rich green glossiness of their fountains of foliage, a sub-tropical quality that draws you to the tree, particularly in winter. The willow-leaf podocarp (*P. salignus*) from Chile and the Japanese member of the family (*P. macrophyllus*) both have this effect. Where they are hardy (zones 7 and 8 respectively) they will eventually grow into fair-sized trees. The biggest willow-leaf one in

At a distance you might possibly take the leaves (above) of Prince Albert's yew (*Saxegothaea conspicua*) for the common kind, but not its sharp-tipped leaves, its pale new shoots, or its elegant droop. It is another rainforest tree from Valdivia in southern Chile.

North Island, New Zealand (right) is home to several wonderful trees that are hard to grow in Europe. In the Pureora rainforest, rimu (*Dacrydium cupressinum*) is a dominant species. Its leaves are awl-shaped, of the cypress persuasion. Sadly, New Zealand is more interested in fast-growing *Pinus radiata* these days.

the British Isles at present is 78 feet. It enjoys Irish humidity – although there is a luxuriant dome of one in a key position in the Savill Garden in Windsor Great Park. Smaller sizes are reported from the southern States. In the extreme south, in hardiness zone 10 at the tip of Florida, the one to grow is the east African *P. gracilior*. (Its seed arrangement – a hard stone in a plum – has had it reassigned to *Afrocarpus*; gardeners will not be concerned.) *Podocarpus macrophyllus* appears in almost every Japanese garden, tonsured or not, under the name kusamaki. In China it is a feng shui tree – and frequently stolen, I'm told, for good luck.

Like yews or cypresses

Botanists in their wisdom are no longer happy with the broad concept of *Podocarpus*, so break it up into several smaller genera scattered across the southern hemisphere. One of the short-leaved species, the Chilean *Prumnopitys andina*, is remarkably like the yew for such a distant relation. To make matters worse it has been christened the plum-fruited yew, to the chagrin (one imagines) of the original (so-called) plum yew.

Another yew-like podocarp from Chile was made memorable by being called after the husband of Queen Victoria: *Saxegothaea conspicua* (Prince Albert's family name was Saxe-Coburg-Gotha). Prince Albert's yew is certainly worth planting. It looks like a rather slender weeping willow with black foliage – or at least it does in Edinburgh, where it is perfectly hardy. Again, humidity and patience are the secret; in Devon 60 feet is the limit so far – and it has had a century and a half to grow.

The podocarps have their cypress types as well as their yew types and willow types. The *Dacrydium* branch of the family has scale leaves that suggest cypress (or perhaps juniper). The New Zealand rimu (*D. cupressinum*) is another small, soft-foliaged, weeping tree, happy in the moderate warmth of San Francisco. The Tasmanian huon pine (*Lagarostrobus franklinii*) has shown itself a shade hardier. If you can imagine a cypress weeping willow you are somewhere near it. Such a tree, if only one could find it (few nurseries have them), would be a superb acquisition for a garden in zone 8. A tree from the wild would probably be more cypress in shape than willow.

Ripe seed of the big-leaved podocarp (*Podocarpus macrophyllus*) (below) could almost be mistaken for the familiar European yew, but its long leaves betray it. As kusamaki it seems to be almost an essential ingredient of Japanese gardens. To the Chinese it is a valuable feng shui tree. One of the exotic Asiatic conifers we have hardly started to appreciate as a garden tree, its leaves are as big as six inches by half an inch.

New Zealand's champion podocarp (above) is this totara (*Podocarpus totara*) in the rainforest north of Auckland, 1,800 years old, more than 40 feet around and 130 high. Totara timber is the favourite of Maori sculptors.

The California nutmeg (centre right) (*Torreya californica*) takes its name from the shape and appearance of its abundant fruits. The ripe fruit is streaked with bands of purple. The fleshy part is thin and resinous.

The Chinese plum yew (above) (*Cephalotaxus fortunei*) is the best-looking tree of its genus. Its single-seeded fleshy "cones" look like plums but are totally inedible. As the "cowtail pine" it sounds less appetizing.

The Yews

I AM NOT SURE that one could rightly describe the yew as beautiful. My feelings, at least, are more of respect than admiration. The yew is old and wise. It is the colour of ancient widows' shawls, as unkempt and wispy as their hair. Yet senile as it seems, its berries are bright red and there are lusty new shoots from the old trunk. From some secret place in its dark boughs it shakes bright yellow dust to catch the springtime sun.

Folk must always have felt like this about the yew. They have so lavished legend on it that there is not a 200-year-old churchyard yew which isn't reputed to feature in the Domesday Book. My favourite fable is built round the old tree (what is left of it) at Fortingall in the Scottish Highlands, frequently claimed to be 5,000 years old. It appears that Pontius Pilate's father was an Imperial official in that area and the stripling Pontius (or so they say) romped beneath those very branches. He scratched P.P. in the bark of the tree and he even put the date: 15 BC.

Sober report has it that this particular yew is no more than 2,000 years old, but even so is one of the oldest living inhabitants of Europe, with the olives of Gethsemane as approximate contemporaries. There are no annual rings to count, though. For yews, unlike redwoods, have no thick rind to protect them. Their heart wood, immensely durable as it is, eventually decays. The Fortingall tree has no centre left at all: it is a palisade of living fragments. Two hundred years ago it measured 52 feet around, but it would be hard to know which bits to measure today.

Venerable versatility

The association of yews and churchyards is so widespread that everyone has their own explanation of it. I find it easy to believe that people have always respected this ageless evergreen. In Britain, moreover, it was the only conifer (save the bushy juniper) throughout the Middle Ages. Along with holly and ivy (both equally associated with Christianity) it was the only green thing in the brown winter world. A practical consideration is that the yew's leaves are poisonous and the churchyard is one of the few places fenced against stock. Another is that its branches make the best longbows and deserve protection.

For garden purposes it is not the species of yew that matters, but the form. Some authorities, indeed, have taken the view that there is only one species, and that it alters slightly in leaf size and colour from place to place. The European species, *Taxus baccata*, is in any case the yew of yews. If there is any argument for growing the northeast American native one (*T. canadensis*), the Pacific coast version (*T. brevifolia*), or the Japanese yew (*T. cuspidata*) it is only a question of local conditions. North of New York, for example, the

Massive yew hedges (right) are the pride of many ancient gardens, and Powis Castle on the border of Wales is as proud as any. The current fashion is for free-form pruning. The 18th-century lead statue is Hercules slaying the Hydra.

The Fortingall yew, near Aberfeldy, Scotland, is an ancient yew tree that could be 5,000 years old and therefore the oldest tree in Europe. What is left is only fragments of the outer bark and wood, well over 50 feet around; the centre has disappeared.

The male yew pollen-producing cones are usually borne in abundance on the undersides of the twigs, here of European yew. In late winter and early spring they scatter yellow clouds of pollen on the wind.

Japanese yew is recommended; *T. baccata* suffers in winter. But the biggest and best-looking yews are all European – and it is the European yew that spawned the most celebrated garden forms.

Stiff or drooping

It is one of the oddest coincidences in gardening that the two most distinguished cultivars of the yew arose within a couple of years of each other at the end of the 18th century. In 1777 a Mr Dovaston at Westfelton in Shropshire in the English West Midlands spotted a weeping yew among the seedlings offered by a pedlar. We don't know the pedlar's name; the yew is called *T.b.* 'Dovastoniana' – or, in the vernacular, the Westfelton yew. And in 1778 or '79 a farmer called Willis saw two female versions of what is now the famous Irish yew (*T.b.* 'Fastigiata') growing in the wild near Florence Court in County Fermanagh. They were moved to the Earl of Enniskillen's garden nearby and from them, by cuttings, are descended the millions of upright yews all over the world. In a fairer world they would be called 'Willisiana'.

One sees far fewer Westfelton yews – largely, no doubt, because they take up so much space. What the ordinary yew tree lacks in poise and harmony this graceful one provides. It makes a tree like a huge pavilion, rising to a conical centre. All around hang weeping swags of near-black branchlets and often a good crop of red berries.

Both the weeping and upright yews, at the Great Gardener's command, have done the sporting thing and produced golden cultivars. The golden upright (*T.b.* 'Fastigiata Aurea') is much the brighter and more effective, especially when spangled with red berries. For a quieter effect *T.b.* 'Semperaurea' is lighter yellowish-green. Both the golden and green Westfeltons sometimes fail to develop central leaders and flourish as immense spreading bushes, recommended only for the largest gardens.

Yews are unusual among conifers in being either male or female. The male tree has much bigger flower buds on the new shoots. It is important to distinguish a tree's sex if you either want the berries or don't want the prodigious clouds of pollen that gild the countryside around a virile yew.

Irish yews are ubiquitous in English churchyards, a suitably solemn presence among the gravestones, as here at Great Saling in Essex. If they are not clipped and supported with wire bands they eventually grow stout and start to fall apart.

There are two stages in the development of a yew "berry" or single-seeded fleshy cone. The first one is a acorn-like cup, which later swells up with sticky red pulp to form the second stage, above. Both the leaves of the yew and the seeds are poisonous.

The leading hedge

Yew is incomparably the best hedging plant. Regularly shorn it is so dense and even textured that one can easily think of it as architecture. Many of the greatest gardens have their structure defined in clipped yew: it makes a perfect background for colours in shrubs or borders. It answers the shears perfectly, yet somehow always contributes its own weight to the effect: there is no such thing as a straight line in yew, yet you can hardly perceive just where it sags and softens the design. Stately homes grow fiercely competitive about the scale and sagginess of their prize hedges: at Powis Castle in Wales they sell rooted cuttings with the assurance that they will grow, given 300 years or so, to the immense size of their famous hedge. If the hedge is grown from seedlings there is a constant play of different colours where one plant merges into the next: no two are ever exactly the same green. When the new growth starts in spring a warm coppery or cooler olive bloom spreads in broad jigsaw pieces over the surface.

If yew hedges are planted much less today than hedges of *Thuja* or Lawson cypress it is partly a question of price but more, I suspect, a question of impatience: yew has a name for growing at

The astonishing collection of yew topiary (above) at Levens Hall in Cumbria was reputedly begun in the 17th century by Guillaume Beaumont. It encompasses scores of more or less absurd shapes, all perfectly trimmed but all gradually evolving.

a snail's pace. We should learn from the Hon. Vicary Gibbs, who in 1897 planted some seedlings which he fed liberally with nitrate of soda. Eight years later, in 1905, he had trees 12 feet high and 16 inches in girth. Better to wait four years for a six-foot yew hedge which needs clipping only once a year than two for a *Thuja* which bristles with new growth after every rain. The hedge I planted around our pool grew to seven feet in 10 years and proves the perfect baffle to swimming pool noise and swimming pool blue.

If Gibbs had waited another three years before planting he could have tried the hybrid between the European and Japanese yews, raised in 1900 in Massachusetts. *Taxus × media*, as it is called, was reported at the Royal Horticultural Society's Conifer Conference in 1972 as being "slightly coarser in texture than English yew, much quicker growing; easily rooted from cuttings". The last part is important: it could bring down the price of yew hedges, or so you would think. The hybrid I planted has grown no faster, though, and merely has slightly bigger needles.

The True Cypresses

TREES OF THE PINE FAMILY have enough in common with the broadleaves for their parts to be easily recognizable. A twig is a twig, a bud is a bud, a leaf, though exiguously narrow, is a leaf. The trees of the cypress family are so different that they seem to have no way of growing, as we usually understand it, at all. You can't see the twigs for the leaves – or, in a sense, the leaves for the twigs. And buds there are none.

"Them ferns" is what an Essex countryman calls the 50-foot Lawson cypresses in my garden. Tree-ferns, if it didn't already mean something palm-like and entirely different, would be a good term for them. They seem as different from other trees as bracken is from a rose bush. So much so that when big trees show their trunks, as stout and straight as any fir's, they look quite incongruous.

All the difference, in fact, lies in the foliage. The little leaves are so closely applied to the twigs they spring from that on most of the more recent shoots – leaf green is all you can see: no wood shows through. Without twigs or buds or any of the usual points of reference the foliage is oddly homogeneous. Instead of forming the next year's shoot in miniature during the summer, and protecting it during the winter resting period in a covering of scales, the cypresses merely pause in their tracks during cold weather – and set off again in the spring. A cypress spray in time becomes a

cypress branch simply by intermittent expansion. There is no difficulty, then, in recognizing a member of the family. Certain parts of it, particularly the junipers, distinguish themselves by unambiguous signs – in their case, by having what seem to be berries instead of cones. With the two genera of cypress – proper (*Cupressus*) and false (that note of perfidy again) cypress (*Chamaecyparis*) and their near relatives – we catch a glimpse, however, of the botanical backroom, where all is by no means so cut-and-dried. The *Thuja*, too, are a group of easily confusable trees, not only among themselves but also with the false cypresses.

And as though this weren't enough, along comes the woodsman and calls half of the cypress family, indiscriminately, "cedar".

On the whole the cypress family is less important to the forester and more important to the gardener. The wood is excellent: durable, light, and often sweet-scented. Hence the name "cedar": a luxury rather than a staple. The Japanese pay huge sums for a big Port Orford cedar (alias Lawson cypress) from Oregon, which they apparently use for lining coffins. There are no great forests of it.

The Monterey cypress is limited in nature to the gale-swept promontory of Point Lobos in southern California. The winds make stunted but still thriving trees with characteristic antler-like branch formations. It is only its vigour that enables it to resist the constant buffeting wind – like its neighbour the Monterey pine.

Nurseries are full of funny cultivars. There is hardly a gawky shape or a gaudy colour which doesn't find an admirer somewhere, particularly among the dwarf trees and little bushes with which the family abounds. On the other hand some of the finest species, fast, hardy, and beautiful trees, get overlooked.

Proper Cupressus

Cypresses proper, in the genus *Cupressus*, the nub of the whole family, are trees whose twigs, closely covered in tiny scale-like leaves, branch again not only sideways in one plane, as it were east and west, but north, south, and points between as well. This, together with their much bigger cones, distinguishes them from the false cypresses which (like *Thuja*) keep their sprays flat. If you hold a false cypress spray horizontal, you will find no twiglets heading upward or downward – only on the level.

The classic cypress (*Cupressus sempervirens*), the cypress of literature, is the dense dark column of the Italian garden, of the Greek olive grove – of all the sunwashed landscapes of the Mediterranean where its point of exclamation, alone or in combination with umbrella pines, contributes the essential architecture of the scene. There is no better columnar tree where it is hardy – and we are probably too ready to assume, because of its southern connotations, that it won't be hardy in our gardens.

It was commonly used in English gardens until a generation of very cold weather late in the 17th century, which killed or damaged so many that it went out of fashion, to be replaced by the newly discovered fastigiate Irish yew. The Mediterranean cypress grows well in the Edinburgh Botanic Garden; the Arnold Arboretum puts it in hardiness zone 7, which brings it as far north as mid-Tennessee and, on the coast, Atlantic City. If it is planted out very young in a reasonably sheltered place it is certainly worth a trial even farther north – although only southern nurseries are likely to stock it.

It may have been because the Mediterranean cypress didn't take to the Po valley in northern Italy that the Lombards took up their perpendicular poplar as a substitute. There is no real substitute, though, for the peculiar glossy blackness of the Mediterranean cypress with its gleaming cones in constellations, towering 50, even 100 feet above the nymphs and gravel of an Italian garden. On condition, that is, that you plant the right variety. The same species contains two varieties, *sempervirens* (the column) and *horizontalis* (a spreading branchy version of no special merit). *Cupressus sempervirens* var. *sempervirens* comes true from seed but still needs selection for the best form. The ideal clone, if you could find it, would be male and hence have no cones; fertile females can have so many that they pull down the branches out of their trim arms-at-their-sides position. The best I have ever seen were in the Pope's garden at Castelgandolfo, towers of immense height clipped smooth to the top, finished off with scissors.

The next best tree for texture, though not for form, is the Monterey cypress (*C. macrocarpa*) from California. In hardiness it

Cypresses dressed for the Pope in the garden of his summer palace at Castelgandolfo seem very distant cousins of the ruffians opposite. The Mediterranean cypress is the trump card of Italian formal gardening, often in conjunction with umbrella pines.

The cypresses have produced a wide range of ornamental garden trees, including some good gold ones. *Cupressus macrocarpa* 'Lutea' is a bright colour but easily becomes a messy shape, especially near the sea.

The tiny male organs on a spray of the smooth Arizona cypress. These flowers will develop within several months into the cones, also shown. Cypress twigs branch radially, making an open conic shape like tiny Christmas trees.

may have a marginal advantage where winters are wet as well as cold – though zone 7 is given as its limit, as well. Tree-talk returns again and again to Monterey. The same wind-battered bit of Pacific coast, which gave us the fastest-growing pine and the best-looking silver fir (and more besides), has the only two natural groves of the Monterey cypress, at Point Lobos and Cypress Point.

The extraordinary junkyard tangle of the trees which have grown up in the teeth of a gale is no indication of how a Monterey cypress will behave in a garden. What is fairly constant is that it will have several, perhaps many, stems. Whether these all go upwards, or spread to an oak-like head, may be a question of breeding, and is certainly a question of light, for which the tree is conspicuously greedy. Shaded branches die without more ado. The biggest and best of the species I have seen are in the Melbourne Botanic Garden in Australia: they are immensely broad domes. Often, on the other hand, one sees narrow, plume-like specimens, or ones with three or four of these plumes, all playing the leader. Trees used as shelter belts in New Zealand are some of the ugliest, snaggiest monsters you could imagine.

The two-year-old Monterey cypress you buy in a pot is curiously misleading if you expect the mature tree to look like that only bigger. It is a much paler green than the adult, which is very dark indeed. Transplanting of older trees is not a good idea, nor is it necessary, as cypresses grow at least three feet a year. The temptation to use them for hedging should be resisted; they take badly to hard pruning and can be surprisingly tender. After one bad winter in Bordeaux all the Monterey cypress hedges were dead.

The Monterey cypress has spawned various golden forms. 'Lutea', 'Donard Gold', and 'Goldcrest' are all well known. Of these 'Goldcrest' is the best shape and 'Donard Gold' the best colour. It has also produced a weeping form (very rare) and two or three pygmy ones – even one which lies flat on its front.

Odd pockets of cypresses with just enough differences to count as species proliferate in California. Often there is only one stand, reached by a botanist-beaten path. The Santa Cruz cypress (*C. goveniana* var. *abramsiana*) is one such. It lives on the southwest slope of Ben Lomond, about seven miles east of Bonnie Doon School in the Santa Cruz mountains. Out of curiosity I planted one in Essex in 1985. It is now a smoothly textured, perfect ovoid cone to 50 feet, if not a terribly interesting one. The similar Gowen cypress (*C. goveniana*), companion to the Monterey cypress in its refuge by the sea, is similarly rare but worth trying. More to the point for gardeners north of zone 7 is *C. bakeri*, the "Modoc" cypress, from 5,000 feet up in the very north of California, which the Arnold Arboretum has found totally hardy in Boston, and as fast as any tree there. The dark green of all these is an excellent background colour for flowering trees and shrubs.

By going inland into Arizona one finds, as in the other conifers, the lighter colours, the grey-greens and grey-blues, of drought-resisting species. The smooth Arizona cypress (*C. arizonica* var. *glabra*) is one of the best; an extraordinary tree. The twigs look no more leafy than lengths of sash cord, covered (like sash cord) with a fine coating of wax. The cultivar 'Pyramidalis' is pale grey with a hint of blue, 'Blue Ice' so cold a colour that you could almost use an enclosure of them, with perhaps an icy eucalyptus, for shooting snow scenes in mid-summer. Their bark, in contrast, is dark red, flaking in patches of purple and cream, the whole tree almost artificially ovoid-conic until branches start to lean out in old age. It is understandably popular, but I confess I got tired of its artificial look and cut mine down.

The Mexican cypress, which owes its Latin name, *lusitanica*, to its early popularity in Portugal (Lusitania) is the south-of-the-border version. There are massive grey-green specimens in the monastery garden on the misty hill of Bucaco, where the Duke of

The foliage of the Arizona cypress develops a heavy wax coating, which protects it from drying out in parched desert conditions. Some of the best blue garden conifers come from arid areas of the Rockies or the southwestern States.

Kashmir cypress (*Cupressus cashmeriana*) is the usual name for what should more properly be called Bhutan cypress, the most graceful and glamorous of its family. The hunt continues for forms with ever-bluer foliage and lengthier hanging sprays.

Europe's biggest Kashmir Cypress grows on Isola Madre in Lake Maggiore. A gale blew it down in 2006, but all 70 tons of it have been winched up again, severely pruned, and richly fed. Whether it will ever recover its previous beauty is sadly doubtful.

Wellington scored his first win against Napoleon's army. The name cedar of Goa (the Portuguese colony on the west coast of India) makes it the only true cypress to be called a cedar.

Of the numerous true cypresses of Asia only one has really found favour in Western gardens: the Kashmir cypress (*C. cashmeriana*), not only silvery-blue but softly weeping. Europe's best specimen grows on Isola Madre in Lake Maggiore, though not quite the tree it was; a storm blew it over in 2006. Eventually, after massive lopping, its 70-ton weight was winched upright again and it remains under intensive care. Moderate cold does not seem to be the enemy of this tree so much as exposure and dry conditions. It is recommended for the southern California coast but has been spotted in the Cotswolds. Clearly more experience is needed.

Further possibilities

There is also room to wonder why nobody grows the alerce (*Tetraclinis articulata*), the North African relative of the true cypress. The ancient Romans apparently rated its scented, bird's-eye-figured wood highly, and called it citrus (a change from cedar, but still further from the truth). A little grows in Spain … Why not in Italy, if the Romans prized it so highly?

And few people grow any cypresses from the southern hemisphere – though the Chilean *Fitzroya* is distinctly decorative, and the dainty, plumey Australian *Callitris* handy if you have a parched plain to furnish; or even, on the evidence of one or two trees in southern England, without that facility.

Fitzroya (named in honour of Captain Fitzroy of HMS *Beagle*) makes a big tree in its native Andes, where it has been exploited almost to extinction, and has reached 70 feet in Scotland. It is like a broad cypress whose hanging branchlets seem to be made of green plastic chain … except that it is really very pretty. Its possible role in modern gardens has yet to be explored.

The False Cypresses

FROM THE FOUR COMMON SPECIES of this minor branch of the cypress family (minor in nature, that is) so many decorative cultivars have arisen that false cypress might be called the flowering cherry of the conifer world.

In gardens in many parts of the world (the south of England is one) false cypresses are such a cliché that it is hard to see them with fresh eyes, or to use them effectively. Yet good specimens are graceful and satisfying plants. The commonest in gardens, by far, is the Lawson cypress (*Chamaecyparis lawsoniana*). Lawson was a famous Edinburgh nurseryman, and fitting it is that such a nurseryman's staple should bear a nurseryman's name. If I occasionally address my trees as Port Orford cedars it is only to reinforce their dignity a little. For, though botanically way off mark, this is their forest name in their Pacific Northwest home.

Port Orfords are common, but never dominant, in the forest which starts on the foggy coastline of Oregon and works back into the Siskiyou mountains, up to that Eden populated by the Brewer's spruce. As far inland as the sea's influence is felt in mist and moisture-laden winds, the Port Orford is dotted, singly or in small groups. It stands out as a paler, more finely detailed tree than the spruce, the Douglas fir, and the western hemlock. On higher and drier spots its place is abruptly taken by the brighter, more tufted, incense cedar. Its love of moisture is a point gardeners should bear in mind. New England and most of California are too dry for it to flourish; it will grow there, as it will all over Europe, but it is a drab thing where it is unhappy. Sadly, today, this includes much of its native zone, where a phytophthora root rot is spreading out of control. American gardeners are rightly wary of it.

Master of disguise

What more than anything has singled out the Lawson (to come back to its nursery persona) for intensive cultivation is its tendency to pop up in the seedbed in eccentric – sometimes quite unrecognizable – forms. No conifer has so many faces. Den Ouden and Boom, authors of one of the standard reference books on conifers, list 200 cultivars of tree or big bush size and some 50 dwarfs. Strange that this happens in cultivation, when in the wild it hardly varies at all. The main variations are in colour, habit of growth, and in leaf shape and arrangement. Some of the best are particularly soft and graceful, by virtue of having permanently juvenile foliage – fronds which never reach the size, firmness, and flatness of the adult tree.

This state of affairs, which also occurs in several other members of the family, was misunderstood when it was first noticed and examples were placed in a separate genus, *Retinospora*. Victorian gardening books are full of *Retinospora*. As a genus it was an

Normally a droopy, dour, even depressing tree, Lawson's cypress is capable of some cheerful moments. Its cultivar 'Wisselii' has perky up-jutting branches. In spring their male cones are positively festive. A full collection of Lawson variants could be astonishing.

illusion. On the other hand it is a perfectly valid concept, and all these trees do share an attractive quality.

Two other fairly frequent aberrations of foliage are a sort of congestion, in which twice the usual ration of twigs and leaf scales crowd into the same space; and the opposite, where they are freakishly spaced out along hanging sprays as thin as strings. The name 'Filiformis' (string-shaped) is given to this type of foliage.

There are other differences it is harder to classify: ones that hold their sprays upright but side-on to the audience, or loose and nodding, or ones with odd, twisted branch tips.

The first and still one of the oddest of the cultivars actually arose from seeds shipped from California to Britain in 1855, so it has been almost as long in cultivation as the species itself. It is called 'Erecta Viridis'; the first name describes it exactly: bright green upright. Up to a certain age, it is one of the best sentinel trees, as formal as a footman and as groomed. When most green cultivars turn grey in winter, it seems greener than ever. But accidents happen to older trees: from footmen they rapidly degrade to tramps.

You can find Lawsons to fit any decorative plan. One can't help thinking of them as garden wallpaper. One of my favourites is a

This rare chance combination of the stringy and the droopy tendencies is called *Chamaecyparis lawsoniana* 'Imbricata Pendula'. Carefully pruned it could make a strange green veil to half-hide some theatrical garden feature.

very early cultivar called 'Intertexta', whose lolling leader and exaggeratedly open-textured, pendulous sprays are hard to ignore.

The best of the gold-leaved Lawsons are 'Lutea' and 'Lanei'; of the blue (for brightness) 'Spek', 'Pembury Blue', and the ultimately huge 'Triomf van Boskoop'; for fluffy greenness 'Pottenii'; for combination of blueness and narrow upright form 'Columnaris' – at least up to 10 feet or so, when it mysteriously changes its act, like an actor forgetting to use dialect. For a broad upright habit like an Irish yew 'Fletcheri' and 'Ellwoodii' are the ones to choose.

It is easy to confuse the ordinary Lawson cypress with other false cypresses or with *Thuja*, which is often found in the same gardens. The *Thuja*'s sprays of foliage are generally greener, simpler, and flatter, but the best way to recognize the Lawson is to look at the underside of the leaves. A faint but distinct pattern of little white crosses marks the lines of pores.

Two species of false cypress grow on the east coast of the United States. The white cypress (*Chamaecyparis thyoides*) has by far the biggest range of any; happy in wet ground from Maine to Mississippi. (Florida has its own species: *C. henryae*.) The white cypress came into cultivation 150 years before the western and

oriental species, but today they have more or less eclipsed it. Its best-known forms in modern gardens are the dwarfs, *C. thyoides* 'Andelyensis' and 'Ericoides'.

... and from Japan

The Japanese sawara cypress (*C. pisifera*) has identity problems like the Lawson. Indeed, if the species itself rather than one of its cultivars were grown few people would recognize it. In two areas of aberration it is conspicuously successful; the string-leafed and the juvenile (or retinospora). It goes so far in the latter that two different degrees of juvenility have separate terms: if it is merely childish it is "plumosa"; if it is positively infantile it is "squarrosa" – vernacularly, "moss". Seventy feet is not an impossible height for a "squarrosa" with the finest and fluffiest foliage. One often sees hacked-about specimens ruining the rockeries where they were planted as pigmies years ago. (The deadpan Alan Mitchell paid them a back-handed compliment when he called them the coniferous equivalents of the spotted laurel, *Aucuba*, in their resistance to the fumes of Jeyes fluid, to judge from their use round toilets.) A compromise, and probably the best cultivar of all, is a "squarrosa"-type called *C.p.* 'Boulevard' which grows slowly to 15 feet, and eventually a little more. Its leaves, if you can call them that, are like a silvery froth all over the plant. The "filiformis" ones, which come in green and gold but not blue, are even less vigorous than the retinosporas. None the less you finally get a huge gold (or green) bun – as much as 30 feet high and 20 feet through.

The other Japanese false cypress, hinoki (*C. obtusa*), is more consistent and in many ways a better tree, keeping dense and healthy-looking when many a Sawara has tired of life. The hallmark of the hinoki is the blunt-ended spray of deep shiny green, which gives the whole tree a compact, well-nourished look. Both the species and its golden version *C.o.* 'Crippsii' are first-class garden trees.

The most popular cultivar, and probably the best, is *C.o.* 'Nana Gracilis'; a big bush comparable with *C. pisifera* 'Boulevard' both in size and in its remarkable finish. It consists of a dense random heap of small vivid-green scallops; an unmistakable plant in the millions of gardens which give it pride of place.

The cones of the Lawson cypress (or Port Orford cedar) are full grown but still unripe here. Usually a few of last year's cones, brown and woody, still hang from the drooping spray.

The full name of this beauty is *Chamaecyparis pisifera* 'Filifera Aurea' – a fact that has not discouraged its enthusiastic public. It is the best of the string-foliaged versions of the Japanese sawara cypress; a broad bright dome of unique texture.

The squarrosa or "moss" types of *Chamaecyparis pisifera* are as different from the string or "filiformis" types as they could be. 'Boulevard', perhaps the best of them, arose in an American nursery in 1934, a slow-growing cone eventually 15 feet high – though prettier as a young plant.

'Rubicon' is a miniature conical sport of the eastern American white cypress (*Chamaecyparis thyoides*, an improvement, many think, on *C.t.* 'Ericoides'. The attraction is the colour switch, in winter, from green to this purplish tint.

Compactness goes berserk in some of the other cultivars; the foliage becomes a congested jumble which would look like a nasty spot of virus trouble if hinoki cypresses weren't such sturdy, bright-coloured plants. *Chamaecyparis obtusa* 'Filicoides', *C.o.* 'Lycopodioides' and *C.o.* 'Tetragona Aurea' seem to make whole trees out of witches' brooms. The last is a good gold colour, but the others, to my mind, have no great virtues to balance their essential freakishness.

Botanists' Bedlam
George Sudworth, my forester hero reporting from horseback on the western forest trees, was at pains to point out all the differences between the two "cedars" he classified as cypresses and the western red cedar (*Thuja*). It seems a rather unnecessary muddle – but a

hundred years later it has grown worse. Port Orford cedar remains Lawson cypress, if you follow me, and is still *Chamaecyparis*. Its forest neighbour, the Alaska or yellow cypress (or cedar), then *C. nootkatensis*, is a battleground in which the only sure thing is that *Chamaecyparis* it ain't. You can pencil in either *Xanthocyparis* or *Callitropsis* as its name – but keep your eraser handy.

The Nootka (false) cypress (or indeed false cedar) has never caught on in cultivation to the same extent as its rival. Its real claim to fame in modern gardens is in being one of the parents of the Leyland cypress. In nature the Nootkas are the northern neighbours of the Lawsons. They are common among sitka spruce along the coast from Alaska to Vancouver, and further south (on Mount Rainier for example) at an altitude much higher than the Port Orford cedar. The Nootka of their name is on the ocean side of Vancouver Island, where Archibald Menzies first caught sight of them.

The usual garden form of Nootka is different from the Lawson chiefly in the hang of its foliage. The sprays are heavy and coarse to the touch. They droop as though the tree were desperate for water; and their rather dingy yellow-green suggests it may be too late. The form 'Pendula' is relatively rare in cultivation but superior in every way. Where Nootka just mopes, 'Pendula' is heartbroken. Each long branch bears a cascade of weeping fronds.

The Leyland saga
Up to 1925 the groups of trees we now call the cypresses and (with some hesitation) the false cypresses were considered one. So when in 1888 the

EIGHT GARDEN VARIETIES OF ONE TREE

All are flourishing specimens of 15 years or so. Age is not always kind to them. 'Columnaris' tends to revert to nature and grow coarse 10 feet up. 'Erecta' falls apart. 'Stewartii' constantly renews its bright gold new leaves around a bright green core. All are capable of becoming very large trees in the fullness of time. 'Intertexta' has immense coarse character. 'Wisselii' is the perky one illustrated on page 138.

'Alumii'   'Columnaris'   'Erecta'   'Fletcheri'   'Intertexta'   'Lutea'   'Stewartii'   'Wisselii'

pollen from a Monterey cypress lit on the female flowers of a Nootka "false" cypress in the grounds of a Welsh country house the affair was perfectly legal. Hindsight (in the shape of a taxonomic dictat of 1925) then declared the two parents members of separate genera and their offspring, therefore, a mule. Do such taxonomic niceties have any resonance? They do when they result in an offspring as sensational as the Leyland. Hybrid vigour is a familiar concept but the Leyland cypress (currently called × *Cuprocyparis leylandii*, but watch this space) carries it to extremes. In normal British gardening conditions the Leyland is the fastest-growing conifer bar none. Cuttings, which root easily, are ready for planting out in one year. Thereafter growth only pauses in really cold weather. The first year out the plant will double in size; three years from a cutting it should be at least six feet tall. It will make a 30-foot specimen in 10 years or so, and it has reached 100 feet in 55 years.

There are faster trees for milder climates. Eucalyptus can go like a train. Even in the north spruces have made six feet in a season. But Leyland is not only tough: it also has qualities that appeal to instant gardeners. It makes a uniquely rapid hedge or tall screen, without gaps or bare patches; it gives a desirable even grey-green background for coloured trees or flowers. Even at 100 feet it keeps its smooth shape like a great green flame with a slightly leaning tip. The result? It has been planted with reckless abandon where its vigour is a potential menace. Already it gives rise to litigation. Leylands are time-bombs all over suburbia.

The Leyland of the name was C J Leyland of Haggerston Hall, Northumberland. It was he who first spotted the hybrid at his brother-in-law's home, Leighton Hall, in Powys on the Welsh border. In fact the cross happened both ways: in 1888 it was the Nootka which was the mother, but in 1911 (by the strangest coincidence on the same estate) the Nootka pollinated the Monterey cypress. Leyland's first find is now referred to as × *C.l.* 'Haggerston Grey'. It is the common Leyland of commerce. The second cross bears the name of × *C.l.* 'Leighton Green'; being harder to propagate it is not so often seen.

To every nurseryman's delight the Leyland has obliged with "golden" (or at least yellow-green) sports. In 1970 × *C.l.* 'Castlewellan' appeared, followed five years later by the – slightly brighter – × *C.l.* 'Robinson's Gold'. Both seem to grow almost as fast as the green forms. When × *C.l.* 'Robinson's Gold' made its appearance I planted two in Essex and gave plants to my brother in Sussex and friends further north and west with a modest wager on their height after the first 10 years. The winner seems to have been the one I planted near the church at Great Saling, which was nearly 80 feet high when I cut it down 30 years later.

Much less well-known are two other Nootka hybrids, respectively the Mexican cypress (*Cupressus lusitanica*) and the Arizona cypress (*C. arizonica* var. *glabra*). The first of these hybrids is a vigorous but dull tree, × *Cuprocyparis ovensii*, unfortunately more like its Nootka parent than its Mexican one, but the second, × *C. notabilis*, is handsome: light grey-green, with branches that gesticulate and droop like a languid heroine.

Hero or horror? The hybrid Leyland cypress can hardly be blamed for its express growth or remarkable shape, but it calls for understanding neighbours. This tree in the national pinetum at Bedgebury was planted in 1935 and last measured at 150 feet.

The Junipers

READING ACCOUNTS OF DESIRABLE CONIFERS I sometimes begin
to despair that so many want the same (to me unattainable)
conditions: acid soil and lots of moisture. What hope of these
arid and alkaline acres ever having anything worthwhile to show?
And then I remember the yews – and the junipers (*Juniperus*).

The juniper in particular desires and does almost the exact
opposite of the general run of conifers: even of the rest of the
cypress family. Where these grow relatively fast, the juniper grows
at a snail's pace. Where they love shelter it loves the south slope of
the hill – unmitigated sunlight. Where they need damp luxurious
leaf mould it likes to get its wiry roots into mineral soil. Where
they have both male and female flowers on one tree, the juniper
(again like the yew) has either one or the other. They have woody
cones; the juniper and the yew have fleshy berries. The juniper's
slow growth rules it out for the forester, except for such small-scale
and specialized uses as providing the scented wood for pencils.
The same quality makes the juniper a basic stand-by in planning
gardens – particularly small gardens. But the great majority of
garden forms of juniper fall far short of tree stature: they are
probably more grown as ground-cover than in any other role.

"Tenderly mysterious beauty"
The common juniper (*J. communis*) is the only tree species which
occurs naturally in Europe, Asia, and America. It is usually dismissed
as scrub; its only claim to fame is that its berries provide the flavour
for gin and do wonders for a stew. To give it a fair hearing, however,
I must quote Gertrude Jekyll, whose sensitive eye made her one of
the greatest gardeners of the 19th century. She feels differently. "Its
tenderly mysterious beauty of colouring … as delicately subtle in its
own way as that of cloud or mist, or haze in warm, wet woodland.
It has very little of positive green; a suspicion of warm colour in the
shadowy hollows and a blue-grey bloom of the tenderest quality
imaginable on the outer masses of foliage." One is driven by such
words at least to take a second look.

Few gardeners plant the common juniper today except in one
or other of its cultivar forms. Three of these are pale grey-green
columnar plants of considerable merit: the tallest *J.c.* 'Suecica',
as much as 40 feet ultimately; the commonest the Irish juniper
(*J.c.* 'Hibernica'), perhaps 15 feet; and the baby, which has no
business here but is such a gem I can't leave it out, *J.c.* 'Compressa':
about 18 inches of drum-tight fuzz – one of the best of all miniature
trees for a sink garden or rockery. I have used Irish junipers as a soft
grey alternative to yew to make an avenue down the central path of
our walled garden.

All common juniper incidentally could (archaically) be called
retinospora, having nothing but juvenile foliage. Some other
junipers tend to be very variable, having some branches adult (with
scale leaves, quite cypress-like) and other branches, or individual
sprays, the juvenile form (with fine sharp-pointed leaves).

The Rocky mountain juniper (left) is a survivor of inhospitable conditions, drought, and exposure high in the Rockies. Its many garden variations include the slimmest of all upright ones: *Juniperus scopulorum* 'Skyrocket'.

The pale green berry-like female "flowers" (above) of the common juniper are borne in the angles of the whorls of spiky leaves. Here they are mature and have released their pollen.

The Himalayan juniper (*Juniperus squamata*) produces some peculiar bush-trees. This is 'Chinese Silver', with evidently vigorous soft blue scaly shoots. This individual seems undecided, growing horizontally, then rearing up its branches with weeping tips.

The Phoenician juniper can live for 1,000 years and has spread from the East throughout the Mediterranean and to the Canary Islands. Its foliage is often entirely scale-like and adult, with few or no prickly juvenile leaves.

The toast of the west

The two junipers which between them give more garden varieties than all the rest are the eastern red cedar (*J. virginiana*) and the Chinese juniper (*J. chinensis*). Both have this characteristic of variable foliage. The eastern red cedar is widespread in the wild throughout the eastern United States. Occasionally it makes a 100-foot tree of a battered conical outline; but only in the south of its range, south of (say) Pennsylvania. About 100 cultivars of eastern red cedar have been named; nurseries looking for shapeliness concentrate on the narrower, smaller, northern ones. *Juniperus virginiana* 'Glauca' is a good blue-grey upright one; *J.v.* 'Canaertii' is a selection made in a Belgian nursery: dark green and firmly conical, with the added attraction of regular masses of blue-bloomed berries – though in the USA it suffers badly from rust. The cultivar *J.v.* 'Corcorcor', usually sold as 'Emerald Sentinel', forms a broad column of dark green, while *J.* 'Grey Owl' forms a spreading mound of silvery-grey.

There is not a great deal of difference between the eastern and the Rocky Mountain junipers (*J. scopulorum*), or between the Rocky Mountain and several other local western forms. According to *Sunset Magazine*, which caters exhaustively for western gardeners, the juniper is the "most widely used woody plant in gardens in the west". I can make no attempt to sort out the Utah from the one-seed or the Californian; the varieties of the worldwide ones are enough to be getting on with. The Rocky Mountain contributes more than 100 garden varieties, of which the best tend to be columnar or densely pyramidal. The familiar *J. scopulorum* 'Skyrocket', one of the narrowest of conifers, is among them. In the wild, to give you some idea of its unhasty progress, it has been measured at 14 feet in 40 years, 18 feet in 80 years, and 30 feet in no less than 300 years.

The western juniper (*J. occidentalis*) is something different, by the way; a rough, big-limbed, unshapely tree from dry sites in eastern Oregon and western Idaho.

Most original of the American junipers is the one known as the alligator (*J. deppeana*) and its variety *pachyphlaea*. All other junipers have thin bark which peels off, if at all, in brown vertical ribbons. The alligator's cracks up into square patches like the carapace, if that's the word, of the terror of the Everglades. Its leaves are light blue, fine, and spiny as befits a tree from the near-desert of Texas.

Multi-tasking oriental

The Chinese juniper (*J. chinensis*), like the common juniper, is the raw material of dozens of cultivars. It is not often seen in the form nature intended. There is no striking difference between it and the eastern red cedar unless it be the considerably larger berries of the Chinese plant. Both carry foliage of two types: fine-prickly, and cypress-like with scale leaves. Some of the selected forms break this rule, however; the size of the berries, if any, is probably the most consistent trait to watch for. One cultivar, *J.c.* 'Femina', was selected on the strength of its profusion of pretty berries.

Of the other selections the golden one, *J.c.* 'Aurea', is probably the most popular, as one of the most reliably bright gold conifers for a small garden. It may reach 40 feet, but it will take 60 or 70 years. 'Keteleerii' is a tidy, solid, dark-toned tree, a broad cone in shape – again very slow. 'Kaizuka' (also known as the Hollywood juniper) is a harum-scarum sort of bush, launching long limbs of bright green in unexpected directions: one for a mid-lawn position where it can freely yawn and stretch as wide as 15 feet. In straightforward columns most shades from bright green through grey to blue are available under one name or another. Nursery names often depart from the strictly botanical; the great thing is to see the tree and like it.

One occasionally sees a very different Oriental juniper, a tree with great character which could be used much more: the temple juniper (*J. rigida*). Not column- or cone-shaped, the temple juniper puts out pine-like branches to make a small but venerable-looking open head. Spiky dark foliage droops densely from the branches

The **most valuable** form of *Juniperus communis* is called 'Hibernica'. Like the Irish yew it is fastigiate and can be kept, by clipping and wiring, to a perfectly trim pillar.

The **ground-hugging** *Juniperus x pfitzeriana* range (above) includes green, blue, and "old gold" forms. They make totally effective if uninspiring ground-cover, but not trees.

with an effect like Spanish moss on an old swamp cypress. Thirty feet would be a big one.

In 1930 another drooping juniper was discovered in the Burmese foothills of the Himalayas: *J. recurva* var. *coxii* (after Euan Cox, its finder) has almost the effect of a weeping willow at a distance. The long camouflage-green sprays, so dense and soft to the touch that you can squeeze them like sponges, hang from arching branches. It is more of a forest tree than most junipers; shade, especially near a woodland pool, suits it well.

To stick strictly to the trees of the family I should ignore the much better-known Himalayan juniper (*J. squamata*), whose form 'Meyeri', a prickly, icy-blue bush with drooping shoots, is familiar in so many gardens. I should also fail to mention the ubiquitous *J. × pfitzeriana*, child of a union between the Chinese juniper and

the shrub *J. sabina*, the savin of southern Europe. The cultivars of *J. × pfitzeriana* and other spreading junipers (especially *J. virginiana* 'Hetzii') are vigorous V-shaped bushes able to cover a remarkable amount of ground. Being capable of growing in pure chalk, dry as a bone and in deep shade, they are the workhorses of the family, inevitably seen in problem corners and clichéd planting schemes and generally traducing the family's good name.

In southern gardens one sometimes sees the Greek juniper (*J. chinensis* 'Pyramidalis'), a tight grey column. From the best accounts the Syrian juniper (*J. drupacea*) is much hardier and more tolerant, as well as being (I quote Alan Mitchell, in his *Conifers in the British Isles*) "a fresh light green colour unique among junipers". One I planted in 1975 has kept a sturdy regular column shape eight feet across to 25 feet.

ADULT AND JUVENILE LEAVES

Of all conifers junipers seem least committed to growing up. Juvenile, pointed, owl-like leaves can occur at any time alongside the scale-like adult ones comparable to cypresses. Trees are most often of one gender, females bear bluish berries which share the pungency of the leaves.

J. communis
The common juniper has little pointed "juvenile" needles and berry-like cones.

J. chinensis
Has both juvenile and adult foliage. A plant is either male or female.

J. virginiana
Eastern red cedar is the tallest juniper – a proper tree. Young trees have only juvenile leaves.

The Incense Cedars

IF THE INCENSE CEDAR grew only in its native woods in the Pacific Northwest it would rate little more attention than a cousin of the better-known *Thuja* with the peculiarity of holding its branchlets on edge instead of flopping horizontally. It is a common tree in the lower mountains of Oregon and California, known to botanists as *Libocedrus decurrens* until someone changed the name to the now current *Calocedrus*. The "incense" of the name is a sweet resinous scent in the leaves and wood – but no more so than in many other conifers. But there is more to it than this. It is one of only two conifers with very similar relations (under the name *Austrocedrus*) south of the Equator. With a little exaggeration you could say its range runs from Oregon to New Zealand. And in cultivation it has given northern parts what is easily their grandest columnar tree, a formal upright more magnificent in size and richer in colour and texture than the Mediterranean's cypress.

As far as the northern hemisphere is concerned the incense cedar is the sole representative of its genus. Colonel Fremont of the US Corps of Topographical Engineers (an officer whose career embraced everything from botany to mutiny, and who eventually found himself the owner of one of California's richest gold lodes and with a San Francisco hotel to his name) discovered the tree in 1848, flourishing in the drier and sunnier spots of the mountains of Oregon and California.

The tree he found is very distinct in colour and texture. Its firm, bright green tufts, vertically set, are a cheerful part of the character of the sunny slopes. It is never one of the tallest trees; rather it makes its presence felt by keeping more branches than most of the forest trees; often being clothed for most of its length in foliage, giving the woods a dense, well-furnished look. Only in old age does it begin to grow gaunt.

The incense cedar of the mountains doesn't, on the other hand, have the habit which makes the garden version so outstanding. In nature it is a bushy, irregular tower of foliage, broadest about its middle reaches. Whereas the same tree grown in gardens in Europe and the eastern States soars up with completely parallel sides, the most regular and densest column to the greatest height of any tree. The best have reached 120 feet in 100 years or so, the only unevenness being a faint billowing of the rich greenery: the embonpoint of a courtier.

A variegated version also exists, but it is a poor thing, merely sprinkled with yellow branch tips as though something were lacking from its diet. Its cordilleran cousin, *Austrocedrus chilensis*, in contrast grows happily, if slowly, in my garden, similar in shape but softer and less striking in colour and texture.

The incense cedar of the Sierras (far right, in Sequoia National Forest, California) superficially resembles the redwood in its well-furnished formality and grows in similar conditions. The bark is cinnamon-brown, peeling vertically.

This incense cedar (right) came home with me from Oregon in 1976 and is now 40 feet high. In English gardens incense cedars maintain this bright green tight column shape better and to a bigger size than any other tree.

The Andes in Patagonia are very different from the Sierras, with far fewer conifers. They have an incense cedar (here in the Valle Encantado in Argentina), but it was known to early explorers as Patagonian *Thuja*.

The leaves of the California incense cedar are stiff and enduring, tilted on edge, deep shining green, and quite strongly aromatic. Their cones are singular: pistachio-shaped, opening into three wings.

Patagonian incense cedar has paler, softer, and prettier foliage with less scent. The trunks of both trees are notably straight and tapering. *Austrocedrus* is still rare in Europe.

The Thujas
or Arborvitae

IT IS COMFORTING TO KNOW that in their day the great Bentham and
Hooker, authors of *Genera Plantarum*, one of the definitive systems
of arranging plants, lumped *Thuja* and false cypress together. One
doesn't feel quite such an idiot. The same flattened fronds, the
same well-furnished conical form make the species infinitely more
like each other than their own varieties. In the garden, the western
red cedar or giant arborvitae (*T. plicata*) – for it has nearly as many
names as branches – and the Lawson cypress are distinct enough.
If the trees have cones it is easy to tell which are *Thuja*: the scales of
their tiny, rosehip-shaped cones all hinge at one end. False cypress
cones open around a central point: the scales point up and down
as well as sideways, forming a sort of skeletal sphere. Perhaps the
other most telling point of recognition which applies to all the
Thuja, except one, is their powerfully pungent smell; between
balsam and turpentine. The leaves fairly exude it; you become
aware of it just by standing nearby.

The two American species of *Thuja* are best known. The familiar
pattern repeats itself: a small, immensely hardy, not particularly
attractive tree shows the flag in the east while out west its great
brawny country cousin luxuriates in perfect conditions, developing
to a size and beauty which makes it almost a match for Douglas fir.

Quirks and foibles

The eastern tree (*T. occidentalis*) ("white cedar" to the forester) –
the roll call of "cedars" is still not exhausted – was the first
American tree to be grown in Europe. There is a record of one in
Paris in 1553, almost 200 years before the trees of North America
began to make their great impression on the world's gardens.

Its importance today, however, is based on its quirks and foibles
in the seedbed rather than its intrinsic value. It is second only to
the Lawson cypress in the number of offbeat seedlings it produces.
Professor C S Sargent of the Arnold Arboretum, once said that if
anyone sowed a quantity of seed he "would be sure to find forms
among the seedlings as novel and interesting as any now in
cultivation". The competition, as represented in nurseries today,
includes mainly dwarf and juvenile-leaved varieties, but one of the
classic golden conifers, 'Rheingold', is a bush form of *T. occidentalis*.

Thuja on the whole are strong on gold cultivars, shifting from
bright gold to dull, bronzy old-gold at the onset of winter. The
billowing flanks of a 12-foot 'Rheingold' (12 feet is about its
maximum) can be one of the best things in a bleak winter
landscape. The winter shift of colour is much less effective on a

Western red cedar is a giant of the same forests as Douglas fir, here in the state of
Washington. Its swooping lower branches frequently touch the ground, take root, and
grow into more soaring trunks. Its leaves are strongly fragrant and its timber light,
workable, and enduring.

The Chinese thuja (now called *Platycladus* to make sure we're not nodding off) has claws on its round cones, which make it easy to identify. As the cones ripen the waxy white bloom fades to brown and the scales gape open.

green tree. Hardy though it is, the white cedar tends to go drably brown in very cold weather. A number of cultivars have been selected for constant colour, among them 'Wintergreen', 'Lutea' (an excellent full-size golden form), and 'Spiralis' (which is slim, compact, and dark green, with distinct formal possibilities).

Patriarch of the family

In all except very northerly or badly drained gardens the western red cedar (*T. plicata*) is a much better tree; greener and more glossy, leafier, and of course larger. Old lawn specimens are sometimes creatures of extraordinary character. Their lowest limbs rest on the ground and take root, to spring up and surround the tree with a grove of green buttresses, a tabernacle cavernous and perfumed within, floored and ceilinged in bracken-brown and raftered with branches. In Washington and Oregon they grow to 200 feet with vastly buttressed bases 35 feet through. One sees old red cedar butts grounded along the shores of Puget Sound, great wooden whales which decay, apparently, can never touch. The Native Americans hollowed out 50-foot canoes from them and carved them into totem poles. As "cedar" in commerce today they provide outdoors all-weather wood for shingle roofs and greenhouses.

The gardener can hardly avoid using such an adaptable tree. It is happier than almost any conifer on chalk soil and in drought conditions – strange when you consider its moist, luxuriant origins. It will start life slowly but surely in quite heavy shade. Given better conditions it can grow three feet a year. It makes an excellent hedge. I have always wondered why it is not grown more in Europe for its excellent strong, light, weatherproof timber.

'Zebrina', one of the forms of *T. plicata*, is the biggest of any golden conifer. It has reached 90 feet: a remarkable sight in its summer colouring. In winter it settles down to a mixture of gold and green which is still, in such bulk, extremely eye-catching. Why such conifers change their plumage with the seasons like birds is an interesting speculation.

Chinese cousin

Golden varieties are also the main attractions of the Chinese thuja – as it was known until they changed it to *Platycladus orientalis*. The green tree is small, often a multi-stemmed, round-headed,

substantial bush, for some reason often seen in Essex churchyards, as it is said to be in temple grounds in China. One in my garden has been reduced to two of its original five stems, leaving a neat dome on bare poles which is rather effective. Its sprays, fine and small scale for a thuja, have a certain springy way of turning on their sides, which gives it a tousled look. It is not glamorous, but its green is bright, and one grows fond of such things. Unlike true *Thuja*, Chinese thuja lacks a pungent smell and has little down-turning hooks or claws on the scales of its cones: an easy tree to identify. 'Conspicua' and 'Elegantissima' are two of its best-known golden forms of small-tree size. 'Aurea Nana' is a particularly good golden dwarf.

The Japanese thuja, *T. standishii*, is a rarer tree than the related *Thujopsis*, another native of Japan, where it is known as Hiba. Many conifers flirt with green plastic artificiality. Hiba goes the whole way. Its scale leaves, firmly (and very prettily) marked with silvery-white below, are a conifer in kit form from a cereal box: the precise, shiny, yellow-green of a stamped-out plastic conifer. (I can think of worse ideas). Though with a possible height of more than 70 feet it could not be classed as a dwarf conifer, it grows slowly in its early years so that you could happily plant it in a small garden. Young plants may grow only an inch or two a year for as long as 10 years.

FROND VARIATIONS

A whole class of conifers, not all closely related, have flattened fronds in place of identifiable twigs and leaves. It is intriguing to watch them develop, slowly replicating with no buds, just adding more of the same scale-like greenery.

Thuja plicata
The western red cedar has strong, flat hanging fronds of deep green, with a smell combining pineapple, apple, and turpentine. Touch the wood and you will have resin on your hands.

Thuja occidentalis
Why is eastern red cedar called "western" in Latin? Because America's east coast was once Europe's Far West. Its foliage is thin, yellow-green above, greyish below, without white patches.

Thujopsis dolabrata
This is Japan's Hiba, an important ornament of Japanese gardens. Its leaves are bright green with a distinct pattern of white stomata markings on their undersides. It grows slower but roots well from cuttings.

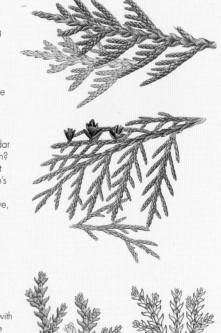

Swamp Cypresses and Dawn Redwoods

CONIFER NAMING HAS BEEN more confident with the nuts and bolts of varieties and cultivars than with the bafflingly broad horizons of families that have been in existence 100 million years or more. The swamp cypress, dawn redwoods, sequoias, and their allies were, until recently, grouped in a family of their own, Taxodiaceae. Since this was based on a combination of characters that they have with those that they lack, modern botanists have found this an unsatisfactory alliance. Recent research (including DNA) has shown that most of the group are better included with the cypresses.

One thing the swamp cypress (*Taxodium*) and its classmates have in common is great age. Another is colossal size. To this group belong the redwood and the giant sequoia, the Japanese "cedar" and Chinese "fir" (the biggest timber trees of their respective countries), and the biggest conifer of the eastern United States, the swamp or, as it is known in America, bald cypress – so laymen had already rumbled the cypress part.

An amphibian

It may seem odd to recommend a conifer for being deciduous, but the swamp cypress, like the larch, makes a virtue of its seeming disability by being the freshest-looking of all the needle trees while it is green, and turning colour superbly in autumn. It complements the larch beautifully, in fact; they should be planted together. For when the larch turns from green to autumnal gold, the swamp cypress shifts from pale green by deepening stages to a ripe ginger-brown. The two colours together in my garden, interspersed with the steel-blue of Scots pine, are one of autumn's most compelling compositions.

Why is the swamp cypress not planted more? Partly, no doubt, on account of its size. Yet of all tall trees in a small garden it can make the most effect while casting the least shade. Partly, I suspect, because everyone assumes it needs a swamp. Curiously, it seems indifferent. For a tree whose home ground (if that's the word) is the stagnant water of swamps in the sub-tropics it does remarkably well in dry ground in the sub-arctic winters of New England. It needs a good warm summer – it fails on the mild west coast of Scotland, needing warm summers for growth, but it freezes without flinching, as the 80-foot tree at Boston's Arnold Arboretum shows.

The only attraction you will probably lose by planting swamp cypress in ordinary garden soil is the eventual growth of its unique "knees" from the roots. Science is puzzled by these spongy knobs, which arise over the points where its deep sinker roots descend

Not many trees are happy to spend their whole lives in water or a marsh. The swamp (or bald) cypress is an exception, seen here in the Big Cypress National Preserve, between Miami and Naples, Florida. Their trunks are powerfully buttressed for stability. With age they will grow the peculiar "knees" seen on the following page.

into the mud. Sometimes they form a whole village of odd mounds like ant hills, coming out of the water or out of the waterside ground. It seemed reasonable to think that they supplied air to the roots, which would otherwise be in danger of drowning. Yet they seem to have no means of doing this. Are they like the camel's hump: a store of nutrients? Perhaps like ear lobes or little toes they are just there.

The swamp cypress includes two other trees in its rather special genus. The pond cypress (*T. distichum* var. *imbricarium*, formerly – and better – known as *T. ascendens*) is a local form, restricted to the southeastern part of its huge range, which runs from Delaware to the swamps of eastern Texas. Although it is not so hardy, the pond cypress is in some ways an even better tree for the garden: smaller and upright, with its leafy shoots set spirally all around the branches instead of in two ranks. The "leaves" in fact are short shoots bearing leaves – longer in the species, short and scale-like in *T.d.* var. *imbricarium*.

The other *Taxodium*, the Montezuma cypress (*T. mucronatum*), is evergreen at home in Mexico, where it is the national tree, only losing its foliage where it is planted near the limit of its hardiness. One specimen can safely be called the world's fattest tree: it is 115 feet in girth. Postcards from Tule, Oaxaca, show nothing but El Gigante: there is no room for anything else.

Enter the fossil

For a long time botanists thought that Asia's nearest approach to a *Taxodium* was a tree of great rarity, *Glyptostrobus pensilis*, reported as living a swamp cypress-like life in Guangdong in the south of China. Whether there is any *Glyptostrobus* left in the wild nobody is sure. A few indigenous trees may persist in southern China and Vietnam, but it is widely planted as a riverbank tree, so the distinction between wild and planted is blurred. It likes a hot summer and water at its toes: in California it has grown 40 feet in 11 years, and there is a 30-foot tree in southwestern Ireland, but elsewhere it is rather tender and slow-growing. Where *G. pensilis* survives it distinguishes itself by turning a lovely warm pink-brown in autumn. After a long search I planted a very small one in my water garden, but looking out at the ice and snow I don't have much hope of seeing it growing in spring.

Are there, or were there, any other relations of this ancient race? In 1941 a Japanese scholar found the fossil remains of a similar tree in Tokyo. He called his find *Metasequoia glyptostroboides* – an indication, perhaps, of his state of scholarly indecision, rather than of his barbaric ear ("Meta" means close to, and so does "-oides") Could this be a link, or at any rate a cousin?

What seems almost incredible is that in the same year, 1941, thousands of miles away, half way up the Yangtze River in eastern Sichuan near Chongqing, a botanist stumbled on three trees of this same unknown genus. It was 1944 before specimens were collected and 1946 before a thorough search of the region was made by the National University. By September that year the Arnold Arboretum, for which E H Wilson had done most of his collecting in that very area, had joined the search. There were plenty of these trees; they were known to the people as water larch and fed to the cattle. In 1948 seeds germinated in Boston and Britain and within a year the new discovery was being planted in gardens all over the world.

These strange protuberances are unique to swamp cypress, coming out of the ground (or more often the water) up to a considerable distance from the tree. There is no accepted view of their *raison d'être*.

The cones of swamp cypress and dawn redwood are similar. Both ripen and fall in their first year and bear fertile seed, but both trees equally root quite easily from cuttings taken from leading shoots.

The dawn redwood is as readily identified by its trunk as the swamp cypress by its "knees". Mature trees have deeply fluted bases of a colour and texture that recalls their distant relationship to the redwoods.

The leaves of both the dawn redwood (left) and swamp cypress (above) are ambivalent. Are they just the little needles or the whole feathery frond? The needles are the leaves, but the shoot is also deciduous and falls with them.

Like the ginkgo, this survival from among the fossils seems to have left its enemies behind. It usually grows much faster than the swamp cypress, at least to start with; there are dawn redwoods (the name that was coined for it, though I thought water larch was pretty good) more than 100 feet high in collections already. In my garden both were planted together in 1959 and have achieved exactly the same height 50 years later. So far it has grown as a regular, pointed tree with very light, indeed rather twiggy branches. It remains to be seen whether it will broaden out with age, as swamp cypresses do, to the splendid, branchy, irregular head it possesses in its native habitat. It already develops a flaring, fluted bole and has a special fondness for water: both typical swamp cypress traits. In some specimens the bole goes far beyond mere flaring; it erupts in a landscape of writhing ridges and hollows.

The ready way to distinguish between the two is by the arrangement of the leaves. Though both have green branchlets in two ranks along the main shoots (which fall in autumn with the leaves) those on dawn redwood are opposite each other, while the swamp cypress's are placed alternately. The same is true of the tiny soft green needles themselves. The fall of what appear to be entire little branchlets (but in fact are just leaves) is more noticeable in the dawn redwood. Perhaps easier to see is the dawn redwood's unique habit of having the next year's buds underneath the branchlets.

RELATIVE DIFFERENCES

The ancient branch of the cypress family containing *Sequoia* and *Metasequoia* have a strong family likeness in their structure whether they are evergreen or deciduous. The free-standing trees shown look more distinct than they may be in real life.

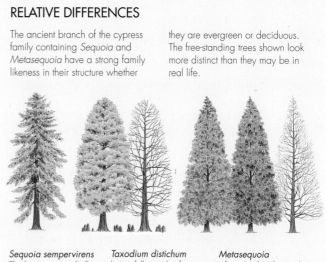

Sequoia sempervirens
The biggest tree of all tapers strongly. Its heavy branches tend to hang down.

Taxodium distichum
has a fuller and softer outline with lightweight branches until it forks with age.

Metasequoia glyptostroboides tends to keep a pyramidal outline with light, wandering branches – so far.

Giant Sequoias and Coast Redwoods

SUBJECT TO AN UNCONCLUDED DEBATE about whether early Australians were just bragging when they reported eucalyptus trees 470 feet high, the redwood is the tallest tree we know and its cousin, the "big tree" or giant sequoia, is the biggest. Just how big they are in relation to more everyday trees comes across well in the comparison made by the American dendrologist Rutherford Platt. The lowest branch of the biggest of the "big trees", in his report, was 150 feet from the ground, but this branch was six feet in diameter and 150 feet long – a branch alone bigger than the biggest elm tree in the world, held horizontally more than an elm tree's height above the ground. It took even 19th-century Californians in all their vigour and rapacity quite a time to think of a way of cutting down such formidable creatures. There is a note of hysterical triumph in the account of five men working for three weeks to fell a tree 100 feet in circumference. It never occurred to them to leave it where it stood.

Before the Ice Ages, redwoods grew in many parts of the globe. Fossil remains of them have turned up in Cornwall, among other places. But like so many trees they have retreated to the one place on earth that offers them ideal conditions, and as so often that place is California.

The fog belt

There is a recipe for instant confusion in the fact that there are two trees in this story, not one. To make matters worse the tree known to botanists as *Sequoia sempervirens* is called the redwood (or coast redwood) while the tree known to laymen as giant sequoia or "big tree" is botanically separate, in the genus *Sequoiadendron*. It is the only species: *S. giganteum*.

Superficially the two are similar enough. Their foliage in fact is quite different, but what distinguishes them even more is their tastes and habits. The coast redwood reaches its preposterous heights in dense stands in the coastal fog belt of northern California; the "big tree" builds its bulk in open groves with white fir and ponderosa pine and incense cedar in the relatively dry and extreme climate of the western Sierra, where most of the precipitation comes as snow.

A redwood forest is like no other. The best stands are on rich alluvial flats sheltered by the coast range but still within 30 miles of the sea. Bull Creek Flat in the Rockefeller Forest near Highway 101 is one of the best remaining, with about 10,000 acres of superb trees. The biggest surprise on seeing them for the first time is not

The 10,000 acres of the Rockefeller Forest in Humboldt County, northern California is the largest remaining area of old-growth coast redwoods. They are guarded from logging companies by the Redwood League, one of the earliest conservation groups in the United States.

so much the height of the trees – you can't stand back and survey one – it is the way they stand shoulder to shoulder, sometimes scarcely leaving room for a man to squeeze through between trunks each 20 feet thick. Unlike any other conifer they sprout profusely from suckers from their roots; cut one down and 10 more appear.

The whole grove seems to grow from one giant skein of roots. Nothing else except a few ferns can grow in that half-light. It is a solemn and simple environment: the soft brown floor of litter, the surrounding trees more like walls than pillars, the dusky brown of their bark lit occasionally with a needle of light from far above. The style of the architecture is distinctly Gothic: the bark splits in deep grooves which end in a narrow Early English arch every 10–15 feet.

Monstrous vegetable

The "big trees" in the Sierra are even more impressive to visit as a spectacle. There is all the fairground stuff of being able to drive through holes in them, either laterally or longitudinally – though no longer laterally; that attraction is closed. You can stand back and see the whole monstrous vegetable: its untapering shaft going up like a road to the tangle of heavy branches that make its head – perhaps 100 feet of lively sprouting above 200 feet of simple, unadorned log. The tallest, a towering 311 feet, is only (only!) 65 feet short of the tallest redwood. A tree 100 feet in circumference at the base is 3,500 years old. There is no question of "second-growth" trees ever supplanting these, except for logging purposes. Yet they are still prolific with their seed and seedlings, aided by an enthusiastic population of squirrels.

In nature, because the trees never grow together there is no opportunity to compare them. In cultivation their foliage – if you

can reach it – is the simple way of telling them apart. The "big tree" is essentially scaly: somewhere between a cypress and a monkey puzzle in size of scale but inclined to the monkey puzzle's reptilian look. The redwood has two kinds of foliage: a little on new shoots like the "big tree" but most of it more like a yew with regular rows of flat, taper-ended needles. The needles grow steadily shorter toward the end of the shoot, so that each individual shoot is roughly boat-shaped. The texture of both trees is stiff and coarse.

In cultivation the "big tree" has proved the hardier and more adaptable. Its arrival in Europe in 1853 caused a sensation. The craze for the cedar of Lebanon was almost over: the "big tree" took its place. It was christened "Wellingtonia" in honour of the Iron Duke, and is still known under this name in Britain. There was not a landowner in Britain who failed to plant one. British conditions suited it perfectly: the trees flourished so that today there is one on every estate; the tallest in Scotland stands 177 feet high.

In the United States there was some attempt to counter-christen the tree "Washingtonia" – a name already pre-empted by a palm. At one time it was called *Taxodium washingtonianum*. It is hardy as far north as Rhode Island on the east coast. But it has never been regarded quite as rapturously as it was in Europe.

Open-ground cultivation gives the drooping lowest limbs of the "big tree" the opportunity to layer themselves around the base. Some of the most magnificent specimens have formed vast pyramids in this fashion. The siblings moulded into the general mass are all the more beautiful because the younger growth is noticeably more grey, or silvery-green, than the dark spire above.

It is the nature of the coast redwood to dwell in groves. Single trees tend to resent the exposure to frost and wind. The best specimens outside California, where it is widely used, are in northern and western Britain where it has reached 157 feet. But in central France, for example, where the wellingtonia grows well, the coast redwood is not happy.

I feel obliged to mention what to me is the ugliest tree in the world – the weeping wellingtonia. This freak from a French nursery has branches that grow straight downward, as near the trunk as they can get. As it gains height it begins to topple from the tip. I have seen them compared to a brontosaurus emerging from a swamp, but they cannot be said to look like trees.

THE BIGGEST TREE: INTIMATE DETAIL

When you see a full-grown "big tree" you have no need for a botany book. Nothing else resembles its thick soft red bark, its tapering tower of a trunk, or the dark, heavy reptilian foliage on its hanging branches.

Sequoiadendron giganteum
Above are its heavy shoots, its leaves, and its cones, fresh and mature.

The cypress-like spray (far left) of Sierra redwood (*Sequoiadendron giganteum*) shows its leaves awl-shaped, pointed, and rough to the touch; they have a strong fresh smell.

The coast redwood (*Sequoia sempervirens*) (left) has more fir- or yew-like needles, dark green above, with two white lines beneath, crowding forward in parallel lines on each side of the shoot. When they first emerge at the tips they look more like the Sierra redwood.

The biggest known tree of all was christened General Sherman by one of his cavalrymen. It stands in the Sequoia National Park, California, 275 feet high and 103 feet around, with a volume (the accepted measure of size) of 53,000 cubic feet, weighing more than 6,000 tons.

The Dwarf Conifers

WHEN THE NURSERYMAN SAYS sport he doesn't mean violent exercise: nor does he mean jest. He means the tendency of plants and parts of plants to have a mind of their own; to deviate from the norm. The nurseryman's business is to follow up the most promising deviants to see how they develop. It happens in the seedbed: out of 100 cypress trees one or two may be blue or yellow instead of green, or thin and narrow instead of pyramidal or bushy. It also happens occasionally on one branch of an otherwise blameless tree. There is suddenly a confusion of twisting, twining little shoots. The nurseryman pounces. If he can propagate from this oddity, either by getting a cutting of it to take root, or by grafting it onto other roots, he has a tiny new tree to patent. In the past he has celebrated the fact in pseudo-Latin; today he must use a vernacular name.

Sports of conifers which grow very little, or at least slowly, are particularly valuable. They are permanent, often pretty, and they take up practically no space. In town gardens which have no room for real trees they offer, pollution permitting, the chance to enjoy the lovely variety of the conifer world in miniature.

There is no clear dividing line between conifers and dwarf conifers. Natural smallness may or may not be permanent; but it can always be made so. By a simple borrowing from the technique of bonsai – just lifting and replanting every few years – it is possible to keep any small conifer (or even any big one) from getting the bit in its teeth and growing away.

On this page are some of the better miniatures. The broadleaf world has no equivalent – only relatively few stunted Arctic forms. A certain amount of tact is needed in using dwarf conifers effectively; their scale works best if they are kept to themselves, in a bed or on a rock garden of their own, perhaps with low heather to cover the ground between them.

A wild form of the Balsam fir (right) of New Hampshire: *Abies balsamea* 'Hudsonia' has short broad leaves and snail-like growth.

The scale of accompanying plants (below) is critical in a collection of dwarf conifers. Even grasses can be overwhelming.

1 *Pinus mugo* 'Mops' is a miniaturization of the already small mountain bush pine of the Alps. It is not fussy about soil or climate.

2 This light green candle-snuffer is *Picea glauca* var. *albertiana* 'Conica', a happy sport of a natural Rocky Mountain hybrid. It can grow to six feet.

3 'Spaan's Dwarf' is like a natural bonsai form of *Pinus contorta* from the state of Washington. Unlike most dwarf forms it has an open branch structure and grows wider than high, suggesting Japanese-style pruning.

4 The Noah's Ark juniper is an exquisite miniature of the common Irish juniper, a minute grey column. It is of the slowest-growing of all conifers but needs care and moisture to thrive.

5 Pines are as apt to sport as other conifers. The shining Bosnian pine (*Pinus heldreichii* aka *P. leucodermis*) has a tiny version, 'Schmidtii', with the hardiness of its parent.

6 *Picea glauca* 'Pixie' is an even slower and smaller sport of *P. glauca* var. *albertiana* 'Conica', bright green in growth and upright in habit.

7 A false cypress with needle-like foliage. *Chamaecyparis pisifera* 'Cumulus' forms an airy little bun.

8 *Pinus mugo* 'Moppet' is a sport of a sport, originally a witch's broom on *P.m.* 'Mops'. It grows about an inch a year. Drying paint is an alternative diversion.

Perhaps the most perfect of several dwarf Alberta spruces is *Picea glauca* var. *albertiana* 'Laurin'. In 10 years it will reach two feet high and one foot wide, dense, and shapely. It is completely hardy but like all miniatures it needs some maintenance. Aphids can be a problem; so can drought.

The Palms

THE PALMS, THE POTENT SYMBOL of the tropics, occupy a byway of evolution. In the remotest past they parted company with the family which contains the water lilies and the magnolias. In strict botanical terms, being properly called flowering plants, which the conifers are not, they come closer to the broadleaves. But unlike any broadleaf tree they come into the class of plants with one-seed leaf, the monocotyledons. The implications of this abstruse detail are far-reaching. From it springs the totally different design and construction of the palm trees.

Palms don't and can't have any real branches. They have no annual rings of growth: they grow taller without growing thicker. All their leaves are produced from the top of the stem, and their flowers and fruits from among their leaves. The workings of a palm tree are shown in the illustration on the right.

There are certain crucial disadvantages in their design. The biggest is that they depend utterly on the cabbage-like bud at the top of the stem … which happens to be a very tasty morsel, much sought by the evolving beasts of every age up to and including man. The palm's defence has therefore been an important part of their evolutionary story. Witness the fact that mainland palms tend to be more or less furnished with spikes to make them uncomfortable to climb, whereas on remote and mammal-less islands in the Indian Ocean they have no spikes at all.

A hundred and twenty million years ago most of the globe suffered the same tropical climate. Then came the cooling that ended with the ice ages. The palms retreated before the cold in just the same way as the rest of the trees, but very few were left in the temperate zone and not many in the sub-tropics; they needed the tropical extreme. Most of the 3,500 species of palm need heat and moisture as a way of life.

To palms, therefore, even the sub-tropical climate of southern California or Florida is chilly. A worldwide group of a very few species venture into the world of frost. Only one, a Chinese species (*Trachycarpus fortunei*), is at home in a truly northern climate.

The study of palms is relatively new. The first and still the most considerable work on them was done by the Dutch in the East Indies *c*.1800. The classic palm drawings of that period have not been surpassed. For the plain fact is that you have to be a blend of botanist and steeplejack to get very far. Not surprisingly, most botanists stick to trees with branches.

The hardiest palms

Only one palm has come to terms completely with snow and ice: the Chusan or windmill palm (*T. fortunei*) of the Far East. It is no beauty. Each year it grows three or four fan leaves, three or four feet across, and sheds their yellowed forebears. The old leaf bases remain, covering the trunk with brown furry fibre. As a young tree, a cluster of green fans on a short stalk, it has definite merit, but in age, a 15- or 30-foot column of tatty fur with a tuft of leaves on

The Canary Island date palm (top) is a squatter, broader tree with a much thicker trunk than the date palm of Egypt. Its dates are inferior but its range much wider. Few palms are more tolerant of poor conditions.

The needle palm (*Rhapidophyllum hystrix*) (above) maybe be a bush rather than a tree but it represents the palm – for those who can't do without – in New York winters. The needles are vicious spikes between the huge and handsome leaves.

HOW PALM TREES WORK

1 Before a palm can begin upward growth it must form a growing point or bud on a base as broad as the trunk will be: which means remaining as a little tuft of leaves at ground level for as long as six or seven years. During this period the bud is very vulnerable. Animals would eat it if it were not densely surrounded by stiff leaves and often spikes.

2 When the leaves eventually fall off, their bases often remain for longer. Finally the stub of the leaf goes and only the ring around the trunk remains. The space between the rings varies from an inch to a foot according to the amount the trunk has grown between each new leaf.

3 Palms often send down roots from above ground level and in some cases send up suckers from the earliest and lowest of the circular leaf scars almost at ground level. Superior forms of date palms are propagated from these suckers.

4 Most palms have a single unbranched trunk. The size of the trunk varies considerably from species to species: some trees have a very slender trunk, no more than two inches across; others such as *Jubaea* are very stout with a diameter of up to six feet.

5 A palm trunk has no annual growth rings. Its core is bundles of softish fibre or pith. Round the circumference actively growing fibres drawing water from the outer layer of woody vessels come to the surface at the points where leaves grow.

6 The yellow flowers (on the date palm shown here) mature into a heavy cluster of berry-type fruit. The leaves in time die off. For a whilte they hang, dead, from the trunk.

7 The single growing point of a palm forms in the centre of the crown. From it emerge the leaves, whose bases, encircling the trunk, add to the tree's height. Among the leaves appear the flower buds.

8 There are two major types of palm leaves: palmate or fan shaped, and pinnate or feather shaped, as here. In the fan leaves the leaflets radiate out from a central point on the leafstalk. The feather leaf has a stem running through it to the tip, with leaflets set at intervals on it.

The coconut palm (left) epitomizes the holiday-brochure image of the race, of white yachts moored in blue waters. Coconuts soon germinate where they fall in the salty sand, sending up a first rush-like leaf as their roots take hold.

Admirers of the little Japanese fan palm (*Trachycarpus wagnerianus*) (above), often impeccably groomed in Japanese gardens, are apt to refer to them as "Waggies", growing and grooming them almost in a spirit of competition.

top, it is a sad creature. Still, a palm is a palm, and the raffish Riviera air it can lend an innocent rose garden is not to be underestimated. Nor, these days, need it be tolerated.

An altogether more domesticated version, *T. wagnerianus* (known in the fancy as a waggy), has made its appearance (it is a Japanese garden plant) with leaves half as wide and more apt for grooming. If you must have a palm in a frosty garden this is the one. If your garden has New York-style winter conditions there is even a shrubby *Trachycarpus* relation, glorying in the name of needle palm (*Rhapidophyllum hystrix*), which resists winters no palm should have to contemplate (on condition it has hot summers). Moreover it has shiny leaves six feet long. The drawback? It has vicious nine-inch spines between the leaves in self-protection.

The French Riviera, indeed the whole Mediterranean, has its own native palm, a similar fan-leaved affair. *Chamaerops humilis* is not quite so hardy and rarely grows so high as *Trachycarpus*. It often forms a bush with a cluster of short trunks, each carrying a bunch of completely rigid fans. At most it will grow, single stemmed, to 20 feet, and look like a better-groomed Chusan palm.

There are two other fan-leaf palms to consider, not as hardy as the first two but certainly acquainted with freezing temperatures – the Caribbean cabbage palm (*Sabal palmetto*) and the Australian (and Asian) *Livistona*. In both, the leafstalk extends through the fan, giving it a distinct midrib. That of the cabbage palm – the common palm of South Carolina and Georgia and many beaches of Florida – curves the whole six-foot leaf outward and downward. In a garden *Sabal minor*, a much hardier but only bush-height version with bluish leaves, might be a better proposition. *Livistona australis* is in some ways the most graceful of the fan-leaved palms: the points of the fan's score of fingers are as long as whips and dangle from the crown.

California's only native palm, the petticoat or California fan (*Washingtonia filifera*), is also fan leaved and moderately frost-proof. The petticoat's peculiarity is to keep its dead leaves indefinitely. They hang around the trunk, eventually making the whole tree look like an Alpine haystack.

Most of the palms on the borderline of the temperate zone belong to the group with fan leaves: in the tropics the feather leaf is much more common.

From date to coconut

The move from fan to feather advances a stage with the date palms. In the date genus (*Phoenix*) leaflets of leaves that are obviously feathers meet the stalk in a pleat with its curling edges facing upward, a characteristic which apparently relates it to the fan leaf. Date palms have magnificent crowns – as many as 200 leaves each 20 feet long. The edible species, *P. dactylifera*, famously fringes desert oases, but is cultivated for its delicious sweet fruit wherever the climate is hot enough, including southern California and central Florida. Left to their own devices certain cultivars of the date palm often prooduce so many suckers from the base that their trunks are lost in fountains of greenery.

The Canary Island date (*P. canariensis*), which has an immensely

The crown of the tropical coconut palm (*Cocos nucifera*) with its feather leaves up to 20 feet long, of which the first yard or so is the stalk, letting the light in and giving the coconuts a clear drop to the ground and climbers a clear route to the top.

thick stem, is hardier and features on many a Mediterranean esplanade: it is also now frequently surviving in southern England.

The hardiest of the feather palms are from temperate South America. The most impressive is the Chilean wine palm (*Jubaea chilensis*). It does very well in California and of all palms has perhaps the most memorable trunk. It can be as much as six feet in diameter, smooth, and elephantine. The nuts of *Jubaea* taste of coconut; the fruits of *Butia* are softly and succulently edible, earning the name jelly palms. The southern Brazilian *B. capitata* is probably the hardiest of all feather palms.

There are two more magnificent feather palms hardy enough for southern California and central Florida: the king palm of Australia (*Archontophoenix alexandrae*) and the queen palm (*Syagrus romanzoffiana*) of Brazil. The king has a worthy crown of short, substantial leaves, rising from a glossy green crown-shaft at the top of a 70-foot trunk. The queen is less than half the height; her leaves droop elegantly.

The royal palm (*Roystonea regia*) and the coconut palm (*Cocos nucifera*) make no pretence of hardiness. The royal palm comes from Cuba, the coconut from almost everywhere in the tropics. Both make enormously stately trees: the royal palm perfectly straight, pale trunked with a gentle swelling at the base and again halfway up; the coconut tall but never straight, characteristically inclined toward the surf over a dazzling white beach.

A CATECHISM OF PALMS

The palms can be memorized by the structure of their leaves: the round obviously palmate ones (like huge maple leaves) being the toughest; the long and feathery ones being the most averse to anything but tropical conditions.

FAN PALMS

The hardiest palms have fan leaves. They will survive in temperate zones but only one, the Chusan palm, is fully frost-proof.

1 The Chusan palm (*Trachycarpus fortunei*) can survive temperatures of 10°F or even lower. It grows slowly with an untidy fibrous trunk.

2 The Mediterranean fan palm (*Chamaerops humilis*) is more often a many-stemmed bush or a hedge.

CABBAGE PALMS

A group of palms with fan leaves formed around a distinct midrib include several moderately hardy species for Mediterranean-climate regions (zone 8).

3 The cabbage palm (*Sabal palmetto*) grows to 90 feet with immense fans on stiff spines curving out and down. Its dried leaves furnish the crosses for Palm Sunday services.

4 *Livistona* is a genus from Australia and China. The slow-growing Chinese form has gracefully drooping tips to the lobes of its fans.

5 *Washingtonia* is a sturdy-looking tree surviving temperatures down to 22°F. The Mexican *W. robusta* is one of the fastest-growing palms, but not as hardy.

MODIFIED-FAN PALM

A third form of palm leaf continues the progression from fan to feather: leaflets are curled almost into tubes where they attach to the leaf spine. Trees with this form of leaf are more tender than fan palms.

6 The date palm of commerce (*P. dactylifera*) is planted worldwide in regions with a climate similar to Egypt's. A temperature of 20°F will defoliate a date palm, but the trunk may survive even lower temperatures.

7 The Canary Island date palm (*Phoenix canariensis*) is the most impressive of its genus. The less hardy Senegal date palm (*P. reclinata*) forms graceful clumps of thin stems.

FEATHER PALMS

The feather-leaved palms are mostly tropical trees with little or no resistance to frost. Their leaves may be as long as 20 feet. The hardiest tree is the Chilean wine palm (*Jubaea chilensis*).

8 The king palm (*Archontophoenix alexandrae*) grows to 70 feet with a smooth green trunk and a broad shapely head.

9 The queen palm (*Syagrus romanzoffiana*) grows rapidly and ramrod-straight to 40 feet with an open crown of long pendulous leaves.

10 The coconut palm (*Cocos nucifera*) with its curving trunk grows only in frost-free regions, on tropical beaches.

11 The royal palm (*Roystonea regia*) is at the borderline of its hardiness in southern Florida. Its pale trunks line avenues in many tropical cities.

Palm-like as it looks (below), the tree fern (here a species of *Dicksonia* in Tasmania) is no relation, but a normal fern with its rhizome or root vertical instead of horizontal in the ground. The illustration is from Backhouse's *A Narrative of a Visit to the Australian Colonies*, published in 1843.

The Broadleaves

THE BROADLEAF TREES BELONG to the earth's dominant group of plants: the flower bearers, in the strict sense of the word. Unlike the conifers the families of the broadleaves include plants that are not trees at all. They stem back to tropical climates (where by far the greater number of species still live). In the course of adapting to temperate conditions – above all to freezing winters – they have had varying degrees of success. Only a minority of species has come through as trees. All our herbaceous plants had tree ancestors in the ancient forests, and many of them still have tree cousins.

Broadleaves originated as an evolutionary improvement on the conifers. What they managed better, above all, was the circulation of sap. The structure of their wood adapted to allow a much freer and stronger flow: this in turn allowed them leaves which could evaporate more water, photosynthesize quicker, and thus get more value out of the summer sun.

Having made leaves that were more efficient for the conditions in which they grew, they were in trouble when (in the course of geological time) colder weather came. Some modified their leaves, giving them a tougher structure and a thicker skin: made them, in fact, more like the leaves of their conifer ancestors. These are our broadleaf evergreens. Some simply became herbaceous, going underground in winter. These no longer exist as trees.

Most adopted a third plan: they went deciduous. By dropping their leaves they could make the best of the sunny season with big, delicate, and efficient leaves, capable of the maximum photosynthesis; but suffer no damage in winter.

As far as flowers were concerned the big step forward by the broadleaves was pollination by insects rather than by the wind. It seems to have started with beetles eating the flowers of members of the magnolia family. The great advantage of insect pollination is that it does not need a great crowd of trees of the same species to have a fair chance of success. There is less drain on a tree's protein supply because less pollen is needed. (The nectar that attracts bees is made not from protein but from more easily obtainable sugars.)

The other development, crucial to the science of taxonomy, but without much advantage for the broadleaves, was the introduction of an ovary to contain the seeds. The term Angiosperm, which covers all the flowering plants, refers to this and distinguishes them from the naked-seeded conifers (Gymnosperms).

"In a large group an imperfect organization is better than no organization at all." So wrote the late Arthur Cronquist of the New York Botanic Garden in his *Evolution and Classification of Flowering Plants*. I am glad I can quote him, because the organization of the broadleaves is a minefield for the amateur – or, come to that, for the professional. How right he was, although his system has been overtaken by one derived from the new DNA-led classification, based on their relative primitiveness or sophistication, as shown (chiefly) by their flowers. Thus the magnolias, which are accepted as being among the most primitive orders of all the flowering plants, come first. The rest are in order of sophistication and modernity.

The Tulip Tree and the Magnolias

EVERY MAGNOLIA IS THE APPLE of someone's eye. To be conspicuous but to manage an air of frailty is a good recipe. To have prominent flower buds and fleshy petals is no disadvantage.

There are over 220 species of magnolia and two of tulip trees (*Liriodendron*) in the magnolia family. Nearly all have flowers larger than any other tree and leaves bigger than most. They range from giant forest trees in the wet forests of the Americas and southeast Asia to café-umbrella size. None comes from Europe; some big ones are natives of the southeastern United States; almost all the pretty little garden ones come from China and Japan. Frailty is deceptive: these last tend to be the hardiest.

Family patriarch

The tulip tree (*L. tulipifera*) is the grandfather of the family. It sounds rather less glamorous under the name American foresters know it by: yellow poplar. As the tulip tree it sounds like a Walt Disney production. Imagine: 190 feet high, all covered with tulips. The only trouble is that the two selling points cancel each other out. It's true about the tulips, but you often need binoculars to see them.

You can hardly grow a more adaptable, less demanding tree or one that grows faster. A good warm growing season is its one requirement; a moist Mediterranean climate is its ideal – but this does not rule it out for cooler zones. In the wild it grows in the Great Smokies and along the Ohio River, but its range reaches from the Great Lakes to the Gulf of Mexico. It should be planted small: like all magnolias it has vulnerable fleshy roots and one of them is a deep taproot. Ideally it likes deep loose soil and a wet spring. A seedling with all its prayers answered has grown 50 feet in 11 years. But tulip trees soldier on in polluted air, mutilated by amateur tree surgeons, in disease (they catch few) and drought (provided there is a moist subsoil).

They offer an individuality that makes one think of the ginkgo. There is no other leaf quite this shape. It sets off like a maple leaf, with pointed lobes at the sides, but where the third and biggest lobe should come to a point it is cut off short. Again ginkgo-like, it keeps up an even, fresh, medium green from spring to autumn, then changes to a clear light yellow. I can never understand the prejudice that autumn colour is not colour unless it is red. A hundred feet of fresh farm butter is a stirring sight. As for the tulips, which open in late spring or early summer, it's a pity they are not often within reach: they are sumptuous flowers, beloved by bees, with pale green petals opening to show a soft orange lining and a noble array of parts.

The tree has a remarkably similar Chinese counterpart; less different indeed than one or two odd cultivars which have cropped

The short-lived green and orange "tulips" are intriguing but often distant, not as important as the lobed leaves with their tips seemingly sliced off. Their seed capsules are seemingly made of wood.

The timber tree of the magnolia family (left) is the hardy, fast-growing tulip tree from the eastern States. From the emergence of its little folded leaves in the freshest green to its golden autumn it commands admiration.

up. It is distinguished by rather bigger leaves, often flushed bronze as they unfold. There is also an upright one, a variegated one, and one without the side lobes so that the leaves are virtually rectangular. In cooler regions, including most of England, *L. chinense* is the one to plant.

When they flower

Magnolias proper can reach almost the same heights as tulip trees, but many stay obligingly near eye level. It is probably in terms of size that most gardeners tend to distinguish them; size and whether they are evergreen or deciduous. That, and the crucial question of when they flower.

For the gardener's purposes, magnolias fall into three broad categories. One is the evergreen, of which the bull bay of the Deep South (*M. grandiflora*) is the prime example. The second is the deciduous magnolia that flowers in summer, so that its leaves set off (or hide, as the case may be) the flowers, which are tulip-shaped, opening to saucers. The third is the kind that flower before the leaves and is therefore by definition of most interest to gardeners. Most of these are small trees, some are shrubs, and all are from the Far East.

When the first of them arrived in Europe it was labelled *M. denudata* because it flowered on bare branches. One botanist described it as "a naked walnut tree with a lily at the end of each branch". Its Chinese name, *yulan*, means lily tree, and Yulania became the name for all this section of the family – which embraces not only these desirable oriental plants but also an American species that for long seemed less interesting, the so-called cucumber tree (*M. acuminata*). Its role in producing a new race of hybrids comes later in the story.

There is no lily, it should be said, that looks like these magnolias. A tulip, perhaps, or a crocus. *Magnolia denudata* has broad pure white petals of the substance and thickness of an ivory

The scented creamy flower (right) of *Magnolia grandiflora*. Some forms of bull bay, notably 'Goliath', have flowers up to 10 inches across, produced in great numbers from mid-summer on.

Magnolia fruit (far right) can be highly decorative too, particularly in the American *M. acuminata* var. *subcordata*. Its flowers are yellow and its seed full of breeding potential.

paper knife – rather too early in the season for the wayward British climate. Frosted, the flowers look tragic. Its Japanese counterpart *M. liliiflora*, a bush with purple flowers, despite its name, also fails the lily test. So do the Japanese Yulanias introduced during the 19th century, of which the best known are *M. kobus* and the closely related *M. stellata*, whose narrower petals stand out like rays. The narrow-growing little *M. salicifolia* with scented leaves is another species ardent in flowering. These and a handful of other species brought almost all the qualities you could want into our gardens – of colour and scent, of earlier or later flowering, of different statures and patterns of branching. *Magnolia salicifolia* crossed with *M. kobus* gives us the hybrid *M.* × *kewensis*, of which the clone 'Wada's Memory', named for a much-loved Japanese nurseryman, laden with drooping white flowers in early spring, is one of the glories of the genus.

Most spectacular and on the face of it desirable of all Yulanias is the giant tulip tree of the Himalayas, *M. campbellii*, which covers its bare branches with waxy pink flowers in early spring, often while there is a good chance (in Britain at least) of a frost to wreck the show. I can imagine, but have never seen, one 150 feet high and holding aloft no leaves but thousands of gleaming chalices: there is certainly no other tree that can compare. In Britain the tallest are 60 feet or so, but there is one in Cornwall with a spread of 140 feet I should love to see. There are several forms with flowers from white to pink/purple, but they demand patience; grown from seed, *M. campbellii* has no flowers at all for 25 years. The most planted today is the *M.c.* subsp. *mollicomata*, which gets under way rather quicker at 15 years or so. A gardener lusting after *M. campbellii* could do worse than to plant one of its western Chinese equivalents, the elegant *M. sargentiana* (particularly its rather smaller (to 30 feet) form, *M.s.* var. *robusta*), or the rather more garden-compliant *M. sprengeri*, a dazzling tower of pink goblets even on my humdrum soil. In one of those confusions all too

frequent among magnolias, the usual pink form is properly called *M. sprengeri* 'Diva' (the Italian for a goddess). There are seven varieties of 'Diva' – all glamorous. Mine, I suppose, should be addressed as "*M.s.*D. 'Westonbirt'".

A quick start is obviously a major consideration with a tree which is grown for its flowers. The excellent *M. kobus* has been largely superseded because it takes 15 years or more to start flowering (though a nursery-grafted plant will flower much sooner than a seedling). The ones that start soonest are *M. stellata*, the low bush with white or pink star flowers you see almost everywhere, its more vigorous and splendid hybrid *M.* × *loebneri*, and the most popular and free-flowering of the tulip-flowered kind, *M.* × *soulangeana*.

The trick with the very-early flowering kinds like *M. stellata* is to grow them in a comparatively cold spot where their buds will not be wooed open by the spring sunshine too soon. *Magnolia stellata* and *M.* × *loebneri* are hardy enough to flourish in a north exposure. *Magnolia* × *loebneri* grows twice as fast, however, and has bigger flowers, either pink or white to taste, in abundance.

Enter the breeder

Gardeners in the past had no choice but to plant the species and a limited number of cultivated varieties. You chose by size, by persistent leaves, by flowering period, or other characteristics, but your choice was limited to what nature provides. What could be a more tempting prospect for a nurseryman or breeder than a family with such a range of genes for desirable features? One of the most-planted of all magnolias, indeed, was the product of crossing by a diplomat in Napoleon's service, Étienne Soulange-Bodin, who was at one time in charge of Josephine's great garden at Malmaison. *Magnolia* × *soulangeana* is his cross between *M. denudata* and the lily magnolia (*M. liliiflora*), made in 1820 and immediately seen as an earlier-flowering improvement on both its parents, blending their

Magnolia kobus (above) is notoriously slow to flower but makes a fine tall tree, up to 80 feet in Japan. Its relative *stellata*, known as the star magnolia, is shrubby but has twice as many petals and flowers remarkably young.

Inside the flower (far left) of *Magnolia campbellii*. It evolved to be pollinated by insects devouring the petals. It is clear to see how the centre of the flowers turns into the candle of a fruit on its right.

A magnificent close-up (left) of the fruit of *Magnolia kobus* splitting to disclose one of its bean-like seeds. They germinate the following spring, but the plant takes 10 years or more to flower.

In the forests of the American south it grows to well more than 100 feet and regularly delivers a heavy, creamy, scented flower on every shoot, making sackfuls of seed. Its leaves are beautiful: best when they unfold with a layer of velvet rust underneath, but always glistening and trim. The downside is that fallen leaves blow around the garden, hard enough to rattle, and compost with great reluctance.

Toward the north of its range (Philadelphia is about the limit on the east coast; it is hardy in most of Europe and all of Britain) the custom is to grow the bull bay up a high south- or west-facing wall, which it totally buries in its magnificent evergreenery. Such precautions, and the extra heat of the wall, are scarcely needed now. It prefers no lime (but is tolerant of it) and lots of nourishment. Once it gets going it flowers steadily through the summer – indeed the spring too in mild-winter areas. Some clones start flowering at a much earlier age (four or five years) and have bigger flowers than others. An 18th-century English selection called *M. grandiflora* 'Exmouth' (or 'Exoniensis') is one of the best on both counts. Another good one, with whopping flowers, is called 'Goliath'. But small-garden owners need not despair, nurseries have found smaller models: 'Kay Parris' and 'Little Gem' are two to try.

SOME RECENT HYBRIDS

Breeders have been busy with magnolias since the early 19th century. Progress continues towards new colours, more robust (or smaller) trees, more and bigger flowers, and later flowering to avoid frosts. Much of the work is done in the USA where Todd Gresham, Phil Savage, and the US National Arboretum are well-known names, and in New Zealand, where the names of Blumhardt and the Jury family label many good plants.

The Yulan or lily tree (above) of Japan and China (*Magnolia denudata*) has sweet-scented flowers before the leaves appear. Eventually it makes a tree up to 30 feet high, billowing with big branches. This tree is just north of Rome.

Magnolia campbellii (left) has been the object of some seriously competitive gardening – especially in Cornwall. Its home is in the Himalayas and the fear of its growers is in spring frost.

M. 'Star Wars' has flowers of almost excessive size from the middle of spring far into the summer.

M. x *loebneri* 'Spring Snow' flowers a little later than other *loebneri*'s but its generous flower buds survive bad weather.

'Wada's Memory' is M. *kobus* x M. *salicifolia*. Its petals hang to give it a unique shaggy look.

M. 'Elizabeth' is one of the best of the new yellow flowers, using the genes of M. *acuminata*.

colours. More crossing and selection soon produced more improvements, until now there are 40-odd varieties from this blood line alone. Of magnolias of all kinds there are at least 200 worthwhile crosses and varieties derived from them. Four of the best are shown in the panel right.

For many enthusiasts today these hybrid magnolias, especially among the Yulanias, form the heart of the genus. One species has been particularly important in their ancestry, the bush-size *M. liliiflora*, whose variety 'Nigra' is almost new-wine purple on the outside of the petals. This rich colour and something of its shape has been passed to many favourite selections, including the darker *M. × soulangeana* cultivars, 'Susan' (a hybrid with *M. stellata*) and the magnificent *M.* 'Star Wars' (which brings the size and colour of the other parent *M. campbellii* to a more tolerant garden tree that flowers freely from a young age and can stay in flower for months on end).

The evergreen bull bay

Many would call the bull bay (what a dreadfully hearty name for such a voluptuous plant) the all-round champion of the magnolias.

Magnolia wilsonii (above) from China has flowers that face downwards, with one of the sweetest scents.

'Lennei' (below) is one of the oldest and best *soulangeana*. Some call them the "saucer magnolias".

Of the many forms (above) of the hybrid *Magnolia* x *soulangeana*, 'Rustica Rubra' is one of the darkest pink. The best white is 'Brozzoni'.

Magnolia liliiflora **'Nigra'** (below) is more bush than tree. It brought darker colouring into the breeder's palette.

More big trees

The first magnolia to reach cultivation was the swamp bay (*M. virginiana*) in 1688. It lacks the magnificence of the bull bay, either in size, in leaf – it is only a committed evergreen in the south – or in flower. Its compensations are a delicious fragrance in flowers which persist (at Kew) even into early winter, and almost luminous blue-white undersides to the leaves – but many other magnolias share this. Its name declares it to like damp ground – which I have given it, but still wait patiently.

The bigleaf magnolia (*M. macrophylla*), also from the southeastern States, has its detractors, but I find it superb. This one is definitely not an evergreen, but then its leaves are really startlingly big – getting on for banana size. In the Mexican *M.m.* var. *dealbata*, which seems easier to grow successfully, white waxy undersides make them even more desirable. Its flowers, moreover, are in proportion. They come in summer, beaming through the two-foot, luminous-green leaves. In autumn, if the tree is well

sheltered from the wind (this is essential) and the leaves hang on in good condition, they can turn anything from malted milk to espresso. Not classic fall colour exactly, but arresting.

The bigleaf is, in fact, a member of class two: deciduous, summer-flowering magnolias. The champion for flowers in this department is the Japanese "whiteleaf" magnolia (*M. obovata*). Its flowers are long-petalled, white, and sweet-scented, and the leaves really do set them off: they are long enough to be slightly floppy; their undersides are milky white; and if you are lucky enough to be standing under the tree on a sunny day you will see how they have the translucency of tropical sea water – a pale green glow suffuses everything beneath.

The cucumber tree of the eastern States (*M. acuminata*) is probably the tallest in the flower-with-the-deciduous-leaves department, but it would be misleading to class it as an ornamental flowering tree; the flowers can scarcely be seen in the towering mass of big shiny leaves. In every way the yellow cucumber tree

Magnolia macrophylla (below) has the biggest simple leaves of any American native tree. It grows down the east coast into the cloud forests of Mexico, where the endangered subspecies *M.m.* subsp. *dealbata* can have leaves more than three feet long.

The Arnold Arboretum chose this new cultivar (left) of *Magnolia stellata* to celebrate its centenary in 1972 and christened it 'Centennial'. It makes an upright tree as fine as its flowers. Nurseries sometimes call it 'Centennial Star'.

tatters, and its flowers come and go very quickly – within a couple of days.

For over a century gardeners in Cornwall and California have grown bushy trees from China with scented white magnolia-type flowers and called them michelias or mangletias (while wondering, I dare say, why not magnolias). *Magnolia* (*Michelia*) *doltsopa* and *M. floribunda* are reasonably well known. *"Manglietia" insignis* is esoteric but recognized. Botany has now allowed there is no significant difference. Magnolias they are.

Warmer summers are extending their possible garden range, given moist soil. They are being joined by newly introduced relations, too: *M. maudiae* only came from China in the early 1990s but is already a splendid early-flowering street tree in Portland, Oregon. Apparently even more adaptable is *M. laevifolia*, the evergreen equivalent of *M. stellata*, forming a neat shrubby tree with masses of white flowers emerging from richly russet-hairy buds in spring.

Wilson's (*M. wilsonii*) and the Oyama (*M. sieboldii*), with its Chinese version, *M.s.* subsp. *sinensis*, are a group of Oriental species without significant differences as far as the gardener is concerned: all flower from late spring or early summer onward, sometimes intermittently for two or three months. All have flowers which hang their heads so that you have to be underneath to see the red stamens in the white shades, and to appreciate their delicious lemony scent. All tend to be big bushes rather than real trees. There is a famous hybrid of *M. wilsonii* named 'Highdownensis' which proved, at Sir Frederick Stern's famous chalk garden at Highdown near Worthing on the south coast of England, to be perfectly happy to grow on that uncompromising rock.

The positioning of magnolias is worth careful thought. Wind and rain can soon destroy their fragile flowers. Shelter is the vital factor; light woodland shade provides the ideal conditions, and a position to the northwest of your viewpoint the best chance of a heart-stopping moment of the sinking sun backlighting the petals.

(*M.a.* var. *subcordata*) is more of a garden proposition. It is half the size or less, but has masses of yellow flowers among leaves half the size. The same cucumbery pods appear on both. The real interest of the cucumber tree to gardeners lies in its breeding potential: it can add yellow to the range of colours of the oriental Yulanias. Since the 1960s it has been crossed with *M. denudata* and *M. liliiflora* with brilliant results. Among the breakthrough yellow cultivars are the pale, fragrant *M.* 'Elizabeth' (*M. acuminata* × *M. denudata*) and the darker *M.* 'Butterflies'. *Magnolia* 'Yellow Bird' and *M.* 'Sunburst' are results from backcrossing with *M. acuminata*. There will be more.

More Chinese ideas

Asia of course has its evergreen magnolias too. *Magnolia delavayi* from western China is a compromise between the bull bay and the American bigleaf: it is evergreen, and its leaves are quite big enough to satisfy most people. A windy corner, indeed, can reduce them to

The Bays and Laurels

IT IS AS WELL, FIRST, TO DENOUNCE the common laurel as the usurper of an honoured name. The honoured name, you might say, since the true laurel, to us the bay tree, sweet bay, or bay laurel, is the Latin *Laurus nobilis*, whence poet laureates, winning your laurels, passing your baccalaureate, and all the rest. Wreaths for such distinguished brows were made, in fact, of the leaves chefs use as home-grown spice. The noble tree of the ancients is the one they clip like a poodle outside the Hôtel de Paris in Monte Carlo and casinos and hairdressers the world over. Whereas the common laurel (page 228) is an evergreen cherry.

Laurels have similarities with the magnolias. They have in common striking, rather big leaves, usually simple ovals (and where not, as in the tulip tree and *Sassafras*, quite odd and original shapes). They share an aromatic principle: their chemistry is not identical, yet your nose tells you there is something shared by magnolia, bay, camphor, and *Sassafras* smells. What they lack, of course, is the fancy flowers. They are essentially foliage trees.

The bay tree comes into the same class as the yew and the box as material for vegetable sculpture. It is a slow grower; naturally a dense cone to no more than 40 feet. But it can easily be trimmed to a pyramid, or more laboriously pruned up to make a ball on a stick. It survives in a tub, or anywhere where the drainage is good. Very hard frost can hurt it, but it is by no means tender.

There is an unusual narrow- and rather wavy-leaved variety of bay (*Laurus nobilis* f. *angustifolia*) which is apparently hardier. Like the golden bay (*L.n.* 'Aurea'), it is not common in cultivation – nor, frankly, worth growing. In a mild spot I would much prefer the sweet-scented *Laurelia serrata* of New Zealand or Chile, which seems, curiously, not to be a relation.

Pungent differences

Bay is a Mediterranean tree with a close relation (*L. azorica*) forming the prehistoric laurisilva cloud-forest of the Azores and the Canaries. It has two far-flung relations that are similar in general effect, but curiously different in their (very powerful) smells. The smell of camphor is familiar enough. Camphor comes from the camphor laurel (*Cinnamomum camphora*) of southeast Asia. But there is no short way of describing the smell of the California laurel (*Umbellularia californica*) – or, as Oregonians have it, Oregon myrtle – except as a kick in the nostrils. It is the sort of smell that spells danger to me. I would no more put it in my soup than I would petrol – though Californians (and no doubt Oregonians too) play

La Gomera in the western Canary Islands has a rare example of what is called laurisilva; laurel cloud-forest, perpetually dripping, sheltering endemic *Laurus azorica*, hollies, and other trees sometimes reaching 100 feet high.

The bay tree and true laurel (*Laurus nobilis*) are the same. The hardiest form, the willow-leaf bay, is also the neatest and most amenable to cultivation in tubs, with wavy-edged narrower leaves.

Sassafras is a tree of many properties, including aroma, but its garden attraction is in its eccentric leaves of three different shapes and its pretty rather than strident autumn colours. It revels in a warm climate and tends to sucker rather than make a single big tree.

chicken with it in the kitchen, scorning the pungent bay as a spice for weaklings. "A fragment of leaf", says one author, "is enough for 4,000 gallons of stew." The crushed leaf has a reputation for giving headaches. I'm prepared to take it on trust.

A smell is not a tree, however. The California laurel is a very fine evergreen. The biggest one, near Santa Barbara, California, is 80 feet high and 100 feet wide. It is a tree that will grow in deep shade before its turn comes to cast it. The big ones have rather wandering trunks of wood so hard that folk who up and move house (in the land where moving house means moving house) just put the California laurel rollers underneath and shove. It is not so hardy as bay but even in England grows to a bigger tree. In Essex it has reached 20 feet in 25 years and is always a cheerful fresh green.

In California the camphor laurel is beginning to be fancied as a street tree. At the Cape the first governor adorned his Vergelegen estate with trees that are now famous for their girth, yet still airy and graceful. In Japan a great camphor tree is always the sign of an ancient monastery. It is slow (like all the family) and rather greedy for a small garden, but it supports a wide canopy of lovely pale shiny leaves on thick black branches. It is not hardy, sadly, in northern Europe, and its relative *Cinnamomum verum*, the true cinnamon of the spice rack, is strictly tropical.

The deciduous *Sassafras albidum* is hardy, and perhaps the most beautiful (and the best-smelling) of the whole collection. *Sassafras* bark and roots used to be used to make a fragrant pink tea, a cheap alternative to tea and coffee then known as "saloop", thought to cure innumerable ailments, more familiar today as an ingredient of Ecstasy. Together with sarsaparilla from a tropical vine it made the root beer which was universal in the deep South. It always reminded me of toothpaste, but the active compound, safrole, is now known to be carcinogenic. *Sassafras* is rarely planted in Europe, but in its native eastern States it is common enough – all the way from Florida to Maine and west to Texas.

Foliar confusion

Sassafras leaves can't even agree among themselves: some are plain, some lobed rather like maple leaves, some lobed on one side and looking like mittens. They stay a really edible, succulent green as late as early autumn, then they turn orange and scarlet.

This excellent tree doesn't take up too much room: its branches are noticeably short for its height. Only its very positive likes and dislikes tell against it. It needs well-drained sandy loam; alkaline soil kills it. It tends to throw up a small forest of suckers. It doesn't like being moved except in infancy, and it clearly prefers summers

These elephantine cinnamon trees were planted by the first Governor of the Cape, Simon van der Stel, on his estate of Vergelegen in the early 1700s. His mansion has been beautifully restored and he would be proud of the wines from the estate today.

The California laurel (or Oregon myrtle) is a cheerful and graceful evergreen. The scent of its leaves is legendary, but rumours of its tenderness, at least in southern Britain, have been exaggerated. The tallest is more than 70 feet.

that are summers and winters that are winters; Europe's wishy-washy arrangement gets it confused. The difference shows in the size. The biggest known in Europe is six feet around – although it was introduced nearly 350 years ago. The biggest in America, in Kentucky, is 17 feet round and 100 feet high

The *Lauraceace* flirt with both hardiness and usefulness. The most useful, but definitely vulnerable to frost, is the avocado (*Persea americana*). It is easy to plant the satisfyingly heavy seed, fun to see it germinate, but rare to see the resulting shiny-leaved evergreen (the leaves vary in shape, though less than those of the *Sassafras*) survive a north European winter outside. Which doesn't stop Londoners trying.

The pawpaw (*Asimina triloba*, in the distantly related custard-apple family, Annonaceae) had a great name among the early settlers in America as a food in the wilds. John Bartram sent it over to England in 1736. Yet today it is far from common. To collectors of the curious its pleasures are its big, drooping, oval, deciduous leaves, its small, browny-violet flowers, and its short brown pods which are edible – but not, I'm told, a gastronomic experience. The name pawpaw is often usurped by the much tastier tropical fruit papaya, the product of the somewhat succulent, handsomely leaved tree *Carica papaya*.

The avocado tree is a native of central America capable of reaching 60 feet but more often seen on kitchen windowsills. Its fruit, like the banana, delays ripening until after falling or being picked. Tropical orchard trees can produce well over 100 pears a season.

The Planes

IT WAS CALCULATED IN THE 1920s that more than 60 percent of the trees in London were London planes, most of them planted in the previous 25 years. If that isn't daredevil monoculture I don't know what is. The figure would be rather different today, but the fact remains that the plane is, in the words of the chairman of the tree planting committee, the "most biddable tree" there is.

The London plane is generally believed to be a cross between two widely separated wild species of plane, the eastern plane (*Platanus orientalis*), native to Turkey and Greece, and the American western plane (P. *occidentalis*), which Americans call buttonwood or sycamore. The first record of the London plane was in England in 1663 – which has not prevented the botanists calling it *P.* × *hispanica*, the Spanish cross. It certainly shows the sort of vigour associated with hybrids. Given a good bit of clay it grows two or three feet a year until it becomes, at a century or two, an immense structure that combines elegance and urbanity with something scaly and almost sinister. I can see a plane as the protagonist in a science fiction film involving terrified young women …

Plane trees rejoice in heat and plenty of water (above). They are in both elements here lining the banks of a backwater of the Canal du Midi in the Languedoc, but grow almost as strongly in avenues shading the roads of the Midi and in the southern town squares where they are lopped relentlessly every year.

The one tree that takes its name from London shades almost all of the city's historic squares (left, in Bedford Square) and is perfectly adapted to urban pollution. Some enormous trees have even survived having carparks dug under them.

Familiar as the plane is, it is hard to fault it for style. Its great attraction is a trunk as tall as an English elm's, not getting lost in a maze of branches but weaving purposefully to the top, dappled with big patches where the dark outer bark has flaked away and left the inner layer, green ripening to buff. Its winter silhouette is one of the most graceful of all: a tracery in which the weeping twigs form countless little Gothic arches, diversified with dangling black balls, three or four to a string. The power of this form to save city architecture from itself has been realized for 250 years. It has never been more needed than today. There remain authorities, notably around the Mediterranean but also in the suburbs of most cities, who hack the branches off planes as a symbol of municipal power over nature. This is what is meant by the plane being biddable: it just shoots again undaunted. But streets where it is allowed to fulfil itself, as it is in countless Provençal villages, are among the most stately, be the houses mere hovels. Given a good water supply and a hot summer it seems capable of limitless growth.

A strong family likeness

The three most important planes (there are of course others, of which the Californian, *P. racemosa*, is the most notable) are easily distinguished by their leaves. They all have what are, basically, maple leaves, but the eastern plane's are deep lobed, the button-wood's very shallow-lobed, and the London plane's something between them. What is odd is that they could hardly be further from the maple in the evolutionary sense.

Of the three, the London plane is the most widespread today. It will grow in an astonishing variety of climate and latitude, ranging from North Africa to New Zealand. It seems immune to city pollution. The French Ministry of Forests has suggested planting it in place of the ubiquitous poplar for its quick-growing timber. But to a tragic extent the range of forestry trees is controlled by what the market is accustomed to.

Where the buttonwood is grown for timber in the southeastern States you can see that the poplar analogy is no exaggeration. A trunk 45 feet high and nine inches thick in 10 years is not bad going. Unfortunately this beautiful tree, in many ways even better-looking than the London plane and more handsome than the wide-spreading eastern one, demands the continental climate of the eastern States. Spring frost and cool summers make it hard to grow in Britain. Nor is its cousin from California much seen elsewhere, though in the canyons of Beverley Hills its vast trunks (often snaking sideways) are the most admired natives.

Eastern plane has the widest range, from the eastern Mediterranean to India and beyond. The old tree in the square that shades the village elders is usually the biggest plant of all. In Kashmir they boast of a girth of 60 feet. The emperor Xerxes numbered himself among their admirers, a story remembered by Handel in his opera *Serse* with the aria "Ombra mai fu …" ("Ne'er was dappled shade granted so lavishly, so lovingly.") It is a tree that sends its branches from on high snaking to earth. At Blickling in Norfolk there is a tree that has layered from such branches enough subsidiary trees to fill half an acre of ground.

As in California, so in Greece. Planes grow where water gathers. In the Taygetos mountains in the Peloponnese they grow in the gulleys where there is a little more moisture, picking them out in gold as their leaves turn in autumn. Below them, at this season, flowers Queen Olga's snowdrop (*Galanthus reginae-olgae*), and many a pilgrim paying respects to the snowdrop has tracked it down by looking for the planes above.

In contrast (left) to the London plane, opposite, an oriental plane in the garden of Ninfa near Rome, planted in the 1920s, has grown perfectly straight and flawlessly smooth and white to an extraordinary height. And yet other specimens of the same tree spread their limbs over huge areas.

This is shameless favouritism (right): one of the London plane avenues in Green Park, leading to Buckingham Palace. The pale bark of planes catches the winter light with a lacy effect above the classic architecture of their trunks.

The downside (below): the fruit of the plane is a spiky ball on a long stalk, consisting of many "achenes", each of which has a tuft of hairs at the base. When the balls break up in spring countless hairs are released and blow about. Many say they cause hay fever-like symptoms, but the connection has not discouraged the planting of such a valuable tree.

The London plane (right) tends to become grossly warty in maturity with strange protuberances all over its trunk. Why some trees do this while others remain smooth is an unanswered question.

The flaking bark (far right) of the American buttonwood or "sycamore". The irregular little shards of the brown outer bark are continually falling away, especially in summer, to reveal the creamy inner layer, which changes to a mosaic of green and fawn.

Witch Hazels, Sweetgums, and Parrotias

THE WITCH HAZELS AND THEIR RELATIONS are one of the orders of plants which have been thrown into disorder (or a new order if you have the stomach for it) by the coming of DNA. They used to be considered members of a very primitive company that included many of the biggest broadleaves. Now they have been given a new billet in, of all things, the Saxifragales. I leave the subject there.

The witch hazel's own family of close relations gives us some of the best autumn colour outside of the maples in the sweetgum (*Liquidambar*) and the Persian ironwood (*Parrotia*). The witch hazel (*Hamamelis*) itself is one of the best winter-flowering shrubs – though scarcely a tree. And a closely related family gives us one of the most delicate and beautiful of all big trees for the garden, the Japanese katsura (*Cercidiphyllum*). Nobody seems to have come up with a vernacular name for this. Katsura seems fine; that's what I call it. It might sell well as the Valentine tree: its leaves are perfect little hearts which turn all kinds of sweetheart colours at the end of the year.

None of these trees survived the Ice Ages in Europe. The witch hazel, which gives the sweet-scented soothing lotion for bruises, was the first one known to botany – the eastern American *Hamamelis virginiana*. Hazel because of its soft oval leaves like the nut tree's; witch, presumably, from witch. Or Wych, as in elm? It apparently means "switchy or pliant".

The most tree-like of the witch hazels is the Japanese *H. japonica* 'Arborea'. The best for winter-flowering in gardens are the Chinese species, *H. mollis*, and its hybrids. They all have beautiful pale yellow leaves in autumn and all tend to grow with a broad horizontal emphasis, covering a fair bit of ground. But trees?

The sweetgum

The relationship of the sweetgum to the witch hazel is far from obvious, and is indeed denied by some botanists who put it in a family they call Altingiaceae. It looks much more like a maple, with leaves lobed in just the same way. It is another tree that ended up everywhere but in Europe. The eastern American version of the tree (*Liquidambar styraciflua*) is by far the best known; but the now rare Turkish *L. orientalis* (its woods were cut down to produce storax resin) and the Chinese *L. formosana* are very pretty: especially the Chinese, which produces lavender-coloured young leaves in spring. Then in the 1990s we were introduced to a hardier Chinese sweetgum, *L. acalycina*.

Sweetgum is not all sentiment in America. Any tree that grows to 135 feet and 20 feet around is forestry material. I have seen it grown as the moist bottomland hardwood in *Georgia*, with its masters expecting quite some performance from it.

It is a fairly slow grower in northern Europe (10 feet in 12

years), and furthermore is a tree that must be planted small: it has fleshy roots that can sucker furiously if they are disturbed. The biggest in England (at Syon House near London) is nonetheless nearly 90 feet high and a beautiful sight, turning from the same sort of rich shiny green as the sassafras to an autumn motley of scarlet and crimson.

There are two simple ways of recognizing the sweetgum. All maples have their leaves and shoots in pairs opposite one another on the twig; the sweetgum's are arranged alternately. And the bark, even of the twigs, has deep ridges of corky material. Its (male) flowers are little upright catkins at the branch tips, but much more noticeable are its "gumball" fruits, akin to the balls that dangle from plane trees but hard and well able to damage your mower.

Autumn colour is the chief reason to plant liquidambar. Even run-of-the-mill trees turn cheerful in a harlequin sort of way with leaves in a range from yellow to purple, but *L. styraciflua* 'Worplesdon' and *L.s.* 'Lane Roberts' are more committed cultivars that turn scarlet and deep red as nights grow cold. Californian gardeners grow *L.s.* 'Palo Alto', *L.s.* 'Burgundy', and *L.s.* 'Festival', which colour even in warm weather. In England I have found 'Palo Alto' forgets to drop its leaves; they stay on the tree bright red well into the frosts. There are also variegated forms: *L.s.* 'Silver King' and *L.s.* 'Variegata', and a lush green column called *L.s.* 'Emerald Sentinel'.

Liquidambar acalycina from China, indeed from the same province as the dawn redwood (*Metasequoia*), may be an advance on any of these, and as hardy as any. Its brief period in cultivation has shown its three-lobed leaves to be extremely handsome, emerging and staying deep red for many weeks (especially in the form selected and named after Spinners Nursery in the New Forest).

Sweetgums like damp ground (though not bogs or floodland), which suggests the bank of a pond as the ideal place to plant them. The reflection can be sensational. Unfortunately they tend to look chlorotic and grow slowly on chalky soil.

More of a big bush

Shiny succulence of leaf doesn't always count as a botanical characteristic, but if you see the sweetgum and the *Parrotia* side by side (a good planting idea, by the way) you can sense a relationship. The *Parrotia* is known to gardeners as a low tree or a very wide bush that grows more sideways than up – as does the witch hazel, in fact. I have the word of Roy Lancaster that in Iran he has seen it straight as an arrow and 60 feet high, but the biggest

The leaves of the sweetgum (*Liquidambar styraciflua*) hold a bright green until late in the season, then take on tones of scarlet and orange. It is the biggest of the family, growing to more than 100 feet.

I have seen in captivity is the Arnold Arboretum's tree – a broad fan with a good dozen stems to maybe 40 feet. It may well be that this is an instance where all the plants in captivity come from a limited source or sources. It is a common enough case: an explorer brings home something exceptional but it is a long time before anyone follows up by collecting more and different material to add to the gardening pool.

Parrotia, drooping as it does, at least in our gardens, tends to show you here and there a cascade of rich deep green leaves from above. Other branches will take off like flapping wings of a big bird, making the tree a free-form performance you need to stand back from. Its little blood-red flowers, rather like the elm's, make a lovely early spring incident. In England its autumn colour is patchy, some leaves going orange or red while others stay green, or parti-coloured. The snag is that the ones on the end of the long branches drop off before the inner ones have turned, so you seldom have the whole tree furnished and coloured at the same time. And I have found that the same tree can be yellow one year and red the next.

What with the colours and the flapping I was prepared to believe that it was the parrot which lent the tree its name. Now I read that it was named in honour of a Herr Parrot, the second man to the top of Mount Ararat (Noah being the first) in 1829.

Esoteric evergreens

The witch hazel family also has some choice evergreens, including the oddly named *Exbucklandia populnea*, with shining leathery heart-shaped leaves that make you think "poplar" until you see its eccentric stipules, like extra little green leaves standing straight up from the twigs. Many plants change their names; few have "ex-" added to the old one; more's the pity. Since its introduction from Asia in the 1970s it has become highly appreciated by gardeners in milder areas on both sides of the Atlantic.

Not related to it, or to each other, are two more relatively primitive evergreens, greatly admired by plantsmen. Indeed, they were planted adjacent to each other by the late Christopher Lloyd at Great Dixter. First, with stiff dark green leaves and a mass of greenish-yellow flowers followed by persistent green fruits, is *Trochodendron aralioides*, representing its own family Trochodendraceae in the forests of Japan and Taiwan. The key to its appearance is in the name *aralioides*, or "ivy-like". It certainly is. The other is *Daphniphyllum macropodum*, whose genus stands alone in its family of Daphniphyllaceae. Whether it is just a big shrub or could be a tree is yet undetermined. What enthused Lloyd and enthuses me is its foliage: each branchlet ends in a rosette of lizard-green, broad, rounded leaves radiating outward on red leafstalks. The flowers are minute, but in a good season it can set a fine crop of attractive blue-bloomed berries.

Exquisite grace

Among all this lush evergreen foliage the katsura (*Cercidiphyllum*) strikes a blow for quiet good taste An old tree makes distinct terraces with its swooping, almost-horizontal branches. On them each small, long-stalked leaf stands out precisely; a neat heart in

The sweeping layers of its branch pattern combine with its little heart-shaped leaves to make *Cercidiphyllum japonicum* (the Japanese "katsura") one of the most graceful of all medium-sized trees. It appreciates deep soil and moisture.

pale green. One of its lovely habits is to produce single leaves from the old wood of its inner recesses, so that when they turn almost pure white as they often do in autumn you catch a glimpse of white hearts against dark bark. Overall it can turn anything from deep pink to ivory: the paler colours stressing all the more the marked separateness of each leaf. Few leaves in the woods are more exquisite as they emerge, and in autumn the tree summons you from yards away with its irresistible smell (it is the damp leaves on the ground) of caramelized strawberry jam.

The katsura likes moisture and shelter. I was surprised by how it thrives in a rather crowded part of my garden among taller trees on quite heavy clay; its natural home is among the maples and cypresses and moss of a Japanese shrine. Eighty feet is the tallest it has yet grown in England, but it shows promise in America of reaching 100 feet. If it has pests I have never seen them. This, rather than one of its cultivars (*C. japonicum* 'Pendulum' weeps, *C.j.* 'Heronswood Globe' is half-size, *C.j.* 'Rotfuchs has red leaves turning bluish) is one of the first-choice garden trees.

Ironwood (*Parrotia persica*) (above) colours brilliant red and/or yellow in autumn. It tends to a broad horizontal shape. Mature trees have flaking bark making patterns not unlike that of the London plane.

The sweetgum (*Liquidambar*) (above) has leaves like a maple's that in good specimens can colour almost as well. Its ridged and corky bark and alternate pattern of branch and leaf are the easiest way to distinguish it.

The nearest witch hazel to tree size is the rare *Hamamelis japonica* 'Arborea', a tree with a sideways bent. The best known (left) is the Chinese *H. mollis* or a hybrid of the two.

The flowers (right) of ironwood (*Parrotia persica*) are minute, but a thrilling scarlet against a cold winter sky. The low spreading branches make it easy to reach them to cut and bring indoors.

The pittosporums (below) are evergreen relations of the witch hazel family, making small trees. The seed capsules split to show red seeds. The variegated *P. tenuifolium* is one of the prettiest and most popular.

The rare and tricky *Disanthus cercidifolius* has leaves exactly like *Cercidiphyllum's* but bright red in autumn. A rare member of the witch hazel family, it makes at best a small tree. It needs lime-free woodland soil, moisture, and shade.

The Eucalypts or Gum Trees

AUSTRALIA IS MORE CLOSELY ASSOCIATED with eucalyptus than any other country with any other tree. Three-quarters of all the forest trees of Australia are eucalypts – from the snowy passes of Tasmania to where palms and tree ferns take over in the tropical jungle of northern Queensland. Their ecological range is thus enormous.

The eucalypts are members of the myrtle family. Most of the myrtles (including the eucalyptus) have simple, untoothed, unlobed, straightforward leaves appearing opposite each other in pairs. Very often the leaves have tiny translucent dots which are visible against the light. The eucalyptus is unique in having a flower bud with a lid which comes off when the sexual mechanism inside is ready to receive insects (for whom it supplies abundant sweet-smelling nectar). There are no petals: the whole of the flower display is supplied by the anthers. And the leaves come in two forms, juvenile and adult. Juvenile leaves are round (and pounced on by florists) while adult ones are usually long and narrow like larger-than-life willow leaves, often abandoning their opposite habit and becoming alternate. More often than not they are grey with wax, at least while they are young. They all have a fragrant oil inside. And finally the bark of many eucalypts is as deciduous as most trees' leaves, resulting in patterns of flaking or tattered strips which are half the tree's character.

Typical eucalyptus country, vast stretches of South Australia, Victoria, and New South Wales, for instance, is strangely like the ghost of a great English park. If you half-shut your eyes those soft contours with grazing sheep, those huge hummocky broadleaf trees, here in a clump, there standing alone in massive maturity, suggest a Petworth or a Longleat drained of all colour, condemned to everlasting summer, to bone-grey and bone-brown. Early painters of this landscape were fascinated by the peculiar light and shade these trees engender. The sheep congregating under them are still sunlit: gum tree leaves hang vertically, filtering rather than blocking sunlight.

In the arid bush on the fringes of desert, where grass gives way to bare earth and stone, the scrubby types of *Eucalyptus* known as mallee take over as dense suckering brush. And in the towns and villages their red-flowering relations (especially what most people call *E. ficifolia* (they are wrong; it has been reclassified a *Corymbia*) line the streets like flowering cherries.

Towering giants

There are 900 or so *Eucalyptus* species. Some are very local; others, like the river red gum (*E. camaldulensis*), are native to every State, growing wherever their niche – in this case the riverbank – can be found. The most magnificent are the mountain ashes of Victoria

Mountain ash is what the locals call the magnificent *Eucalyptus regnans* (left) of Victoria and Tasmania. The Dandenong Ranges, just east of Melbourne, are a vast forest of them, above southern beech and tree ferns. In the country around, including the vineyards of the Yarra valley, they rocket up from streamsides in fertile red earth.

Colossal peeling trees (above) people the south Australian bush, not all of them *Eucalyptus* species. This is the prickly leaved paperbark (*Melaleuca styphelioides*) another myrtle relation that grows as broad as an English oak in pastureland in south Australia where its roots can find an underground watercourse.

and Tasmania, *E. regnans*, which has probably provided the tallest trees ever known, and is easily the tallest nonconifer in the world. In the past it reached at least 430 feet, possibly 470, while the current tallest tree stands at 327 feet in Tasmania, only 50 feet short of the highest extant redwood. Western Australians have to be content with their karri (*E. diversicolor*) reaching 300 feet in forests 250 miles south of Perth. But it is hardly typical Australian conditions that produce these giants; they need rich, rain-soaked soil just as much as coast redwoods do. And just as on the American west coast, predatory logging companies see big trees as mining companies see seams of minerals.

There are huge variations in hardiness within a species. In the often-planted snow gum (*E. pauciflora* subsp. *niphophila*), for example, it is essential to know whether the seed comes from Tasmania or New South Wales, and from what altitude, before it is possible to say if it will survive in northern Europe or the eastern States. In southern Europe there are opportunities for many species. Portugal was the pioneer, in the 1860s, with thousands of acres, largely of blue gum (*E. globulus*), which is all too anxious to take over, with greedy roots and eager seedlings, creating a highly flammable landscape where nothing else will grow.

California saw the first invasion of the United States by eucalypts in the 1880s. The blue gum was widely planted with the fanciful notion that it would absorb the "noxious gases" which were then supposed to be the cause of malaria. It did so well it gave the whole gum tribe a bad name. It seeded itself everywhere and filled good ground with its greedy shallow roots. Nonetheless it looked splendid, and still does, its blue-rinsed tresses and chaotically shaggy forking trunk mingling magnificently with the native oaks and pines.

If Californians curse the blue gum it is only fair to say that Ethiopians bless it. Its introduction at the end of the 19th century by the Emperor Menelik II probably saved their capital city. The native timber had all been cut and the seedlings grazed: Addis Ababa was without fuel. No native tree would grow at anything like the speed of this exotic.

How hardy?
In northern Europe we have always regarded eucalyptus as marginal, doubtfully hardy, distinctly exotic. Too exotic, indeed, to fit in easily in any landscape. This is partly because they rarely grew much beyond their juvenile-leaf stage when they (or at least the species we tried) are an eye-catching light blue. If you are looking for evidence of global warming these gum trees seem to provide it; you regularly pass gardens these days whose dear little blue bushes have become bulky swaying trees, noticeable in winter but fitting in much better than we expected to a temperate summer scene.

In England the commonest by far in the past has been the Tasmanian cider gum (*E. gunnii*). The snow gum followed it in popularity, its appeal being bark peeling to reveal a snow-white trunk. Now it is generally agreed that the not very different

E. pauciflora subsp. *debeuzevillei* is the hardiest, but several other species are gaining gardeners' confidence, and even provoking the interest of cautious foresters. A trial plot of shining gum (*E. nitens*) – its leaves glisten and flash in the sun – has rocketed up to 50 feet in seven years in North Wales. Very probably the larger they get the hardier (within limits) they become. *Eucalyptus dalrympleana* has made a 60-foot beanpole, patched white, grey, and brown, in Essex, so eager for the light that it makes no side branches.

Eucalyptus parvifolia is the one I would grow if I didn't want anyone to recognize it as a gum tree; its leaves are always a modest size and deep green. They also grow so densely that branches can be pulled down by their weight – and a falling branch brings destruction beneath; eucalyptus wood is immensely dense and heavy. *Eucalyptus subcrenulata* is another seemingly hardy one with green, rather than grey, foliage. *Eucalyptus neglecta*, in contrast, is not likely to be ignored: its round juvenile leaves, prized by florists, are slow to give way to oversize grey mature leaves, all giving off a strong balsam smell.

This is only the beginning: others will follow. Unfortunately the most elegant I know, the lemon-scented gum (*E. citriodora*; now reclassified as a *Corymbia*, but still a gum tree in any language) will never be what we in Britain call hardy. It is worthy of a conservatory, though. It possesses as trunk as slim and white as a goalpost, but a goalpost with hips.

Most eucalyptus are easy and quick to grow from seed; all need to be planted out before they grow pot-bound. Once their roots have begun to spiral in a pot they will never be stable.

The ghost gum (*Eucalyptus papuana*) is so-called for its smooth almost luminous white bark. In very hot climates (here in the Macdonnell Ranges, in the central desert of Australia) gum trees get by with very little foliage to minimize transpiration losses. Rainfall in this area is limited to 30 days a year and rarely exceeds 10 inches.

Snow gums in the snow. The snow gum (*Eucalyptus pauciflora* subsp. *niphophila*) is one of the hardiest species; a small slow-growing tree worth having in a garden for its beautiful bark.

The only real tree of the true myrtles, the Chilean *Luma apiculata* shows its kinship with the eucalypts in its peeling trunk. It makes a beautiful evergreen in regions without hard frost.

Adult foliage of the blue gum (*Eucalyptus globulus*) (right). The leaves of young blue gums are often round or perfoliate and have an almost silvery hue. The blue gum is an all-too-fertile giant.

The proteas (below) are the other great Australian tree family. Silky oak (*Grevillea robusta*), tender in Britain, has tiny tubular golden flowers, each tube split and curling, in large clusters in spring.

South America's glamorous protea-family contribution is *Embothrium lanceolatum* (above) from the Andean rainforests. The scarlet tubular flowers smother the little evergreen tree, quick growing but short-lived, in spring. A selection called *E.l.* 'Norquinco' is often grown in the milder and damper parts of Britain.

The elaeagnus family (left) appears here more by sympathy than bloodline. Its leaves are emphatically silver-scaly. Sea buckthorn (*Hippophae rhamnoides*) is a useful little silver tree, tough enough to grow on a beach, heavy with orange berries in autumn if both sexes are present.

More distant relations

Is it a coincidence that most of Australia's trees come in a rush at this stage of evolutionary history? Between the *Eucalyptus*, the *Acacia* which came just before them, and the *Grevillea* and *Hakea* which turn up now in their order of *Proteales*, we have accounted for a high proportion of all Australian trees.

The silky oak (*Grevillea robusta*) is more familiar in the northern hemisphere as a pot plant or summer bedding plant than as the big tree (to 120 feet) it makes in Australia and California. Australians saw a similarity with the oak in its very narrow, but curvily lobed leaves. The comparison is nothing like as far-fetched as with the tree they called the she-oak (*Casuarina*) which has no leaves at all but only bamboo-like twigs. In California the silky oak is a very useful tree, growing fast, looking lush, and thriving in desperately dry areas. What they call the pincushion tree (*Hakea laurina*) is the other member of the family Californians grow, for its pink and gold pincushion flowers.

Gardens as far north as southern and western Britain can grow the proteaceous (but you'd never suspect it) Chilean firebush (*Embothrium lanceolatum*). If there is a more spectacular tree than this in flower I have never seen it. Given acid soil and plenty of moisture it grows rapidly, a lightweight tree with long fresh green leaves, In early spring for several weeks it covers its whole 20 or 30 feet with brilliant scarlet tubes which curl back at the mouth like bells. The effect of flame is almost realistic, and in the Welsh forest where I grow it a touch surreal.

There is nothing so showy in the allied *Elaeagnus* or *Oleaster* family. The family likeness lies in a subtle silver scaliness and a tendency to thorns. In the main they are a collection of the most useful and cheerful of evergreen shrubs, with silver-backed leaves. The one small tree among them (*E. angustifolia*, sometimes known as the Russian olive, or by the French as Bohemian) has the silver scales both sides of its narrow, willowy leaves. The general effect is like a small but highly polished olive tree; even to the olive-like fruit, the many trunks, and the domed top. In the Carpathian mountains in eastern Europe are whole hillsides that glint with their reflective silver. This and its cousin the sea buckthorn (*Hippophae rhamnoides*), which needs a bit of pruning to make a tree shape, are two of the best small silver-leaved trees we can grow; they deserve a place in the sun with a dark background. When the sea buckthorn is covered with its orange berries all along its twigs among the leaves (you need both male and female plants to get fruit) the effect is of firelight in silver hair – a truly autumnal theme.

If you are not intent on a tree but happy with a huge wandering bush *Elaeagnus umbellata* var. *parvifolia* is a wonderful plant. Its leaves are fresh silvery-green and its tiny yellow flowers perfume the countryside with vanilla in early summer. It also has merit as underplanting in forestry, protecting young trees, leading them up, and boosting the nitrogen in the soil.

… and the myrtle itself

Nor should we overlook the myrtles, which betray their eucalyptus affiliation in their bark. Their names have changed more than once, but when you say myrtle everyone knows what you mean. The champion of the genus is *Luma* (aka *Myrtus*) *apiculata* from the Andes, a tree usually with several trunks, happily because the trunk is the best part. It flourishes, and sows its seed generously, in the mild and damp parts of Britain. In Cornwall it can reach 50 feet and we hope for similar results in west Wales. Little white many-stamened flowers, looking faintly rosaceous, spangle the dense mass of small (one-inch max) dark green leaves and become tiny purple edible berries. I trim away the foliage to give a glimpse of stocking, as it were: the highly tactile cinnamon bar, peeling to reveal (can it be?) the white skin beneath.

Once on a lakeshore in remote Patagonia we came on a grove growing in the sand and washed by the breaking waves: silver surf around the glowing rufous trunks: a sight to remember, if a difficult one to reproduce.

The Willows

THE WILLOW FAMILY HAS EVOLVED along a distinct path of its own as a client of the four winds. It stands on its own among trees in being specially adapted to give the wind its seeds to sow. There are only two members of the family in temperate regions: the willows and the poplars. Both have male and female catkins on different trees and offer nectar to the bees as their reward for fertilizing them. Poplars get their American name of cottonwoods from the seed's flying apparatus: all willows and poplars use the same bit of fluff to carry the seed away.

The seed has to be very light, which means it can carry no endosperm, the food supply most seeds include in one form or other. Seed without endosperm has a very short life; germinating conditions must be right straight away – surely the reason why willows and poplars tend to grow in moist ground. For the same reason the new seedlings can't stand competition from other plants. They only colonize bare ground.

Vigour is another family quality, but in this the poplar is more consistent than the willow. For one of the odd things about this unusual genus is that, fast as it can grow, most of its 250-odd species are shrubs. America, Europe, and Asia share most of the tree species of willow – all except the most famous of all, the weeping willow, which grows wild only in the west of China.

A gift from China

Stories abound of the introduction of *Salix babylonica*, the original weeping willow. Before it was established that it came from China it was thought to have been the tree by the waters of Babylon where the Jews in captivity sat down and wept. Its entry into Western gardens came early in the 18th century, from the Middle East, or, according to another story, as a withy used to tie a parcel of tea sent from Spain to Lady Suffolk in London. W J Bean recounts how the poet Alexander Pope, "noticing one of the twigs was alive, begged it, and planted it at Twickenham, where it grew into the celebrated weeping willow of his villa garden". One can imagine his excitement in watching his twig flourish, grow tall, and weep. Why didn't he, I wonder, ever bring it into a poem?

Cricket bat willows (*Salix alba*) (above right) line a stream in an unkempt field on my daily walk in Essex. I always hope for such a dramatic sky; any tree looks good with such a background – especially a silver one.

Jean-Baptiste-Camille Corot was a painter seemingly obsessed with willows (right), as Constable was with elms. Pollarded white willows like these occur in painting after painting, evoking a gentle melancholy *à la mode* in the 1830s.

Did Monet see pictures of the West Lake at Hangzhou (middle right) when he was making his Normandy garden at Giverny? Weeping willows are the essential frame to the bridge in both. They only arrived in Europe from China in the 18th century.

The city of Suzhou (old Soochow) is as famous for its canals as it is for its gardens (far right). This punt-under-the bridge scene reminded me of Cambridge, with weeping willows having the same lulling, dreamy effect.

Only its resemblance to a weeping willow justifies this unrelated tree being on this page. In fact it has no close relations in this book. It is the Chilean maiten (*Maytenus boaria*), beloved of cattle, the nearest thing to an evergreen weeping willow.

The lacquered twigs of *Salix alba* var. *vitellina*. Every spring the pollards are cut back to the short trunk to encourage the trees to produce a profusion of brilliant orange-red stems for the winter. They look like fire when it snows.

The English landscape movement had mixed feelings about this noticeable tree. William Gilpin held that it was "not adapted to sublime subjects. We wish it not to screen the broken buttresses ... of an abbey ... These offices it resigns to the oak, whose dignity can support them. The weeping willow seeks an humbler scene; some romantic bridge, which it half conceals ..."

Most books agree that what really popularized the tree was Napoleon's fondness for one on Saint Helena. So fond was he that he asked to be buried under it, whereupon the poor thing was torn twig from limb by souvenir hunters. Most Victorian weeping willows claimed to be cuttings from that tree.

The blood of *S. babylonica* runs in every weeping willow's veins, but the species itself is rare today. Finding that it was quite tender, nurserymen (or possibly in the first place nature) produced crosses of it with hardy native willows. The commonest of these in Europe is *S.* × *sepulcralis* var. *chrysocoma*, which is *S. babylonica* crossed with the weeping form of the white willow (*S. alba* 'Tristis'). In America other crosses are popular, notably the Thurlow and the Wisconsin weeping willows (*S.* × *pendulina* var. *elegantissima* and *S.* × *p.* var. *blanda*), which are crosses between *S. babylonica* and the crack willow (*S. fragilis*). The Thurlow has reddish shoots, the Wisconsin a bluish sheen in both shoots and leaves. They weep less than *S.* × *sepulcralis* var. *chrysocoma*. *Salix* × *sepulcralis* (which is a cross between *S. babylonica* and *S. alba*) is the most vigorously and grandly ascending of weeping willows. The magnificent trees

with pale yellow stems around the lake at Kew are this hybrid. Sadly though, they are prone to a canker which will soon destroy them. Take notice if the canker is in your neighbourhood.

Today a weeping willow seems almost indispensable to any biggish body of ornamental water. Water is not necessary for the tree, but it mirrors the beauty of the weeping branches. The effect seems so inevitable and right that you instinctively look for a pond or stream wherever you see the tree. The tree itself is indifferent; it will grow in any soil.

William Gilpin went on to call the weeping willow "the only one of its tribe which is beautiful". Evidently he was never invited to Woburn Abbey. There the 7th Duke of Bedford made a complete Salicetum: a collection of every known species. For a time the Salicetum was like the Pinetum: the collection of conifers, a desirable – if not essential – part of the grounds of a great house. The importance of osier beds as a way of using undrainable land provided an economic excuse. But the point of the Salicetum was that willows are pretty – and hugely various. There is a spot in the centre of France, in the soggy wastes of the Sologne, where the river Yèvre winds close to the A 71 and you drive for miles in a watercolour landscape of nothing but willows and poplars, the willows pale and soft, the poplars rigid and smudged with huge nests of mistletoe. It makes an unforgettable impression.

Catkins and brilliant bark

Catkins in the early spring are the first obvious attraction all willows share. The prettiest catkin-bearers tend to be the shrubby kinds, or sallows, which also fortunately have their catkins within reach. It is the male plants that have the bigger and brighter catkins. There are red-catkinned ones such as *S. gracilistyla*; a black-catkinned one, *S.* 'Melanostachys'; and many silvery ones, of which one of the best and earliest is *S. daphnoides*. But it is hard to beat the common European pussy willow (*S. caprea*), with its fat, pearly-furred catkins ripening to saffron-yellow.

Coloured bark is the other willow speciality. Many of the shrubs, and several of the trees, have brilliantly painted new shoots. They come in scarlet, yellow, green, orange, white, purple, and brown.

The colour is brightest on new stems in their first winter. The trick, therefore, is to cut them back hard in the spring, to give them all summer to grow. If you cut them to ground level it is coppicing; to shoulder-level pollarding. Uncut, if the thing becomes a tree, it is scarcely less bright waving in the sky. I can see an ordinary golden white willow (*S. alba* var. *vitellina*) from my desk. Winter morning sun makes it easily the brightest spot in the landscape. The best red is another variety of the white willow (*S. alba* 'Britzensis'); the best purple (overlaid with a soft white bloom), *S. daphnoides*. The Chinese bush *S. fargesii* adds the excitement of buds like painted fingernails on red-lacquered stems – and then deep-veined leaves.

The typical willow leaf is racingboat-shaped: long, narrow, and pointed at both ends. Many are bloomy-silvery, the common white willow (*S. alba*) perhaps most of all. In the landscape, beside streams, they flash in the sun, graceful, light-limbed, almost feathery, perfect against a dark background of oak and alder. One form, called *S. alba* var. *sericea*, a smaller tree more apt for a smaller garden, is like polished silver. There are surprising variations, though, in this too. The English native bay willow (*S. pentandra*) has glossy dark green ovals which look – and even smell – like a bay tree's – and grand yellow catkins with them.

One inspired variation is the twisted leaf. A variety of the original weeping willow (*S. babylonica* 'Annularis') has leaves like little ringlets. In the corkscrew willow (*S. babylonica* 'Tortuosa') the vibrations go right through the whole tree: every inch of twig as well as leaf wiggles. It sounds bizarre: in fact it is elegant – if not reliably hardy. I know of only one willow which has gone too far: *S.* 'Erythroflexuosa' tries to do everything at once: to twist, to weep,

and be cheerful in orange-yellow bark. Result: neurosis.

The white willow, the crack willow (*S. fragilis*, so-called because its twigs snap off if you pull them), and the super-vigorous variety the bat willow (*S. alba* var. *caerulea*) are the common tree willows grown not for their beauty – beautiful as they are – but for their wood. The willow combines first-class timber with rocket-like growth. Whether for cricket bats, the bottoms of quarry carts that have to endure the shock and scrape of stone, or in the terms of the naval timber manual which called it "the best wood for the formation of small, fast-sailing war vessels", willow is strong.

As for speed, a normal rotation period for bat willow, from planting to felling, is 15 years. By then, on a good site with a high water table, the tree is 70 feet high and five feet around. The Essex fields around home are ringed with them, along every running ditch, one of the farmer's most profitable crops. When the corn ripens you can stand in a landscape of pure gold and silver.

Willow growing is made absurdly easy by the readiness of the twigs to take root. The sets used for willows are 15 feet long: simply a young branch driven into the ground for half its length. "Sets" are planted in late winter; by late spring they are in leaf and growing as new trees. You can even stand a young willow on its head; branches become roots and vice versa. As biomass, grown purely for fuel and regularly coppiced, there are few crops to beat it.

This wonderful instant tree must have its drawbacks. It has several, from its liability to the fatal disease of canker to its attractiveness to aphids; also, curiously for such strong wood, its liability to split in a bad storm. But to me they are outweighed by the advantages: its ability to grow so big and beautiful so quickly.

Shrubby willows that will never become trees make a very tempting collection, even in a limited space. This is *Salix* 'Melanostachys', a variety of the Japanese *S. gracilistyla* famous for its black catkins on red stems. It grows to about seven feet.

Pussy willow or "palm" is a stage in the flowering of the male catkins of various shrubby willows that coincides with Palm Sunday before Easter in northern countries, especially in the Orthodox church. If the stems are kept dry, not put in water, they will stay in this state for months.

The purple osier is a valuable shrubby willow for fine baskets, and even richer in salicine (the principle of aspirin) than most. Bean's great reference book has 66 pages of willows. There are 26 varieties of this alone; this is *Salix purpurea* 'Nancy Saunders'.

Poplars, Cotton-woods, and Aspens

THE WILLOW FAMILY includes, in the weeping willow and the Lombardy poplar, the two most famous and distinctive of all ornamental trees: a text-book pair of forms planted together for contrast in almost every park in the world.

The original weeping willow (*Salix babylonica*) is a true species: its seed comes up weeping. The Lombardy poplar (*Populus nigra* 'Italica') is a cultivar, in the sense that its seed may or may not grow bolt upright. It is propagated by cuttings, which luckily are almost as easy to persuade to take root as the "sets" of willows.

But neither tree is typical of its race. They are exceptional not only in shape but in being considered ornamental at all. Poplars on the whole are thought of, like willows, as wood factories. Like willows they grow alarmingly fast. A cottonwood (*P. deltoides*) planted on Mississippi bottomland grew 98 feet in 11 years and I'm told that in China 100 feet in 10 years is as routine, these days, as the four-minute mile.

Like willows, poplars hybridize among their kind with abandon. As racing trees they have also been subject to a great deal of stud activity. Hence, as J L Reed irresistibly puts it in *The Forests of France*: "Their social history is like one of those exhausting novels which describe the adventures and alliances of several generations of an international family established in different capitals." They are social trees in another sense too: their very name, *Populus*, comes (according to the poet Horace) from the same root as populace. He says they were grown to shade squares where people gathered. So closely woven into human history are they that, alone among trees, they have a unique United Nations Commission devoted to them, and the American black cottonwood was in 2006 the first tree to have its DNA decoded.

Four basic kinds

The four branches of the family are the black, white, trembling, and balsam poplars. The black are the biggest branch, containing all the poplars which aren't white under the leaves, don't smell of balsam in bud, and don't tremble (as the aspens do).

The white poplars are natives of Europe, long since established in North America. The basic white poplar (*P. alba*) is an eye-catching tree, even where (which is often on poor land) it grows no higher than 20 or 30 feet. Its bark is like old aluminium and its leaves toss at the merest breeze to show their silver-white undersides. The standard poplar leaf shape is more or less like

The autumn beauty of hybrid black poplars – most familiar in plantations on farms – grown on fertile soil with a high water table. In northern France they are sometimes cropped within 20 years. These trees are grown for match-sticks in Victoria, Australia.

a heart; the white poplar's are lobed like a maple's. In America, furthermore, they often turn red in autumn. In Europe, like other poplars' they turn yellow. There is an upright version of the white poplar (*P. alba* f. *pyramidalis* or 'Bolleana') which is like a rather beamy Lombardy with the distinct advantage of the winking white leaves, and a charming smaller form known as Richard's poplar (*P. alba* 'Richardii'), with leaves yellow on top and white below.

The cottonwood of the western creek beds, *P. trichocarpa*, is the most familiar black poplar in the wilds. A film director is bound to aim his cameras at it or through it when it is the only tree in the landscape, especially in autumn when its leaves turn to gold. Its range runs from the Atlantic to the Rockies; in the east it is less conspicuous until it snows cotton in mid-summer. I fell for its spell when I noticed a strangely formal and very vigorous-looking tree, a stranger in a plantation near home. The cutting I took and stuck without ceremony in the garden was 70 feet high when it blew down 25 years later. I waited for its sweet scent to fill the garden every spring.

The union of the cottonwood with the black poplar of Europe (*P. nigra*) has brought about most of the rapid commercial poplars that in France especially have made their own style of valley landscape. The ordinary black poplar cuts a rugged, almost elm-like figure in valleys in the east of England, though it has become rare. Its varieties – the Lombardy poplar (*P.n.* 'Italica') and *P.n.* 'Plantierensis' (which often steps in to take the Lombardy poplar's place) – are the important ones. Their value as landscape trees is always before one's eyes: either the single huge tower, sometimes 100 feet high or more, dominating some Alpine valley; or the elegant file through fenland, leading to a lonely farm; or the grand approach to an urban institution. There is no tree which creates an architectural incident so quickly and surely.

Aspen is one of the ubiquitous American trees, spreading by suckers to form immense stands. One in Utah, known as "Pando", covers more than 100 acres, weighs an estimated 6,600 tons, and is the biggest living organism on the planet.

The union of these two American and European poplars, known as *P. × canadensis*, has made a multiplicity of hybrids of varied merits, all fast and some beautiful in a light-leaved, summer-meadow way and some of real economic importance. *Populus × canadensis* 'Robusta', for example, leafs late but coppery – an autumnal colour in late spring. Its narrow, light-branched crown (which it shares with *P. × c.* 'Eugenei') can look like a feather on the skyline. *Populus × canadensis* 'Regenerata' and *P. × c.* 'Marilandica' are denser and broader; *P. × c.* 'Serotina' can make a huge cloud-like head. In Melbourne there is a golden poplar (*P. × c.* 'Aurea') a good 120 feet high and built like a sunlit thundercloud.

Trembling leaves

Since all poplars have leafstalks and many of them long ones, the wind always sets them aflutter. The difference with the aspens, the trembling poplars, is that they feel breezes when the whole world is becalmed. You can lie under an aspen on a stiflingly still day and listen to its incessant whisper. The trick is in the leafstalk, a narrow ribbon set at right angles to the hanging leaf blade.

The American quaking aspen (*P. tremuloides*) is the tree with the widest distribution on the whole North American continent. It grows from Labrador to Mexico and to the Bering Strait, leaving out only the southeast and part of the mid-west. Its life is short and vigorous: it demands light and offers bigger and more permanent trees a light shade for their early years; then they overtake it and it dies. In the east of its range it grows with its sister, the bigtooth

The white fluff that carries their seeds in vast quantities gives their name to cottonwoods. This is white poplar. Its leaves are always covered with white down when young. *Populus alba* f. *pyramidalis* is a fine upright form.

The cypress-like pattern of the branching of the Lombardy poplar is seen here in winter. Male trees tend to be very narrow, female ones (here) a little more beamy.

Balsam poplar (*Populus tacamahaca*) gives off a powerful sweet smell in spring but makes a poor tree in Europe. Its western cousin, black cottonwood (*P. trichocarpa*) is bigger and finer in every way and equally sweet scented.

aspen (*P. grandidentata*), which has bigger, toothed leaves. Its favourite place to seed and start a new colony is where fire has been.

According to J C Loudon, the European aspen (*P. tremula*) took over the ruins of Moscow in 1813, the year after Napoleon had reduced most of the city to ashes. It is similar to the American aspen, lacking only its pale smooth bark. Both are (or were when Europe had such things) the beaver's favourite tree. Despite its astringency beavers love the inner bark. Both species have small weeping forms, particularly pretty with their grey catkins in early spring and again in autumn, when aspens turn a shining yellow.

Aspen crossed with white poplar has produced grey poplar (*P × canescens*), a common and on the whole stately tree standing out with its pale trunk beside French roads. It has a bad habit (which it gets from both its parents) of throwing up thickets of suckers. It is a good catkin tree before the leaves come, and the leaves can be almost as silvery as the white poplar's in a breeze.

Poplar wood is not generally sought for special qualities, except perhaps its readiness to split into narrow shavings, good for matches and lightweight fruit baskets. Basically it is grown today for pulp. Grey poplar, however, has smooth, nonsplintering, nonwarping, and lightweight wood which has many uses, from silk rollers to barn doors.

Finally I must mention two Chinese poplars able, in very different ways, to draw people across a crowded arboretum. *Populus lasiocarpa* has stout emphatic leaves over a foot long. "Bull poplar" would be a good name for it if it did not produce such mattress-fulls of white fluffy seed. *Populus simonii* is a complete contrast – its leaves so delicate, on long stalks, that you could almost call it the birch poplar. It is prone to canker, as too many poplars are, and in my garden died back with the infection. I cut it out, however, and the tree has outgrown it to become, well, rather like a handsome birch.

Grey poplar (*Populus x canescens*) (far left) looks rather like white (it may be a hybrid with the aspen) but can grow to a huge size with a gleaming pale trunk, often seen along roads in southern Europe. There is a lovely weeping form, but all grey poplars sucker.

The golden black poplar (*Populus nigra* 'Aurea') (left) is one of the biggest and certainly the fastest of golden trees, a glorious shining sight all summer and autumn. This specimen is in the grounds of Sandringham House in Norfolk.

Locusts and False Acacias

PEAS AND PEA TREES of various kinds grow everywhere on earth. The tropics are their natural base; in the temperate zones they appear more often as herbaceous plants than trees. Peas, beans, and lupins are all examples. Even the big trees of the family, however, share the pea-likeness to a remarkable extent. The *Robinia* (which goes by the name of the black locust, false acacia, or acacia according to where you happen to be) is the first one that springs to mind. Big tree though it is, it has essentially pea-like flowers, pods for fruit, and much the same sort of finely divided foliage as all its kin.

Most of the family are cultivated for their flowers. Of the bigger trees grown for their stature, timber, and shade four are natives of eastern North America. Europe had none like this.

The term "locust", which has stuck with several members of the family, started as the Latin word for lobster. From lobsters it got transferred to the swarming grasshopper-like insect of the Middle East which, with wild honey, was said to be the diet of John the Baptist in the wilderness. Learned discussion produced the verdict that what he was eating was really the fruit of the carob tree, which thus took the name of locust. Thence, via the Bible-reading colonists, the name was attached to the American trees with the same sort of pod: *Robinia* became black locust and *Gleditsia* honey locust.

An American in Paris

The black locust was one of the first American trees to be sent back to Europe. By about 1600 Jean Robin, Henry IV's herbalist, was growing it in Paris. It derives its generic name of *Robinia* from him and its specific name of *pseudoacacia* from its obvious similarity to the sub-tropical acacias of Africa. There are no European acacias. The acacia can hardly have been a well-known plant in the 17th century. Yet for some reason its name stuck and to this day in England acacia means *Robinia*. (Real *Acacia* from Australia we call mimosa.)

Whatever the natural range of the black locust may have been, somewhere in the Alleghenies, today it grows wild in large parts of North America and most of Europe. Greedy though it is, with shallow and competitive roots, it is usually welcome. It grows very fast and makes wood that rivals oak for strength and permanence. In the wine districts of France a copse of "robinier" is a valuable asset; vine stakes from it will not need replacing for 50 years. Copses of it, on the other hand, can easily get out of hand. It both seeds and suckers in abundance. In many parts of France, in particular, it has invaded oak woods, flowering in white sheets, to the delight of the ignorant and the despair of the forester. For when he cuts down the slender young trees they come back again with interest as suckers. I cut down an established *Robinia* in a lawn and am still mowing off the memories 20 years later.

Fresh, enduringly pale green leaves and frothy white flowers contrast with the craggy trunk and zigzag branches of false acacia to make a striking picture. It is a quick-growing tree but can put up crowds of thorny suckers from its roots.

Moved by its performance and its manifest usefulness, the English journalist William Cobbett came home from America in 1823 to sing its praises. In his enthusiasm Cobbett had started a nursery on Long Island and raised thousands of seedlings for the English market. He sold them, and they grew; but unfortunately crooked-trunked and useless.

Crooked or straight, though, from the ornamental point of view the black locust is an original and memorable tree. Its signature in the winter sky is unmistakable: branches like forked lightning zig-zagging out from a pale grey trunk with deep-shadowed furrows. It seems to have no overwintering buds on the twigs; where each bud should be the ridge is crowned with two short sharp thorns.

Like the walnut, it is a black shape for more than half the year. The leaves are late to come and early to fall. They come out tender and yellow-green, deepening to a bluer shade when the white flowers appear. Then for two weeks there is a faint sweetness in the air and the gaunt framework is transformed into as delicate a salad as anything in the garden. But the black locust takes up a lot of space for what it does, and casts practically no useful shade. Nursery work has made a better version of it for street planting: the cultivar *R.p.* 'Umbraculifera', which has no thorns – and alas few flowers – but which forms a dense round head of the most mouth-watering green. In town squares in southern Europe it is an ideal small-scale shade tree, leafing only when the sun grows hot and always keeping a neat shape.

For the garden proper there is a smaller golden version. *R.p.* 'Frisia', which is one of the most desirable of all golden trees – so much so that it has become a town garden cliché. Its annual programme is from springtime pale gold through golden green to golden copper, a sunny sight from urban windows. Its very popularity, in fact, may be making it more prone to disease.

Few other *Robinia* are much entertained in gardens. One that gives me great pleasure is a hybrid raised at Hillier Nurseries between *R. pseudoacacia* and a pink-flowered shrub of the Allegheny Mountains, *R. kelseyi*. *Robinia × hillieri* is a small tree, to perhaps 25 feet, generous with pink pea flowers in early summer.

Seriously fierce

The other American locust, the honey locust (*Gleditsia triacanthos*), has also been called acacia in its time. There is no mistaking this tree in nature: it grows thorns in vicious clusters, even on the lower part of the trunk. Guy Sternberg, who writes definitively on American trees, calls them "weapons-grade devices". A thorn can be a foot long and almost indestructible. Honey locust will grow to twice the size of the black locust: to 130 as against 80-odd feet. The biggest, in Michigan, is 115 feet high and 124 in spread.

The search for an elm substitute has lit on this tree as being tall, shapely, singularly beautiful in its fine, almost ferny, foliage, and not subject to too many pests. Sad to say, though, it has had to be emasculated, disarmed of its thorns and made barren of its long

shining brown pods, to satisfy the parks departments. Overplanting has also encouraged the ravages of insects and fungus. The thornless locust comes by many names from different nurseries: 'Moraine' is perhaps the best-known cultivar, often planted as a substitute for lost elms. 'Sunburst' is a yellow-green one, though no match for *Robinia pseudoacacia* 'Frisia' for sunny effect summer long. *Gleditsia triacanthos* 'Rubylace' is small and reddish and could be used in a subtle colour scheme. Better still, perhaps, for gardens is the smaller Japanese species of the same tree (*G. japonica*) which has leaflets so small (about half an inch long) and in so many millions as to look exactly like a big billowing fern.

Honey locust and its near relation the Kentucky coffee tree (*Gymnocladus dioica*) are unusual in the pea family for having regularly petalled, not pea-like, flowers. Since in both cases they are green they are of no ornamental importance. If the coffee tree is planted it is for its leaves, the biggest and boldest in the family; up to three feet long, easily 18 inches wide. Bare of leaf the coffee tree remains a landmark. For a while the long leafstalks hang on. When they fall the tree seems bereft of all its twigs; what is left is so sparse and each shoot so thick.

The American yellow-wood (*Cladrastis lutea*) is the last of the locust relations from the same astonishingly species-rich forest of the east. It would pass for a specially showy black locust with longer streamers of bigger flowers and fewer, broader leaflets that turn bright yellow in autumn. The way to identify it is to look for its buds. They lurk all summer hidden inside the hollow stalks of the leaves. *Cladrastis* grows slower and smaller than the locusts, making it often a better choice.

Its oriental counterparts, maackias from Manchuria (*Maackia amurensis*) and China (*M. chinensis*) are rare but highly prized little trees: the Manchurian for its upright blue-white flowerheads in summer; the Chinese for ravishing young shoots in spring, emerging dark and bluish with a covering of silver-silk down.

One of the most graceful and decorative of all small garden trees (left): *Robinia pseudoacacia* 'Frisia', a cultivar of the false acacia produced 40 years ago in Holland. Its rich yellow colour from spring to autumn shows off feathery and graceful leaves that seem to cast sunshine into city windows.

Older specimens (right) of the honey locust develop threatening branching thorns on the main trunk. Most barbarously armed of all is the Caspian locust from northern Persia, seen here.

Laburnum, Redbuds, and the Silk Tree

EUROPE'S BIGGEST PEA TREE is the original locust, the carob (*Ceratonia siliqua*) of the Mediterranean. Apart from the occasional experimental bite into the fleshy brown pod (which is mildly chocolaty; not bad at all) nobody pays the carob much attention except confectioners. Because it is evergreen its eight-inch pods often escape notice. It merges into the background: short, gnarled, ageless-looking, scarcely suggesting its unique place in the history of commerce. Its seeds seem so uniform in size and weight that they became the "carat" measure of jewellers.

But experimental bites into pods are not something to encourage. The little laburnum tree is poisonous in every part. Were it not so beautiful, so regular in profuse flower, and so ready to grow anywhere it might well have been banned as a public danger. As things are it is almost as universal in small gardens in Europe (where it is native) as the flowering cherry, with the same drawback: that its big moment is only two weeks of the late spring.

There is common laburnum (*Laburnum anagyroides*), Scotch laburnum (*L. alpinum*, hardier, with shinier leaves, and longer spikes – a better tree), and a hybrid between the two (*L. × watereri*, usually seen as the clone 'Vossii') which flowers even more lavishly than either. Any of them used in calculated masses for knock-out effect (especially pleached into the form of a tunnel), or scattered among light woodland like natural accidents, can be the most telling of all hardy yellow-flowering trees – unless you count broom (*Genista*), another pea cousin. Laburnum is a hard and monotonous yellow, however, and I have often thought how welcome a white-flowered laburnum, with the same graceful profusion of flower sprays, would be.

Most brooms are shrubs. But it needed only a slight help from a wall to windward to egg on a Mount Etna broom (*G.aetnensis*) to 20 feet in my garden. It was planted, with what amounts to horticultural wit, it seems to me, in the same hole as a *Koelreuteria*. Both are multi-trunked; they blended and intertwined in a dome that flowered once in early summer (the broom) and again in mid-summer (the *Koelreuteria*) before the *Koelreuteria* produced its orange seedpods, when the evergreen broom took centre stage again for the winter. On the way up Mount Etna you pass through a belt of this broom growing in black lava soil: a more startling monochrome landscape it is hard to imagine.

The pagoda tree
The best of the hardy Asiatic pea trees, the pagoda tree (*Styphnolobium japonicum*, formerly, and still to most gardeners

The Mount Etna broom (*Genista aetnensis*) (top) tosses clouds of little yellow pea flowers in the air. It scarcely has leaves, but rather thin almost rush-like green stems, limp when new, then making a sparse pretty-much-evergreen canopy.

The weeping form (above) of *Styphnolobium japonicum*, 'Pendulum', might be better named 'Contortum'. The way it twists and turns shows off its pretty fresh green leaves as they tumble in opulent swags. In warm regions its white mid-summer flowering is notable.

The laburnum tunnel (left) in the gardens at Bodnant, north Wales, is perhaps the world's most spectacular use of this almost universal tree. It is made of the hybrid *Laburnum × watereri* 'Vossii'.

Sophora) is more in the vein of the American locusts. Among the curiosities of Kew Gardens there is a specimen that reclines its trunk as though on an invisible *chaise-longue*. It was planted in 1762 – one of the first two or three Chinese trees to reach Europe direct through the offices of M. d'Incarville in Paris. In the same year the garden's 10-storey pagoda was finished. Did the sapling gaze up at the 163-foot tower and realize it couldn't compete?

The pagoda tree remained relatively unknown in the New World, perhaps because the black locust (which was there already) is so similar. Yet a month after the black locust has dropped its leaves the pagoda tree will still be fresh and green. Its chief justification, fine foliage apart, is that it flowers notably late, in late summer. Its flowers are creamy-white, emerging from the most promising-looking creamy buds. It prefers, and flowers much better in,

ALL LEGUMES AT HEART

The range of woody pea-family plants is immense, embracing the American false acacias and locusts as well as the brooms and wattles. Many have pinnate leaves, evergreen or deciduous, and these can take many forms.

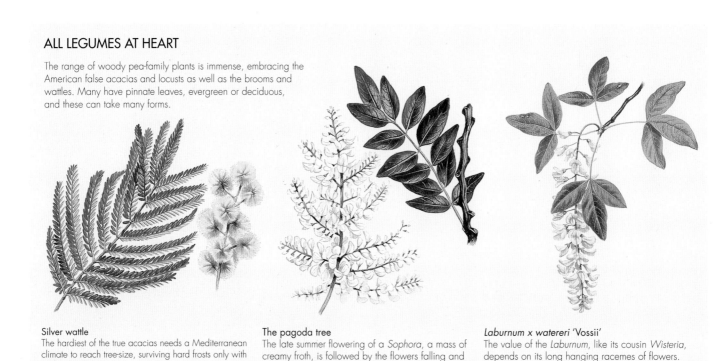

Silver wattle
The hardiest of the true acacias needs a Mediterranean climate to reach tree-size, surviving hard frosts only with artificial aid.

The pagoda tree
The late summer flowering of a *Sophora*, a mass of creamy froth, is followed by the flowers falling and covering the ground.

Laburnum x watereri 'Vossii'
The value of the *Laburnum*, like its cousin *Wisteria*, depends on its long hanging racemes of flowers. The hybrid 'Vossii' has the longest.

regions with continentally hot summers; in Britain 70 feet would be a huge one. The variety *S.j.* 'Pendula', a graceful small tree with long weeping branches, would succeed much better than a weeping willow in very dry soil.

The New Zealand kowhai (*Sophora tetraptera*) is an evergreen pea relative – not a close one – that needs the mild winters of the south to get beyond a shrub. Its flowers are bigger and bright yellow, coming in late spring among ferny young grey-green leaves. More amenable, but also more shrubby, is the hybrid *Sophora* Sun King ('Hilsop') with parents from Chile and New Zealand.

Judas trees and wattles

Yellow is one characteristic pea flower colour. The other is purply-pink. The redbud (*Cercis*) genus, represented in Europe by the Judas tree (*C. siliquastrum*), is probably the most familiar instance of the second kind – aside from sweet peas. The North American redbud (*C. canadensis*) flowers with the dogwood in the woods of the east from Pennsylvania southward, the only native tree with this flowering-cherry colour. Nursery selections are finding new virtues in redbuds of local species from different parts of the States, but the most distinctive up to now is *C.c.* 'Forest Pansy', a cultivar with dark maroon leaves beautifully bloomed as they emerge, small enough in stature for a small garden.

The hillsides of Tuscany in spring are bejewelled by wild Judas trees in flower. Even in England it will grow (as one has in my garden) into an enormous bush 40 feet across. Europe's best display must be the grove of 50-foot Judas trees in the Retiro gardens in Madrid. I remember seeing them being watered one sunny morning by a jet with a great plume like a fire hose. The big trees, arched with rosy blossom, were entirely covered with drops of

water; the sun caught a crystal in each of the million tiny flowers. The tree's peculiarity is to flower not only on young twigs but also even from the most calloused old black bark of the main trunk. There is also an albino version, white-flowering with pale green leaves. China has a similar redbud, and also another whose flowers hang in short streamers (*C. racemosa*). They are all worth having for their glossy green heart-shaped leaves, quite apart from their flowers. The legend that this was the tree from which Judas Iscariot hanged himself marches on. The French name for the tree, however, explains all. They call it *l'arbre de Judée* – the tree from Judaea.

All these trees adhere to the pea's own asymmetrical flower design. It is the mimosa group that rocks the boat. Their ring leaders are the acacias – the true acacias of the tropics. In the sub-tropics they have concentrated in the southern hemisphere. After the eucalyptus they are Australia's commonest trees, known in the vernacular as wattles (and the inspiration for the national sporting colours of green and yellow). Mimosa-type flowers as individuals (they come in heavy bunches) are tiny, radially designed, and symmetrical.

The silk tree (*Albizia julibrissin*) is the one hardy tree of this group that belongs in the northern hemisphere. It was first found in Persia (*julibrissin* is its Persian name) but its range runs all the way to China. Like the true acacias it has doubly divided leaves of the finest texture. Unlike most of them it is deciduous and its fluffy heads of flowers are a rather washy pink. The top of the low, wide-reaching tree is never without a few pink plumes from mid-summer to early autumn. A better form (*A.j.* f. *rosea*) has been introduced from Korea for the north; both hardier (zone 5) and stronger in flower colour, though up to now not happy in Britain's too-moderate climate. Hotter summers are needed to ripen the wood.

The California redbud (right) is very like the the Eastern form *Cercis canadensis*, blooming in early spring with the dogwood in the eastern States, especially from Pennsylvania to Virginia. A white-flowered form is also available from nurseries.

The silk tree (*Albizia julibrissin*) (above) may be more familiar as a young plant as a summer bedding scheme than as the 40-foot tree it can become in a hot climate. Masses of thin thread-like stamens make the flowers look like brushes.

The most promising to try at the moment is a sumptuous purple-leaved Japanese cultivar sold as *A.j.* 'Summer Chocolate'.

Of the 1,000 species of Australian *Acacia*, or wattle, only half a dozen are usually attempted in gardens outside the sub-tropics. Only one has ever made a substantial tree in England: the one the Piccadilly flower sellers call mimosa, now a common sight in London gardens – the silver wattle (*A. dealbata*).

For Californians and Floridians and gardeners on the Mediterranean the wattles to try (apart from the silver) are the Cootamundra, the black, and the Sydney. Cootamundra wattle (*A. baileyana*) from New South Wales is as hardy and adaptable as any, and its grey foliage and creamy-yellow flowers more interesting, perhaps, than traditional "mimosa". It really comes into its own, however, in *A.b.* 'Purpurea', whose leaves are shot with purple; an arresting combination with its flowers.

Black wattle (*A. mearnsii*) and blackwood wattle (*A. melanoxylon*) are very different, the first fluffy-foliaged and the second with phyllodes, or simple strap leaves, but united in being two of the most invasive weeds on earth. The reason to try either of them in a temperate garden would be their flowers. Blackwood is considered a valuable timber tree (as well as a weed) in Australia. Two other wattles worth trying for their foliage are oven's wattle (*A. pravissima*), with curious little triangular phyllodes giving its branches the appearance of grey-green chains (its flowers are bright yellow) and Queensland silver wattle (*A. podalyriifolia*), with paler leaves and darker flowers. They need shelter in Britain but would be happy in the south of France.

The purply-pink flowers of the Judas tree (left) are tiny, but densely massed. The young translucent seedpods, which clearly relate it to the peas, grow to about five inches and ripen purple, then brown.

The Rose Family

BY THEIR FLOWERS YE SHALL KNOW THEM – if you look carefully. They make a big family; so big it was once considered as a sub-class embracing one-third of all flowering plants with two cotyledons. A third of that third were classified as the Order of Rosales. But botany progresses: regrouping has got them down to a mere 6,900 species now. Of these only 3,000 belong to the rose family proper: the Rosaceae. We are almost in manageable numbers.

What matters more to us is their visibility. Most of our fruit trees and ornamental blossom trees belong to this one family. The cherries, crab-apples, hawthorns, blackthorns, pears, apples, mountain ashes, whitebeams, quinces, firethorns, shadblows, laurels, cotoneasters, medlars, almonds, peaches, apricots, and plums are all members – linked by their flowers. The next 34 pages are full of them, and this is only a selection of the best.

Their flowers? Their family likeness (there are always exceptions) lies in five petals and five sepals surrounding lots of stamens, all forming a sort of cup to create a display. Five petals can become 10, 20, or more in flowers bred or selected for doubleness. They can come single, or in spikes, or racemes, or bunches.

There are no great rosaceous forest trees; moderate size is in the genes. Hardiness is another feature of the family: there are few tropical members. There is no nonsense about acid soil either – to this family alkalinity is a positive virtue.

Their blood lines and ancestral relationships are another matter. DNA analysis is helping to elucidate them but many apparent species reproduce themselves perfectly by apomixis, the production of fertile seed without pollination, while there have been some surprising instances of cross-breeding between two very dissimilar trees: the shadblow or snowy mespilus (*Amelanchier*) and the mountain ash (*Sorbus*) for example. On the other hand the crab-apple (*Malus*) will not even graft onto the mountain ash or the pear (*Pyrus*), let alone hybridize with them. There can never be a cross between an apple and a pear.

There are no tree roses, or rose trees. The only way a rose can reach tree height is by climbing, using strong hooked thorns on long shoots as crampons. Some roses can go 50 feet or even higher. The rose perhaps best known for such mountaineering is *R. filipes* from western China – particularly the form called 'Kiftsgate' – but many selections from and crosses of several Himalayan ramblers can be as daring. Any of these will scramble out into the sunlight at the top of an old fruit tree and send down astonishing cascades of a hundred sweet-scented flowers in a spray. If there is an old tree in your garden you consider dull you could do much worse than give it such an adornment as a Christmas present.

But even a 50-foot rose is only a shrub with long shoots, as liable to lie on the ground as to ascend into the air.

'Wickwar' is the name of a rose bred in the village of that name in Gloucestershire in the 1960s. This specimen, planted in 1975, has reached 50 feet and run out of tree to climb. It flowers in early summer and has grey leaves and brown hips.

The fruit of the medlar epitomizes the rose's way of developing its fertilized flower into something birds or animals will want to eat and distribute. What was the little cup forming the base of the flower grows into a more or less fleshy coating around the seeds. In the medlar the end never closes and you can just see the ends of the seeds.

Cotoneasters are a wonderfully versatile group of shrubs not very different from hawthorns but with no lobes on their leaves and no thorns. Their masses of fruit are more important to gardeners than their flowers. The tallest, *C. frigidus*, can reach 20 feet; others are prostrate.

The kinship of cotoneaster with crab-apples is clear in this picture of *Malus x zumi* 'Golden Hornet'. 'Golden Hornet' keeps its fruit hanging well into winter, perhaps indicating that its winged and beaked customers prefer red.

Shadblows, Mespilus, and Cotoneaster

THE TREE KNOWN IN AMERICA (at least to old-timers) as shadbush, or shadblow, or serviceberry, in England as juneberry or snowy mespilus, is one of the most delicately beautiful of the rose's relations, and perhaps the one that gives a small garden best value. *Amelanchier* (here is a case where an international name is obviously a relief) is a tree – or sometimes a bush – to watch carefully, even when it is not putting on its main performances.

The genus on the whole is lightly built and all its parts are cast in a fine mould: thin, pale grey limbs and pointed buds, emerging leaves of pale pink and coppery-bronze, silver-haired (only a few

cherries have the same combination), and accompanying masses of star-shaped white flowers which froth it over lightly (though in warm weather, admittedly, for only a few days). By early summer there are bunches of berries like blackcurrants in shining green foliage. And in autumn, though not for very long, the leaves turn the softest red, orange, and brown: a glow rather than a flame.

North America has 16 *Amelanchier* (at the latest count), but there is often some confusion about which is which – not surprisingly as it occurs in every state except Arizona. When you say shadblow you mean the bush which blows – or blooms – in mid-spring, when the shad are running in the rivers … or so the story goes.

Amelanchier canadensis is the name most often used for the tallest, a tree often as much as 50 feet high. But if you follow the latest edition of W J Bean, which is much the safest course, you must call this *A. arborea*. The Allegheny serviceberry (*A. laevis*) is generally smaller, but has the prettiest unfolding leaves and sweet-tasting berries – as the birds know. And there is a hybrid between these two with pink-tinted flowers. Again this is often called *A. canadensis*, although its real name is either *A.* × *grandiflora* or *A. lamarckii*, according to your informant. The true *A. canadensis* grows as an upright shrub in wet ground: a wildling.

Strictly speaking (to complete the picture) snowy mespilus is none of these but the European *A. ovalis*, a medium-sized to large shrub, which is snowy in its downy new leaves as well as its flowers. When it comes to choosing, perhaps the best of all is a selection of *A.* × *grandiflora* named 'Ballerina'. I have been growing one for 15 years or so on a single stem – it is usually sold as a bush with several stems – and I can't think of a more delicate or graceful little tree (it is now 18 feet high). It has sweet fruit (quickly gobbled by birds) and wonderful autumn colour.

Names of *Amelanchier* are confused, but the tall shrub grown as *A. lamarckii* is one of the best for consistently glowing warm orange in autumn. *Amelanchier canadensis* is better in wet ground and puts up suckers.

Cotoneasters

Here among the rose tribe the shrub barrier is crossed and recrossed so many times that a distinction must be arbitrary. Most of the species of *Cotoneaster* are shrubs without qualification. Indeed, some have a marked fondness for lying flat on the ground. But up rears the excellent *C.* × *watereri*, a more than half-hearted attempt at an evergreen tree. In my garden it loses most of its dark green leaves by Christmas but keeps its incredible crop of red berries: almost a solid dome of scarlet 15 feet high and the same across. True it has 20 stems, but each one carries a tree's baton in its knapsack. For some reason the pheasants are in no hurry to gobble this crop; in early spring there is still a good third of it beaming in the sun. Perhaps this restaurant knows more about serving than seasoning.

The Allegheny serviceberry (*Amelanchier laevis*) (left) produces myriads of delicate white flowers with pink young leaves in late spring. A hardy small tree, it grows best on soil with no lime.

It may be only a fleeting flowering (below), lasting four or five days, but the eruption of a million delicate flowers on *Amelanchier canadensis* is one of the most precious moments of spring. The tender pinkish leaves quickly follow the flowers.

The Hawthorns

"WHENEVER I SMELL A MAY TREE I think of going to bed by daylight." This is William Morris, remembering his childhood on the edge of Epping Forest. Marcel Proust, thinking of Normandy, said the same thing (in a hundred times more words). Hawthorn blossom is the smell of early summer in the country.

To the layman there is scarcely a simpler or more consistent genus of tree than the hawthorn. It is a dense, compact, always wild-looking tree, never a big one, a great place for small birds to nest. It flowers in a sheet of haunting-scented blossom in late spring. Its blossom can be pink, is occasionally red, but (says the sentimentalist) ought to be white. It is fiercely armed with thorns, and covered with dark red berries as the year wears on.

In northern Europe it is the hedgerow tree, or rather it is the hedgerow. Its name means hedge thorn, and long before barbed wire it made it very clear where your property ended and mine began. It was "laid" by being cut half-through with a bill-hook and bent horizontally, to make a barrier that neither man nor beast could pass. Or often where no such fence was needed it grew to make an orchard-shape tree. The meadows (unpurged with selective weedkillers) shone with buttercups just at the season when the hawthorns gave them their dazzling frame.

But if it seems almost elemental to the layman, to the botanist it is a genus to dream of. A thousand species have at one time or another been identified in North America alone. How deceitful that apple-like blossom; those simple, shiny, dark green, tooth-edged leaves … that we should have been happy to think of them all as hawthorns, and never suspect …

The buds of May

The major differences that need concern a planter lie in the colour of the flower, the colour of the fruit, the size and shape of the leaves, and the length of time the fruit hangs on the tree. Many species have red or orange leaves in autumn, too.

The common hawthorn or "quick" of English hedges is *Crataegus monogyna*, whose name means it has one seed per berry (though the proper word is pome). It is not as simple as that, though. England, especially the Midlands, has another common native, formerly known as *C. oxyacantha* (its Greek name), now as *C. laevigata* (literally, the smooth one). Needless to say, the smooth thorn looks no smoother; both have, in spring, delightfully shiny lobed leaves. What is smoother about it is its relative lack of thorns. More diagnostically important: it has two or three seeds per pome.

Also needless to say, the two hybridize. At least their offspring are called *C. × media* – something simple at last. This cross has given us all sorts of ornamental forms: single- and double-flowered, white, pink, and red. But then so has *C. laevigata*.

The common or hedgerow hawthorn has grown to almost 50 feet – and the oldest known is 10 feet in girth. It usually has white flowers, but has them in such profusion that they hide the

The common English hawthorn, otherwise known as "may" or "quick", is northern Europe's commonest hedging material. It is hard to beat for beauty of flower, for its graceful fountaining habit, for scent, or for profusion of flower and fruit.

whole graceful, stooping tree. The date of its flowering has a strange fascination for people. It was a point of pride that it should flower by May Day, the first of the month, in time for the traditional celebrations of spring (and of course socialism). The poor tree has had a hard time keeping to schedule since 1751, when the calendar was altered by 12 days. More importantly, its young leaves are good to eat – known to old countrymen, indeed, as 'bread and cheese'. More importantly still, the whole plant is one of the most useful in herbal medicine. The list of its benefits to the heart and other organs is prodigious.

It is a variety of this thorn (*C. monogyna* 'Biflora') that has earned itself a legend by habitually bearing a precocious crop of flowers at Christmas. The legend goes that Joseph of Arimathea, who visited England to preach Christianity after the Crucifixion, was getting nowhere with a sceptical Somerset audience on Christmas morning when God made his staff (which he had stuck in the ground) burst into leaf and flower. Remarkably, its descendants still do a precocious show of leaf and flower, sometimes in mid-winter.

America has a far greater variety of hawthorns – enthusiasts will tell you they run into hundreds. The most widespread is the

The Washington thorn (far left) is easily recognized by its brilliant scarlet, shining, and translucent little berries, hanging on while its leaves turn a medley of colours in the red/purple range. It is quite a dainty little tree.

The fleshy hawthorn (left) (*Crataegus succulenta*) is a vigorous, thorny eastern American tree with profuse fruit almost the size of crab-apples, glossy leaves, and waxy blossoms that "look like porcelain".

The "may" (far left) of English hedges is strongly sweet-scented but only make its appearance late in the month, when countrymen, in the old phrase, "cast a clout", i.e. took their jackets off.

The American cockspur thorn (left) has the longest thorns (up to three inches) of its prickly genus. Its glossy unlobed leaves colour well in autumn. Its flowers are profuse and its fruit long-lasting.

Some of the most ornamental thorns, especially red-flowered ones, are varieties of *Crataegus laevigata*. This is *C.l.* 'Punicea'. 'Paul's Scarlet' (from the Victorian nurseryman William Paul) is probably the brightest and best known of all English thorns.

red haw (*C. mollis*), with cherry-size persistent fruit but vicious long thorns. Not a tree you would plant in a playground. If none is spectacular, many are handsome and have the advantage of autumn leaves in red or orange. The widespreading cockspur thorn (*C. crus-galli*) has the longest thorns – a good three inches long – and its hybrid *C. × lavallei* slightly less formidable armament but shiny persistent leaves and fruit that hangs on till spring. The combination is almost like holly in late autumn.

The green thorn (*C. viridis*), especially in its cultivar 'Winter King', can be covered with big juicy fruit all winter. *Crataegus succulenta* and *C. pubescens* f. *stipulacea* give heavy crops of large, edible fruits in shades of orange to yellow. My personal favourite (it has been beautiful in my garden for 30 years) is the small, dainty, and shiny-leaved Washington thorn (*C. phaenopyrum*). The Washington implicated is the District of Columbia, not on the Pacific coast – unless of course it is George W. himself. Its berries (sorry, pomes) are tiny but brilliantly translucent, and its leaves slowly turn mixtures of red, purple, and green right into winter.

I also grow two Chinese hawthorns of character, *C. pinnatifida* var. *major* with big leaves almost like an oak's, big bright fruit, and good autumn colour, and *C. laciniata*, aka *C. orientalis*, whose leaves are jaggedly lobed and grey with down and whose fruit is orange. It has taken 25 years to grow to 15 feet, an admirable tree for a small lawn.

Quinces and Medlars

FISH IS ABOUT THE ONLY THING we eat on a large scale today which is not specially bred for our consumption – perhaps the reason why we think of it as a luxury. The relationship of our wheat to the original grass, our heifers to the original wild cattle is pretty remote. As for the unselected, unimproved food which is all around us, we leave it for the birds.

In the rose family you can see every stage of selection and improvement – for every purpose. Selection has produced cabbage roses from sweet hedgerow briars, our big juicy apples from the wild fruit forests of central Asia. In the case of orchard apples it has concentrated on the fruit; in the case of crab-apples it has sought to perfect the flowers.

There remain other trees within the rose family, however, where the fruit has never come up to the standard of interest or flavour (or glamour) that is needed to make it a commercial proposition. It is almost eccentric to eat medlars (*Mespilus*) or to grow quinces (*Cydonia*) today; they are neither flesh nor fowl: that is, neither standard rations nor spectacular bearers of blossom. They come into the same category as mulberries – of garden plants that are more nostalgic than strictly useful – and make an alluring tree to have on the lawn.

The medlar (*M. germanica*), both in flower and leaf, quickly becomes a centre of interest. It grows short and often crooked with a broad spread of generous leaves, hairy underneath (and in autumn a lovely russet colour). In late spring – already leafy – it carries quantities of soft round-petalled pinkish-white flowers. Watching these become medlars you can see exactly how, as in roses, the seed receptacle behind the flower's facade of petals swells into ripeness – without even closing at the end in the medlar's case, but leaving ajar the end where the petals were. The medlar is like a big brown rosehip, perhaps two inches across. In the terms of a healthy Shakespearean exchange of insults, however, "You'll be rotten ere you'll be half ripe". The medlar is not ready for eating until it has been "bletted" by the frost: gone "sleepy", as they say of pears, with incipient rottenness. And then, truth to tell, it is no gastronomic experience. At best, on a frosty morning, it can be like a mouthful of rather sharp sorbet.

The truffle of the orchard
The quince (*Cydonia oblonga*) on the other hand is a tree to grow for the fruit. You can buy all the apples and pears you need, but the quince is the truffle of the orchard: the mysterious savour that makes an apple pie come alive as no apple can. Even to leave a quince in a basket of apples in the larder can be enough to impart the scent. In Portugal quince cheese (*membrillo*), in granular pinky-amber slabs, is a common confection and the inevitable

Petals cover the lawn (above) under a typically tilted and wide-spreading medlar tree beside the churchyard wall at Saling Hall.

The fruit of the medlar (right) show how a flower of the rose family metamorphoses into fruit. The flower base swells and the sepal tips grow into incipient leaves.

accompaniment to salty white ewe's-milk cheese. The quince ripens better there. Britain has historically been a shade too cool for it. The haunting lemony fragrance hangs about the tree, however, and fills the kitchen.

There is not very much to choose between quince and medlar as trees: their habits are similar, although medlars have the habit of leaning over at a tipsy angle. They would make a very peculiar avenue and a whole orchard of them would not appeal to a tidy gardener. The quince has leaves richly white-felted below; flowers a little showier perhaps; fruit, a shining pale yellow pear, much bigger and prettier. The slightly more tender Portuguese quince, *C.o.* 'Lusitanica' is the best for flowers and has softly downy fruit. It was identified by Philip Miller in his *Gardener's Dictionary* as long ago as 1768. *Cydonia oblonga* 'Vranja', from Serbia, is the most commonly grown in Britain. But any quince is a source of pleasure.

They should not be confused, incidentally, with the bushy red-flowered *Chaenomeles japonica*, often just called japonica, however superficially similar, and at a pinch its fruit may even be edible.

The five-petalled medlar flower (above) is white or faintly pink, an inch across. The quince flower is bigger.

The Chinese loquat (*Eriobotrya japonica*) (below left) has some of the biggest leaves of any evergreen tree hardy in Britain. It was found tender until recently, but now flowers and fruits. Edible, but not always delicious.

When given a warm summer and no frost at flowering (below) the quince tree can bear heavy crops. The fruit is densely downy when young. Turkey is the main country of production, although Portugal is the source of our best variety, *Cydonia oblonga* 'Lusitanica'.

Crab-Apples

IN THE MIDDLE AGES the European native crab-apple (*Malus sylvestris*) was important in the kitchen as the source (or a source) of verjus, a powerfully acid liquid that seems to have played the role of vinegar in salad recipes and recipes for preserving. In those days before bottles, when so much wine went bad in the barrel, it is a wonder they needed any other vinegar. But in Elizabethan cookery books there is still a great call for it. In addition, of course, crab-apples make refreshing jelly.

It was not until the end of the 18th century when the Siberian crab (*M. baccata*) was introduced into cultivation that exciting hybrids began to appear with far more telling flowers than the modest white blossom of the old hedgerow tree.

Native American and later Chinese species joined in the criss-cross of breeding so profitably that the pedigree of a modern hybrid would have more "begats" than the Book of Genesis in the King James Bible.

Today crab-apples come second only to flowering cherries for excellence and popularity among flowering trees. Indeed before

The Japanese flowering crab (above) has the simplest little single white flowers. It gains its extraordinarily beautiful effect from the backs of the petals, which are cherry-red before they open. This was one of the first flowering trees to be introduced from Japan in 1862.

Hupeh crab is yet another of E H Wilson's introductions (right). It has a long flowering period that makes it a useful pollinator for apple orchards. It is seen here in Oxfordshire growing in "crates" to protect the trunks and make sheep a viable alternative to the mower.

A big Siberian crab (*Malus baccata*) has been growing at our church gate in Essex since 1930. Local brides time their weddings to be photographed under the arch of frothing white made by its long boughs.

planting a flowering cherry it is very well worth examining the claims of the crabs. They have certain marked advantages that usually get overlooked.

In tough conditions of either soil or climate the crabs on the whole are more adaptable and certainly hardier than the cherries. For north American, and particularly Canadian, gardeners there is nothing to beat them. And under any conditions they are normally longer-lived. Like the cherries they rarely have good autumn colour but in compensation they often have very pretty fruit. What's more, in many cultivars the fruit stays on the tree long after the leaves have gone – even into late witner or early spring. A crab with this peculiarity soon becomes a centre of curiosity and astonishment. I know of one (*M.* × *zumi* 'Golden Hornet') that has nearly caused nasty accidents. It stands, dripping with yellow apples, near a sharp bend in the road. On a brilliant winter day you can't help looking at it as you go by …

Crabs often bear as many flowers as cherries, but few have the weight of petals; the flowers have never quite developed into the same sort of extravaganzas. Some think that with fewer petticoats they have more charm.

The best for flowers

Several of the species of crab-apples that have contributed most to the breed in the stud book are still very much in circulation in their own right. Perhaps the prettiest of them is the Japanese crab

(*M. floribunda*), which is one of the first to flower, a thrilling sight in mid-spring when its flowers open from bright red (the backs of the petals) to white (the front). The new leaves are still shining grass green and very small; the flowers waiting to open look like scarlet berries among the white stars of those that already have. All this happens at eye level: it is only a short tree and the branches tend to arch and hang. Most important, it does its stuff every year; a number of otherwise excellent crabs take alternate years off.

There is another Japanese species: Sargent's crab (*M. sargentii*), which is as beautiful in flower (white flowers with gold centres) and in its bright red fruit, but which is only a big bush eight or nine feet high and wide. One Japanese species, *M. tschonoskii*, was much in vogue a few years ago for street planting, a tallish narrow tree recommended for red, yellow, and orange autumn colour.

The selected, bigger-flowered forms of two native American species, themselves closely related, compete in the flowering cherry league with big, soft, double pink flowers. Bechtel's crab (*M. ioensis* 'Plena') is a form of the prairie crab – no relation of the prairie oyster – which sacrifices fruit for extra petals. The less temperamental tree of the two, *M. coronaria* 'Charlottae', has the great advantage of noticeably larger and toothed leaves, which colour well in autumn.

This is the fruit of the Siberian crab illustrated opposite. It germinates so readily that friends often find themselves hosting its offspring. They even make hedges of them.

Generous fruit

Of the many superb Chinese species – China has the richest natural supply – the cutleaf crab (*M. bhutanica*) is the most unusual on account of its leaves, which not only are deeply cut with lobes but also colour well in autumn to set off the relatively big red and yellow apples. *Malus transitoria* is almost like a miniature version, with clouds of tiny yellow fruit.

Malus halliana 'Parkmanii' has double pink flowers on red stalks; Rivers' crab (*M. spectabilis* 'Riversii') is fairly early with its profuse red-bud–pink-flower combination; *M.* 'Van Eseltine' is almost a columnar version of *M. floribunda* but with double, coarser flowers. The Hupeh or tea crab (*M. hupehensis*), which E H Wilson found, is a strange thing with branches which, when they flower late in the season, are like long tentacles of pink and white.

Different parts of Siberia have given us different improvements on the Siberian crab (*M. baccata*): *M.b.* var. *mandshurica* is the Manchurian candidate, a big early-flowering tree with fragrant white flowers, the hardiest of its race, profuse with its little red or yellow apples. The tree horrifically named *M. pumila* 'Niedzwetzkyana' from Kazakhstan has made an important contribution of red blood: it has a red pigment that shows in its new shoots, its clusters of purply-red flowers, and its plum-coloured fruit. Some of the reddest of the hybrids have this "blood" in their veins. *Malus × purpurea* 'Lemoinei' is a typical example with rich red flowers. The more recent *M. × moerlandsii* 'Profusion' is another and *M.* 'Red Tip' a third. *Malus* 'Echtermeyer' is a tree with the same colour scheme but a weeping habit. I confess to a quiet loathing of a dark red tone in almost anything except wine – but don't listen to me.

There is obvious confusion lying in wait in the two excellent hybrids called 'Red Jade' and 'Red Sentinel', both of which sound like fellow travellers but in fact have white flowers. 'Red Jade' is a small weeping tree; 'Red Sentinel' a bigger one. The great value of both lies in the bunches of red crabs bigger than cherries that hang on and on through the winter.

Of the trees with yellow fruit the 'Golden Hornet' I have mentioned is certainly one of the best, and 'Dorothea' (a product of the Arnold Arboretum) is said to be very good. 'Dorothea's flowers are pink and double.

More recent successes include two from France, with red buds opening to white flowers, 'Comtesse de Paris', excellent for yellow fruit, and 'Evereste', for persistent red. American candidates include 'Snowcloud', bred in Princeton, also in the red-bud–white-flower mould, and the smaller, equally upright-growing 'Adirondack', bred at the US National Arboretum. Red-buds-to-pink-flowers are found in 'Indian Magic' from Indiana and the early-flowering 'Princeton Cardinal', which has red leaves, too.

Even this trugful by no means exhausts the catalogue. 'John Downie' is possibly the best-known cultivar of the old English wild crab (*M. sylvestris*). For a crab it makes a big tree which is generous with substantial red and yellow fruit: several crabs make good jelly material but 'John Downie's fruit is actually good to eat off the tree. Do not grow these bountiful trees, though, if you have no use for the fruit. It can litter the ground for weeks, inches deep.

Breeding continues with the goal of healthy trees, showy flowers, and then maximum quantity of long-lasting fruit. This is a new French hybrid, *Malus* 'Comtesse de Paris'.

Not all crab-apples are big trees. *Malus toringo* subsp. *sargentii* (better known as *M. sargentii*) is a shrub that works well as a dwarf tree. Its disadvantage for small gardens: it only flowers every other year.

A complicated parentage of Chinese and Japanese crabs has produced hybrids with a red tint in both flowers and leaves. One of the most popular cultivars, with reddish-pink flowers, is *Malus × purpurea* 'Lemoinei'.

The Tidal Basin on the Potomac river in Washington DC, where thousands of Yoshino cherries (*Prunus* x *yedoensis*) create a blossom festival in about the first week of April every year. The trees (3,000 of them) were originally a gift from the Mayor of Tokyo in 1912, accepted by Mrs President Taft. It was renewed with a further 3,800 trees in 1965, while Washington reciprocated with flowering dogwoods. The exact date and duration of their flowering is keenly studied, usually starting in the last week of March and lasting 10–14 days.

The Flowering Cherries of Japan

THE FLOWERING CHERRY is not so much a tree as an event: a milestone in the year which even the most ungardening citizen recognizes. It is (as the Japanese have long known) a suitable object for a cult. Would thousands flock out from the city to see the crabs, the laburnums – even the magnolias – in flower? They do to see the flowering cherries.

The cherry alone manages to look both virginal and voluptuous. It is easy (and fashionable) to turn up your nose at it, to consider it tainted by the suburbs which have become its second home.

Yet nature offers no experience of profusion and delight so strong. To stand among the low branches in the fragility of millions of pale petals is in the range of natural experiences that are totally distinct and all-embracing, like standing under a waterfall, or flying through racing scraps of cloud.

It is also, to Westerners, a modern experience. Japanese cherry-worship is a thousand years old, but the garden cherries perfected in Japan were not seen in the West (or had not grown to full size) until the 20th century. In Elwes & Henry's seven volumes of 1907

they are not mentioned at all. Most of the work that has led to their popularization was done by one man: Captain Collingwood (known to all as "Cherry") Ingram.

The Japanese have firm notions about which cherries should be planted where. The elaborately double-flowered forms they keep for important sites, where they plant them singly. For the massed effects of hundreds of trees they use single-flowered species and they stick to one kind. The most famous cherry-viewing district near Kyoto is planted entirely with the hill cherry (*Prunus serrulata* var. *spontanea*). It flowers late enough (the second half of April) for the weather to be perfect when its broad crowns turn white against black pines and the pattern of dark and light is reflected in the Kamo river. A selection of the hill cherry, *P.s.* 'Autumn Glory', adds rich red autumn colour to the repertoire.

The spring cherry

The biggest (to 90 feet in Japan, much smaller in Britain) and longest-lived (to 1,000 years, they say) of these half-wild single cherries is the spring cherry (*P. × subhirtella*). But its little pale pink flowers are not considered up to garden standard. Its cultivated varieties are another matter. *P. × s.* 'Autumnalis' has the inestimable habit of flowering on and off all winter, from late autumn on, as the weather allows, most of the flowers clustering tightly around little wispy shoots from the trunk – ideal for cutting and taking indoors.

Some dismiss the (winter) spring cherry as a flimsy, twiggy tree. In London gardens, though, it is uniquely virtuous. One can forgive the flowers their weak colour and puny size when they are the only ones in the garden, especially as its summer leaves are dainty enough not to cast much shade and turn various colours just before the tree starts to flower again. To me the white-flowered form is preferable to the pale pink, although there is a strong pink double form called *P. × s.* 'Fukubana' that flowers in mid-spring. The weeping *P. × s.* 'Pendula' was the first Japanese cherry to reach the Western world in the middle of the 19th century.

One or two of Collingwood Ingram's Kent-bred hybrids are so early-flowering that they fall into the same useful category: the pink *P.* 'Okamé' and slightly later *P.* 'Kursar' are both first-rate small garden trees with pretty autumn colour.

The most popular cherry in Tokyo and Kyoto today is a relatively modern hybrid, the yoshino (*P. × yedoensis*), which flowers early and fills the suburbs with the scent of almonds. Its advantage for a shifting population is that it grows and flowers quicker than the others. The flowers come in tens of thousands, white increasingly stained with pink, on branches that sometimes droop at the ends as though under their weight. Nurseries often offer selected droopy forms.

Noblest of the common species is the tall great mountain cherry. Picture it on the middle slopes of Mount Fujiyama. It is known to the West by the name of Professor Sargent of the Arnold Arboretum. Sargent's cherry (*P. sargentii*) candies over with bright pink flowers before the leaves, but the leaves catch up and mingle their ruby-brown with the pink. On a tree 40 or 50 feet high the effect is pretty powerful. Cherries as a whole are not great autumn-colour trees. Sargent's cherry is an exception: it is particularly

Kyoto's favourite flowering cherry (left) is the yoshino (*Prunus* x *yedoensis*). Its faintly pink flowers come early and scent the spring air with almonds. Though it grows eagerly to start with it never makes a big tree.

Autumn colour (above) is not normally a feature of Japanese cherries. A number, the Mount Fuji cherry among them, have yellow autumn leaves. The best for red leaves are the big-scale *Prunus sargentii* and this uncommon one, the pink-flowered *P. verecunda* 'Autumn Glory'.

conspicuous for turning orange and red in early autumn while most trees are still green. Its long oval leaves turn their points down and hang rather sadly as summer ends, but when they catch fire the display is all the better for seeing their whole length.

Village cherries

The Japanese use the term *sato-sakura*, which means village cherries, for their cultivated productions. There are 40 or 50 of them in cultivation in Europe and America. When they first arrived they were given Latin names relating them all to the species *P. serrulata*, but the tendency today is to use their Japanese names alone without trying to trace the intricacies of their parentage.

In Britain the most popular of them all, by far, is the variety called *P.* 'Kanzan' (or often, but wrongly, 'Kwanzan'). Its vigour, its freedom with its many-petalled flowers, and above all their piercing pink colour combine to make a fatal appeal. Pink is tricky. There are agreeable clear pinks, salmon-pinks, tawny-pinks, and rosy-pinks, but as soon as pink is stained with blue it starts to shriek. 'Kanzan' is slightly calmed by a white background, but its stridency, a solid mass like a lurid thundercloud, in combination with red brick walls has alone been enough to give the cherries a bad name. As a little tree in the nursery 'Kanzan' looks well adapted for a narrow space. Its branches climb steeply in a narrow V-shape. At 10 or 15 years, however, the narrowness is a thing of the past: 25 feet or more is a normal spread.

Several other cherries have the same vase-shape, at least in youth. *Prunus* 'Ukon' is a very pretty one with white flowers just touched with enough lime-green to give a sulphury effect. *Prunus* 'Hokusai' is another, eventually spreading wide, with hundredweights of pale pink flowers slowly being overtaken by pinky-brown emerging leaves. The only flowering cherry that really fits permanently into a narrow space is *P.* 'Amanogawa', which is built like a stripling Lombardy poplar: a shape which (soldier-like as it is) makes a

The biggest (left) and oldest "waterfall" cherry (*Prunus pendula* 'Pendula Rosea' or 'Takizakura') grows at Mizura in Japan and spans 80 feet with its boughs. This group stooping over a pool are at La Torrechia near Rome.

The Mount Fuji cherry (*Prunus* 'Shirotae') (right) is single flowered when young, then double, with a far-reaching almond scent. Its bright pale green, fringed and pointed leaves emerge just after the flowers in mid-season.

The piercing pink and frilly flowers (below) of *Prunus* 'Kanzan' made it the most popular of the Japanese cherries in the Western world. It is a vigorous, wide-stretching tree, growing quickly to more than 20 feet.

laughable combination with petticoat-pink flowers. 'Amanogawa' is a mistake in the open lawn. At the back of a bed or among shrubs where its shape is not overstressed it could be a useful tree, perhaps, but I got tired of them and threw them out.

What the flowering cherries really want to do, or look as though they want to do, is to reach long branches low across the lawn. The trees that do this offer you a whole universe of coloured petals to swim in at head height – even to wade in at waist height.

Prunus 'Shirofugen' is a late spring-flowering, long-lasting one, pink in bud, and maturing white, which does this. *Prunus* 'Shirotae' is an earlier, scented, pure white one, which makes wonderful use of the bright green of its new leaves to show off its hanging clusters of flowers. The leaves remain a feature of 'Shirotae': their edges are fringed with deep teeth and they have long points. *Prunus* 'Shimidsu' is a third, with pink buds opening late to wide white flowers on stalks six inches long. Then there are cherries grown for the far-carrying honey scent of their flowers: P. 'Jo-nioi' and (later) P. 'Taki-nioi' are two spreading white-flowered trees I love.

The great white cherry
The best of all spreading cherries, and all white cherries, however, is P. 'Taihaku' – a tree with a fascinating history. In 1923 the owner of a garden in Sussex showed Captain Collingwood Ingram a tree without a name. It had the biggest flowers he had ever seen on a cherry: simple five-petalled flowers, shining white with gold stamens, displayed among huge copper-red young leaves. He promptly took grafts and put it into circulation, knowing nothing of its name or where it originally came from. On his next visit to Japan Ingram was shown an 18th-century book of flower-paintings. In one of the paintings he recognized his new white cherry. But, he was told, it had been lost to cultivation since the painting was

made. There was no 'Taihaku' in Japan; the Japanese certainly didn't believe it could exist in England.

How it came to that Sussex garden remains a mystery. All the "great white" cherries in cultivation today, however, are its offspring. Ingram himself recommended planting it by still water where its roots can drink deep and its beauty be reflected.

The Japanese, of course, have by no means given up producing new *sato-sakura*. Several have been introduced by nurseries since the 1960s, and some given names more fitting for a begonia. Do P. 'Chocolate Ice' and P. 'Pink Parasol' even start to conjure up the beauty of a flowering cherry?

FORM AND FUNCTION

Japanese cherry culture has been concerned almost as much with the size and shape of trees as with the form, colour, timing, scent, and durability of their flowers. These are five characteristically distinct varieties.

P. 'Amanogawa'
It must have seemed a good idea to smother a fastigiate tree with large pink blossoms. 'Amanogawa' has a momentary impact but is a poor dull tree for 50 weeks a year. Not recommended.

P. 'Kanzan'
Hugely popular for its crowd-stopping profusion of pink in late spring; a hard colour to mix, soon covering the ground with faded flowers while the tree becomes a heavy inverted cone of big leaves.

P. 'Shirotae'
A mature Mount Fuji cherry is a spreading table top, best seen isolated, preferably against a dark background. Its leaves can turn a bright clear yellow. One of the best where there is the space.

P. 'Ukon'
The name means "yellowish". The only flowering cherry to give a hint of yellow, invaluable in avoiding the usual pink/yellow clashes of springtime. A 30-foot tree, excellent with the "blue" *Rhododendron augustinii*.

P. 'Shimidsu'
A small rounded drooping tree with pink buds opening to blush-white frilly hanging flowers late in the season, as the leaves expand. Some orange autumn colour. The name means "Moon hanging low by a pine".

The Flowering Cherries

THE PRUNUS BRANCH of the rose family has such a strangely various bunch of components that it used to be divided into a number of different genera. The cherries then were *Cerasus*; the almonds and peaches *Amygdalus*; the plums *Prunus*; the apricots *Armeniaca*; the bird cherries *Padus,* and the laurels, or cherry laurels, *Laurocerasus*. The botanical argument, though, is that they all have a single female organ in a five-petalled flower, and consequently a single-stoned fruit.

Japan has no monopoly of the flowering cherries. In fact Captain Collingwood Ingram, enumerating the species, counts 13 for Japan (not counting, of course, the countless varieties), 23 for China, six for North America, and five for Europe. Europe's best have long been known and used both for ornament and for their fruit.

For both purposes the leading species is the gean, mazzard, or simply wild cherry (*Prunus avium*), which flourishes over the whole European continent, in Britain, and as far east as the Caucasus. Its small cherries are sweet or relatively sweet in the wild; in cultivated varieties of course they are delicious. Seduced by the Japanese varieties, however, we have forgotten how beautiful the gean can be. There is more of it growing wild in the woods than you realize until the moment in spring when, with new leaves colouring the woodland canopy, random patches of white pinpoint the native cherry. Its double-flowered variety (*P.a.* 'Plena') not only rivals the sato-sakura in its billowing white flowers but is also hardier. The biggest gean, moreover, is of forest-tree size, almost 15 feet in girth, and specimens 80 or 90 feet high are not unknown. This is the tree, the merisier, whose wood makes so much splendid French furniture and whose fruit makes the potent Kirsch of Switzerland, the Black Forest, and the Vosges.

Of still older use as a garden ornament is the sour cherry (*P. cerasus*), the parent of the dark Morellos from which the Swiss make their marvellous jam. The white double-flowered version of this (*P.c.* 'Rhexii') has been known to gardeners since Elizabethan times at least.

Flowers in spikes

The bird cherry (*P. padus*), which grows all the way from Britain to Japan, is different in having its flowers in spikes; much smaller individually, but almost as effective in being crowded together. Of this too there are selected versions: *P.p.* 'Watereri' with extra-long flower spikes, *P.p.* 'Colorata' with pink ones and leaves darkened with pigment, and *P.p.* var. *commutata*, an oriental form which produces leaves exceptionally early.

The wild cherry of Europe, the mazzard or gean, is a tall woodland tree greatly prized for its timber and very beautiful in early spring. Its double-flowered form is among the best white mid-season cherries for the garden.

North America's native cherries have produced no spectacular ornamental trees; the best of them, the black or rum cherry (*P. serotina*), is the equivalent of the bird cherry. It is very common in the woods of the east, extremely valuable for timber, but neither its short spikes of white flowers nor its little red cherries attract much attention. Black cherry occurs right down into South America in the form of *P.s.* subsp. *capuli*, tall, with sweet and showy fruit. The choke cherry (*P. virginiana*) is another "bird" type with flowers in spikes. The wild red cherry (*P. pensylvanica*) has clustered flowers and fruit like the gean.

E H Wilson brought back a number of cherries from China. The most important in cultivation have been *P. hirtipes*, which he found in Hupeh, whose variety 'Semi-Plena' is singularly weather-proof; *P. serrula* from western China, whose flowers are not up to much but whose bark is in the birch class, glistening squirrel-brown; and *P. campanulata*, which he picked up in Taiwan as a contribution to the gardens of the south. It is one of the reddest-flowered of cherries, but for a warm climate only. A rival to *P. serrula* for irresistible bark was collected in Nepal by the man most famous for his new birches, Tony Schilling. Indeed he catalogued it as a new birch with exciting glossy blood-red bark, until he looked closer and found it was a cherry, *P. himalaica*. Another new find,

from central Asia, *P. sogdiana*, is a generous little tree with sweet fruit classified, confusingly, as cherry-plums.

It was not the great Wilson but an earlier traveller, the Russian Richard Maack, who brought *P. maackii* back from Manchuria to the Arnold Arboretum. It remains rare and is bashful with its little flowers, but rapidly makes a stout tree with bark of a glorious polished golden brown. There is one like a barrel of honey at Wakehurst Place, Kew's country garden in Sussex.

Nurseries meanwhile have been busy with hybrids and selections; far too many to catalogue. Hillier Nurseries described their *P. × hillieri* 'Spire' as "possibly the best small street tree raised this century … flowers soft pink; leaves with rich autumnal tints". It is certainly a usefully narrow shape, but the competition is stiff.

Spare the saw

Given a good start the cherry is such an easy tree to grow in most soils (it prefers a little lime and good drainage) that its performance is more or less assured. Do not expect it necessarily to outlive you, though: in my experience the highly bred Sato cherries rarely thrive for more than 30 years or so. There is one thing they don't like, and that is to be hacked about; they tend to go on producing gum from a bad cut until they die. Their motto is "Don't Prunus."

Bird cherries make their little white flowers conspicuous among their early-appearing leaves by carrying them in spikes like the butterfly-bush buddleia. The American choke and rum cherries do the same.

Cherry bark is easy to recognize by its scattered horizontal scars, or lenticels. A few cherries have satiny, birch-like bark, emphasizing the stripes. One of the most beautiful and striking is the Chinese *Prunus serrula*.

Prunus hirtipes is another of Wilson's Chinese finds. It flowers with soft single flowers in late winter; a fine tree that seems to have been overlooked in the richness of Oriental introductions.

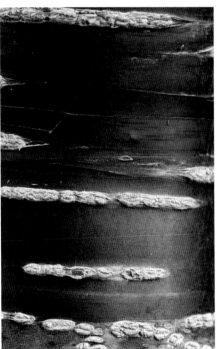

The dusky purple trunk (above) of *Prunus himalaica* shows that cherries can rival birches in the variety of their bark. *Prunus maackii* is another. It grows rapidly with bark the colour of dark honey, but will never be famous for its flowers.

Hillier Nurseries (left) crossed the shrubby 'Fuji' cherry (*Prunus incisa*) with the big-scale *P. sargentii* to create 'Spire', intended for the street-tree market, where a good tight consistent shape and good autumn colour are as important as flowers.

Peaches, Plums, and Cherry Laurels

THE FRIVOLITY OF THE FLOWERING CHERRY has been contagious. Peach, almond, apricot, and plum have all caught it and produced more or less fruitless varieties with a short season of stunning display.

It comes most easily to the almond (*Prunus dulcis*). Even in Mediterranean countries where the almond is grown seriously for the kernels of its fruit, orchards are a tourist attraction. In Sicily, for instance, they are frothing pink in a countryside not given to such femininities – and as early as late winter. The fruit is not up to much in northern parts, but the trees are hardy and flower just as well. *Prunus dulcis* 'Roseoplena' is the almond's nearest approach to a cherry: pink, many-petalled, and one of the first trees in blossom.

An almond is a peach with a lean fruit: or a peach is an almond with a fat one. In any case there is little difference besides the sweet flesh. They are close enough kin to breed together. The most ornamental upshot so far was conceived in Australia at the turn of the 20th century and given the name *P. × amygdalo-persica* 'Pollardii'. Pollard was the nurseryman; *amygdalus* was in those days the specific name of the almonds (*dulcis* is today); and *persica* of the peaches. (From *persica* comes the Italian *pesca*, the French *pêche*, and the English peach – the Persian fruit.) The fruits of this hybrid are like hard green peaches; the main attraction is its big pink flowers – bigger and pinker than either of its parents'. Again, a first-class tree for early blossom.

The peaches flower just after the almonds. A well-planned (though admittedly large) group of trees could use a flowering peach to link the blossom time of the almond to that of the first cherries so that from early spring (or sooner in the south) through to early summer there would always be a tree in flower. Of the peaches selected and sold for their flowers the most telling are perhaps the double-flowered red or rose-pink cultivars. The best-known is *P. persica* 'Klara Mayer'.

Peaches, however, are not the rugged self-supporting individuals that cherries and even almonds are. They have a range of problems, from a curling and crumpling ailment of their handsome long leaves to relatively early decline and death. Growing peaches (or even better their smooth-skinned sisters nectarines) against a wall for fruit is another matter.

Improvements on nature

The fruit-tree apricot from China (*P. armeniaca*) is nothing much to look at, though it is hardy and early in bloom. For its fruit, of course, it is eminently worth growing anywhere warm enough to ripen it. The Japanese apricot (*P. mume*) is the one for flowers. They appear early in spring, often with the snowdrops, and bring warmth to the garden at a chilly time. *Prunus mume* is a shade less hardy than the almond. Double white, double pink, and pendulous

Almond orchards (above) alternate with vineyards and orange groves in the San Joaquin valley in central California.

Sweet almond (*Prunus dulcis*) (left), in Valencia, on the Mediterranean coast in eastern Spain. As in the best olive groves, the soil is kept tilled to minimize competition from grass and weeds in a dry climate.

forms of both are to be had: the deep coral-pink *P.m.* 'Beni-chidori' as lovely as any. Their "floral ardour", as Collingwood Ingram warmly puts it, depends on the sunshine of the summer before.

Most of the domestic plum trees are thought to be forms of an ancient hybrid between the sloe or blackthorn (*P. spinosa*) and the myrobalan or cherry plum *P. cerasifera*, which originally came from eastern Europe and western Asia. Neither is a very impressive tree in its own right – the blackthorn more often a suckering bush with wicked little thorns. Avoid them: they leave an ache behind for days when they puncture you. The blackthorn's flowering, tiny though the white flowers are, is such a feature of hedgerows in mid-spring that the spell of cold weather which often occurs in northern Europe at that time is known as the blackthorn winter. Its fruit, a profusion of bloomy blue berries, are the sloes of sloe gin. Don't be tempted to pick them until the bushes are bare in late autumn; they need time to ripen. Then all you need is sugar – and gin.

I was surprised to discover that there is a fancy purple-leaved variety even of the humble blackthorn. The purple version of the cherry plum is all too well known: the grossly overplanted, purple-leaved cliché of suburban streets. The culprit for this tree was M. Pissard, the French gardener of the Shah of Persia 100 years ago – whence its name of *P.c.* 'Pissardii'. There are other, slightly different cultivars, mostly of dreary hue, but a seedling in my garden caught my eye with its fresh leaves of a rather fetching pinky-brown. Encouraged (it grows beside a silvery willow) it has produced huge crops of little red plums with the perfect acidity for making into jam.

A double-flowered almond (*Prunus dulcis* 'Roseoplena') (above) is the first tree in the garden in spring to make you think of flowering cherries.

In the ruby-pink apricot, *Prunus mume* 'Beni-chidori' (below), ancient Japanese gardeners produced a rival to the almond for late-winter flowering. It is still rare in Britain.

An Australian nurseryman, Barry Pollard, managed to cross an almond and a peach to produce the super-early flowering *Prunus* x *amygdalopersica* 'Pollardii' (above).

Prunus armeniaca 'Rouge de Roussillon' (below) is the name of a variety of red-fleshed apricot grown in the Midi and made into a splendid sweet-scented jam. It is also the name of a breed of sheep.

It is a gradual progression across the genus that brings us from the scarcely edible blackthorn via the cherry plum to the super-succulent plum of desserts and tarts. *Prunus domestica* is the label for a group of crosses between (maybe) *P. cerasifera* and *P. spinosa*, the least edible, with input from ancient Middle Eastern orchards.

In order of sweetness and succulence comes first the highly astringent bullace (*P. insititia*), a small tree with blooming green round fruit that needs a lot of sugar. Second comes the damson (oval fruit, purple in colour, still pretty sharp – especially the skin), reputedly from Damascus. Third the originally Armenian gage, in French the Reine Claude, introduced to France under François 1er (Claude was his queen) and to England by Sir Thomas Gage. You must taste a gage before dismissing it as a small plum. Then you can argue whether the 'Cambridge Gage', the x gage, or the y gage is more engaging. The annual gage problem for the gardener is resisting the temptation to pick them too early (they need a good

long hang) but getting to them before the wasps. It is only a short journey (gastronomically speaking) from gage to mirabelle (*P. domestica* var. *syriaca*), the little yellow plum of Lorraine. Mirabelle is used principally for jam, tarts, and of course the best of *eau de vie*. But ripe from the tree it is ambrosial. Alas two trees we grew in France died of old age (they were not very old).

Bullfinches sadly put an end to our plum growing. The spreading old 'Victoria' tree was regularly stripped of buds right to its branch ends. We used to get one plum per lengthy branch and eventually cut down the tree – since when we have scarcely seen a single bullfinch.

The laurels
It is quite hard to make the mental jump from the deciduous trees of *Prunus* to the big shiny-leaved evergreen shrubs which have borrowed from the bay tree the name of laurels. What could have

The cherry plum (*Prunus cerasifera*) (right) is the earliest plum in flower, with sharp fruit than makes excellent jam. This is a seedling from my garden, possibly a cross with the purple-leaved *P.c.* 'Pissardii', in any case a fine jam tree with bronze leaves all summer.

The Chinese *Photinia serrulata* (below) looks more like a cherry laurel than its nearer relation the hawthorn. Its great value to gardeners is its almost year-round succession of new shoots bearing bright red young leaves. This is *P.s.* Curly Fantasy.

Portugal laurel (*Prunus lusitanica*) will grow into a substantial tree but lends itself well to formal clipping, here in a splendid combination with box. The drawback to topiary with large-leaved trees is the risk of cutting leaves in half, when they may go brown.

less in common with a flowering cherry? Yet *Prunus* species they are.

The cherry laurel is the common one that flourishes in the dankest shade. It has a dozen variants: different leaf shapes, colours, and habits, including a tall one with bold curling leaves, *P. laurocerasus* 'Camelliifolia'. *P.l.* 'Magnoliafolia' has bigger leaves, *P.l.* 'Angustifolia' narrow ones, *P.l.* 'Rotundifolia' round ones, *P.l.* 'Variegata' variegated ones, and *P.l.* 'Zabeliana' throws its branches out at 45 degrees – altogether a most amenable amenity plant.

The Portugal laurel (*P. lusitanica*) is much more of a tree: smart, round topped, and up to 30 or 40 feet, glossy, and a good British racing-green all the year round. Game birds enjoy the little cherries. Forms with broader leaves (*P.l.* subsp. *azorica*), narrower leaves (*P.l.* 'Angustifolia'), and variegated leaves (*P.l.* 'Variegata') can be

found. The capabilities of these trees as the evergreen foundations of gardens on poor and chalky soil have scarcely been explored. They shade with perfect manners into the background or step forward as specimens with equal aplomb. They are not at all averse to clipping and can even, within 40 or 50 years, manage magnificence.

The Chinese *Photinia serratifolia* is more of a performer. Recently, indeed, it has had a vogue as a municipal shrub of choice. To be a carpark plant is the kiss of death, you would think, but this hardy, uncomplaining shrub/tree earns its popularity by constantly producing new shoots of brilliant orange-red leaves, which slowly fade to gleaming green. The effect is of a tree never-endingly in flower. Another Japanese species, *P. glabra*, much clipped in Japanese gardens, has contributed to such cultivars as *P.* × *fraseri* 'Red Robin' – the goal being the maximum of red shoots combined with hardiness. However early the new shoots appear it is rare to see them frost-bitten. The biggest reported – 50 feet – must be quite a sight.

Pear Trees

PEARS, WHITEBEAMS, MOUNTAIN ASHES – they seem strange bedfellows. In the first edition of this book I found myself explaining why botany insisted on lumping them together when common sense said they were different things. Now they are officially different – but still not much clearer. In some ways they remain a close-knit bunch: you can often graft one on the other, for example. But some will interbreed and others won't. In this family the trick of apomixis – a sort of vegetable virgin birth – is relatively common. (It is also very useful; seed of apomictic species always comes true; no need to graft to maintain a clone). On the other hand the usual bedroom inhibitions that keep species distinct seem weaker among this family than among most. There are hybrids between what botany considers different genera: *Sorbus*, for example, with *Cotoneaster* or *Crataegus*. When it comes to *Sorbus*, in fact, botanists can have such fun that they can hardly be expected to reach hasty conclusions.

Age and distinction

What are the real differences, for a start, between apples and pears? Flavour is obviously the most important. The pear tends to have a fleshy stalk, not joining the fruit in a dip but on a bump. Its flesh is granular in texture, gritty when unripe and once ripe, soon rotten.

Pears are wild in Europe and Asia; not in the New World. The common pear (*Pyrus communis*) from which all the orchard varieties are derived occurs frequently in the south of Europe and is not uncommon in the north. It is one of the longest-lived of fruit trees and eventually reaches a remarkable size: there are records of pear trees 16 feet in girth, and 60 feet is not an exceptional height. Domestic varieties can have immensely long histories, too. When I finally identified as 'Autumn Bergamot' a huge pear tree in the garden whose fruit breaks the rules by being apple-shaped I was told it was introduced from Assyria by Julius Caesar (or "has been cultivated at least since the 14th century"). Take your pick.

Old common pears are densely twiggy and often thorny: black and emphatic in winter – which makes their regular early spring covering of delicate white blossom all the more effective. Few (if any) are planted on purpose; where they happen to come up they are nonetheless very much enjoyed. Their wood is the "fruit-wood" of French provincial furniture; the light brown that glows in old armoires and dressers.

Nurseries have investigated dozens of exotic pears for their potential as useful street trees. For streets they need all the amenities except large fruit – something that seems to be overlooked in central Europe. I have walked along roads in Germany that were deep in squishy pears and attendant wasps. Chinese wild pears in particular have a following as ornamental

Of all garden trees not bred or selected for beauty of blossom, the pear gives the best display. Old pear trees are like clouds of white and gentle green in early spring. This is a typical old domestic pear, a cultivar, probably nameless, of *Pyrus communis*.

The tight-packed, pure white flowers of the weeping silver-leaf pear come out at the same time as its silvery leaves.

Pears are easy to grow but crucially difficult to pick at the right moment for perfect ripeness.

The weeping willow-leaf pear (left) (*Pyrus salicifolia*) at Knightshayes in Devon. This little weeping tree with narrow silvery leaves is the most decorative of the pears. Its habit is congested (it needs pruning) but generally pendulous. The form *Prunus salicifolia* 'Pendula' is more common in garden cultivation.

What is known as "fruitwood" (right) in furniture is generally pear wood, though it may be cherry or even apple. It was a favourite of 18th-century cabinet-makers, with a tight grain ideal for carving and a range of subtle colours. In this fancy piano of 1880 fruitwood is contrasted with mahogany and satinwood.

trees. One was the work of the Plant Introduction Station at Glenn Dale, Maryland. From their stock of *P. calleryana* they chose a seedling (known as *P.c.* 'Bradford') which was then strongly promoted as a near-perfect medium-sized shade tree. I planted one in 1976. It is certainly attractive, very early in leaf, a dense pyramid with plenty of flower and good autumn colour. Unfortunately its branch forks split and snap in any strong wind; an expensive experience for the parks department. It has now been superseded by a narrower-growing but excellent cultivar which has become almost a cliché: *P.c.* 'Chanticleer'. Another is the Ussurian pear (*P. ussuriensis*), which has a more northerly range. The near-evergreen (but thorny) *P. pashia* also has potential. None of these gives edible fruit.

The non-fruiting pears one would normally consider planting for their beauty in the garden are those with white or silvery young leaves – the tall *P. nivalis* from the Mediterranean and above all the little willow-leaved pear (*P. salicifolia*) from the Caucasus. For the particular combination of summer-long silveriness and a weeping habit this has no competitors. If anyone is looking for an instant feature, a theme to form the basis for a new garden, or a corner of an old one, they will be tempted by this. Its only fault (and it can be cured by pruning) is a tendency to overbranch and make a tousled head. I made a silver plantation using it beside a weeping silver lime in a thicket of silver sea-buckthorn. Some common juniper (also in part silver) is dotted about. Did I overdo it? Very likely.

A ticklish subject

As for the thousand or so varieties of pear bred for succulent eating at different stages of the season, I must refer you to a specialist book. Many are ancient, many regional, most of them rare. Of many historic catalogues perhaps the most famous is that of Jean de la Quintinye, gardener to Louis XIV at Versailles in the 1690s. He lists 500 varieties, 50 "good", 44 "indifferent", and 66 "bad". 'Winter Bon Chrétien' heads the list as the most historic, winter being the season when it was enjoyed.

The principal point of pears, it seems, was to have fruit that would store and be at its best in winter – although the precise judgment of which pear to store for how long, until its 24 hours of perfection, has always been, to say the least, a ticklish subject. Quintinye judged 'Beurré Rouge' the best pear of his time. Since then landmarks of pear breeding have been 'Williams' Bon Chrétien' (created in Berkshire in 1770 and introduced to America under the name of 'Bartlett'), Doyenné du Comice (from Angers in 1849), and most important (at least commercially) 'Conference' (bred by Rivers Nursery in Hertfordshire in 1894). 'Conference' remains the leading commercial pear.

Pears are near the north of their range in northern Europe; gardeners looking for perfection have always favoured growing them on walls for protection.

The crinkle-crankle or serpentine wall (middle right) is a speciality of Suffolk and the east of England. It supposedly traps warmth for ripening fruit and needs only to be one-brick thick for stability.

Pyrus nivalis (left) from eastern Europe is called the snow pear for the white down that covers its young leaves. It makes a tall tree, up to 40 feet, interestingly pale all summer long, but only giving hard little green fruit.

Pyrus pashia (right) from central Asia has tight flower clusters with rounded, overlapping petals and red stamens. Leaves on vigorous shoots are often three-lobed. It makes a pretty small tree, almost evergreen in mid-winter.

Whitebeams

WHAT WE NOW (or at least till recently) know as *Sorbus* was previously known as *Pyrus*. When it was renamed *Sorbus* it still left many questions: it still muddled up what a layman would call quite different trees. Different in the most obvious way, too: the rowans or mountain ashes have pinnate leaves with many leaflets; the whitebeams have a bigger, undivided single leaf. Happily DNA evidence bears out what we already thought: they are cousins, and not even the closest cousins. The pinnate-leaved rowans continue to be *Sorbus*; the simple-leaved whitebeams now officially have their own genus, *Aria* – easy to remember, at least, because *Sorbus aria* was the whitebeam's old name. The position is still not simple: the two groups hybridize easily, and there is plenty of learned wrangling about the parenthood of hybrids. But we're making progress. Meanwhile if we call them rowans (or mountain ashes) and whitebeams we can't go far wrong.

It is almost a relief, after the frenzied flowering of so many of the rose family, to come to a group of trees which arrange their nuptials more discreetly. Neither whitebeam nor mountain ash is worth planting for the sake of its flowers. They are white or cream, and although there are plenty of them in broad heads they come with the leaves, which not only tend to hide them but are also in themselves more eye-catching. The leaves are the strong point of both, different though they are. And after the leaves, in many cases, the heavy bunches of little coloured berries.

The whitebeams have single, simple leaves; usually just toothed but sometimes lobed, and very often white-felt-backed. The white side of the leaf gives the tree its name. The mountain ashes have compound leaves; anything from three to 30-odd leaflets on a stalk. In the days of more naive botany this was enough to prove them ashes. None of the trees in either class grows to enormous size. Seventy-five feet would be a record. There is another difference, too, that gardeners should bear in mind: whitebeams on the whole tolerate poor and dry soil conditions whereas rowans, with shallow roots, like humus and a good supply of moisture.

The mountain ashes outnumber the whitebeams but remain pretty consistent wherever they grow. Of the whitebeams there are four distinct strains: a splendid one of several species with huge leaves in the Himalayas; the locally variable whitebeam (again several species, but very similar ones) of northern Europe; a couple of trees with leaves deeply enough lobed to remind one of the maple, also in Europe; and a superb Japanese tree with hornbeam-like leaves, which is the best of all for flowers and fruit. America has no native whitebeams.

Beaming white

It is the way in which the leaves appear that first draws your attention to the whitebeam in the spring. They emerge from the bud rather as the petals do in the tulip, forming a chalice-shape with their points up and displaying their undersides. At this stage the underside (in most cases indeed the whole leaf) is covered with a silvery silk that reflects the light like metal. In the wild the tree is characteristic of chalk and limestone hills; it winks from copses of pale beech or the bronzy black of yew. On the North Downs in Kent, where as a boy I first paid attention to trees, it was the first tree I paid attention to. It positively flashed like a signalling light from the downland copses above. When I went to investigate I found the reflective material was not shiny but soft and felty.

By mid-summer the brilliance of the leaves has gone. In many years insects get to work on them and leave mere fretwork. But if all is well they have a second stint in mid-autumn in shades of russet and amber, setting off heavy bunches of red berries.

This is the wild one of Europe, old *Sorbus aria*, now *Aria nivea*. Cultivation has polished up its performance without greatly

In early summer (left) it is hard to resist sitting down to rest and cool off in the shade of the freshest-looking of the whitebeams, *Sorbus aria* 'Lutescens', photographed here in Evenley Wood Garden in Northamptonshire. (Nurseries still have to learn the new official name: *Aria nivea*).

The whitebeams (far left) and rowans have another important role: feeding flocks of birds. A female blackbird makes short work of a cluster of berries.

Whitebeams (left) take their name from the brilliance of their densely white felty leaves as they unfold. The felt wears off through the summer.

changing it in the cultivars *S.a.* 'Lutescens' and 'Majestica' (which has bigger leaves and berries). It has also produced a pale yellow form (*S.a.* 'Chrysophylla'), which to my mind is one of the best of all "golden" trees, and a weeping one (*S.a.* 'Pendula').

The Himalayan whitebeams are recognizably the same tree, but with much bigger leaves, rich glossy green on top, chalk white beneath, and as much as 10 inches long. *Sorbus cuspidata* has broad-oval or long, tapered leaves; *S. thibetica* 'John Mitchell' almost round ones; both are supremely lush and well-nourished-looking trees, whose livery of dark green and white lasts well all summer. Their berries are crab-apple size, but not brightly coloured. The place to plant either would be in a wood where their big leaves are given some protection from the wind. They stand out magnificently among the busy patterns of oak or beech.

When the whitebeams cross with rowans, intermediate forms are found, with broad but lobed leaves. Among these are Swedish whitebeam (*S. intermedia*) and *S. hybrida*.

Service trees

Then there are the service trees. Learned discussion has produced two alternative theories for the meaning of "service". One gives it the same Latin root as *cerveza*, the Spanish word for "beer", on the grounds that a drink was once brewed from its berries. The other, more earthily, reckons it is no more than a corruption of *Sorbus*. In any case people with service trees tend to be rather pleased with themselves, not least because they know there are three – and that one is a sort of whitebeam and the other a sort of rowan (*S. domestica*), and the third, the service tree of Fontainebleau (*S. latifolia*), apparently hybrid.

The "whitebeam" is the wild service tree (*S. torminalis*, or to be bang up to date *Torminaria torminalis*), a rather rare native of northern Europe (England included). It is an understorey tree in forests on clayey ground, surviving drought and putting up for years with low levels of light, waiting for its chance. Its leaves are lobed enough to be taken for a maple until its speckled brown pomes give it away. It colours warmly in autumn. One that I planted in the churchyard in memory of a (rare) farmer who loved trees has grown to 30 feet high and almost as wide in 30 years. It gives us pleasure from its prolific flowering to its autumn tints.

The Fontainebleau service tree (*S. latifolia*) is this crossed with an unknown rowan, years ago in the forest of Fontainebleau. It has eaves both lobed and felted beneath.

The wild service tree (right) is a close cousin with lobed leaves like a maple and quantities of barely edible fruit, once fermented to make a sort of "beer".

The golden whitebeam (right) is a bright, tidy tree with rather narrow leaves. It has wonderful possibilities for both informal and formal planting. (Two flanking a gate would look splendid.) In early and mid-summer no tree is more radiant. (Its official name is *Sorbus aria* – soon to be *Aria nivea* 'Chrysophylla'. A rival called *S. aria* 'Aurea' fades quicker.

The mountain ashes (far right) have brighter and longer-lasting fruit than the whitebeams. It becomes tastier and less tannic after frosts. This is *Sorbus* 'Joseph Rock', a first pick among the mountain ashes for berries that turn slowly from green to white to a mixture of white and amber-yellow with a tinge of pink, among leaves turning scarlet, purple, and crimson, then on the bare tree.

Mountain Ashes

HOW COMMON ARE CROSSES between whitebeams and mountain ashes? Not very. When they occur the upshot is pretty much what you'd expect, compromise leaves with a whitebeam-type oval at the end and a clutch of mountain ash-type leaflets. From the whitebeam they keep white undersides. The best-known white-mountain-ash-beam is *Sorbus × thuringiaca* from Germany, which in its fastigiate form is a popular street tree. *Sorbus hybrida*, on the other hand, which looks and sounds like the same sort of hybrid, is now considered a Scandinavian species. You could be examined on this …

The place of honour, in my mind, goes to a Japanese/Korean species that no one would suspect of either whitebeam or mountain ash blood. You would say horn- rather than white-beam, indeed: the leaves have the same corrugation where the veins go. Botany has described its leaves as alder-like (its name is *S. alnifolia*) which seems a bit farfetched. In any case it is the gayest of the race in flower and the most conspicuous in fruit, keeping its small red berries in large bunches long after the leaves have turned red and fallen. I'm told that *Micromeles alnifolia* is its latest name. There is a rather similar species with even more beautiful pleated leaves, also placed in this new section of *Micromeles*, still usually labelled, in the rare gardens where it grows, *Sorbus caloneura*.

Rowans of many nations

It follows from their leaf shape that the mountain ashes are less substantial, more feathery trees than the whitebeams. They grow to about the same size (or often less) and have the same flowers and fruit, but otherwise they are remarkably different, and one basic model of mountain ash with only stylistic variations in colour of leaf, size of leaf, and colour of berry runs right around the world.

Its most famous representative is the rowan (*S. aucuparia*): a tree, like the Scots pine, identified inextricably with Scotland (though it is native to most of Europe). I have seen landscapes there in which full dress tartans would have simply disappeared into the background as a perfect disguise. The heather gives a purple ground on which the larches and the birches display their yellow-gold. The rowan is gold of a richer tone, set with preposterous carbuncles of scarlet fruit.

It is hard to improve on the common rowan. Various nursery productions offer alternatives: of yellow fruit, for example, in *S.a.* var. *xanthocarpa*; of finely divided, ferny leaves in *S.a.* 'Aspleniifolia'; of a narrow upright head in *S.a.* 'Sheerwater Seedling'; even of edible fruit in *S.a.* var. *edulis*.

North America's native *S. americana* is a similar tree, usually smaller and with smaller fruit. *Sorbus scopulina* is a western version with bigger leaves and fruit on, if anything, a smaller tree.

The most valuable of the innumerable Asiatic mountain ashes are those with different-coloured fruit and leaves. White fruit is particularly effective, making *S. hupehensis* a name well known

The **mountain ash** or rowan of Europe (above) is a small tree of the northern wilds but a willing recruit for gardens, with remarkable quantities of gaudy fruit. This tree is on Loch Osgaig in Wester Ross in the Scottish Highlands.

Sorbus cashmiriana (below) is an easy and reliable white-berried tree, not notable for autumn-colouring leaves but for a good crop of berries hanging on the bare branches, often untouched by birds. Its flowerheads are broad and decorative, too.

among gardeners. A tin-eared botanist has now renamed this
S. glabriuscula, while its pink-berried form has been labelled
S. pseudohupehensis. The upright *S. cashmiriana* and the more
relaxed *S. vilmorinii* are other excellent examples with grey-green
leaves and white or faintly pink fruit, either of them a lovely misty
sight in the gloaming. *Sorbus prattii* is a neat little tree with ferny
green leaves, white fruit, and orange autumn colour. Altogether
bolder in scale of leaf, bud, and bunch is *S. sargentiana*, which is
dark green, has sticky red buds, very long leaves with red stalks,
deep orange autumn colour, and long-lasting scarlet fruit.

We can expect to hear of many new trees (not to mention new
names) in this department. Enthusiasts have reason to study their
sex lives too; prurience can be rewarding. Apomicity in some cases
means that seed from a tree, being effectively self-fertile, will come
up as identical trees. Grafting should in many species become a
thing of the past.

Sorbus commixta sounds exactly right for this genus: you expect
a hybrid. In this case you get a very decorative species from Japan.
S.c. 'Embley' has been much recommended as a street tree because
its branches slant sharply upward out of the way. It has both big
red fruit and red leaves at the same time. Several Asian species in

Sorbus prattii (above)
from Yunnan has up to 30
little leaflets per leaf on a
vigorous little tree turning
warm colours in autumn.
Its berries are pure white
but seem quite popular
with birds.

Sorbus vilmorinii (left) is
a more spreading little
mountain ash from Yunnan,
with delicate foliage, a
rather feminine-looking
tree. Its berries change
from pink to nearly white
as they mature while its
leaves colour briefly.

fact have the same trick of branching: *S. pohuashanensis* from the north of China and very hardy; *S. harrowiana* from the south and rather tender; *S. insignis* from Assam with pink fruit which lasts right through the winter.

I am very fond of *S. scalaris*, whose leaves, dark green above and pale below, have as many as 30 leaflets, giving it a densely ferny look. (I made the mistake of looking up what *scalaris* means, thinking it must have something to do with ladders, which the leaves faintly resemble. The first mention I found was the scalaris or latrine fly.)

Near the top of most gardeners' lists is *S*. 'Joseph Rock'. Whether it is a wild Chinese species or a hybrid is not clear – hence the quotes round its name. It stands out in autumn for its yellow fruit in a unique combination with leaves turning red, orange, and sometimes imperial purple.

More rowans are coming into cultivation all the time. Their success in gardens and profligacy in the wild make them highly collectable. One of the most desired at the moment is the low-growing *S. gonggashanica*, with substantial long-lasting pure white fruits, collected by Roy Lancaster on Gongga Shan, his favourite Chinese mountain.

'Joseph Rock' (above left) wins pride of place in many gardens for its unique combination of yellow berries with a medley of sumptuous leaf colours. The berries often outlast the Joseph's coat of many thousand glowing leaflets.

More than one clone of mountain ash (right) claims the name of *Sorbus pohuashanensis*. Pohuashan is a mountain near Beijing. All are generous with brilliant red fruit among deep green leaves.

The Japanese *Sorbus alnifolia* (left) is not a mountain ash, nor strictly a whitebeam, but leads the *Micromeles* section of east Asian *Sorbus* with simple leaves and decorative fruit. The leaves turn parchment colour and the little berries are a clear red for a long time together. It can reach 50 or 60 feet.

The pure white fruit (above) of *Sorbus prattii*. White berries seem to be fashionable; perhaps because birds usually leave them until they have gobbled the red.

The American mountain ash (below) and the European are very similar. If anything the American has larger leaves and less fruit – though no shortage here in the Blue Ridge Mountains of North Carolina.

The Elms

WHY INCLUDE A CHAPTER on a tree which almost no one plants any more, and many readers will either never have seen or keep only as a distant memory? Because these most majestic trees were the first ones that moved me, that inspired me to study trees more closely. Because I have watched so many of them die. Because in many townships and villages as far apart as New England and East Anglia elms were the landscape. Whether the local species is the American elm (*Ulmus americana*), high-arched and spreading; the English elm (*U. procera*), tall-trunked and balancing a broad fan far above the fields; the wych elm (*U. glabra*) wide-crowned and weeping; or any of a dozen others, the elm set the scale for life in its countryside. In our flat country it did service for the nonexistent hills.

I include it not only for the bitter-sweetness of recollection, but also because it teaches us a lesson about monoculture and the importance of diversity – and another about the importance, and difficulty, of quarantine when a deadly plague arrives.

Dutch elm disease is a fungus carried by a beetle. It grows in the new ring of wood, just inside the bark, on which the tree relies almost totally for its flow of sap. When it blocks the sap vessels, the branch dies – and very soon the tree. Two-hundred-year-old trees can be killed in one season. Because the disease battens on wood that is being created that year little can be done to protect the tree. The only hope is to inject the sap stream with fungicide. I tried it. For summer after summer I knelt at the feet of giant trees, pumping fungicide into their cambium layer hour after hour. That was in the 1970s. But 70 percent of the hedgerow trees in the English Midlands and eastern counties were elms. The task was impossible. We kept hoping that as the disease passes some of the abundant hedgerow sprouts would prove immune and provide our descendants with these irreplaceable trees. There is no sign of it yet.

The English elm

There are about 45 species of elm worldwide: nobody is quite sure how many. There are certainly five that count, and perhaps another dozen hybrids and cultivars. Which you consider *the* elm depends entirely on where you were brought up.

I was brought up in the heart of elm-land, in the southeast of England, where three species and several elusive crosses and crosses of crosses clumped or filed all over the landscape. Our chief glory was the English or field elm which, because it sets no fertile seed but reproduces entirely by new saplings springing up from the roots of old trees, maintains its blood free from taint. In a sense all English elms are parts of one gigantic tree, and look like it – the pattern of branching in a line of them is hypnotically repetitive. They all have short horizontal or slightly drooping branches on the lower part of the trunk, then very often a distinct waist halfway up, then an almost-symmetrical, half-open fan of long branches forming the top third of the tree. William Gilpin, who wrote late in the 18th century about the aesthetics of trees, pronounced

The painter John Constable was almost obsessed by elms (right) – with willows the principal trees of his Stour valley. He made this deeply-felt sketch of a group in Old Hall Park in 1817.

I took this photograph (below) of a tree surgeon high in a dying field elm, beginning to cut off the branches to demolish it, in 1975 at the height of the Dutch elm disease.

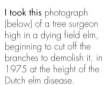

Hoar frost transformed an elm one December morning in central France. This narrow shape of tree without heavy branches is typical of a local population in the Allier, rather like the Jersey or Guernsey elm of the Channel Islands. Most French elms have died since the 1970s.

magisterially: "No tree is better adapted to receive grand masses of light." The grand masses are not exactly of light, but of the dark green, small-leafed foliage whose cloud formations seem to hold it. Constable, a native of this landscape, was constantly trying to paint it: and yet his oaks are better than his elms.

The leaves stay green on the English elm longer than on the others; even into November. Their autumn colour is a fine light gold, which coincides with autumnal mist to create effects I can't begin to describe. It is a savage irony that that lovely colour comes early now, and means death.

The Elms of Europe and America

THE ENGLISH ELM (*Ulmus procera*) is or was essentially the hedgerow, often the parkland tree; important in London's parks but not so frequent in towns. It was as successful in North America as in Europe. At a casual glance, a smooth-leaved elm (*U. minor*) and a wych elm (*U. glabra*) can look very much the same. They usually fork fairly low and broaden out (like the American elm, *U. americana*) into an immense explosion, from the extremities of which the twigs come tumbling in intricate cascades. They are at their most beautiful in winter, when their tracery is on display; not even the grand masses of light can compare with what is surely nature's very best pen-and-ink work.

If you can reach a leaf you have your answer to the identity of the tree: the smooth-leaved elm's leaf is smooth (on top) and shiny; the wych elm's is bigger and raspingly rough. The English elm's has the same sandpapery feeling. The trademark of all elm leaves is a lop-sided base.

The flowers of the elm are often dismissed as being invisible from the ground, and so not worth bothering with. It has no showy catkins, but early in the spring it carries so many tiny red flowers, which are really just bunches of stamens appearing from a bud, that the whole leafless tree, caught by the sunlight, glows misty crimson. Still before the leaves appear come the winged fruit: little discs more or less the shape of a fried egg, which colour the tree green. They ripen brown and fall as the leaves expand. The fertile ones germinate as quickly as they ripen: the whole performance from flower to new elm is a matter of eight weeks or so.

One of the many crosses between the smooth-leaved and wych elms, or rather a whole group of such crosses (known under the collective name of *U. × hollandica*), constitutes most of the elm population of continental Europe, under such names as the Belgian and Dutch elms (*U. × h.* 'Belgica' and *U. × h.* 'Major'). Wandering down to Italy they become the support for many a promiscuous vine, a role they have filled since Virgil's day.

The really weeping forms of elm you used to see on broad lawns (weeping, that is, not just in branchlet and twig, but in the whole branch from the top) were varieties of the wych elm: the rounder *U. glabra* 'Camperdownii'; the taller and more angular *U.g.* 'Pendula'.

The search for immunity
There was not so much planting of the lesser-known American elms, either in the United States or in Europe. The slippery elm (*U. rubra*) with its big velvety leaves (slipperiness, being in the inner bark, is not an obvious character) and the narrow, short-branched rock elm (*U. thomasii*) of the Great Lakes were overshadowed by the far-ranging American elm itself. There are no native elms west of the Rockies. The lack has been supplied partly from the eastern

The American elm, wide-spreading, turning gold in autumn, has been preserved by popular pressure and ruthless hygiene in famous malls in Central Park, New York (here) and in Washington DC. This photograph was taken in October 2009.

States and Europe, but also from Asia. The Chinese elm (*U. parvifolia*) – though not remotely elmy to the untutored eye – is a highly desirable garden tree: all rather miniature, its leaves decidedly so: dark green, shiny, almost evergreen, colouring red before its late leaf-fall and with decorative peeling bark as it ages. In cold places the similar (but less attractive) Siberian elm (*U. pumila*) does service for it.

All discussion of elms now centres on efforts to find crosses or clones that are immune to Dutch elm disease. It is their size and shape, above all, that nostalgic elm-lovers would love to recreate.

The English elms (above) of folklore, memory, and many paintings all conform closely to this shape: a broad fan on top, a narrow waist, and spreading skirts.

Ulmus 'Dodoens' is a hybrid bred for disease-resistance, released in Holland in 1970. It scores four out of five for resistance, certainly no guarantee. No hybrid yet has a silhouette of such strong character as the English elm, above.

There are elms from Asia that have not been affected, but if they are small trees without the splendid presence of our lost ones what use are they? That is the question of the day: can we harness the immunity in their genes to give us trees of the old size and shape?

To a very limited degree isolation has kept populations alive to remind us of what we have lost. Both Washington DC and Central Park in New York have preserved magnificent malls of American elms. In October in Central Park they still provide much of the brilliant yellow of autumn. In England the seacoast town of Brighton, insulated by the bare South Downs and the Channel, and protected by a draconian programme of surgery to any branch that shows symptoms, has kept a large elm population. So has the Isle of Wight. I could take you to half a dozen places in Essex where three or four great elms stand alone in the fields. The two tallest still stand, as I write,

A number of the many strains of elm (below) are identified with their area of origin. The Jersey (or Guernsey) elm (*Ulmus minor* var. *sarniensis*), also called the Wheatley elm, is a fine narrow avenue tree. The number of its names shows its former popularity.

The tiny flowers of elms (above), with clusters of red anthers, are only conspicuous because they come in late winter when the branches are bare. In sunshine they can colour the whole canopy of a tree red-purple.

The fruit of the elm (above) is a single seed in a flat membrane called a samara, looking rather like the yolk in a fried egg. It turns brown and falls as the young leaves appear.

A young elm in Italy (above) in March was the only tree apparently in leaf. There were no leaves, just a mass of its bright green fruit. Rome is full of such trees, giving it an early spring.

Elm leaves (above) are always toothed, always lop-sided at the base, usually oval, and often rough-textured like the wych elm (*Ulmus glabra*) here.

The Chinese *Ulmus parvifolia* is our best hope in breeding trees resistant to elm disease. It has leathery little leaves and flowers in autumn. Its bark cracks and sheds attractively.

ELM SILHOUETTES

The silhouettes of elms are their trademarks, ranging from domes to fountains to combinations of the two, with lighter or heavier branches. Many clearly distinct strains are named from their area of dominance while confusion reigns over their precise species and origin. All the elms below are threatened with extinction except *Ulmus parvifolia*.

U. americana
The American elm can spread to enormous width at great height.

U. minor 'Cornubiensis'
The Cornish elm has a fan top, is bright green, and leafs late.

U. glabra
The wych or Scotch elm forms a broad crown of branches which climb slowly in shallow arcs to 100 feet or more.

U. carpinifolia
The smooth-leaf elm gleams, with many branches and graceful hanging sprays.

U. glabra 'Camperdownii'
The Camperdown elm is a weeping version of the wych elm.

U. parvifolia
The Chinese lacebark elm is small, good in streets, and resistant to disease.

in Queens' College grove by the river Cam, more than 90 feet high. In my own garden, where we lost dozens of big elms, I have succeeded, by ruthless culling, in nurturing several to 50 feet or so.

In the first years of the disease one hybrid from Japan carried our hopes of finding a substitute. 'Sapporo Autumn Gold' was its name. A public-spirited company gave over its carpark to a dozen greenhouses where it was propagated, to my surprise, by leaf cuttings in sand under mist, to be given away to enthusiasts. I still have my SAG, now 50 feet high but sadly nothing like our elms to look at.

In the past 40 years researchers have produced countless crosses, mainly between American or European elms and the Japanese *U. davidiana* var. *japonica*, the Siberian *U. pumila*, Chinese *U. parvifolia*, or Himalayan *U. wallichiana*. Of the American school the most promising is probably 'Morton', from the Morton Arboretum in Illinois, which leads the science. It is sold not as Morton but as 'Accolade'. 'Valley Forge' and 'Princeton' are two other derivatives of the American elm with promise.

European research has been led by the Dutch. It began with the wych elm, or one of its many crosses and variants that go by the name of *U. × hollandica*. Two excellent prospects from Holland are *U. × h.* 'Columella' and *U.* 'Lutèce', adopted by the French and given the Roman name of Paris. Of all these *U.* 'Lutèce' stands the best chance of becoming the giant tree we are all looking for. Its parents are both big: *U. glabra* and *U. wallichiana*.

Zelkovas
and Hackberries

THE ELMS ARE A HOMOGENEOUS LOT, more so even than the oaks. None is quite evergreen; none has wildly eccentric leaves. Their relations also toe the line. The excellent *Zelkova* (if ever a tree deserved a good English name) is thoroughly elmy, though only a cousin. And the more distantly kin nettle tree or hackberry (*Celtis*) has much of the same family look about it.

Very few trees have had such a good press as the zelkova. With no flowers or fruit to speak of and with leaves like those of the elm (only singly toothed; elm leaves have paired teeth) it manages on composure alone. It is a graceful, well-proportioned, short-trunked, many-stemmed, broad-domed, soberly coloured ordinary tree.

Both the zelkovas (the medium-rare ones; there are more, but they are really rare) answer this description. I turned up William Robinson's *The English Flower Garden* to find myself echoed with another forest of commas: "The longer points, sharper teeth, more numerous nerves and leathery texture, together with the fact that they hang longer, may enable anyone to tell the leaf of the Japanese zelkova (*Z. serrata*) from that of the better-known Caucasian tree (*Z. carpinifolia*)."

I would add to the list of distinctions the thin bark of the Japanese zelkova, which in flaking off reveals enticing little patches of quite startling orange. With the wide spread of the long branches from a mere five feet up there is plenty of bark to contemplate.

I also have to add, though, a warning note. Zelkovas are dangerously kin to elms. They may not be immune to Dutch elm disease either. Two 30-year-old Japanese zelkovas in my garden started to show the unmistakable symptoms in late summer. A year later one was dying and one struggling to recover. I felled the former and pruned the latter back to healthy wood. For the moment all is well, but I am not confident.

Related to cannabis
It is the fruit that most obviously distinguishes the hackberries (they are also called nettle trees, but why or by whom I'm not sure) from the elms, but their DNA also shows them to belong to separate lineages, and the genus *Celtis* has now been placed in the Cannabis family, Cannabaceae. Though it is scarcely a reason to grow them, their fruit consists of red, yellow, or blackish berries, compared with the elm's dry, flattened, winged fruit. Certainly birds appreciate the difference. Otherwise in proportions and general aspect they are like small (rarely much higher than 40 feet), spreading, bush-topped, and rather bright green elms, with much to be said for them as street or small garden trees.

One point in their favour is that they root deep. You can grow shrubs under them – if the shrubs can take dense shade, which is what they deliver. Their deep roots are virtually drought-proof.

Leaves of the Japanese zelkova are longer, more tapering, and more toothed than those of the Caucasian species. They turn yellow or bronze in autumn. The fruits are hard little nuts, unlike the elm's flat wafers and the hackberry's berries.

Japanese zelkova (*Z. serrata*) is less vigorous and less distinctive in its growth than the Caucasian tree (*Zelkova carpinifolia*), but potentially useful as a medium-sized street tree. 'Green Vase' is a selected narrower-growing cultivar of Japanese zelkova.

The best part about the common hackberry (*C. occidentalis*) of eastern North America (which will grow more or less anywhere) is its highly original bark. Several elms have corky twigs, but the hackberry handles it differently. By middle age the tree has deep flanges of cork, widely spaced, running more or less north–south but wandering about on its surface. As a tree for public places the common hackberry is not popular for its habit of forming witches' brooms from which the congested twigs fall off and make a mess. This shouldn't put anyone off planting a good and unusual tree in his or her garden. In its place the authorities plant either the Mississippi hackberry (*C. laevigata*) with mainly untoothed leaves, the excellent and even lofty Chinese hackberry (*C. sinensis*) with shiny, bright green leaves, or the southern European hackberry (*C. australis*). *Australis* in this usage simply means "southern".

The *micocoulier* (I love its French name) is common as a street tree in the south of France where there is not enough room for planes – not that tight quarters often inhibit planters. It grows satisfactorily in the warmer parts of England, too. The most unexpected fact about this tree is that its small, dark purple berry is the Lotos of Tennyson's mild-eyed, melancholy Lotos-eaters, no less. I'm afraid the spectacular pollination the poet describes – "Round and round the spicy downs the yellow Lotos-dust is blown" – would be quite beyond it.

Elastic leaves

Eucommia ulmoides is a lone Chinese species in its own family, Eucommiaceae. As its specific name suggests, the eucommia looks like a small elm, but it has rubber in its veins. If you tear one of its leaves in half the barely visible strands hold the two halves together. It is the only hardy tree that does contain rubber; not enough to make it a commercial proposition, but in an attractive and easily grown tree perhaps enough of a reason to plant it.

A fine specimen (left) of the Caucasian zelkova growing in the north of Italy. The short buttressed trunk and extraordinary system of crowded branches are unique to this tree. It was discovered in 1760 but remains little known.

The hackberry (above) of southern Europe garners no superlatives but furnishes Mediterranean streets and gardens with something reliably shady that takes less room and less management than a plane. This is Château Routas in Provence.

The Figs and Mulberries

WHAT CAN YOU GROW that gives a garden such a sense of established well-being as a fig tree or a mulberry? Maybe it is just witless harking back to a three-quarters-mythical past: the fig tree mentioned so often in the Bible, and the mulberry appearing from China in the Dark Ages – and with it the secret of silk.

Nostalgia apart, they are hardly obvious relations. But then we see little of their family in temperate regions: the mere tip of an iceberg sunk (yes, this is an awkward metaphor) in tropical and sub-tropical parts of the world. The huge gloomy trees with roots groping down from the first floor that you see in dusty town squares in hot countries are figs (*Ficus*), and so are the slim-leaved office evergreens that could do with a polish.

What the mulberry family all have in common is the white sap that contains latex. In the fruit department things are more complicated: mulberries (*Morus*), paper mulberries (*Broussonetia*), and osage-oranges (*Maclura*) having essentially raspberry-like fruit, all globules and pips where the old flower cluster was.

The fig has a design of its own in which the flowers bloom (if that's the word) on the inside walls of a little fleshy funnel, which then swells around the pips to become the fig. In most species of fig there is a complex relationship with minute wasps that do the pollinating inside this funnel. Figs love insects.

The hardy fig (*F. carica*) at the northern extremes of its range can scarcely in fairness be described as a tree. A tree in Italy, certainly; in Britain usually more of a sprawling shrub, whose thin grey branches need support from a wall to reach any height. With a bit of care, however, it can be persuaded to ripen fruit in most of Britain and in the United States as far north as New York. The trick to get it to fruit is to give its roots the minimum space to ramble; keep them in a tub, or at least in a confined space, and feed them.

Certain cultivars can set fruit without pollination. Of all the late-summer garden pleasures there is nothing to beat a forage through the heavy, dusty-smelling leaves of the tree to find ripe figs. Their own leaves make a plate for them beside a bowl of cream cheese among the wine glasses. I cut them, not because biting them open isn't a pleasure, but to see the purple-violet-indigo of their livery opening to the silver of the knife

A specialist nursery in Philadelphia lists 150 varieties – not all hardy there, but a indication of how such an ancient and useful plant has been improved, selected, and reselected in countries as far apart as Greece and China. Turkey has long been a great centre of fig growing, and "the best" cultivar is the Turkish *F. carica* 'Bursa'. *Ficus carica* 'Excel' is an excellent yellow variety, *F.c.* 'Panachée' is yellow with green stripes, *F.c.* 'Violette' apparently the hardiest. In Californica *F.c.* 'Black Mission' is said to have been imported by the Franciscan missionaries. In England the most-grown is the quite hardy *F.c.* 'Brown Turkey', pictured below.

Instant history

The mulberry is an easy, though rather slow, tree to grow. You can plant cuttings even five feet long in autumn with every hope of starting something. The great question is where. We rarely get much fruit off our tree, which stands by a pond. The culprits are

The fig as grown in a northern garden (below), pressed up against a wall, is very different from the strong multi-trunked bushes that spring up wherever a fig has been eaten around the Mediterranean.

The fragrant leaves of the fig (above) are almost as much of a pleasure as the fruit. There are two crops: a spring one (often frosted in the north) on old wood and the main crop on new wood in autumn. The best time to prune is after the main crop.

The fig turns a bloomy purple and wrinkles when it is ripe. Fresh figs go beautifully with raw Italian ham or with cream cheese at the beginning or end of a meal, especially in the garden.

The fruit (left) of the paper mulberry (*Broussonetia papyrifera*). Its bark is the source of some of Japan's most precious paper. It is a hardy small tree with odd-shaped grey-woolly leaves. The male plant has yellow catkins.

The fruit (left) of the white mulberry which shares the same structure, ripens from green to cream to chalk-white, pretty in dense clusters among the shiny leaves but lacking the acidity to be really satisfactory eating. This is the silkworm tree.

An almost fully fed silkworm (the caterpillar of the silk moth) feeds on a mulberry leaf. Silk is spun from the very fine thread with which the worm makes its cocoon. Mulberry trees were much planted in England in the 17th century in the hope of starting a silk industry.

Few fruits (above) stain your hands and clothes as effectively as the black mulberry. The chief enthusiasts for the tree by a pond in this garden are the moorhens. The berries, like blackberries, go well with apple purée.

The osage orange has little globular heads of flowers like the mulberry. They fuse together as they develop to form orange-sized green (inedible) fruit that ripens to yellow.

Mulberry trees (left) have a way of looking many times their real age very quickly. Many reputed 17th-century trees are really 19th or even 20th, but they have the presence to inspire such stories, and great affection.

the moorhens, who are adept at running up the trunk and along a branch, popping a mulberry and plopping down again to the ground. It is ultimately a wide, sprawling, low-profile tree with fruit which, unless you eat or your moorhens eat it, falls abundantly to stain everything under it a fine shade of crimson. That is the black mulberry (*Morus nigra*) – the eating one. If you want to keep silkworms you will want the white mulberry (*M. alba*), which grows even quicker. As a tree the white mulberry is often more attractive, with sumptuous shiny leaves in many shapes, if a less characterful trunk. It turns pure butter-yellow in autumn.

Black mulberries are gastronomically neglected. They are somehow not quite right as fruit, combining the extremes of squashiness and pippiness – yet the flavour, searchingly sour and hauntingly sweet, is very special, especially cooked. White mulberries start cream-coloured and turn almost pure white when they are ripe. A bowl of white mulberries makes a pretty sight, but they lack the acidity of the black.

The paper mulberry (*Broussonetia papyrifera*) helps to link mulberry and fig; in terms of leaves, not fruit. It makes a broad, gentle, soft-looking but disorganized tree not unlike an elder that

puzzles visitors. Some of its leaves are more or less heart-shaped like the mulberry's; most are lobed and figgy. The fruit, not edible but pretty, is like a round red mulberry. Some of the earliest Chinese paper was made from the bark of the paper mulberry by soaking it, smoothing it, and finishing it with rice paste. In Japan it is still an important source of super-quality paper.

One thoroughly offbeat American tree, the osage-orange (*Maclura pomifera*), can be dimly perceived to be a relation through the formation of its remarkable fruit. It is limited in nature to parts of Texas and Oklahoma, but widely planted in the eastern and mid-western states as a tough hedge tree with vicious thorns and the very best firewood, rot-proof and long-lasting. The fruit does indeed look like an orange, but its bumpier surface suggests the same composite structure as the mulberry. Each grapefruit-sized fruit, probably once dispersed by now-extinct large mammals, is in fact composed of many smaller ones tightly welded together. I once taught the river pilots of the Gironde to play a sort of cricket with the fruit of an ancient tree on the lawn of a famous château. Should anyone want even bigger fruit (beware of standing underneath) there is a cultivar called *M.p.* 'Cannonball'.

The Southern Beeches

THE WORLD HAS BEEN pretty thoroughly ransacked for trees. It is unlikely that even in China any significant genus remains undiscovered. There is, however, one highly significant and decorative one which is still virtually unexploited and unknown: the southern beeches. But where do you see them? Few nurseries offer them with any knowledge or enthusiasm. To abide by my strictly botanical plan I should place them elsewhere, but to give them the prominence they deserve I link them to the familiar beeches, for comparison and contrast.

The southern beeches are *Nothofagus*, the beeches (*Fagus*) of the southern hemisphere. The botanical differences between the two are relatively minor, though sufficient to place them in different families. They are very different trees, though, to look at. Some of them are deciduous, some are evergreen, but the hallmark of most is a tiny leaf – at biggest, perhaps half the size of a beech leaf, at smallest, smaller than a box tree's.

It is not that the genus is in any sense a new discovery. Species were known from Australia and Tasmania nearly 200 years ago, from New Zealand and Chile only a little later. It was soon realized that this was a close relation to the beech, which alone in its family inhabits the southern hemisphere. Not only that but it straddles the Pacific. It is a race of trees of remarkable singularity and style, fast-growing in many cases and with valuable timber. Why did they not become the rage?

From the first, if seems, they were looked on as too tender for anything except the mildest gardens – and with some reason. There are 34 species and only a dozen are grown more than tentatively outside their natural range. Many attempts have resulted in very slow or poor growth, or death by untimely frost.

The South American branch

One is bone hardy, the rather absurdly named *N. antarctica*, in reality from southernmost Chile. It exhibits all the characters of the genus, but can hardly be called exciting: most cultivated specimens are more bush than tree. The characters? Bark with the little horizontal marks or lenticels you see on cherries, a fundamentally fishbone branch pattern a little like the elm, small, neat unlobed leaves that tend to crinkle, and little beech-like husks for fruit. Their autumn colours are charmingly muted and mixed: one leaf red, another yellow.

Two more species from Chile are big handsome trees that could, you would think, add significantly to Europe's choice, not being too obviously exotic. *Nothofagus obliqua*, known in Chile as roble (or "oak"), has proved hardy in most parts of Britain in the 100 years since it was introduced, though its hardiness varies with the source of the seed. In 1975 I was in Patagonia with a pioneer forester from Cornwall who was frustrated by the lack of seed from high altitude and low latitude (which would have the best chance of being hardy). He tried recruiting young locals to climb some of the best trees to collect the mast. Their upper branches are slender, though, and the boys were reluctant. The only solution was to cut them down.

Thirty years later I appealed in print for seed from trees growing in Britain and received a box of 1,000 seedlings from a wood in Yorkshire – which I planted in Wales. (Foresters have generous natures.) They are growing enthusiastically but rather wildly, with too many forks and deviant branches to make good timber. It may take generations to select the best strains.

Even faster and more handsome is the rauli (*N. alpina* – until recently either *N. procera* or *N. nervosa*, to everyone's confusion). Rauli has bigger leaves, not unlike a hornbeam's. Sadly it is less hardy: we have 50-foot trees in Wales, but in Essex it has been killed in its youth by frost. Good specimens have fine straight trunks with up-sweeping crowns. Roble seems indifferent as to soil, but rauli dislikes alkalinity: definitely a tree for the rainy west.

A cross between the two, potentially you would think a winner, was christened *N. × dodecaphleps* by a botanist who clearly harboured some grudge against it. Another cross, between *N. obliqua* and the rare *N. glauca*, from drier sites in Chile, appears (even from its name, *N. × leonii*) to have a better chance.

One more Chilean beech has been surprising and delighting gardeners: an evergreen of the mid-slopes of the Andes (known locally as Coigue) that forms dense dark-foliaged forests looking at a distance remarkably like cedars. This is *N. dombeyi*, named in honour of a formidable botanist who faced unusual perils. Joseph Dombey came from Burgundy, was self-taught, but reached high rank at Paris's Jardin des Plantes. His first expedition to South America ended in disaster: his valuable collection was captured at sea by the British (and remains in the British Museum). His second

Nothofagus antarctica. One place the antarctic beech won't grow is Antarctica. It is the hardiest, and the only one widely planted in Britain. Its young fruit betray its kinship with the beeches.

Antarctic beech is also the only species of its genus I know to have attractive autumn colour. Its tiny crinkled leaves change to a yellow/green/red medley before falling.

The deciduous rauli (*Nothofagus alpina*) is a big tree from Chile, hardy in western Britain. It has some of the largest leaves, pleated rather like a hornbeam's.

Nothofagus dombeyi (above) from Patagonia (locally "coigue") is a noble evergreen which has reached 100 feet in 100 years in the damp air of Cumbria.

Black beech (*Nothofagus solanderi*) (left) is the hardiest of the New Zealand species, from the rainforests of southern South Island.

was taken by the Spanish, who locked him up as a dangerous rival to their own botanists. Finally he was captured by pirates in the Caribbean, where he died. *Nothofagus dombeyi* was not brought to Europe until 1916, but seems to relish the moderate Atlantic climate. We have no other evergreen tree like it, vigorous, with deeply glossy leaves and a striking but tidy habit. *Nothofagus alessandri* could be equally desirable were it obtainable.

Kiwi cousins

Of New Zealand's largely evergreen beech forest (the bulk of the forest of South Island) only one tree, the mountain beech of the hilltops, is reliably hardy in much of Europe. I was given a little

plant by Ken Beckett, who advised me on the first edition of this book. Some 37 years later it waves its uniquely wispy branches the same number of feet above the garden – wispy because its evergreen leaves are the smallest of any tree's. The misfortune of the mountain beech is not to have been assigned a species. It is *N. solanderi* var. *cliffortioides*. Cliffort? A 17th-century Dutch banker. -*oides*? "Resembling". (A red herring as *Cliffortia* is in fact a South African shrub.) Nothobeeches from the lower, warmer levels include red beech (*N. fusca*), which has more deeply toothed leaves than the others, and silver beech (*N. menziesii*), whose young leaves have a silvery sheen. It would be good to see more of them in cultivation; in forests it is not easy to tell several of them apart.

The Beeches

THE BEECH FAMILY in its broad sense, embracing beeches, oaks, and sweet chestnuts, is the royal family of the broadleaves by any reckoning. It has the oak for king and the beech for queen.

Oaks spread over more of the temperate world than any other broadleaf tree. There are some 450 species, evergreen and deciduous. There are shrubs among them, but most of them are big trees, and it always comes as a surprise to find out how many relations they have in the tropics, too.

There are fewer beeches and less beech forest. Nonetheless, the beech grows in the three northern continents and the related southern beeches circle the southern hemisphere. The chestnut, closely kin to the oak, is the third major member of the family. What marks all three of them as close relations is their fruit. Acorns, chestnuts, and the "mast" of beeches are all formed in the same way: by the woody base of the female flower growing to form a husk round the seed.

Red carpet

If there is a royal seat it is surely Normandy. It is there that the beech grows to perfection, in woods of amazing purity and grandeur. At Lyons-la-Forêt, near Rouen, beeches that seeded themselves during the French Revolution now soar clear and smooth-boled to 100 feet, into a green canopy 30 feet deep shutting out the sky. One cannot avoid the feeling that the inspiration for the great Norman cathedrals came from the ancestors of this solemn forest.

The beech is a much-shorn tree, excellently adapted for making hedges. Appropriately enough, the monarch of all hedges is of beech. It stands at Meikleour, about 10 miles north of Perth in the Highlands of Scotland, a green wall 580 yards long and on average, nearly 100 feet high. It is kept trimmed up to 60 feet – but beyond that ingenuity (and courage, I shouldn't wonder) are exhausted.

It is odd that the beech has not seen fit to diversify in the same way as the oak. The eight or nine species are almost interchangeable, save for a bigger, smaller, longer, or rounder leaf. The beech is clearly the logical end of its own evolutionary road, just as it is often the winner in a struggle for dominance in the forest.

An old beechwood has the longest echo of any woodland: a sound so eerie and disturbing that I remember it vividly from when I was a boy of four living on the chalk hills of Buckinghamshire. (The "Buck" of Buckingham in fact means beech.) The beech thrives on lean chalk, laying down its own red carpet where other trees flag. In due course it takes it over completely: scarcely a bramble, barely a toadstool grows in the moss below the massed beech boles of the Chiltern hills. Their layered canopy lets only

A beechwood in autumn: Burnham Beeches in Buckinghamshire. The floor is bare of all save beech leaves, thanks to the shade cast by the horizontal layers of foliage. Being shallow-rooted, beech is apt to be blown over in a gale.

threads of sunlight through. Hence the long-answering echo; the echo of an empty room. The beech's domineering methods, of filling the surface of the ground with fine roots, and covering it with heavy shade, certainly don't show in its face. Of all the big trees it is one of the daintiest: one of the most graceful in angles and movements, one of the lightest and most feminine in colour, one of the smoothest in texture.

Rather than accumulate calloused old corky stuff to form a protective layer of bark, the beech's cambium stays almost on the surface; its trunk remains smooth and silvery and fresh-looking. By the same token it needs the shade it is so good at making with its layer upon layer of leaves, all horizontally aligned. A beech grown in the open keeps branches right down to the ground to protect its trunk from sunlight. To see the sun playing on its pale curves and hollows, paler in the American beech (*Fagus grandifolia*), like the bloom on old aluminium, is one of the winter pleasures. You could argue that the beech tree is best in early winter. Its leaves turn sometimes to the colour of golden burned butter, sometimes to the clear coppery red of a fox's brush, before they fall. But one in 10 of those on young trees hangs onto the lower branches up to eight or nine feet from the ground. Between the red drifts on the ground and the red spume in the air (and dare I mention the azure firmament behind?), the silver shaft is dazzling. Summer with its geraniums has no monopoly of nature's vivid colours.

Multifarious, adaptable

The American beech is reluctant to grow in Europe. The European (or common) beech (*F. sylvatica*) has no such qualms about the northeastern States, which is just as well, since it is the one with all the interesting cultivars.

Beech's commonest manifestation in gardens is in hedges, for which it is perfectly suited. But next after hedges comes, undoubtedly, the purple or the copper beech — understandably popular as by far the biggest red thing there is to grow.

Red-leaved beeches have turned up many times in the wilds — usually, for some reason, in the middle of Europe: in Switzerland, Bavaria, or eastern France. At their earliest appearances they were inevitably rumoured to be the mark of nature's disapproval of some unnatural crime — the blood refusing to lie down. Redness is a fairly stable characteristic of these trees; often their seedlings are red, too. In forest plantations of beech seedlings a small proportion of seedlings will come up red and flourish as well as their green contemporaries. A

ACORNS, MAST, CHESTNUTS

The fruit of the beech family clearly links its members. Their nuts are all more or less enclosed in a woody sheath which grows up from the base of the female flower. The "mast" of the beech (below centre) has two nuts completely enclosed until they are ripe and the husk splits open as here. The acorn of the oak has one, sitting in an open cup (bottom left). The chestnut has one, two, or three nuts wrapped entirely in a prickly shell (below right).

The acorn is an egg in a cup. Every one of the 400-plus species of oak has a variant of this design.

Beech mast has a hard and enduring little four-part casing, seemingly enjoyed by pigs. The roots of the word *fagus* is a Greek word meaning "to eat".

The prickly overcoat of a chestnut (the only one of these fruits human beings enjoy today) is presumably intended to discourage premature predation of the nuts.

A young beech (above left) grows under the shade of oaks above a torrent in a Welsh woodland. The young tree grows fast, slender, leafing before the oak, using all the light to grow tall. When it reaches the tree canopy it will put on weight.

Adaptable hardy beech (above) is used to make narrow serpentine hedges like "crinkle-crankle" walls in the garden at Chatsworth House in Derbyshire, where Joseph Paxton, designer of the Crystal Palace, was a gardener.

The highest hedge (above right) (anywhere?) is in Perthshire, beeches more than 80 feet high kept in trim since the time of the Young Pretender in the 1740s. It is 580 yards long, along a public road.

number of shades of red are registered in different varietal names. *Fagus sylvatica* 'Cuprea' has been used for the lightest, supposedly copper-coloured; *F.s.* 'Riversii' for the darkest, a full-blooded purple. But nurseries don't often offer a choice, and since beeches all go through the whole range of colours – from new leaves of rosy-red silk to a browny-purple final stage in autumn – it is hard to know which precise colour is the one to put a name to.

Their use in gardens, however, raises passions. To many authorities – notably Alan Mitchell, whose opinion is worth reading on any tree-related topic – they are anathema: "heavy, dark masses, garish without being bright, blending with nothing, contrasting pleasantly with nothing, absorbing light uselessly." They have their (very brief) springtime moment of freshness, but I agree that most are a blot on the landscape all summer long. That said, I have not had the heart to fell the one I inherited and have tried to cheer it up with contrasting neighbours. Its wonderful golden brother, *F.s.* 'Zlatia', which I love, I fear only makes it look darker still. Few trees hold their springtime glamour so well. A golden poplar improves matters a little, but so garishly I feel I have to apologize.

The fern-leafed beech (*F.s.* Heterophylla Group) is a more subtle and, truthfully, far more beautiful tree. (Look for the name 'Aspleniifolia'.) Its leaves are deeply cut into narrow points that somehow give the whole tree a much finer and less shiny texture: a matt effect which soaks up the light. It is difficult to see why this

A hanging branchlet of beech (left) at the point of bud-burst with the silky-hairy leaves just emerging at the same time as the catkins. The flowers of the beech are particularly modest. Beech buds are at an angle to the twig.

The translucent young leaves (above) of beech show more graphically than any how a tree fans its leaves out to catch the sunlight and minimize any mutual shading. The view from below can be like exquisite stained glass.

Fagus engleriana (left) is a rare Chinese species with less vigour but a goodly share of charm. Its leaves are a little narrower and more pointed with a bluish cast. It grows at half the speed of common beech; a fine tree for a small garden with good soil.

Fagus sylvatica 'Aspleniifolia', aka 'Heterophylla', (left) sometimes called the fern-leaved beech, is one of the best of the family, in time making a glorious full-bodied tree. Its leaves are strangely various, from little strips to something so deeply lobed as to be almost pinnate.

tree is quite as effective as it is, and yet if any tree in old parkland has people streaming up for a closer look it is always this one. A totally different, rare, but to my mind excellent variant is *F.s.* 'Rotundifolia', whose leaves are almost small, crinkled, and almost circular. The specimen I had coloured a wonderful deep autumn amber until it was destroyed by a gale.

I am probably alone in thinking the weeping beech (green or purple) an ugly beast. But it comes near the bottom of my list of weeping trees; the ash and the elm (and of course the willow) weep far more convincingly. The beech's are surely crocodile tears, or indeed – as Hillier's *Manual* perceptively suggested – elephant's. I quote: "the enormous branches hang close to and perpendicular with the main stem like an elephant's trunk". Nothing like as bad as a weeping sequoia, which looks like a boa constrictor trying to take off. But still far from graceful.

To attention

Possibly the most useful of all the variants is the fastigiate model,

For a brief moment in spring (above) the golden beech, on the left, and the copper beech, on the right, together make a Spanish flag. Here at Saling Hall the white of a 'John Downie' crab-apple intervenes.

A mature copper beech (above) in spring beside the canal at Combe House in Devon, coming into leaf at the same time as *Malus floribunda* is in flower. Later in the season it becomes a formidable tower of darkest red with a more foreboding feeling.

growing with its arms in to its sides, as it were. It clearly has a lot of planting ahead of it. It was Mr Naismith of Dawyck who found it. Dawyck is an estate in the Tweed valley in lowland Scotland which has been in the hands of tree-lovers since Elizabethan times. Mr Naismith was the incumbent in 1860 when up sprang a beech with the stance of a Lombardy poplar, the perfect compromise for people who want a beech tree but haven't got room. The Dawyck beech (*F.s.* 'Dawyck') is formal when young, but big ones are huge leafy ovals. As a street tree, where shade was not the main need, it would be hard to beat.

One of Europe's great dendrologists, Dick van Hoey Smith of Rotterdam, set out in the 1970s to breed purple and golden versions of the Dawyck beech, paying his children to collect seed in likely spots – with great success. You can choose the shape and colour of your beech with your curtains today. As a piece of décor for a (fairly large) garden a strong – eventually pretty portly – column of leaves in your chosen colour: green, red, or yellow, is a serious proposition.

The upright Dawyck beech (left) can be kept trimmed to a perfect columnar shape like this to great effect. Left to itself it puts on weight, with an embonpoint that can eventually become floppy.

"Atlantic oak woodland" is the term used for ancient woods on thin rocky soils in the rainy west of Britain. Wistman's Wood on Dartmoor in Devon is an example. Mosses, ferns, and a few hollies alone accompany the stunted oaks in an enchanted (but rare and threatened) form of forest. There are others in Merioneth in north Wales and in the western Highlands and islands of Scotland.

The Oaks

"'ROBUR, THE OAK.' I HAVE SOMETIMES considered it very seriously, what would move Pliny to make a whole chapter of one only line." So John Evelyn begins the first chapter of his *Sylva*.

Pliny, and Evelyn, were right. The oak carries such a cargo of symbolism that embroidery is counter-productive.

Robur, the Oak.

"I wonder if you ever thought of the single mark of supremacy which distinguishes this tree? The others shirk the work of resisting gravity; the oak defies it. It chooses the horizontal direction for its limbs, so that their whole weight may tell; and then stretches them fifty or sixty feet … to slant upward another degree would mark infirmity of purpose; to bend downwards, weakness of organization."

This is Oliver Wendell Holmes, *The Autocrat of the Breakfast Table*, chapter 10. I beg you to find the book and read on to where he goes to look for New England's biggest elm.

In Britain it seems as though a sort of race memory still exists of the time when the whole country was oakwood. Even in Henry VIII's time, nearly 500 years ago, one-third of the whole island was still covered by oak trees. Still, today, if good land is left lying fallow, oak will very likely find its way there and cover it before 10 years are out.

Heroic qualities

There are about 450 species of oak in the world. There is one that started the legend. Oak grows in every form: as tree, as bush, as scrub; deciduous or evergreen: in Europe, Asia, and America. But the oak of imagery is the English oak: broad and slow, angular and ponderous-branched: *Quercus robur*.

I should hasten to say that there are two species of oak in England and northern Europe, *Q. robur* and *Q. petraea*; that they are quite distinct (though they often hybridize); and that I have no evidence (or not much) for offering one as the original of the oak image and not the other.

Typically, the English, common, or pedunculate oak (*Q. robur* has all these names) is the broad tree, not specially tall, with the qualities the Autocrat was talking about. Typically the durmast or sessile oak (*Q. petraea*) is a taller, more upward-reaching tree commoner in damper places and regions.

True, it depends above all on where these oaks are growing. In an open field both will be broad and spreading; in close forest both need to aspire toward the overhead light. You have to remember, therefore, that the pedunculate oak has its acorns on peduncles or stalks; the durmast has them sitting on the twig. The leaves are the other way round: the short-stalked leaves of the pedunculate form stiff rosettes at the ends of the twigs, making the canopy broken, with patches of sky showing through. The base of each leaf has two tiny ear lobes. The bolder, thicker, and glossier long-stalked leaves of the durmast oak give an even coverage, a shaftless, uniform shade.

The great oak forests of central France where Europe's tallest oaks grow, above all the Forêt de Tronçais, chosen by Colbert for the French navy in the 17th century, are in the main durmast woods. In the wetter west of England and Wales durmast is in the majority. In an eastern, or midland, or southern English park on the other hand, the huge, broken-headed hulks, reputed (often rightly) to have been there 500 years or more, and all their gnarled and leafy offspring, are pedunculate oak. As far as the navy was concerned, there were never enough of these trees. Their awkward, angular way of growing gave the shipwright the hugely powerful brackets and angle pieces that held a 74-gun ship together as she blundered through the troughs on her way into the line.

Queen Elizabeth's Treasurer, Lord Burghley, was the first man to plant oaks. He sowed acorns and holly berries together, shrewdly giving his saplings evergreen cover that would never smother them.

The common English oak is perhaps more familiar in its open-grown, parkland form with a huge trunk and widespreading branches than as a lofty forest tree. These giants at Windsor Great Park could be the often-pollarded survivors from a Tudor hedge.

It was held that bigger acorns make better trees, so they were sieved and only the biggest were planted.

Variations on the theme

Intensive cultivation produced its crop of ornamental cultivars. An excellent upright version of the pedunculate oak, Lombardy-poplar style (*Q. robur* 'Fastigiata'), is perhaps the best. The rest are rare, but there exist a cut-leafed version (*Q.* × *rosacea* 'Filicifolia'); a purple version (*Q.robur* 'Atropurpurea'); a slow-growing golden version (*Q.r.* 'Concordia'); and a rather anaemic variegated version (*Q.r.* 'Argenteovariegata').

By the mid-18th century most of the oaks of Europe and the Mediterranean region were well known and enthusiastically cultivated. A number of hybrids of permanent value arose in the nurseries where oaks quite foreign to each other were planted side by side. Oak collections, indeed, are hotbeds of miscegenation. Not every chance cross has permanent value.

Turner's oak, the Fulham oak, and the Kew oak (their botanical names are all self-evident) are three examples that have been famous in their time. The most famous and the most used today was a chance seedling in the Exeter nursery of Lucombe & Pince in 1762. The parents were the vigorous Turkey oak (*Q. cerris*) and the evergreen cork oak (*Q. suber*). The Lucombe oak (*Q.* × *hispanica* 'Lucombeana') has the most useful attributes of both its parents short of providing cork. It is a distinguished and huge-growing almost-evergreen with enormous parkland presence.

Perhaps the rarest and most mysterious of all oaks is another hybrid, this time thought to be between *Q. robur* and the Mexican white oak (*Q. rugosa*). There are only two big trees, in the Cambridge Botanic Garden and at Kew. *Quercus* × *warburgii* is often known as the Cambridge oak. Its leaves are hard, bluish, and almost evergreen but emerge in spring almost red at the same time as red catkins: a sight worth going to Cambridge to see. For some reason it is hard to propagate, even by grafting. It has grown fast in my collection; so fast, alas, that the slender top was broken by snow.

The propensity of oaks to hybridize, especially with *Q. robur*, makes life difficult for collectors (like me) with more enthusiasm than specialist knowledge. I am all too familiar with the awkward pause that follows when I show a true quercologist a prize specimen of my own. "I'm afraid it's got a bit mixed up" is the too-frequent verdict.

Not snails at all

Oaks continue to plant themselves, but their qualities are rarely understood and appreciated. And they have a name, like yews, for very slow growth. It is worth examining, therefore, just how fast or slow they are.

Of the oaks I planted when the elms died in the 1970s the best, now 34 years old, is 50 feet high. (More remarkably, one I transplanted at 25 feet has grown to the same height.) You can count on 15 feet in 10 years. It's certainly a satisfying rate of progress to watch.

One should not underestimate young oaks as ornament. The Lammas shoots which make all oaks glow with new life in mid-summer show up most on a young tree. Last year one particular tree was so covered with new pink and golden leaves that no flowering cherry could have been more spectacular, and it took them six weeks to fade through pale green to dark. The same tree, in a sheltered place, has kept its brown leaves on all winter like a young beech. In that tree's future there is really something to look forward to.

The Oaks of Europe and Asia

THE TENDENCY TO KEEP THEIR LEAVES late in the season, dead or alive, is clearly an oak family trait. The oaks as a whole give the impression that evergreenery is never very far away. Many of the oaks of the south, whether in Europe or America, are outright evergreens. Others (especially the big-leafed species) are tardily deciduous. They hang onto their leaves for as long as they can. In northern Europe this is often until mid-winter. In the south it is until the new crop of leaves takes over in spring.

The holm oak (*Quercus ilex*) is by far the most familiar of the Mediterranean species and far and away the biggest and best broadleaved evergreen we can grow as far north as Britain. (In the USA, as far north as Washington DC.) It seems to break all the rules, exposing a mass of leaf to the cruel winter winds. Yet it takes a severe winter to defoliate it, even partially – and even then the tree is often unharmed. Recent warmer years have so encouraged its seeding, in fact, that on the Isle of Wight in the English Channel it has come to be regarded as a weed.

The name "holm" (and also the Latin name *ilex*) means holly. It may have been the shock of seeing another green broadleaf in winter which caused the comparison, because, though its young leaves are sometimes spiny, the adult leaves have nothing like the shine nor the spines of the holly. Holm oak leaves vary from plain ovals to distinctly toothed shapes, from thumb-nail size to three inches or so, and from quite pale green in shade to a dark gloss in the sun. The tree's best moment, like the yew's, is when the young growth of the spring shows against the near-black background of the old. A white moss of hairs covers the fresh leaves; as they open they glow amber and tawny. The broad dome of a big tree, dressed with drooping branches to the ground, and creaming over with these quiet colours, is spring's most solemn celebration.

There is no better tree for the architectural backbone of a garden or park. Pom-pom versions have a vogue, too, in formal gardens. I also grow a fastigiate specimen, a great dark flame shape, which came up in a seedbed but which I have never seen in any catalogue.

The cork oak

The important economic evergreen oak of the Mediterranean is the cork oak (*Q. suber*). It is not quite so hardy as the holm oak, neither is it such a big and good-looking tree. Its interest lies in its bark, which is usually displayed (as the elephant-skin bark of the holm oak is not) by its open and haphazard way of growing. Old cork oaks usually lean at perilous angles which go with the exaggerated cragginess of the trunk to make a convincing picture of age. Cork as a commercial proposition is still pretty well confined to Portugal and Spain, North Africa, and a little in France and Italy. South of the Tagus in Portugal the country is orchard-like for mile

after mile among the spreading, stunted cork trees. There have been scare stories recently that screwcaps for wine bottles will put an end to this landscape and its fauna (notably the Spanish eagle). The landscape was there, though, before the modern craze – it is only 30 or 40 years old – for cork-stoppered wine. In reality I suspect the slow-motion cork industry (one crop every nine years, after an initial 50 to establish the trees) is more at risk from the temptations of planting almost-instant *Eucalyptus*.

Eye-catchers

In many ways the ideal plant for a public place is one that is slightly larger than life. Not a ferocious-looking exotic that makes the native greenery look dowdy, but a close relation with a touch of

The Turkey oak (above) is the fastest-growing and most adaptable European species. Its acorn is distinctive for its densely mossy cup. The rough leaves vary in shape from oval to oblong. Few gardeners can resist this white-variegated form.

The cork oak (left) of Spain and Portugal is still the world's principal source of cork. The thick, rugged bark is stripped off every ten or 15 years.

extra ginger. Just enough to wake people up with the surprise of, let us say, an oak with such big leaves. This alone would be reason enough to bring back the neglected big-leaf oaks of southern Europe, even if they were not in their own right singularly splendid, vigorous, and adaptable trees.

There are three outstanding oaks which once were famous: the Hungarian (*Q. frainetto*), the Algerian (*Q. canariensis*), and the Caucasian (*Q. macranthera*). They all have leaves about six inches long – not as long as some of the American oaks, but long enough to give a distinctively tulgey look.

The Caucasian and Algerian oaks are easily confusable; the Hungarian (excellent for wine barrels, notably for Tokaj) has much deeper lobes – even two inches deep – in its splendid leaves. All are

eventually trees of the biggest size, with plenty of exotic allure.

Another distinctive trio of oaks from the Near East have narrow leaves, not lobed (i.e. with the wavy coastline of most oak leaves) but with a regular row of little points, or teeth: seven or eight a side. The general effect is very much like a chestnut leaf. One of them, in fact, is called the chestnut-leafed oak (*Q. castaneifolia*), another the Lebanon oak (*Q. libani*). When impatient planters find out how quickly the chestnut-leafed oak makes a big tree I suspect its current rarity (surprising, since it has been Kew's most splendid oak for generations) will come to a sudden end. Specimens 30 years old in my collection look twice that age, a cultivar called *Q. castaneifolia* 'Green Spire' in particular, from Hillier's, is a lusty upright of enormous promise, if a bit sparse in winter.

Leaves of the holm (or holly-leaved) oak (far left) are glossy but without the glitter of holly. They fall in early summer and resist rotting, which can make them a problem in gardens. The acorns take 18 months to ripen.

The leaves (right) of the "English" *Quercus robur* are daintier and more intricate than those of the sessile oak (bottom left). Its acorns look like old clay pipes on long stalks that give it its name of "pedunculate".

The evergreen holm oak (left) of the Mediterranean makes massive hedges and thrives under clipping almost as well as yew. It will grow on all types of well-drained soil. These bastions are at Arley Hall, Cheshire.

The sessile or durmast oak (bottom far left), native of most of Europe and of damper western Britain, has broader, often glossier, leaves and bears its acorns directly on its twigs without stalks.

The Lebanon oak has smaller, glossier leaves and the third, the shrubby *Q. pontica* from Armenia, pale scaly buds to watch as they produce bright oval glossy leaves from the tips of its branches.

The only one of this southern group which could be called at all common is the Turkey oak (*Q. cerris*) – though common in Italy, as *cerro*. It has the reputation of being the fastest-growing and could perhaps make in time the tallest tree. At one time the navy had great hopes of it, but its wood proved a disappointment. Its leaves are all points and angles, not oversize but eye-catching. Most important, perhaps, it grows well on unpromising dry and chalky soil. Its white-variegated variety is irresistible to gardeners. In its time it has given rise to countless crosses: always a risk (or hope) where exotic oaks congregate. Most illustrious of its progeny is the splendid evergreen Lucombe oak (*Q.* × *hispanica* 'Lucombeana').

Orientals

The oaks of the Far East in cultivation are not (yet, at any rate) such big trees. The outstanding ones are the sawtooth oak (*Q. acutissima*) from Japan, a very similar tree to the chestnut-leafed oak, and the daimyo oak (*Q. dentata*), which has the biggest leaves of any, sometimes over a foot long. Even in the warmth of Japan, though, it makes a scruffy little tree. The similar *Q. mongolica*, with somewhat smaller leaves, is more frequent and hardier.

Quercus acuta and *Q. glauca*, both valuable small evergreen trees (*Q. salicina* in this group is very tough), and the Chinese cork oak (*Q. variabilis*) occasionally turn up in collections.

Of new introductions of the past 30 years only *Q. franchetii* from southwest China with pink hairy new growth and *Q. schottkyana* with red are really noteworthy. It is curious how Asian oaks compare poorly with the Americans. Most of them, though, are in the sub-genus of the white oaks – and it is the American red oaks that grow so well in Europe.

The Oaks
of North America

THE OAKS OF NORTH AMERICA present a daunting picture at first.
Yet there are certain key features which help one to classify and
recognize the individual species. The chart on page 272 shows the
leaves and acorns of 25 of the most important ones. The trees are
divided according to their red/white affinities, the shape of their
leaf (lobed or unlobed), whether they are deciduous or evergreen,
and their native habitat – whether they belong east or west of the
Rockies. In all there are about 90 species in North America,
without beginning to count Mexico's remarkable contribution –
or the innumerable hybrids which will keep botanists bickering for
ever. Where two populations merge there are often intermediate
forms: should they be given names as species or not?

At times the great American forest seems like an oakwood
with intruders – so many splendid species are there. Ninety grow
in North America and most of them are fully fledged trees. There
can be few people who have mastered the whole catalogue, though
when you start breaking it down into the more obvious distinctions
– region, deciduous or evergreen, whether the leaves are lobed
or not and the colour of the bark – the process of identification
usually leads to a conclusion – even if the conclusion is
inconclusive. The truth is, however, that in most parts of America
there is an obvious oak to plant – the one that grows there already.

American red oak (above) is one that
grows well in northern Europe. Foresters
hoped for good timber rapidly; they
were disappointed.

California black oaks (right) in Yosemite
National Park. Black oaks (*Quercus
kelloggii*) are the most widespread oak
of California, from San Diego to Oregon,
invaluable for timber, fodder, and shade.

The pin oak (*Quercus palustris*) (above) is easy to recognize by the hang of its thin branches: the lower ones sweeping to the ground and the upper ones forming a tidy dome. Deep-cut leaves give it a fine texture in keeping with its almost feminine shape.

Most brilliant of all oaks in autumn (above), the scarlet oak (*Quercus coccinea*) has a natural range of nearly a third of the United States and grows excellently in Europe, given acid soil. Its long, relatively slender branches and open crown cast a light shade.

The black oak of the eastern States (*Quercus velutina*) is one of the tallest of all oaks – a tower of dark glossy green, turning red in autumn. Its young leaves are velvety red, thick, and stiff, and its catkins splendid.

The white oak (*Quercus alba*), most European-oak-like of all the American species with its mighty dome. The leaves turn from autumn purple and reds to golden brown. Sadly, it is rare and unsatisfactory in Europe.

Red, white, and live

The most noticeable distinction is between the live oaks (as the evergreen species are called) and the deciduous ones. The live oaks in fact are few: one in the east and half a dozen in the west.

The real botanical distinction is between white and red oaks. There is no single great giveaway here, but enough clear differences to usually make it quite easy. One rule, though not universal, is a good indicator: white oak acorns form and fall in a single year; red ones take two years on the tree (and have a more bitter taste of tannin). White oaks are therefore a more important larder for wildlife: the menagerie feeding around them can be spectacular.

In the northeast the white oak (*Quercus alba*) has its domain. In what way is it typical of its group? Its name, it seems, comes from the colour of its trunk, which is grey and scaly, contrasting with the blackish and furrowed trunks of the red oaks. The deep lobes of its leaves are rounded at the ends, without points or bristles. Its acorns are reasonably sweet to eat, their cups are shiny inside. And they stay on the tree for only one season; they fall in autumn and germinate straight away – often to be killed by frost before they have got established. In American folklore the white oak plays the part of English oak in England. It is the monumental tree, symbol of the country and its stout-hearted people, generously spreading its arms and living for centuries. Dozens of trees have local status and recall historic happenings – although in regions where other species of oak are the champions they often play the same role.

The red oak (*Q. rubra*) is the protagonist of the complementary sub-section, the former blacks. Its bark, almost beech-like in youth, slowly becomes blackish-brown and corrugated. The leaves have less indented bays, but their lobes are sharp-pointed and their points have little bristles. Acorns last two years on the tree; when they fall they wait until spring before germinating. Their meat is bitter and their cups are downy inside instead of shiny.

The Canyon live oak (*Quercus chrysolepis*), a common evergreen species in California, belongs to the "golden" part of the family. It is a medium-sized, spreading tree with smooth pale bark and gold-fuzzy acorn cups.

The Californian white or valley oak is the biggest oak of the west, the tree that makes part of the Central valley look like a park. The best specimen today is 120 feet high, 103 feet in spread, and 28 feet in girth.

Fennel covers a Washington meadow hillside around an Oregon white oak (*Quercus garryana*). David Douglas called this tree after Nicholas Garry of the Hudson's Bay Company, who helped him transport his seeds back to England.

By and large the rules of identification apply. Often one of the typical features is missing. There are no bristled lobes, for example, on the (black) willow oaks (*Q. phellos*). In the last resort the botanist's criterion has been the wood structure, which divides the two groups perfectly in accordance with the other evidence. White oak timber is, on the whole, much the better. It is an odd sidelight that the white oaks are miserable in Europe and hardly grow at all, while the red have done excellently. What the white oaks lack is the extremes of the eastern climate, above all its summer heat and humidity. It is less clear why the red oaks are happy without.

The big variations in leaf shape that occur within each group help the identification process considerably. The chart on page 272 shows the principal trees – far more red than white, and far more in the east than in the west. The map on page 273 shows the total range of the two groups and the site of the biggest known tree of each species – an indication of where it feels most at home.

Differences in acorns and leaves seem to suggest that there is considerable variation between the species of oaks – but how much, in fact, do they differ in their demands and manner of growth?

In outright hardiness they vary surprisingly little – some of the southern oaks will grow even in New York, though they lose the leaves they might have kept, at home, through most of the winter. They all have their favourite haunts: moist river bottoms, gravelly gullies, or rocky open hills – although the majority need acid or at least neutral soil. In overall size they can certainly vary: from the cliff-like 125 feet of occasional black oaks to the mere 50 feet of, say, the blue oaks of California, not to mention some of the scrub oaks that never make a tree. On the other hand, most oaks with space to spread settle for a medium height: 70 feet or so.

Branch patterns

There are three characteristics in particular that single out certain oaks as trees of real beauty: their habit of branching, their texture, and their autumn colour. The classic oak branch pattern – bold, wide, and angular on a stocky trunk – is best expressed by the

The Virginia live oak (above) is best known for the Spanish moss that drapes its branches in the humid south, from Louisiana to the Carolinas. It makes a huge short trunk with immense long branches and survives drought, even in sand, as well as wet ground.

The red oak (*Quercus rubra*) (left) is more amenable to transplanting than either the white or black and grows faster than any other American oak. It has pale yellow-golden new leaves and red autumn colour.

white oak. The Oregon white (*Q. garryana*) and the burr oaks (*Q. macrocarpa*) are similar. The others in their group are rarely so perfect in form. Overcup oak (*Q. lyrata*), for instance, often has a crooked trunk.

Among the second group of white oaks, those with chestnut-style leaves, the trees are often taller in relation to their spread, with shorter or upward-trending branches. They are less typical of the oaks both in figure and in leaf.

The red and the black oaks tend to be narrower than the white, the total effect more tower than dome. The shumard (*Q. shumardii*) and the Spanish (or southern red) oaks (*Q. falcata*) are typically

taller and narrower trees: the scarlet oak (*Q. coccinea*) more branchy and open. The fastest-growing oaks all belong to this group.

Of these the tree most often planted in America for its profile is the pin oak (*Q. palustris*), which has, instead of stout wandering branches, an abundance of thin and straight ones. The upper branches climb to make a rather narrow crown; the lower ones droop, often as low as the ground, hiding the trunk. For an oak it is a lightweight tree, the more so because of the fine texture of its deeply divided leaves. Being fast-growing and easy to transplant it is the nurseryman's favourite. The live oak (*Q. virginiana*) of the south is the most characteristic of all in its shape: it grows twice as

LEAF SHAPE COMPARISONS

The oaks of North America present a daunting picture at first. Yet there are certain key features which help one to classify and recognize the individual species. The chart below shows the leaves and acorns of 25 of the most important. The trees are divided according to their black/white affinities.

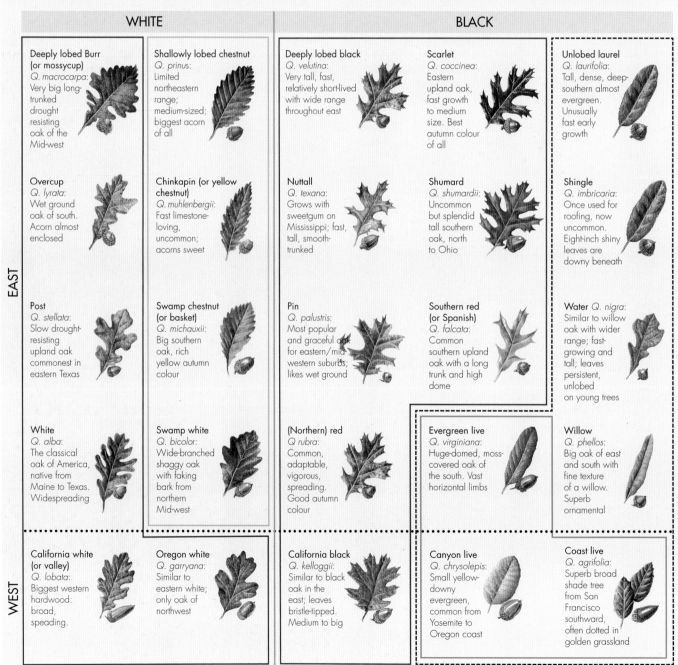

WHITE

Deeply lobed Burr (or mossycup)
Q. macrocarpa: Very big long-trunked drought resisting oak of the Mid-west

Shallowly lobed chestnut
Q. prinus: Limited northeastern range; medium-sized; biggest acorn of all

Overcup
Q. lyrata: Wet ground oak of south. Acorn almost enclosed

Chinkapin (or yellow chestnut)
Q. muhlenbergii: Fast limestone-loving, uncommon; acorns sweet

Post
Q. stellata: Slow drought-resisting upland oak commonest in eastern Texas

Swamp chestnut (or basket)
Q. michauxii: Big southern oak, rich yellow autumn colour

White
Q. alba: The classical oak of America, native from Maine to Texas. Widespreading

Swamp white
Q. bicolor: Wide-branched shaggy oak with faking bark from northern Mid-west

California white (or valley)
Q. lobata: Biggest western hardwood: broad, speading.

Oregon white
Q. garryana: Similar to eastern white; only oak of northwest

BLACK

Deeply lobed black
Q. velutina: Very tall, fast, relatively short-lived with wide range throughout east

Scarlet
Q. coccinea: Eastern upland oak, fast growth to medium size. Best autumn colour of all

Unlobed laurel
Q. laurifolia: Tall, dense, deep-southern almost evergreen. Unusually fast early growth

Nuttall
Q. texana: Grows with sweetgum on Mississippi; fast, tall, smooth-trunked

Shumard
Q. shumardii: Uncommon but splendid tall southern oak, north to Ohio

Shingle
Q. imbricaria: Once used for roofing, now uncommon. Eight-inch shiny leaves are downy beneath

Pin
Q. palustris: Most popular and graceful oak for eastern/mid-western suburbs; likes wet ground

Southern red (or Spanish)
Q. falcata: Common southern upland oak with a long trunk and high dome

Water *Q. nigra*: Similar to willow oak with wider range; fast-growing and tall; leaves persistent, unlobed on young trees

(Northern) red
Q rubra: Common, adaptable, vigorous, spreading. Good autumn colour

Evergreen live
Q. virginiana: Huge-domed, moss-covered oak of the south. Vast horizontal limbs

Willow
Q. phellos: Big oak of east and south with fine texture of a willow. Superb ornamental

California black
Q. kelloggii: Similar to black oak in the east; leaves bristle-tipped. Medium to big

Canyon live
Q. chrysolepis: Small yellow-downy evergreen, common from Yosemite to Oregon coast

Coast live
Q. agrifolia: Superb broad shade tree from San Francisco southward, often dotted in golden grassland

EAST

WEST

wide as it is high. The champion is 132 feet across by a mere 55 feet high. This is the tree of southern allées, ghost-like in its shawls of Spanish moss. Sadly it has never been a success in Europe, even in the south: it needs the steam heat of the Gulf states to strike its extravagant pose. Western live oaks are not so picky – nor so theatrical. *Quercus agrifolia* from California, with small, shiny evergreen leaves, grows rapidly and beautifully in England. Why is it not more often planted?

All oak leaves are lively: the oak is never dull in texture. The burr oak, which has the biggest leaves (sometimes a foot long) has an extraordinarily emphatic pattern. Yet perhaps the finest textured make the most beautiful trees: fine either in having deep-cut leaves like the pin oak, or in having them slim and small like the willow oak. Small leaves on a big tree do not have the busy effect you might expect: they make a perfect patina for the terrific sculpture of crown and bough.

What matters more to the parks department than the exact shape of tree or leaf is the way the tree transplants. Tap-rooting oaks are notoriously reluctant to move: the white, scarlet, and black oaks are all examples. Among the easiest to transplant, and hence the most popular, are the red, water, pin, southern live, and willow oaks. The last three, alas, demand acid soil.

TREE LOCATIONS

The map below shows the range of the major groups of the American oaks and the location of the biggest known tree of each species. The figures of maximum height (h.) and spread (s.) are from the American Forestry Association records.

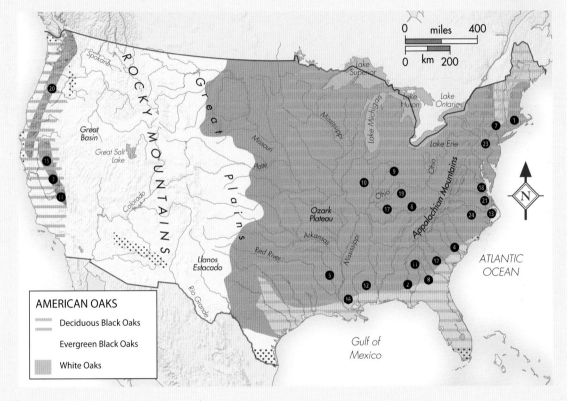

DECIDUOUS BLACK OAKS

1 Black (*Q. velutina*)
80' h. 100' s.
East Granby, Connecticut

2 Blackjack
(*Q. marilandica*)
122' h. 49' s.
Barbour, Alabama

3 Blue (*Q. douglasii*)
71' h. 79' s.
Mariposa, California

4 Laurel (*Q. laurifolia*)
130' h. 81' s.
Conagree NP, South Carolina

5 Nuttall (*Q. texana*)
113' h. 81' s.
Vernon, Louisiana

6 Pin (*Q. palustris*)
135' h. 97' s.
Bell, Kentucky

7 Northern red
(*Q. rubra*)
80' h. 102' s.
Monroe, New York

8 Scarlet (*Q. coccinea*)
119' h. 120' s.
Oglethorpe, Georgia

9 Shingle (*Q. imbricaria*)
105' h. 62' s.
Cincinnati, Ohio

10 Shumard
(*Q. shumardii*)
96' h. 96' s.
Anna, Illinois

11 Southern red
(*Q. falcata*)
123' h. 152' s.
Upson, Georgia

12 Water (*Q. nigra*)
118' h. 108' s.
Jones, Mississippi

13 Willow (*Q. phellos*)
131' h. 130' s.
Chesapeake, Virginia

EVERGREEN BLACK OAKS

14 Canyon live
(*Q. chrysolepis*)
69' h. 121' s.
Tulare, California

15 Coast live
(*Q. agrifolia*)
110' h. 129' s.
Chiles Valley, California

16 Live (*Q. virginiana*)
68' h. 39' s.
St Tammany Parish, Louisiana

WHITE OAKS

17 Burr (*Q. macrocarpa*)
104' h. 93' s.
Woodford, Kentucky

18 Chestnut (*Q. prinus*)
99' h. 98' s.
Arnold, Maryland

19 Chinkapin
(*Q. muhlenbergii*)
76' h. 69' s.
Harrison, Kentucky

20 Oregon white
(*Q. garryana*)
93' h. 65' s.
Douglas, Oregon

21 Overcup (*Q. lyrata*)
109' h. 102' s.
Southampton, Virginia

22 Post (*Q. stellata*)
83' h. 102' s.
Jackson, Georgia

23 Swamp white
(*Q. bicolor*)
93' h. 102' s.
Franklin Township, New Jersey

24 White (*Q. alba*)
86' h. 116' s.
Brunswick, Virginia

The Chestnuts

The deep and regular fissures in the bark of the chestnut have the habit of swirling around the trunk in a spiral, often the quickest way to identify a chestnut tree in a wood or when the leaves have fallen.

IT IS HARDLY A STEP from some of the chestnut-leaved American oaks to the chestnut itself. Botany has made stepping stones rather than a linking path between the two, preferring to give the golden chinkapin (*Chrysolepis chrysophylla*) and the tanbark oak (*Lithocarpus densiflorus*), two American species which come firmly between oak and chestnut, and *Castanopsis*, an Asian one, the status of separate genera. But from the acorn, part in its cup and part out, to the chestnut, all inside, each stage of wrapping is represented.

When you get to the chestnut tree there is no mistaking it. Vivid leafiness proclaims it. Big tongue-shaped leaves, every vein making a valley and every valley ending in a holly-like point. Very commonly the leaves are all the evidence you have of what it is – in Europe because of a still-flourishing industry in coppiced chestnut poles, cut to the ground every seven or eight years; and in North America because disease has in the last 100 years killed every big tree.

At the beginning of the 20th century they were one of the chief components of the eastern forest, bigger than almost any oak, spreading from Alabama to Maine. In 1906 a bark fungus very like the elm disease was introduced accidentally on horticultural specimens from the Far East. By 1940 the chestnut was expunged from the list of American forest trees. Only because of its immense power of sprouting from the base (which makes it ideal for coppice) does it survive at all, as leafy fountains from where huge old trees formerly stood. Every effort is being made to breed blight-proof trees by crossing *Castanea dentata* with the Chinese *C. mollissima* and then back-crossing with *C. dentata*, apparently with encouraging results. The American Chestnut Foundation has hopes of reintroducing an almost-pure America chestnut one day.

Confusing nuts

Uncoppiced and unblighted, standing on a lawn, the chestnut is a dense, substantial-looking tree, forming a billowing pyramid of light-catching leaves. In July it is variegated by its long yellow-green catkins. In them lies its only disadvantage: a sickly smell that haunts high summer.

The genus *Castanea* gets its name from an ancient city in what is now Turkey: the city of Kastanaia, whence the Latin name; *châtaignier*, the French; and chestnut. The sweet (or Spanish) chestnut of Europe is *C. sativa*, from the Latin for "cultivated". The horse chestnut, of another family entirely, has merely borrowed it because its nuts are similar, at least in size and shape. To make matters more confusing, the French call the fruit of the *châtaignier* cooked in sugar a *marron glacé*.

Today it is pigs who eat the nut crop – industrially in such countries as Spain and Corsica. Pigs of the right breed fed on nuts from the right trees produce the caviar of the ham world: *jamón ibérico*. But it is not so long since chestnuts were a major part of the human diet. There are those who believe, indeed, that man's principal diet, before he discovered agriculture, was acorns.

Where the chestnut demonstrably feels most at home is halfway up a full-sized Mediterranean volcano. The most colossal hulk of a tree ever recorded still flourished a century ago on the eastern slopes of Mount Etna. It was 204 feet around, counting all the fragments, and supported a small industry of nut-gatherers, who in their simple Sicilian way eventually killed the goose by cutting off its branches to stoke the fire to cook the nuts. At 2,500 or 3,000 years of age (having been an old tree when Plato lived in nearby Syracuse) it finally succumbed to souvenir hunters.

Stepping stones

The European and American chestnuts are as close cousins as any two trees on opposite sides of the Atlantic. Which means, alas, that the European sweet chestnut is susceptible to the blight and can't be planted in America.

To take its place American nurseries have introduced the Japanese and Chinese chestnuts (*C. crenata* and *C. mollissima*). Of the two the Chinese has the sweeter nuts and makes the bigger and better tree – but still only half the size of the native species. There are beautiful variegated versions of the sweet chestnut, with leaf

The glossy, toothed leaves (left) of the sweet chestnut make it one of the handsomest trees for parkland. In its native Mediterranean it is the longest lived of all deciduous trees.

The tanbark oak (*Lithocarpus*) (left) differs from the oaks chiefly in having erect male flower spikes instead of dropping catkins.

The European chestnut (*Castanea sativa* 'Albomarginata') (below left) proves irresistible to browsers in garden centres. Few variegated leaves are so eye-catching, more so in a young plant than an old one.

The catkin-like flower spikes of the sweet chestnut are much bigger than other common members of the beech family. The smell of the flowers in the cricket season is the main drawback of a noble tree. Plus, of course, the massive litter of husks and nuts that comes with the falling leaves.

margins ivory-white or yellow. But these of course are blight-prone in America too. All prefer acid or neutral soil.

Happily the stepping stones between oak and chestnut are immune. Both are evergreens from the northwest States, Oregon and North California. The more chestnut-like of them in fruit is the golden chinkapin. For one of the most original and charming of relatively hardy broadleafed evergreens, this tree remains curiously uncommon outside its natural range, malingering where the summer is too humid. Its leaves vary from willow-slim to broad as bay, dark green above and beneath permanently upholstered in brown-gold down. Its nuts are very like chestnuts, and can be eaten.

Once golden chinkapins – the name is the Native American for chestnut – grew on a truly Western scale, 150 feet high among the redwoods and Douglas firs in the coastal hills. But their wood was too good to last in a region where hardwood is the exception.

The tanbark oak has less to attract the gardener. It is like a chestnut in leaf and identical to an oak in fruit. Its very tannic bark makes it important for the leather industry. In gardens it has proved wilful and demanding, asking a high price for its handsome foliage and striking flowers.

The golden chinkapin (above) takes on a fluffy bronze look in mid-summer with its thousands of flowers. Where it will not make a full-size tree the chinkapin is worth growing as a shrub.

The fruit (left) of the sweet chestnut: the only beech-family fruit commonly eaten by the human race today. The sweet brown nuts are enclosed in a prickly green case until they are ripe.

The Birches

THE EVOLUTIONARY GROUP that includes the beeches and the oaks ends with the birch family: the birches, the alders, the hornbeams, and the hazels.

Both the birches and the alders, the closest cousins, are lightweight trees of the forest's edge, adapted to rapid colonization of bare, often starved land and extremes of either drought or damp. The birches are the hardiest of all the broadleaves: the only trees native to Iceland and Greenland. Some of the biggest birches grow in the high latitudes, thriving in the almost constant daylight of an Arctic summer – and not minding the winter in the least. Light is the birch's passion; it is miserable in the shade.

There are some 35 species of birch (*Betula*). They circle the globe in the higher northern latitudes without varying very much in stature (save in tundra where it dwindles to scrub) or in stance, remaining always a light-branched, thin-twigged, and graceful tree. Birch species are notoriously difficult to identify. In collections they lose no time in hybridizing – which only makes matters worse. They vary (but not very much) in leaf shape, colour, and texture; more in leaf size up to the six-inch heart-shaped leaves of the monarch birch (*B. maximowicziana*) of Japan. Where they vary most is in the birch's most original feature: the texture and colour of its bark.

The colour range

The common silver birch of Europe (*B. pendula*) has chalk-white bark; the American canoe (or paper) birch (*B. papyrifera*) whiter still; the yellow birch (*B. lutea*) yellowish; the grey birch (*B. populifolia*) white streaked with black; the cherry or sweet birch (*B. lenta*) cherry-like brown; the river birch (*B. nigra*) dark shaggy brown; Forrest's birch (*B. forrestii*) a purply-brown; the Himalayan birch (*B. utilis*) buff or cream; *B.u.* var. *jacquemontii* from Nepal almost goal-post white; the Chinese paper birch (*B. albosinensis*) a range from pink to ox-blood satin.

In all these (except the river birch) the coloured layer is a stage in the bark's development from a reddish-brown to an eventual craggy black. During this stage the outermost layer peels off in the thinnest of ribbons. It has incredible endurance. Pieces of birch bark hundreds of years old have been found intact in peat bogs. In Siberia (at Dworotrkoi, to be exact) it has even been found in its original state attached to wood that has fossilized.

To dwellers in the north birch bark is indispensable. Native Americans make canoes of it; the Sami make cloaks and leggings; Norwegians roofs, covering a layer of bark with a foot of earth. When all is sodden in the forest birch bark will burn. To Russian leather (which is tanned with it) it gives its peculiar musty scent

There is a certain kind of heathland where birch is the first and sometimes the only tree, colonizing ground that only heather and bracken enjoy. This is silver birch country in southern England. The trees are stunted and age quickly, the white of their bark soon cracking and turning black.

and astonishing powers of endurance: rolls of it from a ship sunk in the Baltic were recovered and made into shoes a hundred years later. I confess to being a peeler. It offends birch addicts, but I find a quiet hour picking at the delicate curls of exfoliating bark and unwinding them as far as they will go to reveal the fresh bark underneath is one of the calmest and most absorbing garden pastimes. With a granddaughter to help it can be grandpa heaven. Others I'm told use a power-washer – perhaps those with grandsons.

Bark apart, the birch's beauty lies in its poise: the way it puts dense swarms of lacy twigs in the air with the flimsiest engineering. (There is a downside to this: the flimsy twigs constantly break off and litter the lawn.) Perhaps its best time is in spring, when its natural droop is emphasized by the weight of catkins. Beside an oak or a pine – its natural companions – it looks unmistakably feminine. Coleridge called birch "The Lady of the Woods". There is no avoiding the inevitable image.

But then there is autumn. Birch leaves turn the purest, most sovereign gold of any tree's. It begs to be planted against a background of dark conifers, where the slender pallor of the trunk, fountaining up to fall in shining leaf tresses can really be seen. Plant birch to the north of your house: not only is it a tree that wants the light shining on, rather than through it, but the moss which tends to grow on the trunk and hide its colour concentrates on the north side of the tree.

Birches are trees to plant in clumps rather than singly, for the maximum effect from their bark. Or you can cut a sapling back right to the base: it will sprout up again as a cluster of stems but they will take a few years to turn white. The American grey birch, which is perfectly white with interesting dark marks at intervals on the trunk, forms clumps naturally. They are not, in my view, trees to plant in avenues – or indeed in any formal or urban setting.

They are essentially nature's stopgaps, quick-growing and short-lived, graceful rather than dignified. It is their nature to lean slightly from the true; their strength is in their wayward femininity.

Stiff competition

There is no need to look for exotic species. The common downy birch (*B. pubescens*) of damp ground in Europe is not so beautiful as the more drooping silver birch of sandy heaths, above all in Scotland. The downy birch has downy twigs; the silver birch rough and warty ones – you can tell them apart even as young trees while their bark is still brown. The cut-leafed Swedish birch (*B. pendula* 'Dalecarlica') is a striking tree, but is it an improvement on the whole-leaved silver birch? The cultivar *B.p.* 'Tristis', narrowly soaring and weeping, and Young's weeping birch (*B.p.* 'Youngii') with a low domed head both draw attention to themselves. Young's is peculiar for having branches as bare as curtain rods with all the foliage hanging below them. But can one honestly say they are more beautiful than the species itself? Or than the straight-forward canoe birch? Or the Japanese silver birch, catalogued as *B. platyphylla* or *B. mandshurica* var. *japonica*? (Its tribe of *B. platyphylla* is now considered just an Oriental *B. pendula*.)

The coloured-bark species compare with the snake-bark and paper-bark maples as trees with a special value in winter: trees to plant near the house, or at any rate near a path where you will see them at close quarters. Their satin glow seems to reach its warmest on a cruel snowy day. Exactly which is which of these imports from the Himalayas is a science in itself. The two names to learn are *B. utilis*, whose varieties and nursery selections have some of the whitest bark, and *B. albosinensis*, which tends more to the brown and pink. To the gardener the name of the clone is more important than that of the species. Of the former, *B. utilis* var. *jacquemontii* is the best known – indeed becoming something of a carpark cliché these days. Of the latter, *B. albosinensis* 'Hergest' is a good example, with a trunk tending to the beetroot.

Eastern American gardeners know that white-trunked birches from abroad are caviar to a borer which starts at the top and wrecks the tree within a few years. Their best recourse is a native birch; *B. populifolia* 'Whitespire Senior' seems as close to a borer-proof birch as they will find.

I have lived with all sorts of birches and enjoyed them all for a while. The purple-leafed form of silver birch, a seductive idea at first, grows slowly and tends to revert to green. The Japanese silver birch has been the fastest to make a solid tree. I have a *faiblesse* for the odd bush called *B. medwedewii*, looking halfway to an alder with its stiff branches, big buds, and cones and corrugated leaves that infallibly turn butter-yellow. One yet to try is *B. insignis* from western China, which grows by the lake at Kew. It has such long catkins that you might almost think it a very early laburnum.

The birch of the western Himalayas, *Betula utilis* var. *jacquemontii*, is variable: its bark may be white, cream, or brown. Our knowledge of the birches of Asia is far from complete. This extra-white cultivar is called 'Jermyns'. A group is more effective than a single tree, especially by water for reflections (and especially with a hammock for contemplating the trunks close to).

BIRCH BARKS COMPARED

The outermost layer of bark of a birch is in a state of constant replacement. Few other trees have peeling bark, certainly not in the range of colours and characters that make birches so appealing to gardeners. Interesting as they are to collect it is still a group of one kind (and generally the kind that flourishes locally) that makes the most effective plantation.

The catkins of birches are not the first reason to plant them, but they come before the leaves and open the season with a swing. The male catkins are longer, up to four inches in some species. The shorter female catkins become the papery cone-like fruit.

B. pendula
The "silver" of the European birch is an ever-changing mix of chalk-white and grey. Its grace of growth is its strongest suit.

B. papyrifera
The American canoe or paper birch has some of the whitest bark of all, peeling away in thin layers usable as paper.

B. lutea
Yellow birch is yellow more in its magnificent autumn leaves than its peeling bark, which is amber or golden brown.

B. populifolia
The grey birch is a short-lived, many-stemmed tree colonizing derelict land. Its grey-white trunk is marked with black lines.

B. lenta
The American sweet birch has reddish bark like Tibetan cherry and yellow autumn leaves. Its sweet sap makes "birch beer".

B. nigra
The North American river or black birch has crackly dark brown bark very different from the usual silky sheath.

The monarch birch of Japan (*Betula maximowicziana*) has the biggest leaves of the genus: heart-shaped and up to seven inches long by five across.

B. utilis
The Himalayan birch is so variable in its bark that a sample is difficult. It can be chalk-white, peeling in horizontal sheets.

B. albosinensis
The Chinese paper birch is generally amber or copper to pink, with a sheen and a creamy-glaucous bloom.

B. ermanii
The Russian rock birch is a tall tree with pinky-white peeling bark on the trunk and orange-brown branches.

The unconventional Caucasian *Betula medwedewii* is a big shrub, perhaps 15 feet high and twice as wide, with glossy brown stems and green winter buds. Its roundish leaves turn light yellow and can stay on the plant a long time. The catkins stand erect.

The Alders

Alder is the streamside tree of neglected places, tough, undemonstrative, a good foil to feathery willows. At Rousham in Oxfordshire it plays its part in one of England's greatest landscape gardens, designed by William Kent in the 1730s along the River Cherwell.

ALDERS MUST BE THE EASIEST TREES to overlook, or simply to accept as part of the scenery. I lived for 30 years before I became aware of them. Perhaps that's why they give me so much quiet pleasure now.

Alders have their moment in late winter. They are the dark-brown silhouettes at the water's edge, often in single file along a stream bank, which gradually become more and more cluttered with dangling bits and pieces as the winter wears on. By spring their silhouette is noticeably denser, with a purplish bloom; the transparency of a winter tree is gone. In its place is an abstract of clustered catkins and cones. Cones on a broadleaf? Only on an alder.

Having spotted the alders in winter I watched them into full leaf and discovered that their silhouette is still their hallmark. By coincidence they tend to be neat and narrow, as short and lightly branched as conifers and with their branches almost as regularly arranged. Their leaf shapes vary in detail from species to species, but all are round or fat-oval in general outline and all tend to be held horizontally. From below, therefore, an alder floats black discs across the sky.

Thoroughly worthy

There is little call, it's true, to plant the common alder of Europe or the American speckled (or grey) alder from the east or red alder from the west – unless to vary the endless willows and poplars that tend to be stuck in wherever ground is boggy. But the genus has a dressy member in the Italian alder (*Alnus cordata*), and a tree, moreover, that seems completely impartial about site and soil. As to soil, indeed, it actually improves what it finds, having (like the pea family) the trick of introducing nitrogen with its roots. It is the tree to plant on the slag heap of an old coal mine to get vegetation started.

I first saw the Italian alder growing by the terrace of a mansion that had been burned down and its ruins demolished. The garden was rank and dispiriting: the ponds choked, the once-trim columns of yew toppling and coming apart. What were those gleaming dark green trees still formal and polished as butlers in the chaos? They made a great impression on me. Latterly they have had their virtues recognized by councils looking for smart street trees. Unfortunately this exposure has revealed what we all know about paragons: they can be a bit dull.

To be specific: a tall (to 90 feet) and narrow tree, though with branches more horizontal than upright. Substantial (up to four inches) heart-shaped leaves like a birch's, but darker green and glossy. Remarkable little cones: black eggs standing up on the branch tips in trios. Tolerant for dry and chalky conditions; always on best behaviour.

Natural variation of the common alder of Europe (*A. glutinosa*; so-named for the stickiness of its buds and twigs) and the grey alder (*A. incana*, or in America the very similar *A. rugosa*) – grey of leaf – has produced a few garden forms of these unpretentious trees. The most successful are cut-leafed versions (*A. glutinosa*

'Laciniata' and *A g.* 'Imperialis') and golden ones (*A. incana* 'Aurea').

There are more alders to explore, even if not many gardeners do. *Alnus maximowiczii* is as robust as its maxi-name, like its birch equivalent, the largest-leaved of all the birches. Noble catkins are the attraction. Maximowicz was a Russian botanist when the pick of east Asian plants was ripe for naming. *Alnus formosana* is more beautiful (I am always surprised when a plant from Taiwan – formerly Formosa – turn out to be hardy; but then Taiwan has high mountains and hardiness runs in the family.) *Alnus pendula* from Korea has shining leaves with acuminate points to make most alders feel dowdy. True, some of these grow more as tall shrubs than as single-stem trees, but the alnophiliacs is still a club with few members, and there is much to find out.

Alders (left) are often coppiced and spring up again as characteristic fan shapes in the landscape. In winter their silhouettes thicken and begin to glow orange or purple in the low sunlight as their catkins expand – one of the quiet pleasures of the short dark days in the country.

It is tempting (below) to refer to the fruit of the alders as their cones: they look so coniferous – especially when they become woody and brown. This is *Alnus glutinosa*, the common alder of damp ground in Britain.

ALDER CATKINS

Alder and birches are obvious relations in their catkins and subsequent fruit. Alder catkins are generally longer and heavier than those of birches; like the fruit, more eye-catching. The fruit of birches is similar but longer, thinner, and papery when dry.

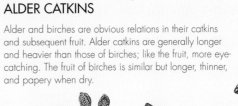

A. glutinosa is the common green-leaved alder of Britain. The pink female flowers will be the "cones"; three-inch male catkins open in early spring.

A. cordata
The Italian alder is the dressy street tree of the family.

A. incana
The European grey-leaved alder has three-catkin clusters in late winter.

Early in the year (above) the alder puts on its most telling performance with its catkins and fruit on its bare branches.

This is *Alnus maximowiczii* from Siberia producing its three-inch catkins in the coldest weeks. I bring them indoors.

The Hornbeams and Hazels

THE ALDER IS NOT THE ONLY SLEEPER in its family. It's a long time since there was exactly a craze for hornbeam (*Carpinus*). But serious tree-lovers hold them in high regard.

Hornheam's fate is to be passed over as beech, with the feeling that it is a bit short of beechiness – the silver trunk and silky leaf we prize beech for. In Germany it is commonly known as "white beech", as opposed to "red beech" for *Fagus*. It is the tree for stiffer, clayey soils, whereas beech likes them light, even sandy.

Certainly there is only one superlative that applies to hornbeam. It is the hardest. Ironwood is another name for it. Before iron became cheap and plentiful the load-taking parts of machines – cogs, axles, spokes – were made of hornbeam. So was the sweet and simple furniture of the dairy: the yoke, the scrubbed white milkpails, and the churn. It was also the premium firewood with the most calories, dense, heavy, and slow, better even than ash. Pizza parlours still use it in their ovens if they can get it. But its engineering role past, what part has it to play today?

The distinctive quality of hornbeam is its texture. An old parkland hornbeam is simply a good broad-headed tree of something less than terrific height, colouring well (the American hornbeam. *Carpinus caroliniana*, splendidly) in autumn. At close quarters, though, the subtle corrugation of its surface gives it special interest. Where the bole of the beech is smooth and round, the hornbeam looks as though muscles were straining within making irregular flutes and ridges. Musclewood is one of its American names. And where beech leaves are slips of well-ironed silk, hornbeams' are corrugated with ribs between the veins.

Where this texture can best be seen and appreciated is in a hedge. Not even the beech takes to the shears so well as the hornbeam. It makes a sturdy windbreak on almost any soil, like the beech keeping many of its dead leaves through the winter. One couldn't call their shade of brown a festive colour, but it has warmth.

Biddable as it is, the hornbeam will do more interesting things than a simple four-square wall. Its most effective use is at Hidcote, the Gloucestershire garden where hedges attain the status of art. Hornbeam here makes a hedge-on-stilts, pleached, as the old word is, into a perfect box shape carried on wide-spaced immaculate stems four feet off the ground. The European *C. betulus* is best for hedge work. Its upright form, *C.b.* 'Fastigiata', is sometimes used for avenues, though their proprietors may get a shock when the lissom things they planted succumb to middle-aged spread. For specimen trees there is far more choice; something like 40 species in all, the majority from Asia.

In hedging hornbeam you lose, of course, its natural shape. You also lose its fruit. The ordinary hornbeams of Europe and North America have little clusters of hanging winged nuts which

LEAVES AND FRUIT

It is not hard to see a kinship to birch and alder in the hornbeams, and a link with the hazels is not far-fetched. Botany keeps them apart, and gardeners have very different roles for them.

Carpinus betulus
Hornbeam fruit comes in short chains of little three-winged (inedible) nuts.

Corylus avellana
The common European hazel is typical of the genus with its roundish, hairy leaf and nut in a frill of leaf-like material.

Corylus colurna
Turkish hazel has a bolder, hairier leaf and bizarre whiskery convoluted nuts, not worth eating.

Fifteen years before this picture (above) was taken the site (the Prieuré de Notre-Dame d'Orsan in central France) was a farmyard. No hedging material could have achieved these architectural effects so easily or rapidly.

catch the eye in autumn. But the Japanese ones (*C. cordata* and *C. japonica*) seem to be covered in bunches of drying hops when the leaves fall. *Carpinus japonica* is more bush than tree. These reach their ultimate development in the glamorous Chinese *C. fangiana*, with distinctive long green buds, where the fruits form monkey tails up to 20 inches long.

Hornbeams have a close relation which carries the hop theme far enough to be called the hop-hornbeam. The hop-hornbeams (*Ostrya* species) flourish in America, Europe, and Asia, all looking remarkably alike and like hornbeams, save for their shaggy bark, which recalls the shagbark hickory. Like hornbeams, texture is their selling point. Their long catkins add spring to their season of display, and like hornbeams they colour yellow in autumn.

With added nuts

Perhaps the most familiar catkins of all are those of the homely hazels and filberts, species of *Corylus*, a genus associated with many trunks rather than one, hence more bush than tree. In their lowly ranks there are some very attractive individuals. There is no more velvety leaf, for example, than the Californian hazel's (*C. californica*). And the purple hazel (*C. maxima* 'Purpurea') is one of the best purple-leafed shrubs, its catkins deep red velvet.

The one tall tree the hazels can boast is a splendid one. It was

brought to Europe from Turkey at about the same time as the horse chestnut. Vienna, on the fringe of the old Turkish empire, has always specialized in it. The Turkish hazel (*C. colurna*) is a shapely, many-branched tree with a ramrod trunk, forming a regular narrow pyramid, ideal for avenues. One example leads through the Bois de Boulogne to the race course at Longchamps. Its foot-long catkins appear in late winter, to be followed by trios of nuts enveloped in weird convoluted husks with sticky whiskers. By all accounts the Chinese *C. fargesii* may even surpass it, adding the attraction of birch-like bark.

Hornbeam woods (above) are frequently seen coppiced back to permanent stools every 30 or 40 years, often under standard oaks. Their dense shade leaves a bare woodland floor.

The Chinese *Carpinus fangiana* (right) is the recently discovered horticultural star of its genus, with long green buds, charmingly pleated leaves, and long catkins and chains of fruit.

The Walnuts

WHAT MAKES A GARDENER proud and happy about having a walnut on his or her lawn? There can scarcely be a tree that puts on less of a show. It is one of the last bare trees of spring and the first of winter. Its leaves fall a sullen brown. Its catkins are by no means ornamental, coming with the leaves, nor is its fruit. It is stocky, heavy of detail, and thick of limb and twig.

Can trees communicate extra-sensorily? A walnut can. It speaks of fatness and fertility even while it stands looking like a puritan about to close the theatres. What it offers is of the highest quality: the best timber and the best of all nuts.

Of the 21 species of walnut which inhabit Asia and America two are widely planted – the American (or black) (*Juglans nigra*) and the European (or Persian) (*J. regia*). Hickories – those quintessentially American trees – are distantly allied. There is an Asiatic branch, too, the wingnuts, with very similar character and in some ways more beauty than either.

What all the family have in common are catkins; compound, often aromatic, leaves with anything from five to 25 leaflets along a stalk up to two feet long; bold buds above pronounced scars from the old leaves; and nuts. Their twigs are distinctly knobbly and substantial, full of slight changes of direction that give them character compared with, say, the ash, which has similar leaves. The scent of the leaves is not only diagnostic, it can also make an excellent cordial. You crush young walnut leaves into a jar and cover them with *eau de vie*, keep it for six months, then drain it, and mix the spirit with red wine and sugar.

With so much in common it is often a job to tell which member of the family is which. The way to know a walnut from a hickory (in the absence of a nut) is to slice a twig longways: if the pith is solid it is hickory; if divided with air pockets, a walnut or a wingnut. If it's a wingnut you're in an arboretum.

Fruit and timber

The Persian or European walnut has a history of cultivation as long as the fig's. Its Latin name was concocted for it by the Romans when it came to them from Persia. *Juglans* was derived from *Jovis glans* (Jupiter's acorn). Our name for it means simply "foreign nut", from the old German word, *Welsh*, for "foreign".

As a fruit tree the walnut has had a good deal of attention paid to it: there are a score or more of named clones, chosen for bigger fruit or more of it, or thinner shells. The best known, perhaps, is *J. regia* 'Franquette', a vigorous widespreading French tree with long, oval nuts, not specially big or in great numbers but full of meat, and the meat full of flavour. The biggest orchards of the European walnut are in central California (there are no less than 250,000 acres).

In Europe orchards are rare and it is planted in a more desultory way, in hedgerows and gardens. Grenoble in southeastern France has the great name for walnuts as fruit; the Dordogne in the

The black walnut of the eastern States is the most vigorous of the walnuts, whether it is planted in America or Europe. Its deeply furrowed bark is dark grey.

southwest for walnuts as timber. In the Dordogne a walnut tree can be an important family asset. Walnuts are indispensable trees for gunstocks: for weight, elasticity, and smoothness of touch there is apparently no comparable wood. Today the famous gunsmiths of London make special journeys to the Dordogne to find the best; the rest, one imagines, goes to make 17th-century furniture for the antique shops of Paris. Circassian ("Persian") walnut from the Caucasus has always had the name for the most beautiful figure of all. It is like a troubled, swirling pool of browns, from parchment to near-black, on the brink of a weir that draws the ripples toward it.

The common European walnut grows with almost alarming speed at first, then slowly. It never makes such a big tree as the American, either in height or spread. The only attraction it has which the American lacks is a gleaming pale grey bark: the black walnut's is blackish-brown. You can distinguish them easily by their leaves: European walnuts have up to a dozen leaflets with a terminal one to round them off; black walnuts have up to twice as many, can be two feet long, and often have no leaflet at the end.

The American nut can be huge – on some commercial varieties even orange-size – and good to eat if you can get in. So hard is the shell that industrial abrasives are made of it. America's biggest walnut stands in Oregon, 130 feet high and 125 feet across, with a trunk 21 feet around, a hybrid between *J. regia* and *J. hindsii*, a

The leaves (right) of European walnut emerge from pale grey twigs with a red tinge (and a sweet spicy smell) just after the catkins. The catkins are the male flowers; the female ones are tiny.

The European walnut (right) in summer has a green outer coat over the maturing nut within, which is still soft. At this stage it is good for pickling. The leaflets are in odd numbers, with a terminal leaf; while the black walnut's are in even numbers.

European walnut is a widespreading tree with a pale trunk and smaller leaves that flush reddish in spring. There is fierce competition with the squirrels for its nuts.

black walnut from northern California. Black walnut grows almost as well as this in Europe, compared with our native's best performance of 80 feet or so. So prized is its timber, indeed, that fine trees are at risk from "log-poachers". A rare but excellent form with fine-cut leaves, *J. nigra* 'Laciniata', is a show-stopper.

There is a cross between the two, *J. × intermedia*, which has no net advantage. More interesting are three Asiatic species, *J. ailanthifolia* from Japan, *J. cathayensis* from China, and *J. mandshurica* from Siberia: you are most likely to spot them by their oversize leaves – in the case of the Japanese tree up to three feet long.

Aggressive roots

The European walnut in fact is first choice for nuts. The American is the choice for ornament or timber. The American walnut has one nasty habit, however, that you should watch: it is capable of poisoning neighbouring trees and shrubs, particularly fruit trees (including its own offspring) with a substance called juglone in its roots. Apple trees near walnuts are often known to die mysteriously. So are birches, pines, and azaleas. It is a sinister development in the battle for survival – happily a secret the walnut can't impart to other trees. One other *Juglans*, *J. cinerea*, is not a walnut but a butternut – and potentially as delicious as it sounds. It comes from the northeastern States, is extremely

Walnut wood has some of the most beautiful colour and "figure", or patterns of graining, of any tree. It was the height of fashion for furniture in the late 17th and 18th centuries and is still indispensable for gunstocks.

hardy, but sadly subject to a blight which is destroying the native population; a good reason to plant it here out of reach of the blight. You would know it by its brown-felty shoots and many leaflets. Like all such eastern American trees, though, including the black walnut, it does better with a hotter summer than Britain provides.

Hickories and Wingnuts

IF THE WALNUTS INHERITED the dignity in their family, it was the hickories and the wingnuts which inherited the looks.

The hickories (*Carya*) are hardly known in Europe; not even for their superlative wood whose smoke goes up like incense from the barbecue pits of America. Yet the vast forests of the eastern States, with their vitality and variety undreamed of in the Old World, have hardly a more characteristic ingredient.

It is not hard to characterize the hickories as a race. They are like taller and more graceful walnuts; more finely textured; in brighter colours. Their catkins, instead of being a single tassel, fork to become three-pronged. Their wood is the finest of all for the traditional offices of ash – for tool handles and firewood.

Plant them small

The most distinguished of them is the pecan (*C. illinoinensis*). It is also, alas, the least hardy: a native of the Mississippi basin, in Europe not happy north of central France, but ripening fruit in Bordeaux. If the walnut (nut, not tree) has a superior it is this. Given the soil it likes (deep and moist), and a hot summer, the pecan is the fastest of its race and makes the biggest tree – taking spread into account as well as height. A pecan in Tennessee is 143 feet high, and not far off in total width. The leaves are composed of up to 15 leaflets, the longest of all the hickories', and every leaflet is gracefully curved. Bulk and grace combine to make a most impressive tree.

Its economic importance makes it by far the most planted of its race. There are apparently 900 square miles of pecan plantation in the States, and 500-odd varieties chosen for their fruit. The pecan has only one problem. All the hickories have it and so to some extent do the walnuts: they hate being moved. They depend on a long taproot, which resents disturbance. Nurseries don't like them, and often don't stock them, for this reason. At best they sulk; usually they die. The only thing to do is to plant them out tiny (and protect them from squirrels) – or bury some nuts.

The easiest hickory to identify is the shagbark (*C. ovata*). If ever there was a case where the botanist who christened the tree in Latin should have seen it first it is this. But he was Philip Miller, 3,500 miles away at the Chelsea Physic Garden, and all he had was a few leaves and a few (oval) nuts. Had he seen the amazing trunk of the tree he would have ransacked mythology for some old witch in moth-eaten rags to express its appalling look of wear and tear.

The pecan has the longest leaves and the sweetest fruit of all the hickories. But it is less adaptable than the other species, demanding hot summers to grow well. In this orchard in Texas the trees are irrigated for heavier cropping.

The hybrid wingnut (*Pterocarya x rehderiana*) (left), with Caucasian and Chinese parents, makes a massive and handsome deeply leafy tree in record time. The drawback is a strong tendency to sucker.

The pignut (*Carya glabra*) (right) might be happier to be called the smooth hickory. Its trim downy buds produce relatively small leaves. The nuts are not big, but the tree has reached nearly 200 feet and colours glorious yellow in fall.

It is when the shagbark comes to fruiting age, at 30 or 40 years, that its trim trunk starts to tatter. It may well be an evolutionary adaptation which has proved successful in keeping squirrels away from the nuts. Certainly the sharp-edged strips of bark, still rigid although apparently ready to fall, must present them with a problem. Since squirrels are their best agent for dispersing their seeds the Darwinian logic is hard to see. The shagbark has fewest leaflets per leaf; only five; the end one bigger than the rest. It is an upright, tower-headed tree. The tallest, 132 feet high, is in Texas.

The bitternut (*C. cordiformis*) has a pale yellow bud, curving and without covering scales, which is unlike any of the others. This is the most widespread of the family and the commonest, reaching as far north as Minnesota and Maine, where it turns from its gay green to a lovely straw yellow in autumn. Sadly, it has totally inedible fruit.

The mockernut (*C. tomentosa*) (its fruit, almost empty, is the mockery) has the alternative name of bigbud, which tells its story precisely. The velvety grey end bud is twice as thick as the twig behind it. This is the commonest southern hickory; not a big tree compared with the others, but excellent for timber and liked, also, for having sweet-smelling leaves.

One more, the pignut, is important and relatively distinctive for the smoothness of bud and twig, and glossiness of the undersides of its leaves, which give it its Latin name of *C. glabra* (literally, smooth hickory). It has the same range as the shagbark.

The question remains why all these species evolved along their different paths. Each has its preferred location, of damper or more sharply drained soil, but all like fairly rich feeding. Their ranges are fairly distinct, but they largely overlap; you often see two or more species in the same forest. Any of them would make a magnificent addition to a European garden in places with hot summers.

Impatient?

Only those with a large garden need consider the *Carya*'s more distant cousin, the wingnut (*Pterocarya*). Its botanical difference is not so important as its performance: the late Alan Mitchell memorably described it as "the fastest thing on roots". The Caucasian wingnut (*P. fraxinifolia*) is like many trees from the Caucasus, almost supernaturally vigorous and healthy. Crossed with its Chinese equivalent, *P. stenoptera*, it went ballistic. Either of them only asks for moist ground to create a huge dome within a few years. It has the liveliness of millions of shining leaflets (21 or so per leaf) on long stalks. By mid-summer its winged nuts are developing from the catkins in long yellowy-green streamers all over the tree, remaining summer-long as they develop into strings of little nuts, a sight to stop and gaze at, and at a time when performing trees are few. The tree I planted in 1975, not in a damp spot but on gravel, is 70 feet high with a trunk people take to be almost centenarian. There is, however, a snag: it suckers. Its roots run almost on the surface and put up countless shoots. Mowing to suppress them is difficult because of the roots. I need sheep.

The trunk of a tall shagbark hickory is covered with stiff curling and crackling bark – possibly a defence against nut-hunting animals. The shagbark covers a vast territory and grows as well in Europe as in North America.

HICKORIES AND WINGNUTS COMPARED

Hickories are recognized by their nuts and the number, size, and shape of the leaflets that make up their handsome leaves and turn various shades of yellow in autumn. Pecan leaves can have as many as 17 leaflets up to six inches and be up to three feet long in total. The map of where they grow, reflecting their tastes in soil and climate as well as their heritage, is complex and covers most of the eastern USA. Planters yet have to do justice to them in Europe.

C. tomentosa
Planted chiefly for its drooping sweet-scented leaves; its nuts are often empty.

C. illinoinensis
The pecan has the longest leaves and the sweetest fruit of all the hickories.

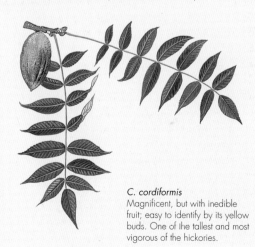

C. cordiformis
Magnificent, but with inedible fruit; easy to identify by its yellow buds. One of the tallest and most vigorous of the hickories.

Limes, Lindens, or Basswoods

HOW MANY OF THE GREAT country houses of Europe do you approach along drives of rearing twiggy monsters: England's limes; France's tilleuls; Germany's lindens; the basswoods of America? It must have been in the 17th century that they first came into fashion as the show trees of the great. One can see why they were thought the very thing for avenues. They take the form of a tower; a huge rectilinear tree, not spreading wide but with the upper branches going up and the lower ones down. Of all the biggest class of trees they are the most softly leafy, with fine-textured, heart-shaped leaves as big as the palm of a (lady's) hand. And in mid-summer they sweeten the air most mellifluously with the scent of their flowers. Yet despite its benefits the lime is one of the few trees whose vices must be taken into account.

The common lime (*Tilia × europaea*) is a hybrid, a cross between Europe's big-leaf and small-leaf limes (*T. platyphyllos* and *T. cordata*). It has all the vigour associated with hybrids – it has reached 150 feet in England, making it the tallest broadleaf in the country. Its problem is its suckers, which spring up all round the base and grow furiously in ungainly competition with the crown. The same exuberance produces great whiskery nobs on the trunk. Meanwhile it rains sticks from its crown. Its wide circulation is blamed on the Dutch nursery trade in the 18th century, which found the royal lime (*T. × e.* 'Koningslinde') as they called it, one of the most profitable trees to grow. The parks where it was planted were constantly grazed, it is true. Nonetheless it is a mystery why they didn't select a sucker-free clone.

Both parents of the common lime are widespread natives of Europe, and both are more satisfactory trees. The big-leaf lime is often the bigger in stature as well as leaf; otherwise the differences between them are unimportant. Today these are the ones that are commonly planted, particularly the red-twigged form of the big-leaf (*T. platyphyllos* 'Rubra'), which is distinctly warm-looking in winter and makes an excellent hedge. *Tilia cordata* also has a red-twigged version, 'Winter Orange', and a very useful upright one, *T.c.* 'Greenspire'.

Easy to spot; hard to tell apart

The botany of limes is far from easy. "When you've seen one you've seen the lot" was the reaction of one frustrated student. Even Linnaeus was fooled: his definitive or "type" species turned out to be the common hybrid. Poor John Gerard, 150 years earlier, illustrated his chapter on limes with an elm leaf – fooled, perhaps, by their both being lop-sided. They belong to the mallow family, the Malvaceae, which on the whole have hairy, alternate leaves and flowers in which bracts play a conspicuous part. The characters limes have in common are more or less heart-shaped, often hairy leaves (bring a magnifying glass; the position and even the shape of the hairs can be diagnostic); soft timber with a smooth grain that makes it ideal for carving; fibre (or "bass") under the bark once used for making rope; and an odd design of flower that springs on a stalk from a leaf-like bract. The sweet, far-carrying, somewhat citric scent of the flowers is one of the reasons for growing them. With a collection of five species you can be bathed in it from early summer to mid-autumn.

The search for a fault-free lime continues, however. All the foregoing have another problem. They are the regular diet of aphids which cover their leaves all summer with honey-dew. Honey-dew is not the pleasure it sounds; it encourages fungus that blackens the leaves and drops stickily on to everything below.

Sad to say the American basswood (*T. americana*) is just as much a victim. It should be a spectacular tree: it has by far the biggest leaves of the family – up to a foot long. But by the end of summer they can be a sorry sight, sucked dry and turning brown.

The most aphid-proof of the limes has turned out to be another hybrid, known as the Crimean lime (*T. × euchlora*). So far this has good reports, if not for its overall shape, which is too squat, at least for its glossy leaves; though my experience is unhappy: the wind round the corner of a barn tore the top out. All limewood is soft; this seems softer. *Tilia × euchlora* also has narcotic flowers – to bees that is. Limes draw bees in swarms. Feebly buzzing bees cover the ground below the tree. An alternative explanation: the bees simply eat too much and die of indigestion.

The bee problem is the only possible reason for not planting two of the most beautiful trees of the genus, the silver lime and the weeping silver lime (*T. tomentosa* and *T.* 'Petiolaris'). Both these trees have the backs of their leaves richly silvered with fine hairs so that when the wind ruffles them they glisten and wink. The silver

Lime flowers (above) hang on a stalk from the centre of a pale oblong bract (in fact a modified leaf) which presumably advertises them to bees – if the scent is not enough. This is broadleaf lime. Any lime flowers make a delicious infusion.

The weeping silver lime (right) (*Tilia* 'Petiolaris') makes one of the grandest parkland trees. Its wild origin is unknown, perhaps the Caucasus. It must be grafted, but it is predictable in form and in its glorious autumn colour.

Lime takes to hedging or pleaching, hedging on stilts, as readily as the traditonal hornbeam. This formal arrangement in the French style is at the early 17th-century Ven House in Dorset. The red-twigged varieties when pleached look well in winter.

lime is the most stiffly upright of the limes, and a more familiar sight in continental Europe than in Britain. The weeping version, which really only has pendulous branches, has leafstalks twice as long, which makes its leaves stir more readily and gives more glimpses of silver. As weeping trees go it is not the full cascade that, say, the willow can be but a tower usually with three soaring limbs from which its foliage hangs like robes. Trees don't come much grander than this.

Judgment of Paris

Does this leave us any lime totally free of faults? There are 20-odd species, and some of the Asiatic ones are still little known. At least four have outstanding qualities. The tree usually sold as *T. insularis* (aka *T. japonica* 'Ernest Wilson') has profuse bunches of flowers in mid-summer. *Tilia mongolica* has almost birch-like, small, slightly lobed leaves. *Tilia henryana* has big leaves that come out red with an extraordinary fringe of long points. But I have kept the best to last: there is one lime which clears up all the points. E H Wilson brought the Chinese silver lime home in 1900 and called it *T. oliveri*. In late spring and early summer when its pale soft leaves stir to show their silver backs, and its flowers grow from almost white bracts to perfume the garden, it is simply the best tree you can grow.

Limes are great role players, though; it is not simply a question of which is best, but which is most culturally appropriate. For the avenue through the park to a great country mansion the common

lime, for all its sprouts, offers the greatest height. Given the space and time you could make an even grander show with the weeping silver lime. The Chinese silver lime would be right for an *haute couture* villa, *T. henryana* for a romantic garden (it flowers long after the others, scenting the autumn air), the small-leafed lime or *T. mongolica* for no-nonsense public planting. The red-twigged big-leaf lime makes a fine specimen and looks perfect pleached. The straight-up silver lime risks looking naff and overdressed. Oddly enough its natural home is flanking a farmhouse door.

As an appendix to the limes is the right place to put their distant malvaceous relations from New Zealand, *Hoheria glabrata* and *H. lyallii*, or in Kiwi the mountain ribbonwoods. As sweet-scented small trees or large shrubs, in mid-summer stooping under bushels of soft white flowers, they deserve to be better known in the northern hemisphere. A selection from Wales called *H. glabrata* 'Glory of Amlwch' is best known. Their relations the lacebarks, *H. sexstylosa* and *H. populnea*, may not be so hardy but can grow into bigger trees. "Lacebark" refers to the fibrous bass they have in common with the limes. Roy Lancaster speaks warmly of the almost columnar, generous-flowering *H. sexstylosa* 'Stardust'.

SLIGHT DIFFERENCES

Even a specialist can be left pondering when faced with a collection of limes. The size of the leaf is hardly constant enough to be a guarantee. Experts bring out a hand lens and start examining the hairs under the leaves. Some contrive to be star-shaped – not easy for a hair. Toothing of the edges and length of leafstalk are other diagnostic features; so is time of flowering.

T. platyphyllos
The leaves of the big-leaf lime are sometimes six inches long, but those of the American basswood (*Tilia americana*) can be twice the size or more.

T. x euchlora
The leaf of the Crimean lime has the darkest and glossiest upper surface. Its branches are more or less pendulous and the whole tree rather squat.

T. cordata
The leaves of the small-leaf lime are paler underneath and less hairy than those of the big-leaf. Its flowers stand more upright and show up more clearly among the leaves.

T. tomentosa
This tree and its weeping counterpart, *Tilia* 'Petiolaris', are the best known of a number with silver-backed leaves which glint in the breeze. They flower toward the end of the season, after mid-summer.

One of the rare but very pretty Chinese species of linden or lime, *Tilia henryana*, has strikingly long teeth on its leaves, which emerge blushing. It flowers later than the rest, still fragrant in mid-autumn.

Tilia mongolica has some of the smallest leaves, toothed and distinctly lobed, rather like a silver birch. It is a moderate-sized tree and flowers at the same time as the small-leaf lime.

The most beautiful of the silver limes (*Tilia oliveri*) in flower comes from China. Its white leafstalks give each leaf a twist to catch the light with its white downy underside. As the tree gets taller its leaves grow bigger. It flowers early in the season.

Horse Chestnuts and Buckeyes

THERE ARE NO IDENTIFICATION problems with the horse chestnut. It peddles a whole catalogue of its own patented paraphernalia. Item: one sticky bud – take indoors for an instant spring. Item: one huge fingered leaf, with no competitors for boldness of design. Item: one tall candle of flowers, as original as orchids. Item: one prickly green container, holding the most succulently shiny of inedible fruit – the conker. Botanists have recently discovered that it is a distant cousin of the maples. If they say so … After all they both have opposite buds, twigs, and leaves. Also some pretty fancy leaves. But I don't think they're about to hybridize.

Despite its manifest entertainment value, the horse chestnut gets a bad press. It is often proclaimed "dirty": which means it drops things. Its foliage is described as "coarse": which means its leaves are big. And those responsible for public trees get themselves into terrible knots over the favourite sport of every nine-year-old: chucking sticks up at the conkers.

Recently it has been in the papers for health reasons, too. There could hardly be a more prominent and public tree; the showpiece of avenues in cities and parks throughout Europe. The public is rightly worried. It is being attacked by a leaf miner that turns its leaves a miserable brown in mid-summer and a canker that appears as alarming – bleeding of sap from the branches or trunk. The first is ugly but not fatal; chestnuts leaf early and make all their growth in early summer. By the time the leaves are ruined they have performed their function. The second can kill – though it is a mistake to assume that it will.

The buckeyes

The horse chestnut (*Aesculus hippocastanum*) is a rather rare cliff-dwelling native of northern Greece, Albania, and Macedonia. It was brought into Western cultivation from Turkey in Elizabethan times. It soon proved not only to be the biggest of all flowering (or rather ornamentally flowering) trees, but also to be completely hardy, to grow fast, and to grow in any soil. Also to grow old. The oldest dated trees in Britain were planted in 1664 and one of them (in Surrey) is still 125 feet high.

Yet it remains essentially an ornamental tree, the sign of human habitation. There are no woods of it: its timber is hardly worth having. We plant it because we love its paraphernalia, but also its billowing, drooping outlines, its bright shade of green in early summer, and its early autumn colouring of a warm yellow. It is one of the first trees to hang out its limp green flags in the spring.

In northern Europe horse chestnut is assigned almost as many avenue roles as the lime. This is a relatively young avenue in Windsor Great Park, leading to the grand formal entrance of Windsor Castle. Autumn colour can be fine in yellows and browns but is often anticipated now by unsightly leaf damage.

At some point in its evolution the horse chestnut must have been bothered by some animal attacking its unripe nuts and developed the not very formidable prickles on the encasing outer husk. As the nut ripens the case gapes open.

The sticky buds of the horse chestnut swell and begin to open in early spring, among the first trees to come into leaf. The bud scales fold back as the growing tip emerges. Six compound leaves and a spike of flowers are all packed into this little parcel.

America has its own set of close relations of the horse chestnut: the buckeyes – so-called because their conkers have a light spot where they are attached to the husk that supposedly looks like a deer's eyeball. There is the red buckeye (*A. pavia*) of the south; the sweet buckeye with yellow flowers (*A. flava*) from the Appalachians; the Ohio buckeye (*A. glabra*) with creamy flowers; and the California buckeye (*A. californica*) with pink or white flowers. The first three of these are medium-sized tree, the last a low widespreading one, and *A. parviflora* from the south an elegant lounging shrub.

Red buckeye is a true bright red, and pollinated by hummingbirds; in Britain, though, it makes a feeble, top-heavy tree. Sweet buckeye is splendid – perhaps not as eye-catching as the common horse chestnut, but head-turning with its yellow flowers in summer all the same. My favourite is California buckeye, a doughty performer with a unique sense of timing.

I brought back conkers from the Napa valley, where they hang in dense clusters from the bare branches in early autumn. They germinated almost instantly. In spring the leaves come so early you are sure they will be frosted; neat leaves of dark slightly metallic green already spread by the mid-spring. Frost doesn't harm them. It flowers profusely, white and pink, intricate and showy flowers, for weeks in early summer, forms its mass of dangling, drop-shaped fruit, and sheds its leaves in late summer. By year 12, I already had the second generation of trees from my Napa conkers in flower.

The inevitable cross between the imported horse chestnut and the native American buckeye (in this case the red one) has given us the red-flowered horse chestnut, *A. × carnea*. It is fashionable to regard this as the ideal horse chestnut for parks, its red (or rather dark pink) flowers and its smallish stature being regarded as advantages. But truth to tell this hybrid horse chestnut is a bit of a runt, and distinctly dowdy beside the majesty of a full-scale horse chestnut flashing white flowers from ground level up to its cloud-like crown. If you must have a coloured version the Swiss cultivar *A. hippocastanum* 'Baumannii' with double flowers (and no conkers) is a better choice.

Bigger, better, different

Some of the common horse chestnut's oriental relations offer it serious competition. The Japanese one (*A. turbinata*) has leaves as much as three feet wide. And the Indian horse chestnut (*A. indica*) has an indefinable grace and polish. The leaves are a little shinier, with perhaps a little more emphasis on the centre leaflet of the seven. It is different enough to catch your eye as an extra-handsome specimen, rather than a different and exotic species. So far, moreover, it appears immune to leaf miner. A particularly fine clone was selected at Kew and named *A.i.* 'Sydney Pearce' after a former curator.

The Sapindaceae, this far-flung family, now also embraces one of my favourite small trees and a jewel of my Essex garden, *Koelreuteria paniculata*. Its popular (rather than common) name of pride of India is no help; it comes from China. Its entertainment value starts with elaborately pinnate leaves, increases in mid-summer when its broad head is yellowed over with 10,000 little flowers, changes when the flowers turn to showy orange bladders all over the canopy, and comes to a climax with fiery autumn colour. It should be on the short list for small gardens.

A leaf miner attacks horse chestnut leaves on the inside, chewing out the green to leave ugly brown patches. It makes them fall early but is only marginally damaging to the tree.

The sweet (or yellow) buckeye is a tall tree in central and eastern States with yellowish flowers. Hybrids between this and the red buckeye have red and yellow flowers.

Pride of India (from China) is an overperforming relation; a glorious small tree whose yellow summer flowers, spectacular flourishes of orange bladder seedpods, and the bright orange of its fretted pinnate leaves is each a major garden event.

The California buckeye (left) is a wonderfully floriferous big bush on a short trunk, bowed down in early summer by masses of its elaborate and delicate white and pink flowers. It grows fast and flowers early; ordinary dry soil will suffice. (This is by way of being a recommendation.)

The sweet buckeye (*Aesculus flava*) (above) of the eastern States with yellow flowers is unaccountably little known in Europe for such a beautiful big tree. Crossed with the (less glamorous) Ohio buckeye (*A. glabra*) it has produced a bright yellow-flowered shrub: *A. x marylandica*.

The Maples of North America

THE OAKS, THE HOLLIES, the mountain ashes, and the southern beeches get through to the semi-finals. But in the finals there is no competition. The most varied and the most beautiful leaves belong to the maples.

The maples' own rules are fairly lax. Among the 150 species there are shrub-sized maples and very big trees. Though there is a tendency to a hand-shaped leaf, interpretation runs through everything from a simple oval to a filigree of 15 fingers. Bark is almost as varied. Nearly all are deciduous. Though autumn colour is very much part of the game, there are maples whose leaves just turn crisp and dismal brown, rattling to the ground with the first good frost.

A number of maples make a very pretty show of flowering. Where their family loyalty is strongest is their fruit – they all have "keys", consisting of two little nuts with one wing each, linked together by their bases – and in their almost universal rule of branching: two twigs at a time, opposite each other on the shoot, and opposite pairs of leaves. Often in summer it is the unripe fruit, bright bunches of keys, that attract attention.

Maples group curiously well by geography. America provides big and handsome ones, given to brilliant autumn colouring. Eastern Asia contributes small and intricate ones, carefully shaped and often beautifully coloured all summer. Europe has the strong silent ones – the bull of the family, the sycamore, for instance, or the workhorse, the field maple.

The New England palette

Europe has never seen anything like New England's maples in mid-autumn. There is no describing, nor photographing satisfactorily, the trumpet-pitch of red their leaves achieve. And strangely you never seem to see two trees with the same tone side by side: the whole gamut of the highest-frequency colours is in use.

Sugar maple (*Acer saccharum*) and red maple (*A. rubrum*) grow together through most of the eastern States (though sugar maple not in the south). Both colour supremely well in the northern part of their range, where they get the right combination of sunny days and freezing nights. Over part of their northern range, they are joined by the black maple (*A. saccharum* subsp. *nigrum*). None, unfortunately, gives quite such a performance in Europe.

It is easy to confuse the sugar and red maples. They are in the same size range: medium to big, but rarely giants (though the tallest red maple, in Michigan, reaches an exceptional 179 feet) . They are dense trees, close-branched, and rather upright; oval

Sugar trapped in the dying leaves of maples turns them an infinity of fiery colours. New England, where cold autumn nights alternate with sunny days, has the most pyrotechnical palette of anywhere on earth. An inevitable, perhaps even a corny, photograph …

shapes in silhouette. Extremely upright forms of both (*A. rubrum* 'Columnare'; *A. saccharum* 'Temple's Upright') have been propagated. *Acer rubrum* 'Autumn Spire' combines a narrow form and extra-hot autumn colour. The diagnostic clues are the red buds, flowers, and leafstalks of the red maple. Its leaves have more acute angles; the sugar maple's leaf lobes are separated by curves. The red maple grows faster, flowers earlier, and often turns red earlier in autumn. Given moisture it will stand city conditions; sugar maple is essentially a countryman.

Silver maple (*A. saccharinum*) is the third one with a similar range. In summer it is a more obviously decorative tree. Its leaves are a cheerful light green backed with a soft silver lustre; their lobes are longer, parted with deep indentations almost like the red oak's. You can spot it at a distance: its swooping branches turn up at the ends. Unfortunately there is a history of accidents with the silver maple. Evidently it was overcommended for its beauty and speed (it is very fast) at the turn of the 20th century in areas where conditions were too tough for it. Because, unpruned, it makes too many acute and fragile branch angles it got a name for splitting in storms. Yet of all the big American maples this is the happiest in Europe. It makes a beautiful light-shaded wide-crowned tree remarkably quickly and turns a happy pale yellow toward the end of autumn. There are also cultivars with yellow leaves and deeply cut, almost shredded leaves (*A.s.* 'Lutescens' and 'Laciniatum').

Possibly better than either red or silver maple for the future is a hybrid that happens occasionally in nature, now called *A. × freemanii*. Nursery selections are giving strong, tolerant, and autumnally brilliant trees: *A. × f.* Autumn Blaze and the narrower 'Armstrong' are apparently two of the best at present.

Coast to coast

One has come to expect the west coast to produce something bigger. In maples it is the leaves. The tall-tree maple of the west coast is the big-leaf or Oregon maple (*A. macrophyllum*), which has whoppers. In the Pacific forest, where broadleaved trees (even narrowleaved ones) are far outnumbered by conifers, the foot-wide foliage of the big-leaf stands out everywhere. In autumn when it is yellow or orange a single leaf shuffling to the floor among the dark fronds of Douglas fir is an incident worth watching. Big-leaf maple, like many western trees, thrives in Europe (though prey to grey squirrels) but not in the eastern States.

The native maple of the mid-west, so common there that it is often regarded as a weed, is the Manitoba maple or box elder (*A. negundo*). "Elder" is easy to understand, it grows to about the same size in the same disorganized bushy way. "Box" is more difficult. Box elder departs from the expected maple leaf shape, simply by treating three (or sometimes five) lobes as three (or five) separate short-stalked leaves. It is the only American maple with this character; in the East there are several more. There is scarcely an easier or more undemanding tree to be had. It can, and often does, grow in the dustiest and driest places. On the other hand it is dull in form, dull in colour, and makes no special effort in autumn. It is one of the rare species in which the variegated forms are superior to the normal one in almost every way. *Acer negundo*

Sugar maple (left) is one of the two great players in the New England fall. The other is red maple. Both are variable in their colouring, even on one tree, from yellow to any shade of red. Oaks, hickories, and birches are among the supporting cast.

Maple flowers (top) are underestimated ornaments, ranging from the tiny red bells of vine maple in the West, here, to entire canopies coloured red (the red maple), or yellow-green.

Leaves of red maple (above) and sugar maple compared. The leaf on top is red maple, with sharp-angled lobes. Under it are the simpler leaves of sugar maple.

'Elegans' has yellow edges to its leaves and *A.n.* 'Variegatum' has white. *Acer negundo* 'Flamingo' goes the whole way with pink shoots and white-edged leaves. Both are extremely conspicuous trees. A "blush" variety from California called *A.n.* var. *violacea* has flowers like pale pink beards – for a short while.

The west coast has its shrubby maple too: the vine maple (*A. circinatum*), so-called because in the tangle of the forest it often fails to make a real trunk, but wanders vine-like on the ground and up other trees. It is possible to make a tree of it, but more interesting to let it sprawl. The vine maple is one of the most worthwhile small maples for a garden – as good a choice as most Japanese maples and less demanding, excellent in autumn, and very pretty in spring with its quite showy red flowers.

America has one striped-bark maple, a ravishing tree which the settlers christened, with inimitable earthiness, moosewood (*A. pensylvanicum*). A few Asiatic maples have the same characteristic: the constantly smooth, fresh green bark of a tree with its living cambium almost on the surface, vertically striped with bold chalk marks. In this case the leaves are like ovals with wings: a totally different and memorable design. Moosewood is a tall thin tree, as though in a hurry to get out of the moose's reach. Its rather spindly branches are a beautiful clear green like the trunk, or in one cultivar (*A.p.* 'Erythrocladum') a pretty pale pink, which turns red in winter. If it were easier to grow (or say?), 'Erythrocladum' could become a craze. It can be too sparse a tree to plant in the middle of a lawn; one would tend to see right through it. Its place in nature is in light woodland, where its pale snake bark blends with the fleeting shadows.

Happily dendrology advances like other sciences, if more slowly. A cross between *A. pensylvanicum* and the Chinese *A. davidii*, known as *A.* × *conspicuum*, grows better and has even better-coloured bark. *Acer* × *conspicuum* 'Silver Vein' was the first introduction, in the 1960s; *A.* × *c.* 'Phoenix' is said to be even more conspicuous.

The variegated box elder (above) is in a different class from its dull green brother. In the gardens at Ninfa, south of Rome, the warm climate has made it a resplendent picture of freshness and vigour.

In the variety *Acer pensylvanicum* 'Erythrocladum' the stems are coral-pink: coppicing to multiply them is a good plan, but it is not the easiest tree to grow well.

The vine maple (*Acer circinatum*) (left) of the Pacific Northwest gets its name from its scrambling habit. It fills forest openings with yellow in autumn, but is very much worth growing as an alternative to some Japanese maples.

The big-leaf maple (*Acer macrophyllum*) (right) of the northwest is the finest broadleaf of the vast coniferous forests, growing to 100 feet with leaves a foot or more across. In spring it is dense with scented yellow flowers and in autumn a tower of gold, scattering leaves like plates.

The Maples of the East

THE MAPLES, WITH THE FLOWERING CHERRIES and the pines, are the cornerstone of Japanese gardening. They are the second annual event: mid-spring brings white and soft pink to the gardens and the woods; mid-autumn the flame colours of ripening maple leaves. Composing tone poems with maples, however, is a less exact art than it is with cherries. For you never know exactly what colour a maple will turn. And cherries are constant year after year; maples louder or fainter, earlier or later, according to the season.

None of the maples of the East is a big tree. Fifty feet is about as high as any of them go. But they have an incomparably rich range of variations on their elegant theme. Central China is the great reservoir of species. Many of the Japanese trees were imported from China in the distant past.

Nature's delicate caprice

As far as the West is concerned the Japanese maple is the little shrub-sized but tree-shaped plant with a domed head and fine-cut, sometimes red, foliage. This particular tree, the most popular of all garden maples, is *Acer palmatum*. 'Atropurpureum' is the dark red variety. Nurseries have concentrated on two variables: leaf colour and leaf shape. There are bright green, pale gold, and bronzy green, as well as various mottlings and marginations, besides the red form. There is a second group , with their leaves cut right down to the stalk in a series of almost needle-like lobes and generally referred to under various combinations of the epithet "Dissectum", depending on the colour of their leaves.

These are the weakest-growing kinds, never reaching tree height. I used to think all Japanese maples were reluctant growers, and it is true they need certain specific conditions to thrive. My infant collection suffered from overexposure; what they want is a degree of shade, protection from wind, and moisture in the air. What they hate is sun-baked drought. As soon as they found themselves in a woodland atmosphere they grew lustily, to 15 feet and as wide, not in the least objecting to my alkaline clay soil. They do have a disconcerting habit of shading out their own inner branches, which die off and catch your eye as conspicuous pale sticks. The mysterious little coral spot fungus can make a deadly attack on such dead wood. I enjoy snapping off these rejected twigs to burn them.

Then there is a third group of relatively vigorous small trees with bigger leaf lobes, normally seven of them, each with a serrated edge. Three of the best of these are *A.p.* 'Sango-kaku' (formerly

For a combination of elegant shape with brilliant colour there is no small tree to compare with the Japanese maple, photographed here at Miyajima. Scores of its cultivars have different leaf shapes and colour which seems to seize and intensify sunlight or even twilight.

Acer palmatum var. *dissectum* 'Seiryu' is one of the hardiest and most reliable of the cut-leaf maples and challenges the best for fiery autumn colours. It is vigorously upright, like another of my favourites, *A.p.* 'Katsura', whose leaves flush orange-yellow in spring.

The best of the golden maples is *Acer shirasawanum* 'Aureum', a tree that slowly builds up horizontal plateaux of foliage. This specimen has been allowed to grow too many branches; it looks better simplified to a few distinct stems separate from each other.

'Senkaki'), which has red twigs and colours a thrilling clear yellow in autumn; *A.p.* 'Linearilobum', whose green or red leaves are shredded as deeply as those of the "Dissectums"; and *A.p.* 'Osakazuki', which simply turns furnace-red for two or three weeks in mid-autumn. The original 'Osakazuki' tree, imported from Japan in 1886, is at the British national arboretum at Westonbirt, 30 feet high with half a dozen stems. Although it is shaded by a huge red oak it lights its dark corner like a lantern. A glade at Westonbirt is given to seedlings of *A. palmatum* (which, being seedlings, have no more exact name). The range of shapes, sizes, and colours is extraordinary.

More variations
The maples which botany knows as Japanese (i.e. the *A. japonicum* group) are not very different. The species tends to have more leaf lobes (from seven to 11) and to grow a bit bigger, but the detail of a downy leafstalk is the only crucial distinguishing mark. Two named cultivars of *A. japonicum* are outstandingly beautiful small trees: 'Vitifolium' (vine-leaved) for its broad, fan-shaped leaves; 'Aconitifolium' for its deep-cut leaves. Both colour magnificently in autumn. Very similar in shape and size of tree, *A. shirasawanum* 'Aureum' is precious for its pale warm-yellow leaves, especially as they unfurl in spring. It needs careful siting, in light shade, as hot sun scorches it. With a little guidance all these trees will make wonderfully harmonious shapes, putting on a terrace of leaves here, a terrace there like the hand movements of a Japanese dance.

The botanical section called Palmata, to which all these belong, has other members which have hardly been discovered yet by gardeners. *Acer sieboldianum* colours as well as any and is considerably more rugged, putting up with a Scandinavian winter, which would kill *A. palmatum*. It should be popular in the eastern States. Its name remembers Philipp von Siebold, the third of three

celebrated physicians (the others were Kaempfer and Thunberg) who used their post in Japan to collect and export precious plants, the lovely *A. shirasawanum* 'Aureum' among them. Siebold was expelled for spying. Would he be so pleased to have his name involved in *A. pseudosieboldianum* subsp. *takesimense*, however beautiful and useful? There are limits.

Japan has one totally different maple, with simple oval leaves. Only the keys and the opposite branching give away the hornbeam maple (*A. carpinifolium*). It has the same corrugations between the leaf veins as the hornbeam. Simple shapes, but smooth surfaces, also characterize the surprisingly numerous evergreen maples from the tropical forests of southeast Asia, whose occasional appearance in milder gardens often provides an identification challenge. Remember the opposite leaves.

Leaves in threes
Some excellent species share the characteristic of the box elder (*A. negundo*): having the lobes of their leaves completely separate and on stalks. One of the most beautiful of all small trees, the paper-bark maple (*A. griseum*), has little three-leaf leaves, greyish underneath as the name suggests. The bark of this tree is much more like a birch than a maple: it peels away in tatters of rich red-brown. Even the bark of quite thin twigs starts crackling and coming away. As it does so it catches the light, so that the interior of the tree is always full of reddish lights and shadows. The leaves turn scarlet, and so do its thousands of tiny keys lining the branches. Unfortunately very little of this harvest is fertile.

The Nikko maple (*A. maximowiczianum* syn. *A. nikoense*) is not common anywhere, even in Japan. Among the three-leaf species it is easily distinguished by having densely furry stems. The whole of a young tree bristles like a moss-rose. It grows into an upright fountain shape and colours brilliantly.

No maple turns true scarlet more consistently than the seven-lobed maple from Osaka, *Acer palmatum* 'Osakazuki'. Does being conspicuous in autumn have any evolutionary advantage? 'Osakazuki' seems to be a tough little tree when many get the vapours.

The Japanese maples with very finely divided leaves (here *Acer palmatum* 'Dissectum Atropurpureum') never grow higher than low mounds, which perfectly display the beauty of their almost ferny texture.

The paper-bark maple adds to the delicacy and autumn brilliance of its foliage the beauty of birch-like peeling bark in the warmest tones of red-brown. It is among the bigger oriental maples, growing as high as 40 feet, though slowly.

Another group of a good half-dozen have snake-bark more or less like the American moosewood. *Acer davidii* and *A. grosseri*, both from central China, are probably the best known. *Acer davidii* is a conspicuously coloured tree all year, with green and white bark, shiny green leaves which are heart-shaped and toothed rather than lobed, and bright red leafstalks. Its genes have also given us *A.* × *conspicuum*. One of Westonbirt's examples is a widespreading tree of a beautiful terraced shape. It colours rich yellow in autumn. *Acer grosseri* and its variety *hersii* have botanists in a tangle about whether they are two or one, or even just forms of *A. davidii*. Joseph Hers was a Belgian railway engineer who improved his leisure, while hunting up timber for his tracks, by some very distinguished botanizing. For the gardener's purposes, rather than the botanist's, it is fair to consider his, *A. davidii*, and *A. grosseri* as one group: immensely decorative and valuable trees. *Acer* 'White Tigress' is one of the best of a growing choice. With the related *A. forrestii* and the *A.* × *conspicuum* group they provide more and more wonderfully coloured bark.

None of these trees is by any means common. We are likely to see more of the lush, deep green Chinese *A. pubinerve* and maybe the Kawakami maple (*A. caudatifolium*) from Taiwan with glorious autumn colour. I grow *A. capillipes* from seed I collected in Japan and love its striped bark and coral-red shoots. There are endless possibilities. The trident maple (*A. buergerianum*), with leaves that suggest the name, is popular in California. And the Amur maple (*A. tataricum* subsp. *ginnala*), from the north of China, whose leaves are similarly three-pronged but with teeth and a long central lobe, is hardy and has a good press, though its promised autumn orange has never appeared in my tree.

The Maples of Europe

EUROPE HASN'T MANY MAPLES. What she has, on the other hand, are exceptionally useful. Two of Europe's maples are now basic planting material far beyond her frontiers. The sycamore (*Acer pseudoplatanus*) and the Norway maple (*A. platanoides*) are among the fastest, hardiest, and least demanding of all big trees. And both of them are prolific in coloured versions for ornament. One would not claim exquisite grace of design for any of them, but they are easy, colourful, and on the whole tough and vigorous.

The sycamore is the giant of the maple family. It is also the beanstalk: it reaches full height in 60 years. It makes a great girthy tree, longer in branch than in trunk, with pale rough bark breaking into plates like rhinoceros armour. Its green winter buds break to produce so many dangling yellow-green flowers that a tree of mine that flowers with a neighbouring laburnum almost rivals it in beauty. The big leaves of the sycamore darken during the summer to yew-green, often overlaid with black by "tar-spot" fungus. They have no autumn glory, but then they soon fall after a frost.

You can tell a sycamore in spring without raising your eyes from the ground. It sows its seed with such abandon that the ground below is a lawn of its little strap-shaped seed leaves uncurling. The town-dweller as a result often sees sycamore in the most unlikely and undesirable places. He comes to regard it as a tree weed.

Scotland (where it is known as the "plane") is where the sycamore grows in its greatest beauty. On the windswept moors of the border country wind-shaped sycamores are drawn like hoods around the cowering farms. In the "policies" (as they call the parks) of mansion and castle the "plane" is the dark green cumulonimbus, and in winter the doughty silhouette that defines where pleasure-grounds end and country begins. Above all the sycamore is the front-line defence against the salt-laden sea wind. It may crouch, but at least it grows, and provides that minimum shelter in which other plants can start to grow.

Useful all-rounders

Dark as the sycamore is, a purple pigment comes naturally to it and suits it very well. The purple sycamore (*A.p.* 'Atropurpureum') is a subtle tree; green at first glance, but with crimson flashes where the wind lifts a leaf. On top the leaves are green; underneath they are dark purply-red.

The golden sycamore (*A.p.* 'Worley') also works well, particularly in the spring. It is not such a big tree as the green or purple. But the doll of the family is the almost-miniature *A.p.* 'Brilliantissimum', a little mophead with leaves of a lovely faint

The European sycamore is one of the biggest of the maple family; fast-growing, immensely hardy, and prolific with its seedlings. It makes a broad tree in the open, noble in winter, casting a heavy shade with its big dark green leaves.

The Montpellier maple (*Acer monspessulanum*) is the common hedge maple of southern Europe. Its neat three-lobed leaves distinguish it from the five-lobed hedge maple of further north. The evergreen Cretan maple has leaves a similar shape.

Acer pseudoplatanus 'Prinz Handjéry' is not so well known as *A.p.* 'Brilliantissimum', nor so showy, but its odd yellow-bronze approach to spring, with multiple flowers the same colour, is unique in the tree world.

pink at first, then turning a rather chlorotic green. Maples as a race are shy of the hottest sun. 'Brilliantissimum' should be kept in a light shade. *A.p.* 'Prinz Handjéry' comes out a curious bronzy-green and stays that way.

The Caucasus, home of some of the most vigorous trees of all, offers in the sycamore line the frighteningly fast Van Volxem's maple (*A. velutinum* var. *vanvolxemii*). I planted one to admire its odd sharp brown buds, its flowers which stand erect, and its huge leaves. I'm ashamed to say that after 20 years I had to cut it down; it was 50 feet high in the wrong place. Its seedlings were all around, though, like a sycamore's, so a replacement was no problem.

The Norway maple is looked on as a more refined tree and often used for street planting. Any doubt about whether a tree is Norway maple or sycamore is easily resolved: the sycamore's buds are bright green, the Norway's brown. It has many of the sycamore's qualities of speed and indifference to conditions, but less of a liking for the sea. Its leaves are of thinner fabric and a paler colour, and they turn splendid gold, rather briefly, in autumn. A Norway maple should be encouraged to branch as low as possible: its yellow-green flowers are early and conspicuous. They repay a closer look.

Hardly any broadleaf tree, at least big tree, has given us such an imaginative catalogue of forms. There are mophead, upright, cut-leaf, variegated, and red forms. The red pigment is not altogether satisfactory in this tree. Either there is too much or not enough. Some forms (*A. platanoides* 'Goldsworth Purple' and *A.p.* 'Crimson King' are examples) start wine-red but turn a depressing dull purple-brown by summer. Far too many of them are planted; in parts of the States they are a purple plague. Some on the other hand – and these are much preferable – run out of red and end the season an original kind of bronzed green. *A.p.* 'Schwedleri' and *A.p.* 'Reitenbachii' are of this school. None of the fancy ones is as good as the straightforward species. *A.p.* 'Drummondii', on the other hand, promises to be a riot of green and white with its almost

The common field maple (*Acer campestre*) is too common it seems to have a fan club. Its tidy foliage (yellow in autumn) and moderate size make it extremely useful.

The design of keys varies endlessly. The Montpellier maple's (usually in thousands) have red wings pointing down so their edges overlap.

The Norway maple is one of the fastest-growing and most adaptable trees, beautiful in flower and in autumn colour. It has produced a dozen coloured varieties, including some very dark purple ones that should have been put on the compost at birth.

gaudy variegated leaves but usually reverts, out of reach and very conspicuously, to plain green.

Smaller sizes

America appreciates Europe's little hedge (or field) maple (*A. campestre*) more than Europe does. On the continent it is used to form hedges, but in Britain its growth is mainly wild. When I see it growing on the chalk downs in Kent I almost wish it came from Japan so that someone would make a fuss of it. Its yellow-gold is one of the best in a British autumn and it makes a neat little tree with its very small leaves, and answers the shears as well as a beech. No doubt with a good choreographer it could adopt poses as graceful as any. By chance I have the national champion of its golden form, *A.c.* 'Postelense', in my garden – by no means a monster but a fresh yellow all summer long.

Toward the Mediterranean its place is taken by the Montpellier maple (*A. monspessulanum*) which has three lobes to its leaves, a little like clover, instead of five. Most intriguing is the Cretan version (*A. sempervirens*), which is reliably evergreen, only bushy in England, but a great talking point.

The Italian maple (*A. opalus*) and the Turkish (*A. cappadocicum*) are both well worth considering, the Italian maple for its brave show of yellow flowers, as early as early spring, and the Cappadocian for its dome of pale yellow in autumn. The variety of *A. cappadocicum* called 'Rubrum' has startling blood-red new shoots in summer. Unfortunately, though, *A. cappadocicum* puts up suckers from its roots to make a thicket.

The medium-sized maples include the Caucasian maple (*Acer cappadocicum*), often grown in its red-leaved form, *A.c.* 'Rubrum', whose new summer leaves emerge scarlet and very decorative. Sadly it tends to sucker.

Cashews and Sumachs

THE SINISTER PRINCIPLE of poison links the members of the cashew family – for the main part inhabitants of the tropics, but with some desirable plants for our zone. They all have to some degree the faculty of giving you a nasty rash. North America's appalling poison ivy is the extreme example; Europe happily has nothing like this. But even the unroasted husks of cashew nuts can be an irritant. Even mangoes (another tropical relation) can give people who eat them unripe an extremely sore mouth.

The cashew is a Caribbean tree, a native of Haiti. The hardier nut-bearer of its race is the pistachio (*Pistacia vera*) of the eastern Mediterranean. Anyone who has sat in a café in Athens knows that the pistachio is a very profitable item of commerce in that part of Greece: old men and little boys are waiting in the wings to sell you a tiny bag for a huge price. There is another pistachio tree in the same area that produced the chewing gum (mastic, hence masticate for chew) which was issued to the inmates of harems to give them

white teeth and sweet breath. Californians and Italians could have a shot at growing either of these useful trees. There is also a very hardy pistachio (*P. chinensis*), but it has no comestibles to offer. It is grown for its upright shape, moderate size, and first-rate autumn scarlet. It will perform in the same way at Palm Springs and Kew Gardens.

The graceful pepper tree (*Schinus molle*) from South America is also limited in Europe to parts with a Mediterranean (or hotter) climate and in America to zone 9. The pepper part of it is its little red berries. But what makes it so popular is the combination of thick gnarled trunk and branches with droopy, almost weeping branchlets and fine feathery foliage. It reaches its ultimate height and equal spread of 35 or 40 feet in only 20 years.

People are apt to install pepper trees in brand-new gardens, only to find that they have commandeered the whole place. They line the streets of Riviera towns, strewing them with berries, and shade yards from Los Angeles to New South Wales. I owed my tree-cred entirely to a pepper tree once in a TV studio in Hollywood when the first edition of this book came out. "But you're a wine man", they said. "Bet you don't know what this [their yard tree] is."

The Chinese pistachio is fully hardy at Kew with fine glossy foliage and brilliant autumn colour, but inedible nuts. Nothing daunted, the Chinese apparently eat the young foliage cooked as a spring-time vegetable.

Staghorn sumach (*Rhus typhina*) (right) makes a gaunt little tree, wider than high with a flat top, whose female flowers (you need a female tree) are covered with crimson hairs while the pinnate leaves, two feet long, turn fiery colours. It comes from North America. *Rhus typhina* 'Dissecta' is a cut-leaved form.

The smoke tree (*Cotinus coggygria*) (right) envelopes itself in a haze of minute flowers. Their stems show up clearly against the foliage, which colours red in autumn.

The pepper tree of Peru (above) is a feathery evergreen for a warm climate. Its autumn cascades of fruit like red peas make it popular in southern Europe, Australia, and California.

The tree (right) of the true pistachio (*Pistacia vera*) can reach 20 feet or so but has only its richly oily nuts to recommend it. It demands a continental climate. The orchards of Asia and the Near East consist of selected female clones.

It may have been the first one I'd seen, but there is nothing else like it. "Don't you know a pepper tree?" I said.

Stagshorn and smoke trees

The two members of the cashew family important in northern gardens and generally considered small trees are *Rhus* and *Cotinus*. Rhus is best represented in the species known as staghorn sumach, from its antler-like growth with supple bends and very few twigs – and also for its antler-like texture; when stag's horns are "in velvet" in the spring they have the same mossy covering. The species (*Rhus typhina*) is a native of the eastern States; a hardy and tenacious plant despite its exotic, sub-tropical appearance. Plant it in a border where its roots get pronged when you are forking around and it suckers freely. But it is a little tree people love for its scarlet lollipop fruit and its fiery autumn leaves.

Chinese rhus (*Rhus verniciflua*) is the source of lacquer for furniture. E H Wilson gives the recipes for the various colours in his account of his journeys. The natural colour of the dried sap (which can also give you a rash) is black. Oil from the tung or Chinese wood-oil tree (*Aleurites fordii*) is added to make it brown; mercuric sulphide to make it red. Apparently only in a hot climate does the sap give the right results.

Cotinus is the smoke tree. The European *C. coggygria*, common in gardens with its often purple leaves and fawn fuzz of flowers, can hardly claim tree status. America's *C. obovatus* is more upright, and by all accounts one of the best-colouring plants even in the firebox of an eastern autumn. It attracts attention every year by the Victoria Gate of Kew Gardens. *Cotinus* 'Flame' and 'Grace' are two good cultivars produced by crossing the European and the American.

The Tree of Heaven and the Cedrelas

IT IS POSSIBLE FOR A TREE to be too easy and prolific – to devalue itself by appearing whether it is invited or not. The tree of heaven (*Ailanthus altissima*) makes this mistake. "Urban weed" is the phrase most often used about it; even "ghetto palm". No other tree in this book comes with a "Don't plant at any cost" label. But before dismissing it we should at least look closer. Its leaves, often dismissed as "coarse", are among the most impressively tropical-looking of any hardy tree's: plumes, sometimes a yard long, of as many as 30 substantial pointed-oval leaflets. Yet with these splendid leaves it contrives to remain open and light-shaded; often forked and wide-spreading; rarely more than 60 feet high.

Trees of heaven are (usually) either male or female. There are disadvantages to both; the male flowers smell nasty, but then the female has offspring. The female, however, is the more interesting: its fruit, bunches of propeller-like keys ripening bright red, is as good as a flowering season.

There is said to be no limit to the range of soils and other substances (ash, gravel, garbage) in which the tree of heaven will seed itself and thrive.

The way many people take advantage of such noble leaves without having to accommodate a whole tree is to cut the whole thing back every year as coppice. The effect in a border is of some ferocious fern, in which you will observe a curious habit: the leafstalks, usually in such compound leaves as deciduous as their leaflets, seem undecided as to whether they are leafstalks or permanent twigs. Often they let their leaflets fall some time before they let go themselves.

Close to mahogany

All the tree of heaven's close connections are tropical: the West Indies is the home of several species of *Quassia* that are relations. The West Indies is also the headquarters of the mahogany family,

Ailanthus is the "tree that grows in Brooklyn", associated above all with the seedier parts of New York. These shapely and decorous trees are managed by the Parks Department of Potsdam near Berlin.

AILANTHUS IN DETAIL

The long leaves are the main attraction of *Ailanthus*, though they have a foetid smell. Another is the heavy load of apparently red fruit that weighs down a female tree. Avoid male trees: the flowers stink.

The mature bark is grey with long vertical splits.

Ailanthus leaves can have from 15- up to 36-inch leaflets.

The fruit is a samara or key with a twist that makes it spin as it falls.

The variety *Ailanthus altissima* var. *sutchuenensis* has shining reddish brown shoots and purple leafstalks.

Ehretia dicksonii (right) develops rapidly into a short broad tree with 10-inch leaves and intensely scented summer flowers. Its corky bark is another talking point.

Three-foot leaves (below) reminiscent of ash are perhaps the strongest point of the toona. They can lend an exotic sub-tropical air to any group of trees. Long panicles of little white flowers come in summer and the autumn colour is a fine yellow.

of which one branch – the West Indian cedar (*Cedrela odorata*) and its Chinese counterpart (*Toona sinensis*) – is remarkably like the tree of heaven, and also hardy. The West Indian cedar gets its inappropriate name from its scented red-brown wood; the wood of cigar boxes.

The Chinese cedrela, now rebaptized the toon or *Toona sinensis*, has little-known gastronomic potentialities. Its young shoots and leaves, which are delicate shades of pink and cream, have a smell and taste of onion (to me closer to rubber). To the Chinese they are a vegetable. I can't help wondering how long the ones in the streets of Paris would remain so leafy if this were generally known. The Chinese toon is in every way the tree to plant if you like the look of the tree of heaven. Its fault is that its roots are apt to sucker. Its advantages are its pink flush as it comes into leaf and an autumn parade of yellow as close as we will easily get in England to the colour of hickories. The tree I planted in 1980 is as wide as it is high – about 35 feet. In summer open-branched racemes of little white flowers three feet long hang among the splendid leaves.

Only in the south of Britain, and then only as a shrub, can we grow the chinaberry or bead tree (*Melia azedarach*), its near-eastern sub-tropical relation, but it is worth the effort. Melia is fast-growing and shady, with fragrant lilac flowers; pretty in leaf but prettier after leaf-fall, when the yellow berries stay glistening in its crown. In each berry is a hard little bead.

Melia has insecticidal properties, but they are dwarfed by the virtues of its Indian relation the neem (*Azadirachta indica*) – otherwise known as "the village pharmacy". The virtues of the neem as a tree are its shade in the desert (Saudi Arabia has half a million shading haj camps) and its fertilizing roots. As a drugstore its list runs: acne, contraception, diabetes, fever, fleas, lice, malaria, scabies, soup, toothpaste … A transnational company tried to patent its chemistry and was seen off by the Indian government.

Worth it for the scent

I risk a taxonomic shootout by putting it here, but must find a spot for the distantly related *Ehretias*, *dicksonii* and *acuminata*, from Taiwan. They have no grace and are rare in cultivation, but those who have smelt their white flowers in summer strongly recommend that we plant more.

The Citrus Family

RUE IS A LITTLE HERB WITH A STRONG SMELL. Few of the 1,600 plants that sail under its flag as Rutaceae are trees outside the tropics, but the few include the highly decorative and useful Mediterranean-hardy genus *Citrus*, the oranges and lemons; also the eastern American hop tree or wafer ash (*Ptelea trifoliata*) and the Amur (which means, roughly speaking, Manchurian) cork tree (*Phellodendron amurense*).

What singles them out as associates of rue is in their having translucent glands in leaves filled with aromatic oils. Fastidious Frenchmen of the 18th century "stopped their noses" with rue (don't try this at home). These "pellucid-punctate" glands are the best way of identifying a member of the family: held to the light they show up as translucent spots.

Where oranges and lemons grow

The orange tree (*Citrus sinensis*), prettiest of the citrus family and the most able to stand relatively low temperatures, came into cultivation in Arab lands from the Far East at least 1,000 years ago, with a complex history of hybridity already in its genetic baggage. In 17th-century Europe the orangery, a room with big windows and a stove for overwintering orange trees, became a craze. The huge white tubs containing the trees were manoeuvred into it in the early autumn and out into the garden again in the spring. The trees barely existed through the winter and would have died without their summer out of doors. They must have looked pretty hung-over in the spring. It was an essential part of the upkeep of the ornamental oranges in those days to hose them down regularly – even daily – to keep pests and disease at bay. Oranges are singularly prone to both.

A winter temperature in the upper forties Fahrenheit (10°C) is enough for an orange or lemon to thrive and set fruit. Spring is the season of blossom, so sweet-smelling that in orchard areas it can become too much of a good thing. The fruit takes a long time to come to maturity. The following winter is the normal harvest period in sub-tropical conditions, but in cooler parts oranges are not ripe till the following summer – hence the charming combination of fruit and flowers on the same tree together.

The citrus list is formidable. A wander in the citrus collection of (for example) the Hanbury Gardens at La Mortola on the Italian/French border is bewildering. Go to Dominica, the home of Rose's lime juice, and the varieties of lime in the orchards tells of a long history of experiment. Apart from the sweet oranges there are Seville oranges (bitter; the best for marmalade), tangerines with their reach-me-down skin, lemons (the least hardy trees), limes (which gave their name to all the "limeys" who sailed on British

Orange trees at the monastery of Sant Antonio near Tivoli, east of Rome. Their charm lies in the glossy evergreen foliage against which white flowers and bright orange fruit stand out like decorations: often at the same time, since fruit takes a year or more to ripen.

The orange trees in a Paris couturier's garden on Cap Ferrat are formally dressed.
Oranges are eminently full-dress ornaments in frost-free gardens, and adapt perfectly
to life in pots or Versailles tubs brought into an orangery (or shed) for the northern winter.

SUB-TROPICAL FRUIT SALAD

The limes have their origin in northern India. Their skin and flesh remain green and their citric acid content high.

The grapefruit (or "shaddock") is *Citrus x paradisi*, apparently a West Indian cultivar of the sometimes enormous shaddock of Polynesia.

The lemon tree was probably acquired by the ancient Romans in Persia or India. The little Meyer's, from China, grows well in pots.

The sweet orange, the sweetest fruit of the family, was bred from natives of southeast Asia. *Citrus sinensis* is sweet orange, *C. aurantium*, bitter orange.

ships and drank lime juice to ward off scurvy), grapefruit and its forebear the shaddock, and citron with its thick skin, used for flavouring cakes. The list of hybrids is as long. There is the ugli (a cross between grapefruit and tangerine), the citrange, the limequat, the tangor, the orangequat, and even the citrangequat. The -quats involve the blood of China's shrubby, sour-fruited kumquat *C. japonica* (aka *Fortunella margarita*), which is a hardier plant.

The urge to grow oranges or lemons is strong in those who hanker for a Mediterranean climate – at least in winter. Planting a pip is not a hopeless resource – kept indoors it may well grow into a tree – but it will be a tree with useless, if any, fruit. The best tree you can grow if you have a conservatory and a hankering is the little Meyer's lemon (*C. limon* 'Meyer'). It needs minimum conservatory protection, but it rewards you with unstinting generosity and beautiful small lemons all winter – so mild and fragrant that it is tempting to suck them. They were introduced to the States from China 100 years ago by Frank Meyer, working for the US government. He had found them used as a popular pot plant. They are happy in pots (and will fruit well even in a small one) but they will also grow – even from seed – to seven or eight feet in the ground.

In London's Chelsea Physic Garden, meanwhile, where a few years ago it was considered extraordinary to see a mature olive tree, grapefruit now grow and set fruit.

The Japanese bitter orange (*Citrus*, syn. *Poncirus*, *trifoliata*) is hardy even in the country in Britain and up to Boston. It can scarcely be called a tree, but it has plenty of flowers and little oranges that could be made into marmalade in desperation. There is a cultivar with wonderfully contorted stems called *C.t.* 'Flying Dragon' which looks best in winter when you can see the sinuosities of its deep green stems.

All with translucent glands

The same chemistry gives the little hop tree one of North America's best-smelling flowering seasons, and in a smaller degree pervades the Amur cork tree and the Korean *Tetradium daniellii*. The hop tree is occasionally planted as a hardy small tree with mid-summer scented flowers – especially in its yellow-leaved version. You are unlikely to notice the green-leaved one until it sends out its head-turning scent. The Amur cork tree is planted more for its corky bark, which looks impressively bulky on its thick, wide-reaching branches. Both these Asiatic trees have long compound ash-like leaves; those of the Amur cork tree turn a useful autumn yellow.

Euodia (to most of us), now changed to *Tetradium*, *daniellii* is a rare orange-relation from China, notable for its generous sweet-smelling blossom in late summer when flowering trees are scarce, and its purplish fruit. The biggest in Britain is more than 70 feet high.

The hop tree (*Ptelea trifoliata*) shows little sign of its citric blood except in the seductive scent of its flowers, crushed leaves, and bark, which is bitter enough to substitute for hops. Its elm-like fruit hangs on after the leaves. *Ptelea trifoliata* 'Aurea' with gold leaves is popular in gardens.

The soft blackish bark of the Amur cork tree from northern China, here at Ussuriland, Primorsky, southeastern Siberia. The trunk often forks low and massive furrowed branches take over, making a most impressive wide-spreading shade tree.

The Dogwoods

THE NAME DOGWOOD (*Cornus*) covers more different kinds of shrub and tree than one name seemingly should. It covers creepers ankle-high, shrubs waist-high, little trees window-high, and big trees of the high forest. There are almost as many designs of flowerheads (the actual flowers are all pretty similar) as there are sizes. The leaf, always an oval with veins curving to meet the centre axis at both ends, is the only clearly visible link between them all.

Although the dogwood trees have two very different types of flowerheads they are all so whole-hearted about it that the effect comes to much the same thing: a spread of warm white, of branches obliterated under tablecloths of massed flowers which have no equal – not even among the cherries. The common flowering dogwood (*C. florida*) of eastern North America, the Pacific dogwood (*C. nuttallii*), and the Japanese dogwood (*C. kousa*) are the proponents of one type of flowering. The giant dogwood (*C. controversa*) of China leads the other type (its junior members are the suckering shrubs we grow for their winter stems).

What seem to be petals on the flowers of the first type are really no such thing: they are the modified leaves known as bracts, which usually play a humble supporting role protecting the unopened flower. Inside that ring of four (sometimes six) white pseudopetals the real petals are tiny: miniature parts of a cluster of miniature flowers. The second type simply has bigger clusters of bigger flowers, and no fancy bracts.

It is the way the dogwood grows almost as much as its floral ardour that makes it such a spectacle in spring. Its tendency is to throw out long branches at 45 degrees. Every flower tells because it is held face upward on these branches.

"Flowering" dogwoods

From the point of view of habit and flower there is little to choose between the flowering dogwood of North America and the (normally rather smaller) Japanese dogwood – except their flowering times, which are staggered … making it worthwhile planting both. Gardeners in the eastern States have seen a fungus disease, anthracnose, play havoc with native dogwoods and hope that Asiatic species will be immune – which is by no means sure.

Unfortunately the American dogwood demands a continental climate: Britain is too temperate for it, and so is the northern Pacific coast. We therefore miss the pink varieties which only occur in this species. The Japanese dogwood does triumphantly in England, however, and in France both succeed as well as they do in New York or Pennsylvania. The Chinese variety of the Japanese (*C. kousa var. chinensis*) is if anything even better, more vigorous, and ardent in flower, than the species, and is the form now most

The wild flowering dogwood (*Cornus florida*) of the eastern States has had plenty of attention from nurserymen, with results like these perfect trees. In nature the flowers are white or pinkish. This is in Illinois.

frequently planted. Cultivated varieties have proliferated and crosses have produced such class acts as *C.* Venus (see below), but the species themselves are so beautiful that it is largely oneupmanship to plant them. I have three mature plants of *C. kousa var. chinensis* and each is distinct in the form and deployment of its bracts. It would be easy to invent names for them – but is there any point? In a good summer the flowers are followed by intriguing pink or red fruits on stalks rather like bumpy cherries, sweet and good to eat by sucking the flesh out of the skin.

Having seen dogwood in flower in Connecticut I was prepared to settle for that. It would have been absurd to ask for more. More is what they get on the west coast, though. The most memorable of many stirring forest sights is the Pacific dogwood in mid-October high in the hills of Oregon, not only lighting the depths of the Douglas fir and hemlock with flurries of golden and scarlet leaves, brighter even than the big-leaf maple on the same hill, but also lighting its own brilliant branches with a second crop of its huge white clematis-like flowers.

Alas this, the biggest of the broad-bracted dogwoods, able to grow to 100 feet in its own woods, is less happy abroad. In the east it suffers from the cold. In Europe it is good, but short-lived and never a big tree. Its qualities have been partially assimilated in such English crosses (with *C. kousa*) as *C.* 'Porlock' and *C.* 'Norman Hadden'. Recently they have been triumphantly subsumed into a

new cross that seems to tick all boxes. *Cornus* 'K30-8' Venus has genes from both the Japanese and Chinese forms of *C. kousa*. They have combined with the Pacific dogwood to make a broad plant with huge creamy-white bracts resistant to anthracnose. Acid soil and ample moisture are the formula that suits it best. The problem of what to call these ingenious crosses remains. 'K30-8', indeed!

The American east coast stock has been improved by breeding with the Pacific dogwood in such varieties as the famous *C.* 'Eddie's White Wonder' and *C.* 'Ormonde', which seem to grow better than either of their parents, even in Britain.

Alternative models

The biggest tree of the dogwoods we can grow in either Europe or the eastern States is the giant dogwood of China. This is the one without big white bracts, but with broader clusters of little flowers which at a distance have almost the same effect.

The horizontal branching pattern is most marked in this tree. There are specimens, perhaps slightly aided and abetted by the gardener, which have made perfect wedding-cake tiers year after year on which the flowers lie like a covering of snow. Its freer form is perhaps even more dramatic. The branches go like the swirl of a dancer's skirt or the tipping horizons of distant alps. By 50 feet with most trees a degree of round-headed anonymity is creeping in. Not with the giant dogwood: it builds another majestic plateau for its 50th birthday.

Gardens where such an empire builder would soon take over have the alternative of an exceedingly pretty variegated-leaved version of the same tree, *C. controversa* 'Variegata'. Like most trees with a short share of chlorophyll it grows slowly, but with the same deliberate pattern of whorled branches. The leaf margins are white, so whether the tree flowers much or not makes little difference: you have your wedding cake all the summer. A less emphatic and smaller but charming alternative is *C. alternifolia* 'Argentea'.

The dogwood family is by no means exhausted with these. There is the cornelian cherry (*C. mas*) of Europe, which flowers yellow before the leaves in late winter, and like all dogwoods gives a good account of itself with autumn colour. The white-variegated form of this is one of the prettiest and daintiest of all variegated plants, and can have masses of edible (at a pinch) red berries. Its larger Chinese cousin *C. chinensis* has large clusters of yellow flowers in spring (it has been described as a cornelian cherry on steroids), and beautiful glossy leaves with long drip-tips in summer. When it was first brought over from India it proved too tender for England – there are now hardier forms from China circulating in cultivation.

Meanwhile for privileged southern gardens the new challenge is *C. capitata* from western China, an evergreen tree that smothers itself in wide creamy-white bract-flowers and follows them with fruit the Chinese consider a feast. To be one up on a neighbour growing *C. capitata* you have to grow *C. hongkongensis*.

VARIATIONS

Flowers with big bracts, usually white or cream, sometimes pink, are the main attractions of the "flowering" dogwoods; in others habit of growth, autumn colour, colour of bark, variegation, and/or fruit are individual and are often eye-catching.

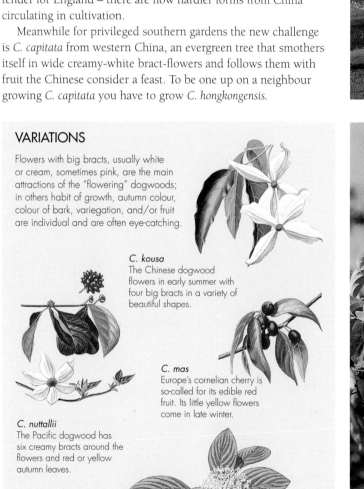

C. kousa
The Chinese dogwood flowers in early summer with four big bracts in a variety of beautiful shapes.

C. mas
Europe's cornelian cherry is so-called for its edible red fruit. Its little yellow flowers come in late winter.

C. nuttallii
The Pacific dogwood has six creamy bracts around the flowers and red or yellow autumn leaves.

C. controversa
The true tree dogwood of China can reach 60 feet with noble tiers of branches.

The giant dogwood (*Cornus controversa* 'Variegata') (above) of the Orient has a superb horizontal habit of growth, coming in time to look almost like a cedar in the snow when its branches are covered in dense masses of creamy flowers in late spring.

The eastern American *Cornus florida* f. *rubra* (far left) is an old variety but a wonderful one. Sadly it is rarely this good in Britain, even on the acid soil it needs.

Strange strawberry-like fruits (left) follow the yellow flowers of *Cornus capitata*. In very mild winters this dogwood is evergreen, although some leaves will turn purplish and fall.

Dove Trees and Black Gums

SO MANY OF THE TREES OF NORTH AMERICA have turned up again as survivors of the Ice Ages in China that it makes headlines when a family is found that holds out in only one of these refuges.

Leaving aside the ginkgo, and with due respect to the wollemi pine, no individual tree has caused quite so much wonderment as the *Davidia*. The idea of the dove tree, the ghost tree, the handkerchief tree (it has been called all these) stirs popular imagination – and the sight of it is sufficiently rare to keep the buzz going. Just don't say handkerchief tree when you're in China. Apparently there are sensibilities …

Father Armand David, the French missionary who first reported the giant panda and the deer that bears his name – a water-loving beast bred only by the emperor of China but extinct in the wild – was also the first to report, in 1869, a beautiful new tree in the mountains of western China. He told of huge white flowers, hanging like handkerchiefs (or fluttering like doves in the wind) among its branches.

Not until 1897 did another missionary, Father Farges, collect seed from the tree and send it home to France, to the great tree-collector Maurice de Vilmorin. Of the 37 seeds he sent only one germinated, but M. de Vilmorin planted it out in his arboretum at les Barres, near Fontainebleau, and it did splendidly. It flowered in 1906 and the West saw its first dove (or handkerchief).

But Veitch's Exeter nursery, the most enterprising in Britain at the time, knew nothing of this Gallic transaction. They had just engaged the young E H Wilson and the first job they gave him was to collect the *Davidia*. He set off in 1899. His only information about the tree came from Dr Augustine Henry, the famous amateur botanist who had lived in China for nearly 20 years. First Wilson had to reach Henry, who was in Yunnan in the southwest. Then, with Henry's instructions, he had to travel the 900 miles to Ichang in central China where the tree was. Henry drew Wilson a map – to show a solitary tree in an area the size of England.

Wilson was boat-wrecked in rapids. He was travelling through country that was notoriously unsafe – it was the time of the Boxer Revolt – and his Chinese guide was an opium addict. He found the tree, though. Its stump was standing by a house that had just been built from its wood. "I did not sleep that night," he wrote in his diary. Eventually, by combing the neighbourhood, he found a grove of *Davidia* and collected their seed. When he sent it off to London, England he was sure it was the first ever to leave China.

Wilson was bitterly disappointed when he came home to find that de Vilmorin already had the tree in his collection. But on closer examination it turned out that there were two different varieties. Farges's was named *D. involucrata* var. *vilmoriniana*, Wilson's was the typical *D. involucrata*.

The two varieties are almost identical, the only difference lying in the colour of the buds and the undersides of the leaves, but more significantly in their chromosome numbers. They are not compatible; there is no hybrid between them. And there is a big enough ecological difference for Farges's to be fully hardy and thrive in Boston, at the Arnold Arboretum, while Wilson's tree struggles.

The botanists' first impression of the dove tree was that it belonged to the dogwoods, and some still say that is where it belongs.

Tupelo or black gum (left) is one of the events of the eastern American autumn. It is a swamp-loving tree, oak-like in construction, with lustrous leaves which blaze as bright as any maple when the cold nights come. Its flower is totally inconspicuous; its fruit a small blue oval. Neither (any more than the leaves) suggests its relationship to the dove tree. In Europe its autumn colour is less certain.

The dove tree (above right) looks rather like a linden (or lime) with oversized dogwood flowers. The two "petals" of each flower have a leaf-like construction and one is four times as long as the other. The "doves" (or "handkerchiefs") grow along the whole length of its long branches in late spring, mingling with its bright green to make a most refreshing sight. The tree grows to about 60 feet and as much across: a light-limbed and airy tree.

The fruit (right) of the dove tree is a green ball, purple-bloomed at first, containing a single, hard, ridged, inedible nut. E H Wilson brought enough of these back from central China to recoup the costs of his expedition to find the tree.

Berries of the tupelo or black gum (far right). The tupelo has clusters of blue berries for fruit. It was formerly classed as a sort of dogwood. The tallest American "black gum" is in Texas, more than 100 feet high. In Britain the best is less than 70 feet. It needs higher temperatures than we can provide.

For just as in the dogwoods, the apparent petals of the huge white flowers turned out to be bracts. Where the dogwood has four equal ones, however, the dove tree has two – and one is four times as long as the other. The leaves, moreover, are heart-shaped, more like those of the lime tree (though symmetrical, unlike the lime's). It was not a dogwood, but it was a close relation. So it was given its own family of Davidiaceae. Now it belongs to the Nyssaceae.

Distant American cousin

Nothing but a botanist's word would have convinced me that the nearest relation of the dove tree, nearer even than the dogwoods, was the tupelo or black gum (*Nyssa sylvatica*). Tupelo is an eastern American tree – common, though scattered, on moist ground throughout the whole of the eastern States. It has relations in China … other tupelos. But it has neither bracts nor heart-shaped leaves; the evidence for its connection with dove tree lies in the intricacies of the little flowers and fruit.

When tupelo is planted as an ornamental tree it is for its generally glossy and wholesome appearance – a strong and shiny green – and for its whole sunset of autumn colours. It goes with *Liquidambar* as one of the trees for a damp site on the sunny side of a pond. All summer long (and it needs warm summers) it deepens the water with its dark reflection; then suddenly in autumn it lights

the surface with the reds and yellows of a flaming evening sky.

Not that this is enough for an ambitious gardener. He or she will – if British – insist on the even fierier *Nyssa sylvatica* 'Jermyns Flame' or *N.s.* 'Wisley Bonfire'. In the US *N.s.* Red Rage is said to be best, though *N.s.* 'Wildfire' has showy summer shoots coloured purple/bronze/red, and *N.s.* 'Autumn Cascade' weeps and turns orange and yellow. Gardeners in areas with cooler summers are probably better off with the Chinese *N. sinensis*.

The Trees of the Tea Family

THE LEAST SPECTACULAR MEMBER of the tea family is the tea tree itself (*Camellia sinensis*). Its little nodding white flowers hardly prepare you for the glory of its sisters the camellias of the garden, nor the curiously named trio of *Franklinia*, *Gordonia*, and *Stewartia* which represents the family in North America. In Europe, ice-ravaged as ever, there are none.

Absurd, indeed, to call tea tree a tree at all (it is no more than a 10-foot bush at most) – but not to salute its history as an economic plant of the first importance or the take-no-prisoners botanist who stole it from China (China had the monopoly of tea) for the British Empire would be churlish. Robert Fortune introduced some 150 new plants to the West from China, but his industrial espionage in tea country is his most famous exploit. On one occasion he was lying in a fever in the hold of a junk when pirates came alongside. He heaved his 12-bore to the gunwale and fired both barrels. The pirates fled. He eventually shipped 20,000 tea plants and eight skilled workers to initiate the tea industry in India.

Flowers in winter

Nurseries have developed a huge range of garden camellias from the original species. The process began centuries ago in China. *Camellia japonica*, the common species with single red flowers, is totally submerged in a sea of hybrids and cultivars – which is not to say it is not itself well worth growing. Two other species, however, are still important, both in their own right and for the magnificent hybrids they have generated. *Camellia reticulata* from Yunnan in China is one; *C. sasanqua*, from Japan, the other.

All camellias are capable of making small trees (or very big bushes) in time, given the summer heat of Virginia and southward (in Europe, of Bordeaux, or the Italian lakes), and given the right humus-rich soil that doesn't dry out and the right degree of woodland shelter. *Camellia reticulata* (which flowers in early spring) and *C. sasanqua* (which flowers from late autumn onward) are or can be the most tree-like of them. In Portugal there are bushes 60 feet high planted in the 16th century. Their flowers come in such a range from white to scarlet and from single to double and so convoluted that they can match almost any human emotion.

But even in regions where growth is slow and flowers are often spoiled by frost it is worth thinking of camellias for their foliage; not just the permanent polish of their leaves but their varied tints, from light green to reddy-bronze, as new leaves unfold.

The *Stewartia* are by comparison committed trees.

The modest white flowers of the tea bush seem to have little in common with their cousins the camellias. The tea plant is kept small by the repeated drastic pruning of the harvest – here at Limuru, 7,000 feet up in the Kenyan highlands. High-quality tea was introduced from China to India in the 19th century by the British botanist Robert Fortune.

The *sasanqua* (above), the autumn-flowering camellia, is a relatively recent import from Japan, where it is grown in almost every garden for winter flowers. It can flower on into spring in soft weather, and makes a tree 20 feet high. There are 60-odd varieties available in Britain.

Stewartia has only white flowers (left) or it would be more popular, especially as it follows on from camellias in mid-summer. *Stewartia pseudocamellia* is Japanese; *S. koreana* is best; *S. sinensis* makes less of its flowers but more of its beautiful flaking bark, cream, buff, and purple. All have good autumn colour.

The flowers of both *Stewartia* are obviously of the camellia family. *Stewartia sinensis* (below) is smaller but later in the season. *Stewartia malacodendron*, the shrubby American equivalent, has bigger flowers with purple stamens.

Camellia japonica has produced hundreds of large-flowered varieties – this and other camellias having been coaxed by breeders into every size, shape, and colour since the Chinese emperors began to collect it more than 1,000 years ago. The process still goes on – there are about 500 kinds in British nurseries today.

(The "Stewart" in question was John Stuart, the Earl of Bute who was unofficially the first Director of Kew Gardens.) *Stewartia* are native to both North America and the Far East, but the ones grown to such effect in gardens are Japanese (*S. pseudocamellia*) and Korean (*S. koreana*). Both are small-to-medium trees, often low-forking and branchy, with white flowers like single camellias coming in summer, often for eight weeks on end, and leaves that turn episcopal shades of crimson and purple before they fall. Their intricately flaking bark is my reminder of the connection with tea: its colours are the shades of gentle brown in strong tea just clouded with milk. A mature *Stewartia* (they grow slowly) gives such an air of settled distinction to a garden that it is worth planting one without delay and forgetting any idea of moving house.

The missing species

Franklinia and *Gordonia* are both prizes of the great John Bartram, who collected them for his (and America's first) botanic garden in Philadelphia in the 1760s. He found them down in Georgia, where the *Gordonia lasianthus* is still to be found, known as the loblolly bay, but whence the *Franklinia* has totally disappeared. It has not been seen in the wild since the 18th century.

All the *Franklinia* grown now are descendants of the plants found by Bartram and his son William. He named his find *Franklinia* after his friend Benjamin Franklin and *alatamaha* after the Georgian river (its name is Altamaha, but Bartram is not the only botanist to have made a spelling mistake). The tree is not unlike a *Stewartia*, but smaller, less hardy but less fussy about soil – the only member of the family in fact which is happy to grow on lime, demanding only the hot summer it is guaranteed in the eastern States. They flower rather later, with creamy flowers, and they, too, colour richly in autumn. To celebrate Bartram's 300th birthday in 1999 the botanic garden that bears his name launched a census to see how many *Franklinia* are now growing worldwide: 2,000 were reported, with the greatest number in Pennsylvania, 92 as far north as Massachusetts, 50 in Great Britain, and 36 in New Zealand. No one responded from France, where Franklin was America's first ambassador.

Gordon was a London nurseryman, James Gordon of Mile End, supplier of many plants to the infant Kew. His "loblolly bay" is perhaps the least distinguished of a very fine bunch. It is grown in the southern States for its evergreen foliage like the bull bay magnolia's, but its flowers are out of their class in such company.

The story does not end here. Sub-tropical southern China has species which were added to the genus until they were taken away again and named *Polyspora*. (Rather unkindly, it seems to me; there are several nasty fungi with many spores that use this name.) *Polyspora* forests, tall and evergreen, full of white flowers "like fried eggs", are full of *Gordonia*-compatible genes.

Southeast Asia has another genus called *Schima* that brings more possibilities of hybridization with both *Gordonia* and *Franklinia*. Already there are crosses between the different genera called × *Gordlinia* and × *Schimlinia*. A resurgence of *Franklinia* (whatever they call them) is a distinct possibility, for deep-southern gardeners.

A *Franklinia* **flower** is a rare sight; there are none you could call a tree in Britain, but given enough warmth it flowers at the end of summer as the leaves turn rich tones of red and orange.

Persimmons, Silver-bells, and Snowbells

THE PERSIMMON IS ONE of many trees which the chances of history have left in China and eastern North America and nowhere else. The Chinese tree (*Diospyros kaki*) gives the edible persimmon; a succulent yellow fruit the size and shape of an apple. The American tree (*D. virginiana*), which is much more impressive, has fruit that is edible only late in the season, after a frost (when it is soft and sweet, but still without flavour). As ornamental trees, however, both have merit. The smaller Chinese tree has long leaves of a glossy green that can turn orange, yellow, or purple in autumn. The American is tall and narrow to 60 or 70 feet and its bark is one of the most arresting of any tree's: clear and regular fissures mark it both ways, dividing it into rectangular blocks of sooty grey. In Washington's National Arboretum it grows well in swampy ground by a stream, among dark-barked river birches. There were no leaves left when I saw it, but I picked the squashy yellow fruit which hung on the dipping boughs.

The fruit of the Chinese or kaki persimmon ripens orange as the leaves turn to a noble shade of purple. These are the sweet persimmons or date-plums of the fruit basket, grown in Japan, the southern States, and in the south of France.

Snowbells and snowdrops

Persimmons claim no beauty of flower. Their fruit is their calling card. But three neighbours of theirs on the evolutionary tree

Despite the daintiness of its flowers hanging in rows below the branches, the Japanese snowbell tree (*Styrax*) is hardy and uncomplaining. Its only enemy is lime in the soil. It needs a little shade from early morning sun to avoid frost damage.

The kaki or Chinese persimmon is deeply symbolic of autumn on the Japanese table. It looks so pretty hanging like tomatoes from the bare tree that it seems a shame to pick it, although the flesh is sweet – and slightly glutinous.

The Japanese snowbell tree (*Styrax japonicus*) is well worth planting in a woodland garden or glade for its graceful habit. Its American sister, the shrubby *S. americanus*, is called storax – probably to give horticultural editors a moment of self-doubt.

(neighbours rather than close relations) are among the trees with the prettiest flowers to grow on any tree: the American snowdrop or silverbell tree (*Halesia*), the Japanese snowbell (*Styrax japonicus*), and the American-named Chinese fragrant epaulette tree (*Pterostyrax hispida*).

There are two silverbells (called snowdrop trees in Britain): the shrubby Carolina version (*Halesia tetraptera*) and the much more noticeable mountain one (*H. monticola*) from the Great Smokies, which grows to 90 feet and wears big bunches of its bellflowers on its bare shoots like cherry blossom in the spring. The flowers are succeeded by fruits – also bell-like but with four little wings – which can hang a whole year. Silverbells were once common forest trees in the Great Smokies of Tennessee, growing to more than 11 feet in girth. The Carolina silverbell is a smaller version. It flowers in spring before the leaves.

The Japanese snowbell tree seems more decorative in intention, for when it flowers in early summer it contrives to put the flowers just where you can best see them, in long rows along the long, low branches. It is a tree with a strong horizontal inclination, often twice as wide as it is high. The leaves perch in pairs like butterflies along the tops of the branches while the dainty little white flowers hang beneath. The fragrant snowbell (*Styrax obassia*) has scented flowers on erect spikes as an added advantage, but its bigger leaves are less neatly disposed.

As for the fragrant epaulette tree, the lavish way it covers itself with creamy cascades of sweetly smelling flowers could make it a mid-summer successor to the laburnum. It is the least particular of its group in the matter of soil – the others all shun lime.

In spring the mountain silverbell or snowdrop of North America (*Halesia monticola*) is a cascade of whitish-pink bells. In nature (and in gardens) it is taller and even more handsome than its sister *H. carolina* (or *H. tetraptera*). It grows fast in the right conditions and flowers young.

The creamy, fringe-like flower panicles of the fragrant epaulette tree are remarkably spring-like for mid-summer flowers. After them come little spindle-shaped fruits with five ribs. The leaves are like those of the related silverbell tree, above.

Heather and Strawberry Trees

THE HEATHS ARE PART OF A RACE derived from the tea family, which through the ages has developed cooperation with soil organisms called mycorrhizae. For some heaths this has gone so far that they are totally dependent: without their fungus friends they can't make their own food at all. The mycorrhizae attached to most of the familiar heaths need acid soil; therefore rhododendrons do too.

The most tree-like of the family are also those that are least particular in this regard. They are the tiny-flowered *Arbutus* from California, the Mediterranean, and Killarney. Southwest Ireland is one of the warmest points in the British Isles and thus a logical outpost for a Mediterranean plant, but in fact the strawberry tree (*A. unedo*) is generally hardy (give it good drainage; it doesn't like a bog) in much colder places: almost anywhere in Britain and to zone 7 at least in North America. The flowering of the strawberry tree is a charming piece of understatement: the clustering of

Fruit of a strawberry tree (*Arbutus unedo*), in mid-autumn (above) turning from yellow to red. Flowers can be seen at the same time as last year's ripening fruits.

The madrone (*Arbutus menziesii*) (right) is the western American version of the strawberry tree: a magnificent evergreen with a red trunk and red fruit, both contrasting splendidly with the dense green foliage.

Few trees (above) have as much to offer as the sourwood (*Oxydendrum arboreum*) from the eastern States. Brilliant autumn colour combines with its long racemes of flowers, drooping from the tips of its shoots.

The sourwood (its flowers are above) is a slender graceful little tree, demanding acid soil and sunshine but seeming happy in the neighbourhood of London. Its leaves are acid to taste, hence the name.

A hybrid (right) between the strawberry tree (*Arbutus unedo*) and its Greek sister (*A. andrachne*) is popular in gardens for its smooth cinnamon-red bark that peels rather like *Acer griseum*. In my experience it can suffer from an unsightly mildew in spring that strips it of leaves.

"Pitcher-like" is the usual description of the heath family's flowers. You might say "Chinese lantern". These are *Arbutus menziesii*, the monarch of the family (not counting the rhododendrons).

Clethra arborea of Madeira is known as the lily-of-the-valley tree. Again, the heather relationship is clear from the flowers; the rhododendron kinship is suggested by the leaves. *Clethra* needs a mild oceanic climate.

The great sweep of the heath (or heather) family is summed up by the flowers of *Erica arborea*, those of the only true heather (*Erica*) to grow to tree size. Despite the huge difference in size the family resemblance is clear.

hundreds of tiny near-white pitchers among handsome bay-like leaves. It happens in the autumn when flowering trees are few. But the best thing about it is that the fruits from last year's flowers ripen red at the same time. They are more like cherries with roughened, pimply skins than strawberries. Everyone tries eating one – but only one; it is a dry and disappointing mouthful. Hence apparently the name unedo: "I eat one." The overall effect, which is more important, is of a quietly prolific plant in glowing good health; a happy thing to have around.

The California version, the madrone (*A. menziesii*) flowers more conspicuously, at the branch tips, in spring. It is altogether much more of a tree: to 80 or 90 feet. The strawberry tree's maximum is 30 or 40 feet.

The madrone's real beauty lies in its combination of rich green foliage and beautiful smooth red bark on a graceful curving and forking trunk and limbs. A big tree in evening light can show up on a hillside two or three miles away. In flower, in fruit (which is red, yellow, or orange), or in mid-winter, when its evergreen foliage is all the more telling, it is one of the finest trees hardy enough for northern Europe. There is a 55-foot specimen in the glorious collection at Hergest Croft on the Welsh border.

The strawberry tree, and also its Greek cousin *A. andrachne* and the splendid hybrid between the two, *A.* × *andrachnoides*, are content to grow with a modest amount of lime in the soil; the madrone would much rather not.

Clear floral links

Arbutus flowers betray the relationship of the genus to the heathers (*Erica*): the same little pendent pitchers cluster among the conifer-like foliage of the biggest of the heathers, *E. arborea* (a big bush, seldom a real tree), and the broad leaves of the madrone.

They appear again in long streamers on the sorrel tree (*Oxydendrum arboreum*) of the eastern United States. Sorrel tree is a deciduous member of the family, the only one of its kind and a plant of very decided character. It wears a sort of Oriental grace, its shoots falling away rather like the multiple eaves of a pagoda to form a tall pyramid – as high as 70 feet in the wilds. It flowers very usefully in late summer, is hardy to Cape Cod, and colours fiery red in autumn. Given good rhododendron country, every garden needs one. Their rarity argues they are not easy to grow; they certainly like a hot summer.

And again the tree *Clethra arborea* from Madeira has the same flower design – to which it adds a wonderful sweet smell. Most *Clethra* grow only to large shrub size but *C. arborea* can make a very attractive, multi-stemmed tree, up to 25 feet. In late summer the tree is a cascade of fragrant, white, bell-like flowers on delicate spiky stalks. Its little American shrub relation, the sweet pepper bush (*C. alnifolia*), is worth growing in a damp place for its smell and its late-summer flowering. But you must live in Madeira or its climatic equivalent (which includes southwest Ireland) to succeed with *C. arborea*. Ideally it likes a rich, acid soil.

The Rhododendrons

FOR A NON-FANATIC to write of rhododendrons at all is foolhardy, but to try to distinguish those that should be called trees, at least without long experience in the Himalayas, is almost suicidal. There never was a rhododendron with the sort of long straight trunk that would tempt a forester, that is certain. Yet who would call a plant 90 feet high, however curving and many-stemmed, a shrub?

Until 1820 the European and American rhododendrons were the only ones known in cultivation. Of these the rosebay rhododendron (*R. maximum*), from the eastern States, was the biggest – though scarcely a tree.

In 1820 the first seeds of *R. arboreum* arrived in England from the Himalayas, packed by the prudent Nathaniel Wallich in brown sugar. With them arrived the glorious red blood that transformed the relatively tame and dowdy colours of the rhododendrons then known. For though *R. arboreum* was tender at first, and needed a conservatory in Britain, it soon hybridized and started hardy strains. Waterers, the famous nurserymen of Surrey, were the pioneers in adapting this tender giant to cultivation. Most modern rhododendron varieties, even small ones, have the blood of *R. arboreum* in their veins.

Sir Joseph Hooker's expedition to the Himalayas from 1847 to 1850 set the seal on the rhododendron as the supreme flowering evergreen. He introduced 43 species, including the tree-size *R. falconeri*. His Himalayan journeys conjured up landscapes that became the model for large tracts of British woodland. Leonardslee, Bodnant, Borde Hill, Exbury, Wakehurst, and many gardens in Cornwall became famous for their collections, but none went so far as the armaments tycoon, Lord Armstrong, who at Cragside in

Complex pedigrees (right) can produce astonishing shrubs. This is *Rhododendron* 'Hydon Dawn', with the blood of the Japanese *R. yakushimanum*, in the Savill Gardens at Windsor.

The nearest thing to a blue rhododendron (left) is *R. augustinii* from China, seen here at Bodnant in north Wales. There are a dozen selections, with *R.a.* 'Electra' the brightest, but all are beautiful.

The purplish *Rhododendron ponticum* (right) has become an invasive weed. Here it has taken over a hillside at Beddgelert, Snowdonia

The discovery of the red-flowered *Rhododendron arboreum* (left) in Kumaon in the Himalayas changed the history of the genus. Its red-flowering genes were introduced to Britain in the 1820s.

From the same country as *Rhododendron arboreum* comes the scarlet-flowered *R. barbatum* (below). Its smooth red-barked trunks peel in blue-grey patches making a glowing cavern under the canopy.

Northumberland planted 17,000 acres of rhododendrons and conifers. The tales of plant hunters vying with each other for grander, more exotic, or more abstruse species is an epic involving jungle journeys, vicious rivalries, and Edwardian sums of money.

Magnificent as they are in flower, a yet greater virtue of the bigger species is their leaves: great glossy tongues of darkest green, often backed with thick brown or light grey felt.

The curse of ponticum

Over the century or so since their planting many have grown to tree height, if not tree shape: their planters certainly wouldn't recognize their woods today. Unfortunately in their eagerness they also sowed seed of the easiest rhododendron to grow, *R. ponticum* from Turkey. They loved its fast growth and prolific flowering in shades of purple and lilac, without taking into account its astonishing fertility. *Rhododendron ponticum* is now Britain's most serious forest weed. Dense ineradicable thickets of it choke forest plantations. The government spends hundreds of thousands of pounds poisoning it, to not much avail. Tourists, meanwhile, turn out to admire it – and I'll admit that it makes an alluring opponent.

Among the Himalayan species Sir Joseph Hooker introduced was *Rhododendron falconeri* (left), whose big stiff leaves have a thick buff felt beneath. The flowers are creamy-yellow bells splashed with purple.

The Ashes

Not even the oak carries quite such significance out of the murk of legendary time as the ash. In which cold land of myth-making was it such a conspicuous tree? Somewhere on the Baltic shores where the Vikings' ancestors spent the dark evenings scaring themselves with sagas and dreaming up dragons, primeval ash woods must have held them in awe. To them the ash was Yggdrasil, "the greatest and best of all trees. Its branches spread over the whole world, and even reach above heaven." At the top (so they said) sat an eagle; at the bottom a dragon. In between a squirrel scrambled up and down relaying endless threats and counter-threats.

Was it size alone that promoted the ash to this position? Certainly it is one of the tallest of Europe's broadleaves. It has reached 150 feet in modern times. Linnaeus's name for it, *Fraxinus excelsior*, stresses its height. I once asked a French friend why one never sees an avenue of ashes forming the grand approach to a chateau. He replied that the ash is not a noble tree. Not noble?

Ash is not exceptional in leaf: best perhaps when its pale grey-green leaflets stream in the wind. It is more conspicuous in winter: notably austere, grey-barked, and sparse-branched. Dryden wrote "nature seems to ordain the rocky cliff for the wild ash's reign". Certainly it plants itself in some pretty bleak places. It appreciates deep soil and moisture, though, as much as any tree.

A versatile family

Botanically the ashes belong to the olive family, which also includes (among common garden plants) privet, lilac, and jasmine. The family traits are by no means obvious (unless you count opposite branching, which is common to most of them, but to thousands of other plants too). The fruits of the ash and the olive could hardly be less alike.

Most of the world's 40 or so ash species conform to the general style of *F. excelsior*. They have compound leaves; the leaflets arranged in opposite pairs (except for the odd one at the end) on a green stalk seldom more than a foot long. They have simple, single-winged keys for fruit. They are vigorous, dense rooted, and demanding of the soil; easy to transplant but bad to grow near shrubs; liable to commandeer all the rations.

All North America's common ashes follow this general pattern. The white ash (*F. americana*) is the tallest, best for timber, and most commonly planted. The basic hardiness of the race comes out in the fact that it grows as well in Europe as in America; not true of American oaks, beeches, or elms. White ash is white and downy beneath the leaves: in autumn purple in the north (there is a cultivar called 'Autumn Purple'); yellow farther south.

Autumn colour is not a feature of European ash. Red ash and green ash (both *F. pennsylvanica*) are varieties of one species; the green one (green beneath the leaflets) the commoner of the two, with the biggest natural range of any of them. These, and several other American ashes, have a great advantage as street trees: they

The black bud (top) instantly identifies *Fraxinus excelsior*, the ash of northern Europe, with its rather flattened twig, opposite buds, and flowers in short dense bunches either just before or just after the oak. The ash of southern Europe has brown buds.

The fruit of the ash (above) is a "key", rather like half a maple's, and as prolific and as fertile as a sycamore's. Autumn leaf colour is not guaranteed; yellow/brown and brief.

The ash is an airy tree (left), often paler green than its neighbours, looking profusely leafy with its many leaflets. Here it grows on a farm in the Yorkshire Dales, undoubtedly self-sown in a rocky outcrop. Its wood is the very best for the fireplace, burning almost equally well fresh or dry.

are fast-growing at first, but slow down and form round crowns at about 60 feet. Male trees are the one to plant to avoid proliferating seedlings – there are plenty of varieties chosen for autumn colour and narrow form. I am particularly fond of a white-variegated one, such a refreshing sight among darker trees in summer that I'm surprised it is not planted more often. Authorities say beware reversion; mine has not reverted in 30 years.

The same size range includes: black ash (*F. nigra*; a far-northern swamp tree), blue ash (F. *quadrangulata*) from the southeast, and velvet ash (*F. velutina*) from the southwest. Blue ash has twigs with a square cross-section, and bark which makes blue dye. Velvet ash has a downy layer on both twigs and leaves, together with a dense fashion of leafing that makes it a good shade tree. Only the Oregon ash of the northwest (*F. latifolia*; a David Douglas discovery, named by him for its broad leaflets) grows to the same sort of height as the white ash.

More ornamental

There are European ashes that give relief from this generally business-like approach. The default ash of southern Europe is *F. angustifolia*, narrower in leaf than *F. excelsior*, more graceful in growth, and lighter in shade; a good shade tree beside a Mediterranean swimming pool. Its Caucasian version (*F. angustifolia* var. *oxycarpa*) produced a variant for an Adelaide nurseryman which must (or should) have made him a fortune; a seedling whose leaves turn such a fine red in autumn that it became known as the claret ash (*F. angustifolia* var. *oxycarpa* 'Raywood'). Everybody loves it and many plant it, especially as a street tree. Unfortunately they often discover that it grows too fast for its own good. Its wood is weak, and big branches in the crown crack in a strong wind. Once there is a substantial branch broken and tangled up in the top of the tree it is far from easy to reach it to bring it down. Another variety of *F. angustifolia* called 'Lentiscifolia' (what a mouthful) has an abundance of extra narrow leaves that give it a wonderfully lush appearance

Fraxinus excelsior has some useful variants too: an odd one with only one leaf at a time (*F.e.* 'Diversifolia') instead of nine or 11; one with dandelion-yellow bark (*F.e.* 'Jaspidea'), which is remarkably conspicuous in winter (especially if you encourage new shoots by

The claret ash is fast-growing with a dense canopy of small leaves turning many shades of red, orange, or purple. It is a form of the southern ash found in South Australia and baptized *Fraxinum angustifolia* var. *oxycarpa* 'Raywood'.

pruning): and above all the weeping one (*F.e.* 'Pendula'): to me the best of all weeping broadleaf trees. The weeping ash manages to be both stiff and graceful together. It is made by grafting weeping branches on a stem already grown to the required height; this has been done, successfully, to a tree 90 feet high. It is a particularly fine sight in winter when its hanging ramrods of branches are like wooden rain.

In my enthusiasm for ashes (they also make the most luxurious firewood, burning almost as well green as dry) I planted a grove of them in central France. A few years later I noticed die-back in their tops and some curious white marks on the bark. When I investigated I found that hornets were mining it, eating channels in it, and ringing the twigs – presumably as material for their nests.

But the biggest departure from what we expect of ashes comes in the group that flaunts its flowers. Of these the manna ash (*F. ornus*) of southern Europe is by far the best known. The flowers are showy, creamy-white, in late spring. But they are only half the story. The whole tree is so voluptuous in glossy leafage that others look quite abashed beside it. It stands out in staid parkland like the velvet-framed bosoms of Nell Gwyn and her contemporaries on ancestral dining-room walls.

One excellent ash (above) has ornamental flowers. The manna ash (*Fraxinus ornus*) from the eastern Mediterranean is never big but always luxuriantly leafy and lovely in late spring. The "manna" is sugar obtained from the sap by slitting the bark.

The weeping ash (left) is an artificial construct, made by grafting weeping branches (which can only grow downwards) on an upright trunk. This was famously done in the 19th century on a tree 90 feet high. The effect is a giant arbour of stiff rods.

The American white ash (*Fraxinus americana*) (below) is very similar to the European in its tastes and the quality of its timber but has fewer, bigger, broader leaflets which often turn to charming pale shades in autumn.

Olives and Fringe Trees

IF YOUR FAIRY GODMOTHER would let you grow just one tree from outside your climate zone what would it be? For me the answer is not the coconut palm nor even the lovely weeping blue cypress of Kashmir. I would choose the olive. And it looks increasingly as though my fairy godmother, disguised as climate change, is granting my wish. Will it be a desire that brings regret?

Olive trees have Homeric style. I cannot see flowering cherries in the Elysian fields. I can see grey olive orchards, black cypresses, and pines. Grace and gravity the olive has, living as long as yew, growing as gnarled, yet still leafing with a silver spray. The olive tree is given to branching low, and its branches to wandering. Down into them through leaves like the white willow's pours Mediterranean light.

The range of the olive in Europe has always been taken as the limits of the Mediterranean world. Wherever olives are planted they bring their mood. No Doric columns, no transplanted temple ever evoked the classical world as powerfully as an olive grove.

I am not so certain that the transplant works in an English garden. It can be as aestheticaly troubling as a rhododendron among roses or a bulrush in a rockery. Other exotics can enrich the picture without provoking a clash of cultures; the olive carries too much baggage. And yet of all trees it is the easiest to transport. By diligent digging and packing of its roots you can move even an ancient specimen. Fashionable gardeners easily succumb to the temptation. There are even instances of criminal tree rustling.

Olives love the sun and dread humidity. They like the soil deep but dry, and love limestone. Florida is not such good olive country as California, and of California the south and the central valley have in the past been best. As olive culture has grown more profitable, though, and varietal olives have joined varietal wines as fashionable desiderata, Oregon has joined in, moving the Homeric world into new territory.

Oil and wine

No one knows where the fruit-bearing olive came from. "The peoples of the Mediterranean", wrote Thucydides in the 5th century BC, "began to emerge from barbarism when they learnt to cultivate the olive and the vine." Oil and wine moved society beyond hunting and growing corn to crops that demand settlements and stimulate trade. Olive trees grow slowly but live for centuries. Wild olives are as decorative, but edible ones (there are hundreds of varieties) must be grown from cuttings or layers, and

Olive trees can live for 1,500 years or more. They share with yews the distinction of being Europe's oldest trees. In age they become grotesquely gnarled, in wonderful contrast with their light silvery foliage.

Olives for oil are first crushed with a stone roller, then pressed like grapes in a huge vice. The ancient process is shown here in a 16th-century Flemish copper engraving from *New Inventions* by Johannes Stradanus.

are better grafted or budded. There are more than 20 million acres of olive groves worldwide, with the biggest production in Spain, Tunisia, and Italy. A cultivated tree, well-pruned and fertilized (the soil must be kept bare of grass and weeds), produces about five pounds of fruit a year, or alternate years, which is a tendency with many trees. Before eating or oil-making it has to be fermented or cured. It is not a culture without patience and labour.

Each country and region has its favourites. Tuscany's best olives are the mild *Olea europaea* 'Leccino' and the stronger 'Frantoio'. The principal olive of Spain is 'Picual', with the 'Manzanillo' the favourite of Andalucia. The most famous olive grove of Greece is at Amfissa, near Delphi, with more than a million trees. The black 'Kalamata' is the most popular olive to eat. Turkey specializes in the 'Gemlik', a black olive whose flesh comes off the stone when it is cured. In Lebanon and Palestine the 'Souri' and 'Nabali' both give excellent oil. France's best-known olive is the 'Lucques', from the western Midi, with long green nutty-flavoured fruit. California still grows the 'Mission', introduced by the Franciscan brothers.

And for the north?

What should a northern gardener plant as a substitute? The traditional best imitation, used at least since the 16th century, is the olive's cousin, *Phillyrea latifolia*. It puzzles me how this admirable and not uncommon Mediterranean tree has kept such a low profile for centuries that it has no common name. Even in France it has

got no further in familiarity than *Filaria à larges feuilles*. *Phillyrea* can be remarkably olivey in shape, with wide-spreading, curving branches: instead of silver-grey, though, it is very dark glossy green. It also takes gamely to topiary. The tree I planted 35 years ago seems hardly to have grown in the past 10. It spreads its rather solemn dome over a space 30 feet across, attracting little comment (people ask if it is a holm oak; it is shinier) but certainly adding gravitas to the glade.

To turn from olive and the dignifed *Phillyrea* to their cousin privet (*Ligustrum*) sounds at first like a descent from the sublime to the ridiculous. And is privet a tree? Common privet isn't: just one of the necessary evils of gardening. But its Chinese counterpart with glossy leaves (*L. lucidum*) is one of the best hardy (to zone 7) evergreen trees. *Vogue* would no doubt call this the wet-look privet. Its leaves are camellia-size, dark green, and glossy. In summer it carries upstanding batons of white flowers with the sickly privet smell. Thirty feet (fairly fast) is as high as it grows, but 30 feet of such intense dark green gloss can be quite a showpiece – especially with a heavy load of flowers – followed, in hot summers, by heavy bunches of blue fruit. There is a good variegated form too, with the resounding name 'Excelsum Superbum' suggesting that its discoverer held it in rather exaggerated regard.

One of the stranger tales E H Wilson told of backwoods China is the story of the wax insect. It seems almost as though the insect sensed that the privet and the ash are of one family. In one valley in

the west (at Chien-ch'ang to be exact) the wax insects breed on the leaves of the Chinese ash (*Fraxinus chinensis*). In late spring they are collected in boxes and carried by the porters once called coolies, travelling only at night (it being too hot for the insects during the day) for 200 miles to their feeding ground – the leaves of the privet trees of Kiating. Such extraordinary pains are rewarded by three months of unrelenting wax-laying, when the insects weigh down the privet with hundredweights of the finest wax – used for candles, for coating paper, and polishing jade. But how did the business start?

The olive-ash family roster also includes the lilacs (*Syringa*). The only *Syringa* rated a tree is the Japanese *S. reticulata*, which might reach 30 feet or so. Like the privets, however, the flowers smell bad enough to discourage its planting.

With a bit of imagination one can see the fringe tree (*Chionanthus*) as the missing link in the olive-ash family; close enough to the ash to be grafted on its roots, flowering rather like the flowering ashes, and with fruit not unlike the olive. But a fringe tree in flower, more fringe than tree, needs no botanical footnotes. Like the ashes it comes late into leaf, but almost at once, in early summer, puts on a covering of white flowers: five spidery-thin petals joined at the base (whence "fringe"). It is one of the most stunning small trees (or big shrubs) that a northern garden – on any soil – can grow. *Chionanthus virginicus* is the American one, *C. retusus* the Chinese.

This *Phillyrea latifolia* (left) on Trumpington Road south of Cambridge is the finest I know in Britain. Its sombre shining green remains consistent all year, like a holm oak of the finest pedigree. They were better known in the 17th and 18th centuries.

The olive harvest (left) in the Mediterranean takes place in the winter. Olives for oil are left on the trees to ripen; eating olives are picked both half-ripe (when green) and ripe (when black, swollen, and wrinkled).

The Chinese fringe tree (above) carries its dainty and fragrant flowers in erect clusters – by the million. Both this and its American counterpart do best in a continental climate.

Empress Trees
and Indian Beans

BIG PALE CRUSHABLE LEAVES give a tree the look of being in tropical kit. In Florida or Cap Ferrat you wouldn't give the Indian bean tree (*Catalpa*) a second glance: which makes it all the more effective in Philadelphia or Frinton. The *Catalpa* and the empress tree (*Paulownia*) are two trees northern gardens can grow as a reminder of the *Jacaranda* (their South American cousin) and other tropical extravagances. They are so similar in most ways that one is surprised to find them in different botanical families – respectively the Bignoniaceae and the Paulowniaceae. What they have in common are flowers like foxgloves, weak and pithy young wood, and their oversize leaves. The most obvious difference is that *Catalpa*'s flower clusters hang while *Paulownia*'s give the appearance of standing erect. And that *Catalpa* flowers are white spotted with purple, while *Paulownia*'s are a pale wisteria-like purple all over.

Catalpa's common name of Indian bean tree only reinforces the tropical impression. But Indiana bean tree would be nearer the mark. It is a mid-western native; another of the trees that turn up only in North America and China; nothing to do with India – only with Indian beans; its pencil-like pods.

The *Catalpa* choice

There are two American *Catalpa*: the southern *C. bignonioides*, more common in gardens, and the northern *C. speciosa*, a far bigger tree, making (surprisingly, considering its pithy shoots) timber that will

The Brazilian *Jacaranda mimosifolia* (below) that grows so well in Punakha Dzong, in the Bhutan Himalayas, is not hardy enough to warrant its place here, except as a distant relation of the *Catalpa*. Its ferny-fine leaves are almost evergreen, and its habit horizontal, making it a first-class shade tree for tropical and sub-tropical cities.

The northern catalpa (*Catalpa speciosa*) (right) is the giant of the family, a great rough craggy tree smothered in white flowers in summer. This monster is 82 feet high, in the grounds of Swarthmore College, Pennsylvania.

The name of Indian bean tree for *Catalpa bignonioides* (left) comes from the 15-inch seedpods that often hang from the tree all winter. The Indians referred to are, of course, Red.

Indian bean tree (*Catalpa bignonioides* 'Aurea') (right, in the Cambridge University Botanic Garden) is one of the most striking of all yellow-leaved trees with its most voluptuous of mid-summer flowers.

Catalpa needs considerable space, sun, and shelter from wind to do it justice. Such rare spots are not lightly allotted, though. A tree must be more than merely conspicuous to compete: especially one that is inherently messy, dropping not only its big leaves, but also its long thin pods and fragments of twig. Luckily Chinese genes come to its rescue.

The Chinese *C. ovata* is not necessarily a winner in itself, but its hybrid with the southern catalpa (*C. × erubescens*) is an advance on both, faster-growing with bigger leaves, flowering early in its career. Its variety 'Purpurea' is the one to choose, for the purple flush of its shoots and new leaves. Alternatively (or even together to get the neighbours talking) plant the golden variety of the southern catalpa, *C. bignonioides* 'Aurea' – the closest thing you can plant to a sunlit hill.

I rarely see another Chinese catalpa, one with surprising pink flowers, orange-speckled, and entirely smooth leaves with long points, perhaps because it is slow to flower as a young plant. Mine is 20 feet high, a narrow tree, after 25 years and only flowers spasmodically, but *C. fargesii* f. *duclouxii* could become a favourite if nurseries would graft a good form of it onto vigorous roots. Again, sun is essential for a good performance.

Invasive alien

The empress tree (*Paulownia tomentosa*) can be disappointing for flowers in northern gardens. It has the rather illogical habit of forming substantial flower buds in autumn, which winter cold can easily kill. Recent years have been kinder to it, at least in England, but the risk remains. Some gardeners for this reason forget about them altogether and coppice the plant, taking advantage of its natural vigour to force huge furry leaves out of it. On the other hand its round fruit (which could hardly be less like *Catalpa*'s) is attractive and extremely fertile – so much so that parts of the US consider it an invasive alien. On the plus side its timber is almost natural asbestos, resisting fire. It seems that the more recently introduced *P. fargesii* from China may do better with flowers – and also make a bigger tree.

lie on the wet ground for a century and not rot. The biggest northern catalpa is over 100 feet high, craggy in bark, and coarse in foliage with foot-long leaves – a savage monster of a tree.

The southern catalpa is also the commonest in Europe. It tends to spread, even to sprawl, from a short trunk, or from none. Its heart-shaped leaves start hairy in the spring, growing shiny above as they reach their full nine or 10 inches, at flowering time in mid-summer. Young trees produce only leaves, but old ones sometimes perform prodigiously with clusters of their extremely elaborate flowers.

The Chinese *Catalpa fargesii* f. *duclouxii* (above) tends to be taller than broad. Its leaves are glabrous rather than hairy, with longer points and its flowers later and pink with red/brown and yellow spots with an orchid-like effect.

The empress tree or royal paulownia (left) is well-named: a conspicuous but impractical tree with huge leaves, weak and pithy wood, and pale purple flowers (shown on page 23) which appear in bud in autumn and often succumb to frost. Its fruit follows the flowers on tall spikes.

Catalpa flowers (above) would not look out of place in the herbaceous border, generous panicles of them, each flower two inches across, in mid- or late summer. Each frilly white bell is spotted yellow and purple inside.

More exotic ideas

No botanical considerations, only the arresting strangeness of the tree, brings the rare *Kalopanax septemlobus* in here. With the related thorny shrubby Hercules' club and the Japanese angelica tree (*Aralia*) it is botanically nearer the dogwood (*Cornus*), with the same little massed florets and seeds arranged in a ball.

Kalopanax septemlobus (sometimes called the castor aralia) is most like a succulent sycamore or a sweet gum in leaf: a striking pile of glossy greenery. But it is its winter frame that is really distinctive: its few, thick, and forky twigs make memorable skywriting. There could be no better tree to sharpen up bland avenues of plate-glass offices.

Aralia is more often shrub than tree. The most worthwhile of the genus are the variegated forms of the Japanese angelica tree (*A. elata* 'Variegata' and 'Aureovariegata'). The leaves are huge and many leafleted. In mottled green and creamy-white they look both ferociously tropical and innocently pretty. A warning to gardeners, though. Any damage to the roots of an *Aralia* or a *Kalopanax* will give rise to suckers. Your innocently pretty tree will become ferocious – and the suckers will not be variegated.

The Hollies

IT MUST HAVE STRUCK YOU by now that the evolutionary order of things is a pretty funny one. To leap from poplars to laburnums, from pears to elms to dogwoods to hollies seems like no kind of order at all. This, however, as far as we know, approximates the order in which, over tens of millions of years, families of new trees evolved. We can only be thankful that the randomness of evolutionary mutation and selection did produce such a mixed bunch, and wonder humbly how each one of them in its own way manages to be so intricately and originally beautiful.

There is no objective reckoning by which one tree is more beautiful than another. But one can have favourites. One of mine is the holly (*Ilex*). And I find I am in good company: this is what John Evelyn had to say (about his holly hedge): "Is there under heaven a more glorious and refreshing object of the kind than an impregnable hedge about four hundred feet in length, nine feet high, and five in diameter, which I can show in my now ruined gardens at Say's Court (thanks to the Czar of Muscovy) at any time of year, glittering with its armed and varnished leaves?"

Poor Evelyn. His fractured syntax makes it plain that he was distracted by having the Czar in the house. Peter (later to be called "the Great") was his tenant. He had come to England to learn about shipbuilding and needed a house near the naval dockyard at Deptford. Who but a Czar, having discovered that his landlord was a passionate gardener, would have set about wrecking the garden?

The story goes that the Czar amused himself by having his staff wheel him through the holly hedge in a wheelbarrow. But apart

The native holly of Europe (left) is not afraid of exposure. This old tree on the Atlantic coast of Donegal is damaged every winter but struggles on, producing a lavish crop of berries. The snow-capped mountain in the background is Muckish.

A holly hedge (above) is routine compared with these formidable individuals looming though the mist in the garden at Buscot Park, Oxfordshire. One wonders how they are cut.

from the difficulty (not to mention discomfort) of being any part of an assault on five feet by nine feet by 400 feet of holly, Evelyn's own words seem to contradict it. He goes on "… It mocks the rudest assaults of the weather, beasts, or hedge-breakers." Whichever of the three he considered the Czar, it seems the hedge survived.

Evelyn was, of course, talking about the English holly (*I. aquifolium*). There are 400 species of holly, even deciduous ones – the American possum haw (*I. decidua*) for example. But it is a tree that has been in cultivation for ornament so long that the cultivars and hybrids outnumber the species. Most of the best of these are

either variants on the English holly, or forms of a cross between the English holly and the native holly of the Azores (*I. perado*). Indeed more than a thousand different cultivars are recorded.

The typical English holly, left to nature but given some woodland shelter, is a tall tree (the tallest is 74 feet, the thickest 11 feet around) and almost as pointed as a Christmas tree. But such trees tend to be thin and transparent. The best-looking ones are in hedgerows where the wind has a pruning effect: shorter shoots keep the foliage dense and stress the contrast of glitter and darkness that is half the pleasure. For this reason holly is best in summer, when the new shoots emerge so soft and shiny that they look (and even feel) wet. The new leaves glow with pink, brown, and purple tones as well as green.

Strangely, you might think, for a native, English holly is not 100 per cent hardy in England. All broadleaved evergreens, for that matter, are taking a risk. Only a once-in-a-century winter

will actually kill a holly tree, but really sharp weather can strip it of leaves. In consequence, its range in the eastern States is limited to zone 6 – from Connecticut southward.

Northward the American holly (*I. opaca*) is hardy, and can be excellent for berries. But it lacks one of the English holly's principal virtues, what Evelyn calls its "varnished leaves". At its best the American holly makes a distinctively matt, greyish or yellowish-green small tree (the biggest, 76 feet high, is in Alabama) on which a good crop of red berries (or in its form *xanthocarpa*, yellow berries) shows up well.

The search for "varnished leaves" in hardy hollies has borne fruit in the US with a series of crosses initiated by Mrs Leighton Meserve in New York in the 1960s and known as *I. × meserveae*. They involve the European holly crossed with the Chinese *I. cornuta* and the Japanese *I. rugosa*. Perhaps the best is *I. × m.* 'Blue Princess', with deep blue-green spiny leaves.

Holly and Box

MOST HOLLY TREES are either male or female, which means they need a mate if they are to bear berries. Many of the variegated cultivars, unfortunately, are male, and will never have berries. One would hope and expect that their names would give some indicaion of their sexes – especially as the most famous of them are called either Kings or Queens. But the Joker must have stepped in. 'Golden King' is a female; 'Golden Queen' a male. There is no 'Silver King', but 'Silver Queen' is also a male. These are all variegated with a band of either silver or gold around the margin of the leaf and green in the middle.

The two 'Milkboys', 'Golden' and 'Silver' (both male, reassuringly), have the alternative form of variegation: green around the edge and a beautiful mixture of apparent brush strokes of dark green, light green, and gold (or silver) in the centre. It is these plants with centre-variegated leaves that have a built-in tendency to revert to plain green. If you see any plain green shoots on them you simply cut them off. Sometimes, on the other hand, shoots with no green at all – leaves exactly the colour of that Swiss white chocolate – appear, only to wither from lack of chlorophyll.

Leaf shape varies as much as colour. There is hedgehog holly (*Ilex aquifolium* 'Ferox' – the fierce), both green and green and white, which has spines on all the surfaces of its strange hunch-backed leaves. There is *I.a.* 'Crispa', a rare cultivar on which everything – leaf, twig, branch, and trunk – curls and contorts: a slow-growing, but ultimately unmatchable, specimen for a courtyard or a lawn. There is weeping silver-variegated holly (*I.a.* 'Argentea Marginata Pendula') – bright as a button and not in the least doleful. There is yellow-berried *I.a.* 'Bacciflava'; there are purple-stemmed, unarmed, narrowleaved, and broadleaved English hollies. And there are combinations of most of these characteristics. It is hard to think of any tree more worthwhile collecting (Kew

Box cut into a miniature maze in the formal garden of the Pope's summer villa at Castel Gandolfo, just south of Rome. Box has been used for low hedges and crisp small-scale topiary since at least the days of ancient Rome.

Gardens has a famous Holly Walk), or more rarely collected.

Moreover, the crossing of English holly (*I. aquifolium*) and *I. perado* from the Azores 150 years ago has produced another whole range known as the Highclere hybrids (*I. × altaclerensis*). On the whole these tend to be more vigorous than English holly, and less prickly. One of the most magnificent varieties is the camellia-leaved holly (*I. × a.* 'Camelliifolia') which has very few prickles, and those only on the lower branches. 'Golden King' is in fact supposed to come from this stable. 'W J Bean' is famous

AN UNRIVALLED GALLERY

Almost any combination of green, white, cream, and yellow can be found in the leaves of English hollies, variously more or less prickly. Crossed with a mid-Atlantic species it has made another catalogue of even more lusty varieties with fewer prickles.

Ilex aquifolium
Common English holly flowers in early summer, inconspicuously but with a sweet scent. Berries are borne only on female plants. They often last all winter.

Ilex aquifolium
'Golden Queen' (left) has no berries. 'Bacciflava' (right) has yellow berries, often in strikingly large quantities.

Ilex x altaclerensis
The fruitful cross between English and Azores (or Madeira) holly seems to have originated in Hampshire early in the 19th century.

Hollies with gold or white margins, such as *Ilex aquifolium* 'Golden Queen' (far left) often produce shoots with no green in the leaves at all.

Ilex aquifolium 'Silver Queen' (left) is a male (berryless) holly with yellow leaf margins surrounding green and grey-mottled centres.

American holly (*Ilex opaca*) (far left) is similar to English. It has good berries but matt leaves.

Perry's weeping holly (*Ilex aquifolium* 'Argentea Marginata Pendula') (left) is a famous Victorian cultivar with pendulous branches that eventually mound up into a big bush up to 10 feet high.

for his (her) profuse berry-bearing. 'Belgica Aurea' is one of the tallest-growing variegated hollies, and female. 'Purple Shaft' has tremendous deep purple shoots. 'Lawsoniana' has gold centres like the 'Milkboy' …there are 20 or so: another whole collection.

Among the species, which come from almost everywhere in the northern hemisphere, my own favourite is Father Perny's holly (*I. pernyi*) which was found by yet another of the French missionaries in western China. It has small leaves, each one a crisp green diamond. It makes a sparkling little tree. A true gourmet's Christmas pudding has its perfectly scaled fruit for decoration.

At the other extreme the Japanese broadleaf holly (*I. latifolia*) you would take for a cherry laurel, with its long, wide, serrated but unprickly leaves. Crossed with *I. aquifolium* it has given us a splendid, hearty, fast-growing tree that lacks only leaf-gloss to be a champion (although truth to tell its berries are a bit on the brown side for total stardom). Its name is *I. × koehneana* 'Chestnut Leaf'. The Dahoon holly (*I. cassine*) of the southeastern States is also unarmed, but tender. There is a prettier and much hardier one from Japan (*I. pedunculosa*) which carries its berries on stalks as long as the leaves. Uniquely in a white-flowered genus, *I. purpurea* has pink to purplish flowers, on stems densely furnished with spineless leaves.

There is only one word of warning needed about hollies: they do not like being moved after two or three years. It is very risky indeed to move them without a big ball of soil containing as much as possible of the roots. And even then it is better to move them in very late summer or mid-spring, when the roots can make some headway before the full demand of transpiration starts to dry out the plant. If a holly does seem to be under this sort of stress, take off some, or even all, of the leaves. It can recover from being denuded, but not from drying out. Do not assume, by the way,

that a holly is its own protection against grazing animals: rabbits love the green bark. They will eat an unprotected little plant down to the ground.

Cousin by marriage

Defenceless box (*Buxus*), on the other hand, they leave alone. Its smell, which divides opinion as does the smell of old walled gardens in summer (I love it; others stop their noses), rabbits apparently shun. Box (*B. sempervirens*) is more of a social than a botanical connection of holly, but it has some of the same virtues and is too modest a tree to have any vices. It seems equally happy cropped down to a six-inch hedgelet grilling by a pathway in the sun or reaching up in permanent gloom as the undergrowth of thick woods. Wherever it is its neat, almost round, yellow-green leaves give it a satisfying texture. It seems to have been a defining garden plant since gardens began. Pliny gives an account of his box topiary (*topiarius* was the Latin for a gardener) that makes him sound slightly crazy. I know a hill in the centre of France where box spills down into the country around ancient trees and their offspring, from what two thousand years ago was an important Gallo-Roman town. The town has totally disappeared – but not the box its gardeners planted.

Given time (measured in decades rather than years) *B. sempervirens* grows into a straggling tree, usually with many trunks. Its smooth-grained wood is ideal for block-engraving, hence the scarcity of mature specimens. The many varieties – there are at least a hundred – of garden box are well worth exploring. I have a weeping hedge now (after 25 years) seven feet high. The variety that makes the best dwarf hedges is *B.s.* 'Suffruticosa'. *Buxus balearica* is a relatively large-leaved species that will become a small tree in years rather than decades.

The Elders

EVOLUTIONARILY NEW-FANGLED as they are, elders are still ancient in human history. The Latin *Sambucus* takes us back, via Greece, to very ancient things … among them the sackbut that accompanied the Psalms. Elder still makes whistles, as every schoolboy once knew, and wine, as every grandmother did.

The elders and viburnums are close kin, as their flat flower clusters show. The viburnums are unusual in sometimes having two kinds of flowers in one cluster: some working (i.e. sexual) and some purely to attract insects. The hydrangeas do the same. Most insect-fertilized flowers, as W J Bean puts it, "do their own advertising". Yet here in this relatively recent design a division of labour is introduced.

In playing to the gallery like this they have played into the nurseryman's hands. The snowball tree (*Viburnum opulus* 'Roseum' – formerly known as 'Sterile'), if not strictly a tree at all, is a form of the "guelder rose" in which the sterile parts have been kept and the sexual rejected: the flower is all the showier.

The viburnums are a highly collectable crowd with huge variety and flowering times almost year-round. They are scarcely trees, though, even the wayfaring tree (*V. lantana*), the other one with a tree name. The wayfaring tree is a European hedgerow bush with white-wool-backed leaves which have much the same effect as the whitebeam. It was William Turner the Elizabethan herbalist who likened it to a wayfarer come from a dusty road. But then I have never thought of showing a wayfaring tree the sort of courtesy that might make it shine. I recently met the southeast Asian *V. awabuki* and admired it for its broad highly polished leaves. Now I hear that in Portland, Oregon it is trained up and used as an effective small street tree. The eastern American black haw (*V. prunifolium*) and the sheepberry or nannyberry (*V. lentago*) are the principal viburnums of North America which make small trees. Both are worth growing for their autumn colour, apart from their edible fruit and masses of white (or off-white) flowers.

Back in fashion

The elder of elders is the truly tree-like blueberry elder (*Sambucus nigra* subsp. *caerulea*) of the Pacific Northwest. It grows in the wild to 50 feet and staggers under its late-summer load of bloomy fruit. I have never seen it grown in England, and wonder why.

Common elder (*S. nigra*) will be a bush (and very likely a weed, too) if you let it. But such a vigorous and uncomplaining – and immensely hardy – plant can easily be harnessed into service. It will grow in dense shade, on a minimum of soil – and deliver barrow-loads of berries. Its worst point is its smell – not of the flowers (though this is a matter of opinion) but of its pithy new wood. Flies apparently agree: carters used to make chaplets of elder to keep the flies from their sweating horses' heads.

Elderflower has recently become a fashionable flavour. For years we used to harvest the flowers as they emerged in heavy creamy panicles, boil them briefly with lemon juice and sugar, and bottle the result in screwtopped flagons while it fermented to make what we called "champagne". We laid it down on a cool concrete floor and counted the days. Fifteen or so was enough to give us a sweet bubbly and faintly alcoholic drink with a marvellous Muscat fragrance. Leave it too long and the drink turned nasty, smelly, and overstrong. That is if the bottle had not exploded. These days elderflower cordial (unfermented) has spawned a small industry.

Elderberry was not only grandma's grape, it once played a large part in the making of port, giving it dark colour and a certain *je ne sais quoi* – something strictly forbidden these days. The gardener will be more interested in the surprising leaf forms of such cultivars as *S.n.* 'Black Lace' with deep maroon blood, or so it seems, that tinges its flowers pink. I enjoy the parsley-leafed *S.n.* 'Laciniata', the yellow-leafed *S.n.* 'Aurea', and the variegated *S.n.* 'Albovariegata'. Seldom are such useful colours and styles of foliage so easy to introduce into the garden.

My hero John Evelyn must have the last word: the pleasure he took in matching trees and words comes so clearly over 300 intervening years. He is speaking of inflammation: "An extract of elder is efficacious to eradicate this epidemical inconvenience."

Yellow-leafed common elder (left) (*Sambucus nigra* 'Aurea') is weighted down with sweet-scented flowers in early summer. The flowers make a good drink and the broad flat heads make heavy clusters of sweet black berries.

The berries (right) of the "guelder rose" (*Viburnum opulus*) turn and remain ruby-red, hanging on far into the winter. It is a tree or tall spreading bush for wet or boggy ground. The lobed, maple-like leaves colour richly in autumn.

The American sheepberry or nannyberry (*Viburnum lentago*) (below right) will grow up to 30 feet or so with shiny toothed leaves, broad flowerheads, and blue-black bloomy berries. Its best moment is when its leaves turn red in autumn.

The wayfaring tree (right) is conspicuous in spring with its dusty-white shoots and in autumn for its jewel-like fruit, which turns from red to black.

The blueberry elder (*Sambucus nigra* subsp. *caerulea*) (right) is the most tree-like of its tribe and it stoops under massive loads of its apparently bright blue berries. When their bloom goes they are black.

A Twelve-Month Succession of Ornamental Flowers, Fruit, and Foliage

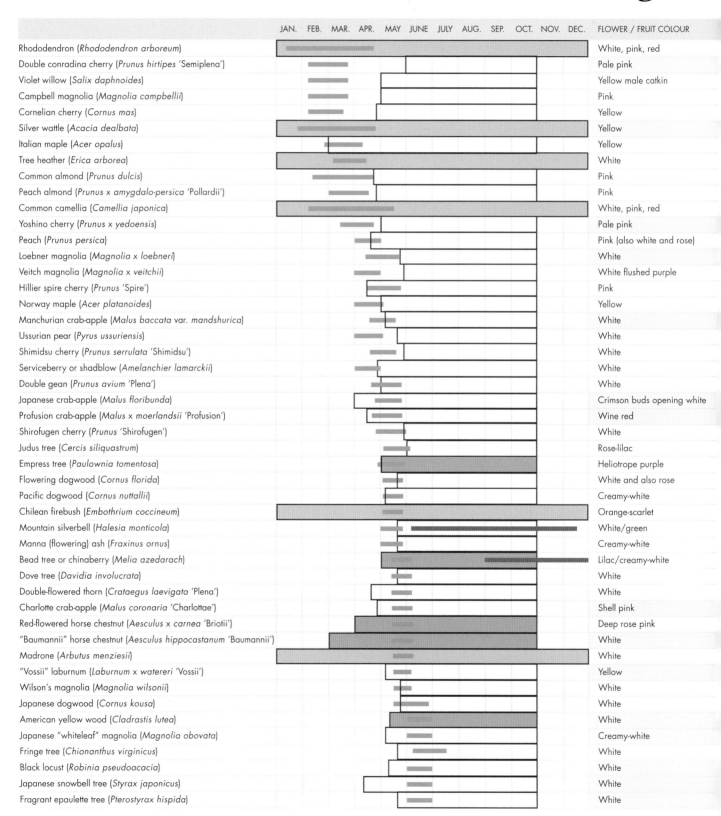

	JAN.	FEB.	MAR.	APR.	MAY	JUNE	JULY	AUG.	SEP.	OCT.	NOV.	DEC.	FLOWER / FRUIT COLOUR
Rhododendron (*Rhododendron arboreum*)													White, pink, red
Double conradina cherry (*Prunus hirtipes* 'Semiplena')													Pale pink
Violet willow (*Salix daphnoides*)													Yellow male catkin
Campbell magnolia (*Magnolia campbellii*)													Pink
Cornelian cherry (*Cornus mas*)													Yellow
Silver wattle (*Acacia dealbata*)													Yellow
Italian maple (*Acer opalus*)													Yellow
Tree heather (*Erica arborea*)													White
Common almond (*Prunus dulcis*)													Pink
Peach almond (*Prunus x amygdalo-persica* 'Pollardii')													Pink
Common camellia (*Camellia japonica*)													White, pink, red
Yoshino cherry (*Prunus x yedoensis*)													Pale pink
Peach (*Prunus persica*)													Pink (also white and rose)
Loebner magnolia (*Magnolia x loebneri*)													White
Veitch magnolia (*Magnolia x veitchii*)													White flushed purple
Hillier spire cherry (*Prunus* 'Spire')													Pink
Norway maple (*Acer platanoides*)													Yellow
Manchurian crab-apple (*Malus baccata* var. *mandshurica*)													White
Ussurian pear (*Pyrus ussuriensis*)													White
Shimidsu cherry (*Prunus serrulata* 'Shimidsu')													White
Serviceberry or shadblow (*Amelanchier lamarckii*)													White
Double gean (*Prunus avium* 'Plena')													White
Japanese crab-apple (*Malus floribunda*)													Crimson buds opening white
Profusion crab-apple (*Malus x moerlandsii* 'Profusion')													Wine red
Shirofugen cherry (*Prunus* 'Shirofugen')													White
Judus tree (*Cercis siliquastrum*)													Rose-lilac
Empress tree (*Paulownia tomentosa*)													Heliotrope purple
Flowering dogwood (*Cornus florida*)													White and also rose
Pacific dogwood (*Cornus nuttallii*)													Creamy-white
Chilean firebush (*Embothrium coccineum*)													Orange-scarlet
Mountain silverbell (*Halesia monticola*)													White/green
Manna (flowering) ash (*Fraxinus ornus*)													Creamy-white
Bead tree or chinaberry (*Melia azedarach*)													Lilac/creamy-white
Dove tree (*Davidia involucrata*)													White
Double-flowered thorn (*Crataegus laevigata* 'Plena')													White
Charlotte crab-apple (*Malus coronaria* 'Charlottae')													Shell pink
Red-flowered horse chestnut (*Aesculus x carnea* 'Briotii')													Deep rose pink
"Baumannii" horse chestnut (*Aesculus hippocastanum* 'Baumannii')													White
Madrone (*Arbutus menziesii*)													White
"Vossii" laburnum (*Laburnum x watereri* 'Vossii')													Yellow
Wilson's magnolia (*Magnolia wilsonii*)													White
Japanese dogwood (*Cornus kousa*)													White
American yellow wood (*Cladrastis lutea*)													White
Japanese "whiteleaf" magnolia (*Magnolia obovata*)													Creamy-white
Fringe tree (*Chionanthus virginicus*)													White
Black locust (*Robinia pseudoacacia*)													White
Japanese snowbell tree (*Styrax japonicus*)													White
Fragrant epaulette tree (*Pterostyrax hispida*)													White

There is no difficulty in having beauty and interest in garden trees in the spring. This chart suggests trees to carry on the interest, with either flowers, ornamental fruit, or eye-catching leaves, all the year around.

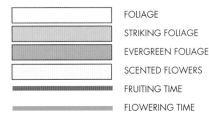

FOLIAGE
STRIKING FOLIAGE
EVERGREEN FOLIAGE
SCENTED FLOWERS
FRUITING TIME
FLOWERING TIME

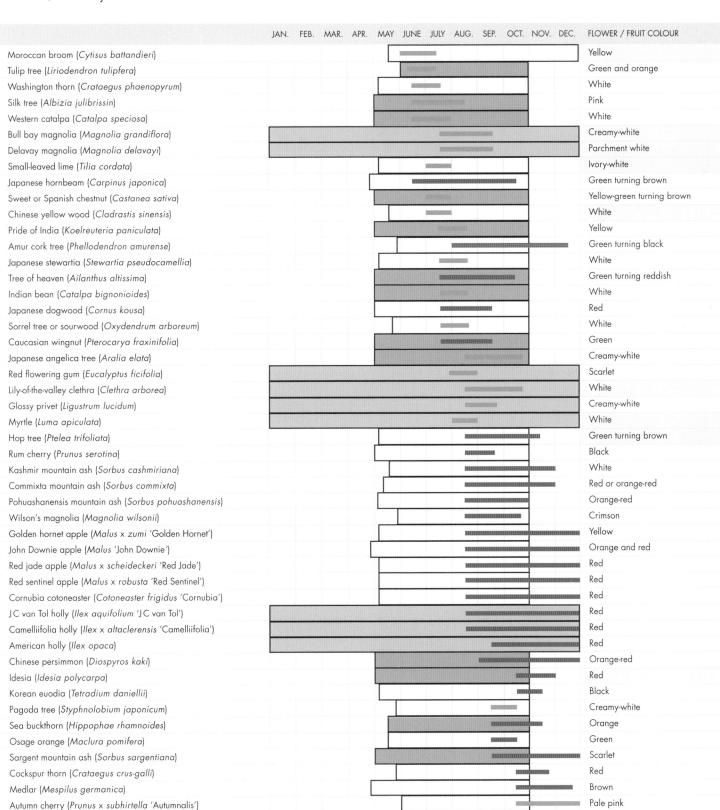

	JAN.	FEB.	MAR.	APR.	MAY	JUNE	JULY	AUG.	SEP.	OCT.	NOV.	DEC.	FLOWER / FRUIT COLOUR
Moroccan broom (*Cytisus battandieri*)													Yellow
Tulip tree (*Liriodendron tulipfera*)													Green and orange
Washington thorn (*Crataegus phaenopyrum*)													White
Silk tree (*Albizia julibrissin*)													Pink
Western catalpa (*Catalpa speciosa*)													White
Bull bay magnolia (*Magnolia grandiflora*)													Creamy-white
Delavay magnolia (*Magnolia delavayi*)													Parchment white
Small-leaved lime (*Tilia cordata*)													Ivory-white
Japanese hornbeam (*Carpinus japonica*)													Green turning brown
Sweet or Spanish chestnut (*Castanea sativa*)													Yellow-green turning brown
Chinese yellow wood (*Cladrastis sinensis*)													White
Pride of India (*Koelreuteria paniculata*)													Yellow
Amur cork tree (*Phellodendron amurense*)													Green turning black
Japanese stewartia (*Stewartia pseudocamellia*)													White
Tree of heaven (*Ailanthus altissima*)													Green turning reddish
Indian bean (*Catalpa bignonioides*)													White
Japanese dogwood (*Cornus kousa*)													Red
Sorrel tree or sourwood (*Oxydendrum arboreum*)													White
Caucasian wingnut (*Pterocarya fraxinifolia*)													Green
Japanese angelica tree (*Aralia elata*)													Creamy-white
Red flowering gum (*Eucalyptus ficifolia*)													Scarlet
Lily-of-the-valley clethra (*Clethra arborea*)													White
Glossy privet (*Ligustrum lucidum*)													Creamy-white
Myrtle (*Luma apiculata*)													White
Hop tree (*Ptelea trifoliata*)													Green turning brown
Rum cherry (*Prunus serotina*)													Black
Kashmir mountain ash (*Sorbus cashmiriana*)													White
Commixta mountain ash (*Sorbus commixta*)													Red or orange-red
Pohuashanensis mountain ash (*Sorbus pohuashanensis*)													Orange-red
Wilson's magnolia (*Magnolia wilsonii*)													Crimson
Golden hornet apple (*Malus x zumi* 'Golden Hornet')													Yellow
John Downie apple (*Malus* 'John Downie')													Orange and red
Red jade apple (*Malus x scheideckeri* 'Red Jade')													Red
Red sentinel apple (*Malus x robusta* 'Red Sentinel')													Red
Cornubia cotoneaster (*Cotoneaster frigidus* 'Cornubia')													Red
J C van Tol holly (*Ilex aquifolium* 'J C van Tol')													Red
Camelliifolia holly (*Ilex x altaclerensis* 'Camelliifolia')													Red
American holly (*Ilex opaca*)													Red
Chinese persimmon (*Diospyros kaki*)													Orange-red
Idesia (*Idesia polycarpa*)													Red
Korean euodia (*Tetradium daniellii*)													Black
Pagoda tree (*Styphnolobium japonicum*)													Creamy-white
Sea buckthorn (*Hippophae rhamnoides*)													Orange
Osage orange (*Maclura pomifera*)													Green
Sargent mountain ash (*Sorbus sargentiana*)													Scarlet
Cockspur thorn (*Crataegus crus-galli*)													Red
Medlar (*Mespilus germanica*)													Brown
Autumn cherry (*Prunus x subhirtella* 'Autumnalis')													Pale pink

	NORTH EUROPE	SOUTH EUROPE & MEDITERRANEAN	CAUCASUS & SOUTHWEST ASIA	INDIA, BURMA, NEPAL & BHUTAN	CHINA	KOREA
Ancient / 1500	U.carpinifolia A.pseudoplatanus Pi.abies	U.procera C.sativa Lau.nobilis Cup.sempervirens Sor.domestica Pl.orientalis	Jug.regia Fic.carica Pr.dulcis Pr.cerasifera Pr.cerasus 'Rhexii' Mo.nigra		Pr.persica Mo.alba	
1550 / 1600	Ab.alba L.decidua	Lab.anagyroides Q.ilex Jun.sabina Pr.laurocerasus Cer.siliquastrum Ph.latifolia P.pinea Lab.alpinum Ae.hippocastanum P.pinaster	Dio.lotus Ced.libani Fr.ornus		Mo.alba	
1650 / 1700	A.platanoides	Pr.lusitanica Pl. x hispanica P.halepensis Jun.phoenicea Q.suber				
1700 / 1750	P.cembra	Pr.mahaleb Os.carpinifolia Arb.andrachne Q.canariensis Q.cerris A.monspessulanum Jun.oxycedrus A.opalus A.sempervirens	Cel.tournefortii Carp.orientalis Liq.orientalis		Plat.orientalis S.babylonica Cam.japonica Sty.japonicum	
1750 / 1800	Pop. x canadensis 'Serotina' Jun.communis 'Suecica' Al.incana	Jun.thurifera A.tataricum P.nigra var.laricio Py.salicifolia O.excelsa Cel.australis	Z.carpinifolia Ti.tomentosa Pt.fraxinifolia Cra.tanacetifolia P.nigra var.caramanica		Ai.altissima Ko.paniculata Jun.chinensis Mal.spectabilis Mag.denudata U.parvifolia Dio.kaki	
1800 / 1850	Al.viridis Mag. x soulangana Jun.communis 'Hibernica'	Fr.angustifolia Cra.orientalis var.sanguinea Q.alnifolia Al.cordata Q.pyrenaica Ab.cephalonica Jun.excelsa P.nigra subsp.nigra Ab.pinsapo Q.frainetto A.lobelii Ced.atlantica Q.canariensis	Fr.angustifolia var.oxycarpa S.daphnoides Pi.orientalis Par.persica A.cappadocicum Ti.'Petiolaris' Q.infectoria	P.roxburghii Q.leucotrichophora Pi.smithiana Ab.spectabilis P.wallichiana Cot.frigidus Ced.deodara A.sterculiaceum Ab.pindrow	Cun.lanceolata Cup.torulosa Jun.recurva Cas.cuspidata Cry.japonica var.sinensis Cat.ovata	G.biloba Mal.baccata Lig.lucidum [all 1750–1800] Q.dentata Pau.tomentosa P.bungeana
1850 / 1900		Ab.numidica P.peuce Mal.trilobata Pi.omorika Ced.brevifolia Ab.borisii-regis P.heldreichii Q.macedonica A.syriacum	Q.castaneifolia Jun.drupacea Ab.nordmanniana Ab.cilicica Al.subcordata Q.libani Ti. x euchlora A.hyrcanum Q.macranthera A.velutinum A.velutinum var. vanvolxemii Ti.dasystyla Q.pontica B.medwedewii	Ts.dumosa L.griffithii Ae.indica Pr.cornuta Fr.xanthoxyloides Sor.vestita Pi.spinulosa Mag.campbellii B.utilis Q.semecarpifolia	Cup.funebris Psl.amabilis Pt.stenoptera Cedrel.sinensis P.tabuliformis I.polycarpa Pts.hispida Fr.sieboldiana Sor.pohuashanensis A.truncatum Ket.davidiana P.armandii D.vilmoriniana Jug.cathayensis	Tr.fortunei Q.myrsinifolia T.grandis Q.variabilis Q.acutissima Liq.formosana Q.serrata A.davidii A.maximowiczianum Sor.vilmorinii Euco.ulmoides Cat.fargesii
1900 / 1950	L.decidua subsp.polonica	Z.cretica Ab.nebrodensis	F.orientalis A.cappadocicum subsp.divergens	A.caudatum Mag.insignis Jun.recurva var.coxii Ab.pindrow var.brevifolia Sor.insignis A.campbellii	Lir.chinense Sor.esserteauana B.albosinensis var.septentrionalis Sor.hupehensis F.englerana Ab.delavayi S.matsundana 'Tortuosa' Jun.distans A.amplum Met.glyptostroboides	Te.sinense Sor.folgneri Ab.squamata Pi.asperata Ab.sutchuenensis Ti.japonica 'Ernest Wilson' Th.koraiensis A.hersii Ti.chinensis
1950 / 2000	Sor.admonitor	Q.aucheri Jun.foetidissima	Il.spinigera Q.brantii Q.vulcanica Sor.takhtajanii	A.wardii Ab.densa Pr.himalaica Cup.himalaica Ex.populnea L.himalaica P.bhutanica Pop.glauca Sor.khumbuensis	A.elegantulum Alan.chinense B.insignis Camp.acuminata Carp.fangiana Catha.argyrophylla Cer.chingii Corn.hongkongensis Cory.fargesii Cup.gigantea Illi.simonsii Jug.sigillata Ko.bipinnata Liq.acalycina Lith.variolosus Mag.cylindrica Mag.maudiae Mell.xylocarpum Notha.cavalierei Par.subaequalis P.hwangshanensis Pitt.brevicalyx Pt.macroptera Q.fabrei Q.schottkyana Sor.pseudovilmorinii Sor.hemsleyi Sor.yuana Tetr.ruticarpum Ti.endochrysea U.lamellosa	Albizia kalkora Cel.choseniana Pr.takesimensis

A Chart of Tree Discoveries

The chart on these pages relates our knowledge of the trees of the world to an historical timescale. It gives the date of introduction to the British Isles of more than 550 tree species. This can be taken in most cases to be the approximate date of their identification and naming. In a few instances where the exact date of introduction is not known, the approximate date is indicated by an arrow.

The astonishing acceleration of botanical knowledge in the 18th century, with the exploration of North America, can clearly be seen. The climax came at the end of the 19th century, with the introduction to the Western world of the trees of Japan and China.

For reasons of space (and since many of the trees have no real common names), names are given in their botanical Latin form, which can be translated by referring to the index.

The chart is the work of Alan Mitchell and is reproduced here with his kind permission.

TAIWAN (formerly Formosa)	JAPAN & MANCHURIA	NORTH ASIA	AUSTRALASIA	SOUTH AMERICA	WESTERN NORTH AMERICA & MEXICO	EASTERN & CENTRAL NORTH AMERICA	

Legend (abbreviations):

A. Acer	Ceph. Cephalotaxus	Gym. Gymnocladus	Mo. Morus	Qui. Quillaja
Ab. Abies	Cer. Cercis	H. Halesia	Myr. Myrtus	R. Robinia
Ac. Acacia	Ch. Chamaecyparis	I. Idesia	N. Nyssa	S. Salix
Ae. Aesculus	Ci. Cinnamomum	Il. Ilex	Not. Nothofagus	Sass. Sassafras
Aex. Aextoxicon	Cord. Cordyline	Illi. Illicium	Notha. Nothaphoebe	Sax. Saxegothaea
Ag. Agathis	Corn. Cornus			Sch. Schefflera
Ai. Ailanthus	Cory. Corylus	Jug. Juglans	O. Olea	Seq. Sequoia
Al. Alnus	Cot. Cotoneaster	Jun. Juniperus	Os. Ostrya	Seqd. Sequoiadendron
Alan. Alangium	Cra. Crataegus	Kal. Kalopanax	P. Pinus	Soph. Sophora
Arau. Araucaria	Cry. Cryptomeria	Ket. Keteleeria	Par. Parrotia	Sor. Sorbus
Arb. Arbutus	Cun. Cunninghamia	Ko. Koelreuteria	Pau. Paulownia	St. Stewartia
Ath. Athrotaxis	Cup. Cupressus		Pe. Persea	Sty. Styphnolobium
			Ph. Phillyrea	Styr. Styrax
B. Betula	D. Davidia	L. Larix	Pi. Picea	
Ba. Banksia	Dac. Dacrydium	Lab. Laburnum	Pic. Picrasma	T. Torreya
Bl. Blepharocalyx	Dio. Diospyros	Lag. Lagarostrobos	Pitt. Pittosporum	Ta. Taiwania
	Dip. Dipteronia	Lau. Laurus	Pl. Platanus	Tax. Taxodium
C. Castanea	Dr. Drimys	Laur. Laurelia	Plat. Platycladus	Taxs. Taxus
Call. Callitris		Lig. Ligustrum	Pod. Podocarpus	Te. Tetracentron
Calo. Calocedrus	El. Elaeocarpus	Liq. Liquidambar	Pol. Polylepis	Tetr. Tetradium
Cam. Camellia	Euc. Eucalyptus	Lir. Liriodendron	Pop. Populus	Th. Thuja
Camp. Camptotheca	Euco. Eucommia	Lith. Lithocarpus	Pr. Prunus	Ti. Tilia
Car. Carya	Eucr. Eucryphia	Lu. Luma	Pse. Pseudopanax	Tr. Trachycarpus
Carp. Carpinus	Ex. Exbucklandia		Psl. Pseudolarix	Ts. Tsuga
Carpo. Carpodetus		Mac. Maclura	Pst. Pseudotsuga	
Cas. Castanopsis	F. Fagus	Mag. Magnolia	Pt. Pterocarya	U. Ulmus
Cat. Catalpa	Fic. Ficus	Mal. Malus	Pts. Pterostyrax	
Catha. Cathaya	Fitz. Fitzroya	Man. Manglietia	Py. Pyrus	W. Wollemia
Ced. Cedrus	Fr. Fraxinus	May. Maytenus		
Cedrel. Cedrela		Mell. Melliodendron	Q. Quercus	Z. Zelkova
Cel. Celtis	G. Ginkgo	Met. Metasequoia		
	Gle. Gleditsia			

Time axis entries (pre-1800, North America columns):

Ancient

1500

1550 — Th.occidentalis / P.strobus

1600

Eastern & Central North America (c.1600–1650):
Pr.serotina   Car.ovata   Sass.albidum
Jug.cinerea   Pl.occidentalis
R.pseudoacacia   Tax.distichum

1650

A.rubrum   Cel.occidentalis   Lir.tulipfera
Jun.virginiana   Liq.styraciflua   Jug.nigra
Pop.tacamahaca   A.negundo   Q.prinus
Cra.crus-galli   Ab.balsamea   Q.coccinea
Gle.triacanthos   Pi.glauca

Western North America & Mexico: Cup.lusitanica

1700

Pi.mariana   Ae.pavia   A.saccharinum
Fr.americana   Q.phellos   Cat.bignonioides
Q.alba   Q.nigra   Mag.grandiflora
Ts.canadensis   Q.rubra   A.saccharum
P.palustris   Ch.thyoides   Mag.acuminata
P.resinosa   B.nigra   L.laricina   P.echinata
P.virginian   P.taeda   R.hispida   Gym.dioica
A.spicatum   N.sylvatica   B.papyrifera
Mag.tripetala   Ti.americana   U.americana

1750

H.carolina   A.pensylvanicum   B.lenta
P.rigida   Ae.flava   Car.cordiformis
Car.tomentosa   B.alleghaniensis   P.banksiana
Fr.pensylvanica   Sor.americana   Mag.fraseri
Q.imbricaria   Tax.ascendens   Car.glabra
O.bicolor   Q.palustris

1800

Body (post-1800):

TAIWAN (formerly Formosa)	JAPAN & MANCHURIA	NORTH ASIA	AUSTRALASIA	SOUTH AMERICA	WESTERN NORTH AMERICA & MEXICO	EASTERN & CENTRAL NORTH AMERICA	Date
Jun.formosana	T.nucifera [1750–1800] Pod.macrophyllus A.palmatum Ceph.harringtonii var.drupacea P.massoniana Cry.japonica 'Lobbii'	L.sibirica P.pumila Ab.sibirica L.gmelinii	Soph.tetraptera Euc.obliqua [both 1750–1800] Ac.dealbata Ag.australis Cord.australis Euc.globulus Euc.coccifera	Arau.araucana [1750–1800] May.boaria Dr.winteri Not.betuloides Not.antarctica Lu.apiculata Sax.conspicua Fitz.cupressoides	A.macrophyllum A.circinatum P.ponderosa Ab.amabilis Pst.menziesii Ab.procera Ab.grandis P.radiata Pi.sitchensis Cup.macrocarpa Ab.religiosa Ab.concolor subsp.lowiana P.ayacahuite Th.plicata Seq.sempervirens Ch.nootkatensis	Mag.macrophylla Q.velutina Car.laciniosa P.pungens Ab.fraseri Ae.glabra Q.macrocarpa Mac.pomifera Q. x heterophylla U.rubra Q. x leana	1850
Pic.quassioides A.buergerianum Ti.oliveri Dip.sinensis	Ts.sieboldii Jug.mandshurica P.densiflora P.thunbergii Th.standishii Pi.polita Ch.obtusa Ab.firma Ts.diversifolia L.kaempferi Mag.obovata Mag.kobus Kal.septemlobus A.japonicum A.cissifolium A.rufinerve 'Hatsuyuki' Q.acuta Pi.glehnii A.argutum Kal.septemlobus f.maximowiczii Pi.jezoensis A.crataegifolium A.pictum A.distylum A.rufinerve A.carpinifolium A.diabolicum A.capillipes B.maximowicziana A.miyabei Mag.salicifolia Pst.japonica B.grossa	Pi.obovata U.pumila Ti.mandshurica Pi.schrenkiana B.costata	Euc.gunnii Lag.franklinii Ath.spp. Not.fusca Not.cliffortioides Not.moorei	Eucr.cordifolia Pod.salignus Pod.andinus Laur.sempervirens	Ab.magnifica Seqd.giganteum Ab.bracteata Ts.heterophylla Pi.engelmannii P.aristata Q.wislizeni Fr.latifolia Q.garryana Q.lobata Ab.concolor Q.kelloggii Cup.guadalupensis Cup.arizonica L.occidentalis Al.rubra Pi.breweriana Fr.velutina	Cat. x erubescens U.thomasii Q. x ludoviciana Cat.speciosa Ts.caroliniana Q. x schochiana H.monticola Q.shumardii	1900
Carp.turczaninowii Ch.formosensis Mag.sargentiana Ab.koreana Pi.likiangensis Ta.cryptomerioides A.triflorum Ab.kawakamii Ts.formosana	S.gracilistyla Pr. x yedoensis A.mandshuricum Ab.holophylla St.monadelpha Pi.koyamae Q.aliena Pr.maackii Carp.laxiflora St.serrata Ti.kiusiana	L.gmelinii var.olgensis		Not.obliqua Not.procera Not.dombeyi	Pop.trichocarpa Jun.deppeana Cup.sargentii Pst.macrocarpa Cup.forbesii Jun.ashei	U.serotina Pi.glauca var.albertiana Cup.arizonica var.glabra Fr.tomentosa Jun.virginiana 'Pseudocupressus' Jun.scopulorum 'Skyrocket'	1950
A.caudatifolium Al.formosana Calo.formosana Carp.kawakamii Lith.kawakamii P.armandii var. mastersiana P.morrisonicola Sch.taiwaniana Sor.randaiensis Styr.formosanus	A.morifolium A.pycnanthum Cas.sieboldii Ci.japonicum Fr.lanuginosa Pe.thunbergii Ti.amurensis	B.tianschanica Mal.kirghisorum	Ac.pataczekii Ba.integrifolia Call.rhomboidea Carpo.serratus El.dentatus Euc.gregsoniana Euc.pauciflora subsp.debeuzevillei Euc.nitens Pse.ferox	Ac.caven Aex.punctatum Arau.angustifolia Bl.cruckshanksii Not.alessandrii Not.nitida Not.pumilio Pol.australis Qui.saponaria S.humboldtiana	Cup.bakeri Cup.abramsiana Ab.vejarii P.cooperi Arb.xalapensis Corn. florida subsp.urbiniana Mag.tamaulipana Pi.chihuahuana P.longaeva P.pseudostrobus Pl.mexicana Q.affinis Q.rhysophylla Taxs.globosa Ti.mexicana	B.lenta f.uber Pe.borbonia Q.buckleyi Q.gravesii Q.leucotrichophora	2000

W.nobilis (Australasia)

Rates of Growth

NO RATE OF GROWTH holds good for any species of tree under all circumstances. A tree's performance is governed very much by the area in which it is planted: climate and soil are all-important. Soils vary considerably in depth, moisture content, and nutrients. These factors in turn depend on rainfall and temperature. Trees, however, have their characteristic patterns of growth. The graphs on the right show how a selection of conifers and of broadleaves reach their maximum growth rates at different ages: some holding back for the first 10 years, before putting on three feet a year; others racing away but soon slowing. The trees shown below and on the following pages are chosen for the wide differences in their speed of growth.

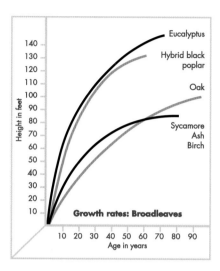

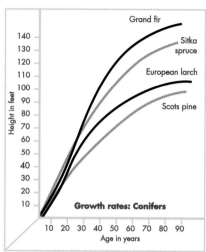

Holm oak
The holm oak starts slowly: it barely reaches 10 feet in 10 years. Then it speeds up to reach 35 feet in 30 years. In the next 70 years it only doubles its height, but spreads hugely.

Wellingtonia
Wellingtonia (or "big tree") can be relied on to grow a steady two feet a year for 40 or 50 years, slowing down as it approaches 100 feet.

Cider gum
Cider gum is immensely vigorous from the start, reaching 40 feet in 10 years but slowing down, scarcely doubling its height in the following 20 years.

White willow
White willow makes a really big tree faster than almost any: its height and spread at 30 years make it look twice its age. Thereafter its growth rate is slow.

Monterey pine
Monterey pine is one of the fastest of all conifers: in exceptional conditions it has grown 20 feet in one year: it can be relied on to do three or four.

Common silver birch
Silver birch gets away quickly but settles down to a slow advance after 15 or 20 years – making it an excellent garden tree.

English yew
English yew is notoriously slow to start with, but grows as wide as it is high – and eventually much wider.

English elm
English elm is always vigorous, but most so in middle age: a 100-year-old tree looks 200 or 300 years old.

After 15 years

After 15 years

After 100 years

After 15 years

After 15 years

After 100 years

After 100 years

Feet

Douglas fir
Douglas fir in good damp conditions is a three-feet-a-year tree for the 30 years from five to 35. It then slows down and thickens its stem.

European horse chestnut
Horse chestnut reaches its full height, like many broadleaf trees, in 50 or 60 years at about two feet a year. Thereafter it just spreads outwards.

Lombardy poplar
Lombardy poplar can grow between two and three feet a year for 25 years before it slows down. Planted at eight feet it should make a 20-foot screen in five years.

English holly
English holly starts as slowly as yew. Only after about 15 years does it put on 18-inch shoots in one growing season.

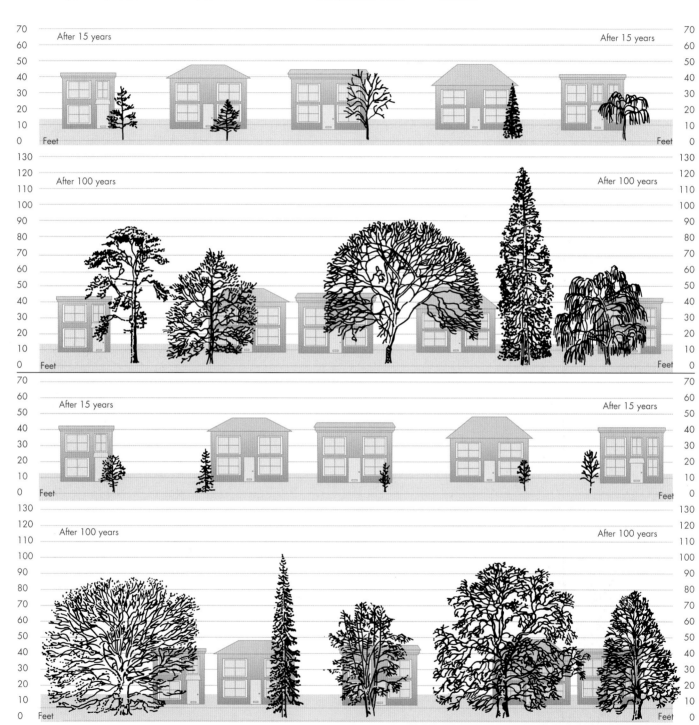

Scots pine
Scots pine grows surprisingly fast after the first three or four years: a three-foot shoot is normal. But it soon loses its upward vigour and spreads out.

Scarlet oak
Scarlet oak in the acid loam it likes is among the fastest of the oaks. The growth rates shown here are about average for any oak.

Wingnut
Wingnut is in the championship class for high-speed broadleaves, growing 45 feet in only 15 years, then building a wide symmetrical dome.

Giant thuja
Giant thuja is one of the fastest conifers: a reliable two-feet-a-year tree in gardens, though much more in the Pacific northwest.

Weeping willow
Weeping willows usually surprise their owners by their rapid spread: at 15 years they are 30 feet across as well as high.

European beech
European beech is rather slow in its early years but makes up for it in its second and third decades. At 100 years it is a fully mature tree: as wide as it is high.

Serbian spruce
Serbian spruce is the best of the spruces for a confined space, never spreading but mounting steadily at two feet a year in a trim spire.

Maidenhair tree
The maidenhair tree or ginkgo is never a fast grower. A one-foot shoot is about normal for its formative years.

London plane
The London plane is slower than its American parent, but the estimate here is conservative: it should grow two feet a year in its early years.

Big-leafed lime
The big-leafed lime (Tilia platyphyllos) is not exceptional for speed but keeps a narrow tower-like shape in maturity, unlike most broadleaves.

Spanish chestnut
Spanish chestnut is
relatively slow in its earlier
years but immensely
vigorous in middle age,
when it will put out 10-foot
shoots in one year if it is
cut back hard.

Silver maple
Silver maple is among
the fastest of the maples,
but has a reputation for
making weak wood which
is liable to split. Its initial
energy fades relatively
early in life.

European larch
All larches are among the
quicker-growing conifers.
The speed of the European
larch is typical: at 15
years it is growing three
feet a year; 100 years
brings it to maturity.

Common juniper
The junipers are a slow-growing
race – hence excellent for small
gardens. Common European
juniper scarcely reaches tree
stature – even in 100 years.

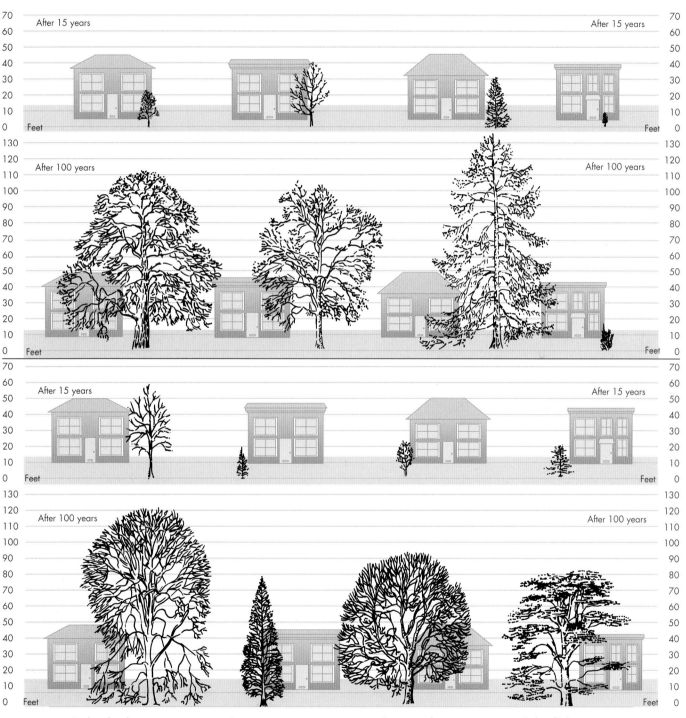

Southern beech
The southern beeches
(*Nothofagus*) are among the
fastest of all broadleaves:
the roblé's 60 feet in 15 years
is comparable to the
performance of much shorter-
lived willows and poplars.

Swamp cypress
Swamp cypress grows
fairly slowly but steadily to
a great age. At 100 years
it is still not fully grown.

European ash
American ash is one of the
fastest-growing American
hardwoods. European ash
(above) is steadier, but still
vigorous, reaching 90 feet
or so in well under 100
years.

Cedar of Lebanon
Cedar of Lebanon grows and
forms its venerable plateau much
faster than you might expect:
100 years is enough to
complete the picture of sublime
old age.

A Guide to Choosing Trees

By their shape, colour, performance, and preferences

The following lists classify a selection of trees from the Conifers and Broadleaves sections according to their outstanding characteristics.

SHAPE

Picturesque/romantic/irregular
Black mulberry Morus nigra
Japanese snowbell Styrax japonicus
Judas tree Cercis siliquastrum
Strawberry tree Arbutus unedo

Araucaria Araucaria species
Nikko fir Abies homolepis
Pines – most with age, especially eastern white pine Pinus strobus
Japanese black pine Pinus thunbergii
Japanese white pine Pinus parviflora
Scots pine Pinus sylvestris
Umbrella pine Pinus pinea

Broad-domed
Algerian oak Quercus canariensis
American walnut Juglans nigra
Black poplar Populus nigra
Blue gum Eucalyptus globulus
Broad-leaved lime Tilia platyphyllos
California laurel Umbellularia californica
Caucasian wingnut Pterocarya fraxinifolia
Chestnut-leaved oak Quercus castaneifolia
Elm zelkova Zelkova carpinifolia
English oak Quercus robur
European beech Fagus sylvatica
European walnut Juglans regia
Hardy rubber tree Eucommia ulmoides
Holm oak Quercus ilex
Hungarian oak Quercus frainetto
Indian bean tree Catalpa bignonioides
Italian maple Acer opalus
Japanese crab-apple Malus floribunda (and others)
Japanese zelkova Zelkova serrata
London plane Platanus × hispanica
Lucombe oak Quercus × hispanica 'Lucombeana'
Norway maple Acer platanoides
Red horse chestnut Aesculus × carnea
Red oak Quercus rubra
Silk tree Albizia julibrissin
Sweet chestnut Castanea sativa
Sycamore Acer pseudoplatanus
Turkey oak Quercus cerris

Cedar of Lebanon Cedrus libani
Monterey cypress Cupressus macrocarpa

Tower-like
Bat willow Salix alba var. caerulea
Bitternut Carya cordiformis
Common lime Tilia × europaea
Cucumber tree Magnolia acuminata
English elm Ulmus procera
Mockernut Carya tomentosa
Pecan Carya illinoinensis
Pignut Carya glabra
Pin oak Quercus palustris
Roblé beech Nothofagus obliqua
Shagbark hickory Carya ovata
White willow Salix alba

Giant cone
Atlas cedar Cedrus atlantica
Big tree Sequoiadendron giganteum
European larch Larix decidua
Japanese cedar Cryptomeria japonica
Japanese larch Larix kaempferi
Western hemlock Tsuga heterophylla
Western red cedar Thuja plicata

Globular-headed
Black locust Robinia pseudoacacia 'Umbraculifera'
Cockspur thorn Crataegus crus-galli
Norway maple Acer platanoides 'Globosum' and 'Summershade'
Sugar maple Acer saccharum 'Globosum'

Horizontally spreading
Black mulberry Morus nigra
Chinese witch hazel Hamamelis mollis
Fig Ficus carica
Japanese cherry Prunus 'Kiku-shidare-zakura' and 'Shirofugen'
Japanese snowbell Styrax japonicus
Laburnum Laburnum species
Persian parrotia Parrotia persica
Silk tree Albizia julibrissin
Witch hazel Hamamelis virginiana

Golden larch Pseudolarix amabilis

Pyramidal: broad
Dove (or handkerchief) tree Davidia involucrata and D.i. var. vilmoriniana
Silver lime Tilia tomentosa
Turkish hazel Corylus colurna

Golden larch Pseudolarix amabilis

Pyramidal: narrower
Alder Alnus species
Cornish elm Ulmus minor 'Cornubiensis'
Jersey elm Ulmus minor var. sarniensis
Sweetgum Liquidambar styraciflua

Caucasian fir Abies nordmanniana
Dawn redwood Metasequoia glyptostroboides
Serbian spruce Picea omorika

Upright, columnar, or fastigiate
Common hawthorn Crataegus monogyna 'Stricta'
Crab-apples Malus baccata 'Columnaris' and M. 'Van Eseltine'
Dawyck beech Fagus sylvatica 'Dawyck' and F.s. 'Dawyck Purple'
English oak Quercus robur
Honey locust Gleditsia triacanthos 'Columnaris'
Japanese cherry Prunus 'Amanogawa'
Lombardy poplar Populus nigra 'Italica'
Norway maple Acer platanoides 'Columnare' and A.p. 'Erectum'
Red maple Acer rubrum 'Columnare'
Silver birch Betula pendula 'Fastigiata'
Small-leaved lime Tilia cordata 'Swedish Upright'
Sorbus Sorbus × thuringiaca 'Fastigiata'
Sugar maple Acer saccharum 'Temple's Upright'
Tulip tree Liriodendron tulipifera 'Fastigiatum'

Washington thorn Crataegus phaenopyrum 'Fastigiata'
White poplar Populus alba f. pyramidalis

Arolla pine Pinus cembra
Atlas cedar Cedrus atlantica 'Fastigiata'
Chinese juniper Juniperus chinensis 'Aurea' and J.c. 'Columnaris Glauca'
Incense cedar Calocedrus decurrens
Irish juniper Juniperus communis 'Hibernica'
Irish yew Taxus baccata 'Fastigiata'
Italian cypress Cupressus sempervirens
Lawson cypress Chamaecyparis lawsoniana 'Columnaris', C.l. 'Erecta Viridis', and many others
Leyland cypress × Cuprocyparis leylandii
Maidenhair tree Ginkgo biloba 'Sentry'
Pond cypress Taxodium ascendens
Rocky mountain juniper Juniperus scopulorum 'Skyrocket'
Scots pine Pinus sylvestris 'Fastigiata'
Syrian juniper Juniperus drupacea
Western red cedar Thuja plicata 'Fastigiata'

Pendulous (not weeping)
Japanese flowering crab Malus floribunda
Silver birch Betula pendula
Smooth-leaved elm Ulmus carpinifolia
Weeping silver lime Tilia 'Petiolaris'
White elm Ulmus americana
Wych elm Ulmus glabra

Brewer's spruce Picea breweriana
Deodar Cedrus deodara
Drooping juniper Juniperus recurva and J.r. var. coxii
European larch Larix decidua
Huon pine Lagarostrobus franklinii
Kashmir cypress Cupressus cashmeriana
Lawson cypress Chamaecyparis lawsoniana 'Intertexta'
Nootka cypress Xanthocyparis nootkatensis and X.n. 'Pendula'
Prince Albert's yew Saxegothaea conspicua
Westfelton yew Taxus baccata 'Dovastoniana'
West Himalayan spruce Picea smithiana

Weeping
Camperdown elm Ulmus glabra 'Camperdownii'
Perry's silver weeping holly Ilex aquifolium 'Argentea Marginata Pendula'
Silver birch Betula pendula 'Tristis'
Spring cherry Prunus × subhirtella 'Pendula'
'Tortuosa' beech Fagus sylvatica 'Tortuosa'
Weeping beech Fagus sylvatica 'Pendula'
Weeping oak Quercus robur 'Pendula Rosea'
Weeping purple beech Fagus sylvatica 'Purpurea Pendula'
Weeping willow Salix babylonica, S. × sepulcralis, and S. × s. var. chrysocoma
Weeping wych elm Ulmus glabra 'Pendula'
Young's weeping birch Betula pendula 'Youngii'
Big tree Sequoiadendron giganteum

'Pendulum'
Eastern hemlock Tsuga canadensis 'Pendula'
Japanese red pine Pinus densiflora 'Umbraculifera'

Weeping – height depending on graft
Flowering dogwood Cornus florida 'Pendula'
Japanese pagoda tree Styphnolobium japonicum 'Pendulum'
Weeping ash Fraxinus excelsior 'Pendula'
Young's weeping birch Betula pendula 'Youngii'

Atlas cedar Cedrus atlantica 'Pendula'

Multi-stemmed
Caucasian wingnut Pterocarya fraxinifolia
Hazel Corylus avellana
Japanese maple Acer japonicum, A. palmatum
Oleaster Elaeagnus angustifolia
Olive Olea europaea
Rhododendron Rhododendron species
Sea buckthorn Hippophae rhamnoides
Stewartia Stewartia species
Vine maple Acer circinatum

Chinese juniper Juniperus chinensis 'Kaizuka'
Eastern hemlock Tsuga canadensis
Lace-bark pine Pinus bungeana
Mountain pine Pinus mugo

Contorted trees
Corkscrew hazel Corylus avellana 'Contorta'
Corkscrew willow Salix babylonica var. pekinensis 'Tortuosa' and S. 'Erythroflexuosa'
"Tortuosa" beech Fagus sylvatica 'Tortuosa'

LEAVES, FLOWERS, FRUIT, AND BARK
Leaves – big
American lime Tilia americana
American walnut Juglans nigra
Bigleaf magnolia Magnolia macrophylla
Bigleaf maple Acer macrophyllum
Burr oak Quercus macrocarpa
Caucasian wingnut Pterocarya fraxinifolia
Chinese poplar Populus lasiocarpa
Chinese toon Toona sinensis
Daimyo oak Quercus dentata
Delavay's magnolia Magnolia delavayi
Empress tree Paulownia tomentosa
Hercules club Aralia spinosa
Himalayan whitebeam Sorbis thibetica 'John Mitchell', S. vestita
Idesia Idesia polycarpa
Japanese angelica tree Aralia elata
Japanese broadleaf holly Ilex latifolia
Japanese horse chestnut Aesculus turbinata
Kentucky coffee tree Gymnocladus dioica
Mockernut Carya tomentosa
Palms All species
Rhododendron Rhododendron calophytum, R. falconeri, R. sinogrande
Southern catalpa Catalpa bignonioides
Tree of heaven Ailanthus altissima

Jelecote pine *Pinus patula*
Montezuma pine *Pinus montezumae*
Pitch pine *Pinus palustris*

Leaves – very small
Antarctic beech *Nothofagus antarctica*
Black beech *Nothofagus solanderi*
Chinese elm *Ulmus parvifolia*
Mountain beech *Nothofagus solanderi*
  var. *cliffortioides*
New Zealand sophora *Sophora
  tetraptera*
Perny holly *Ilex pernyi*

Oriental spruce *Picea orientalis*

Evergreen trees (not conifers)
Acacia *Acacia* species
American holly *Ilex opaca*
Bull bay magnolia *Magnolia grandiflora*
California laurel *Umbellularia californica*
California live oak *Quercus agrifolia*
Camellia *Camellia* species
Carob *Ceratonia siliqua*
Chilean fire bush *Embothrium* species
Chinese photinia *Photinia serratifolia*
Citrus *Citrus* species
Cork oak *Quercus suber*
Delavay's magnolia *Magnolia delavayi*
English holly *Ilex aquifolium*
Eucalyptus *Eucalyptus* species
Evergreen dogwood *Cornus capitata*
Glossy privet *Ligustrum lucidum*
Golden chestnut *Chrysolepis chrysophylla*
Heath *Erica* species
Highclere holly *Ilex × altaclerensis*
Holm oak *Quercus ilex*
Holly *Ilex latifolia, I. pedunculosa*
Laurel *Laurus nobilis*
Live oak *Quercus virginiana*
New Zealand sophora *Sophora tetraptera*
Oak *Quercus acuta, Q. glauca*
Olive *Olea europaea*
Palms All species
Perny holly *Ilex pernyi*
Portugal laurel *Prunus lusitanica*
Rhododendron *Rhododendron arboreum,
  R. barbatum, R. falconeri, R. maximum*
Silk oak *Grevillea robusta*
Strawberry tree *Arbutus* species
Tanbark oak *Lithocarpus* species

Leaves – red/purple
Crab-apple *Malus × purpurea,
  M. × p.* 'Lemoinei'
English oak *Quercus robur* 'Atropurpurea'
Japanese maple *Acer palmatum*
  'Atropurpureum'
Norway maple *Acer platanoides*
  'Goldsworth Purple', *A.p.* 'Schwedleri'
Purple beech *Fagus sylvatica* 'Riversii'
Purple-leaf filbert *Corylus maxima*
  'Purpurea'
Purple-leaved plum *Prunus cerasifera*
  'Pissardii'
Smoke tree *Cotinus coggygria* 'Royal
  Purple'
Sycamore *Acer pseudoplatanus*
  f. *purpureum, A.p.* 'Atropurpureum'

Japanese cedar *Crytomeria japonica*
  'Elegans'

Leaves – silver/grey
English holly *Ilex aquifolium* 'Argentea

Marginata Pendula', *I.a.* 'Silver
  Milkboy', *I.a.* 'Silver Queen', *I.a.*
  'Silver Sentinel'
Olive *Olea europaea*
Russian olive *Elaeagnus angustifolia*
Sea buckthorn *Hippophae rhamnoides*
Silver maple *Acer saccharinum*
Weeping silver lime *Tilia* 'Petiolaris'
Whitebeam *Sorbus aria, S. thibetica,
  S.t.* 'John Mitchell', *S. vestita*
White poplar *Populus alba*
White willow *Salix alba* var. *sericea*
Willow-leaved pear *Pyrus salicifolia*

Arizona cypress *Cupressus arizonica*
  'Pyramidalis'
Blue spruce *Picea pungens* 'Glauca'
Colorado white fir *Abies concolor*
  'Candicans'

Leaves – green and yellow
Box elder *Acer negundo* 'Elegans'
Chestnut *Castanea sativa* 'Variegata'
English holly *Ilex aquifolium* 'Golden
  Milkboy', *I.a.* 'Golden Queen'
English oak *Quercus robur* f. *variegata*
Highclere holly *Ilex × altaclerensis*
  'Golden King', *I. × a.* 'Lawsoniana'
Japanese angelica tree *Aralia elata*
  'Aureovariegata'
Sweetgum *Liquidambar styraciflua*
  'Variegata'
Tulip tree *Liriodendron tulipifera*
  'Aureomarginatum'

Chinese arborvitae *Platycladus orientalis*
  'Elegantissima'
Golden yew *Taxus baccata* 'Aurea'
Lawson cypress *Chamaecyparis
  lawsoniana* 'Lanei Aurea', *C.l.* 'Lutea'
Oriental spruce *Picea orientalis* 'Aurea'
Western red cedar *Thuja plicata* 'Zebrina'
Yew *Taxus baccata* 'Standishii'

Leaves – cream/yellow/gold
English elm *Ulmus procera* 'Louis van
  Houtte'
English oak *Quercus robur* 'Concordia'
European alder *Alnus glutinosa* 'Aurea'
European ash *Fraxinus excelsior*
  'Jaspidea'
European beech *Fagus sylvatica* 'Zlatia'
False acacia *Robinia pseudoacacia*
  'Frisia'
Golden-leaved laburnum *Laburnum
  anagyroides* 'Aureum'
Golden poplar *Populus* 'Serotina Aurea'
Honey locust *Gleditsia triacanthos*
  'Sunburst'
Japanese maple *Acer shirasawanum*
  'Aureum'
Laurel *Laurus nobilis* 'Aurea'
Southern catalpa *Catalpa bignonioides*
  'Aurea'
Sycamore *Acer pseudoplatanus*
  'Brilliantissimum', *A.p.* 'Worley'
White poplar *Populus alba* 'Richardii'

American arborvitae *Thuja occidentalis*
  'Rheingold'
Atlas cedar *Cedrus atlantica* 'Aurea'
Chinese juniper *Juniperus chinensis*
  'Aurea'
Hinoki cypress *Chamaecyparis obtusa*
  'Crippsii'

Lawson cypress *Chamaecyparis
  lawsoniana* 'Winston Churchill'
Scots pine *Pinus sylvestris* 'Aurea'

Leaves – blue-green
Eucalypt *Eucalyptus* species

Bhutan pine *Pinus wallichiana*
Blue cedar *Cedrus atlantica* 'Glauca'
Hemlock *Tsuga mertensiana* 'Glauca'
Lawson cypress *Chamaecyparis
  lawsoniana* 'Columnaris', *C.l.* 'Pembury
  Blue', *C.l.* 'Triomf van Boskoop', *C.l.*
  'Wisselii', etc.
Scots pine *Pinus sylvestris*
Weymouth pine *Pinus strobus*

Leaves – unusual and attractive
Camellia *Camellia* species
Cherry laurel *Prunus laurocerasus*
  'Camelliifolia'
Chinese persimmon *Diospyros kaki*
European alder *Alnus glutinosa*
  'Imperialis'
Fern-leaved beech *Fagus sylvatica* var.
  heterophylla
Golden chestnut *Chrysolepis chrysophylla*
Holly *Ilex* species
Hornbeam *Carpinus betulus* 'Incisa'
Hungarian oak *Quercus frainetto*
Japanese maple *Acer japonicum*
  'Aconitifolium', *A.j.* 'Vitifolium',
  A. palmatum var. *dissectum*,
  A.p. 'Linearilobum'
Katsura tree *Cercidiphyllum japonicum*
Oriental plane *Platanus orientalis*
Paper mulberry *Broussonetia papyrifera*
Scarlet oak *Quercus coccinea*
Swedish birch *Betula pendula*
  'Dalecarlica'
Sweetgum *Liquidambar styraciflua*
Tulip tree *Liriodendron tulipifera*

Maidenhair tree *Ginkgo biloba*

Leaves – divided/cut
Cut-leaved walnut *Juglans regia*
  'Laciniata'
European alder *Alnus glutinosa*
  'Imperialis'
Fern-leaved beech *Fagus sylvatica*
  var. *heterophylla*
Hornbeam *Carpinus betulus* 'Incisa'
Japanese maple *Acer japonicum*
  'Acontifolium', *A. palmatum*
  var. *dissectum*
Mountain ash *Sorbus aucuparia*
  'Aspleniifolia'
Norway maple *Acer platanoides*
  'Dissectum', *A.p.* 'Laciniatum',
  A.p. 'Palmatifidum'
Swedish birch *Betula pendula*
  'Dalecarlica'

Leaves – very early
Bird cherry *Prunus padus* var. *commutata*
Hawthorn *Crataegus laevigata*
Horse chestnut *Aesculus hippocastanum*
Japanese flowering crab *Malus floribunda*
Manchurian birch *Betula platyphylla*
Roblé beech *Nothofagus obliqua*
Southern beech *Nothofagus × alpina*
Dawn redwood *Metasequoia
  glyptostroboides*

Leaves – late to drop
Algerian oak *Quercus canariensis*
Chinese elm *Ulmus parvifolia*
English elm *Ulmus procera*
Hawthorn *Crataegus × lavalleei*
Laurel oak *Quercus laurifolia*
Lucombe oak *Quercus × hispanica*
  'Lucombeana'
Water oak *Quercus nigra*
White oak *Quercus alba*

Autumn colour
Amur cork tree *Phellodendron amurense*
Antarctic beech *Nothofagus antarctica*
Aspen *Populus tremula*
Beech *Fagus* (most species)
Birch *Betula* (most species)
Black cottonwood *Populus trichocarpa*
Black tupelo *Nyssa sylvatica*
Buckeye *Aesculus parviflora*
Caucasian wingnut *Pterocarya fraxinifolia*
Chinese persimmon *Diospyros kaki*
Chinese toon *Toona sinensis*
Chinese witch hazel *Hamamelis mollis*
Crab-apple *Malus tschonoskii*
Disanthus *Disanthus cercidifolius*
English elm *Ulmus procera*
Flowering dogwood *Cornus florida*
Franklinia *Franklinia* species
Giant dogwood *Cornus controversa*
Guelder rose *Viburnum opulus*
Hickory *Carya* species
Japanese cherry *Prunus sargentii*
Japanese dogwood *Cornus kousa*
Japanese maple *Acer palmatum*
  'Osakazuki'
Katsura tree *Cercidiphyllum japonicum*
Kentucky coffee tree *Gymnocladus dioica*
Mountain ash *Sorbus alnifolia,
  S. americana, S. aucuparia,
  S. commixta, S.* 'Embley',
  S. sargentiana
Nikko maple *Acer maximowiczianum*
Norway maple *Acer platanoides*
Ohio buckeye *Aesculus glabra*
Pacific dogwood *Cornus nuttallii*
Persian parrotia *Parrotia persica*
Pin oak *Quercus palustris*
Quaking aspen *Populus tremuloides*
Red maple *Acer rubrum*
Red oak *Quercus rubra*
Sassafras *Sassafras albidum*
Scarlet oak *Quercus coccinea*
Smoke tree *Cotinus* species
Snowy mespilus *Amelanchier* species
Sourwood *Oxydendrum arboreum*
Stag's horn sumach *Rhus typhina*
Stewartia *Stewartia koreana,
  S. pseudocamellia*
Sugar maple *Acer saccharum*
Sweetgum *Liquidambar styraciflua*
Trident maple *Acer buergerianum*
Tulip tree *Liriodendron tulipifera*
White oak *Quercus alba*
White poplar *Populus alba*
Willow oak *Quercus phellos*
Witch hazel *Hamamelis virginiana*
Yellow wood *Cladrastis* species
Zelkova *Zelkova serrata*

Dawn redwood *Metasequoia
  glyptostroboides*
Golden larch *Pseudolarix
  amabilis*
Larch *Larix* species

Maidenhair tree *Ginkgo biloba*
Pond cypress *Taxodium ascendens*
Swamp cypress *Taxodium distichum*

Flowers – early in spring
Alder *Alnus* species
Almond *Prunus dulcis*
Aspen *Populus tremula*
Birch *Betula* species
Camellia *Camellia* species
Cherry plum *Prunus cerasifera*
Elm *Ulmus* species
Goat willow *Salix caprea*
Hazel *Corylus* species
Italian maple *Acer opalus*
Magnolia *Magnolia campbellii*
Norway maple *Acer platanoides*
Silver wattle *Acacia dealbata*
Witch hazel *Hamamelis* species

Flowers – ornamental
Camellia *Camellia* species
Carolina silverball *Halesia carolina*
Cherry *Prunus* (most species)
Chilean fire bush *Embothrium coccineum*
Chinese witch hazel *Hamamelis mollis*
Cootamundra wattle *Acacia baileyana*
Crab-apple *Malus* (most species)
Dogwood *Cornus* species
Dove (or handkerchief) tree *Davidia involucrata, D.i.* var. *vilmoriniana*
European pussy willow *Salix caprea*
Golden-rain tree *Koelreuteria paniculata*
Hawthorn *Crataegus* species
Horse chestnut *Aesculus hippocastanum, A.h.* 'Baumannii'
Indian horse chestnut *Aesculus indica*
Japanese witch hazel *Hamamelis japonica* 'Arborea'
Judas tree *Cercis siliquastrum*
Laburnum *Laburnum* species
Magnolia *Magnolia* species
Moroccan broom *Cytisus battandieri*
Mountain silverbell *Halesia monticola*
Red buckeye *Aesculus pavia*
Redbud *Cercis canadensis*
Red horse chestnut *Aesculus* × *carnea*
Silk tree *Albizia julibrissin, A.j.* f. *rosea*
Silver wattle *Acacia dealbata, A. decurrens*
Stewartia *Stewartia koreana, S. pseudocamellia*
Sydney golden wattle *Acacia longifolia*
Violet willow *Salix daphnoides*
Willow *Salix gracilistyla*
Wingnut *Pterocarya* species
Witch hazel *Hamamelis virginiana*
Yellow wood *Cladrastis kentukea*

Lawson cypress *Chamaecyparis lawsoniana*
Likiang spruce *Picea likiangensis*
Pine *Pinus* (many species)

Fruit – ornamental
American holly *Ilex opaca, I.o.* 'Xanthocarpa'
Black mulberry *Morus nigra*
Blackthorn *Prunus spinosa*
Blueberry elder *Sambucus nigra* var. *caerulea*
Campbell's magnolia *Magnolia campbellii* subsp. *mollicomata*
Caucasian wingnut *Pterocarya fraxinifolia*
Chinaberry *Melia azedarach*

Chinese persimmon *Diospyros kaki*
Choke cherry *Prunus virginiana*
Cotoneaster *Cotoneaster* × *watereri*
Crab-apple *Malus* (many species)
Dogwood *Cornus* (many species)
Elder *Sambucus nigra*
English holly *Ilex aquifolium* 'Baccliflava', *I.a.* 'J.C. van Tol'
Hawthorn *Crataegus* (most species)
Highclere holly *Ilex* × *altaclerensis* 'Camelliifolia', *I.* × *a.* 'Golden King'
Italian alder *Alnus cordata*
Japanese broadleaf holly *Ilex latifolia*
Japanese hornbeam *Carpinus japonica*
Japanese whiteleaf magnolia *Magnolia obovata*
Judas tree *Cercis siliquastrum*
Medlar *Mespilus germanica*
Mountain ash *Sorbus* species
Osage orange *Maclura pomifera*
Perny holly *Ilex pernyi*
Persimmon *Diospyros virginiana*
Quince *Cydonia oblonga*
Sea buckthorn *Hippophae rhamnoides*
Strawberry tree *Arbutus unedo*
Tree of heaven *Ailanthus altissima*
Whitebeam *Sorbus* species

Cedar *Cedrus* species
Fir *Abies* species (late summer)
Pine *Pinus* species
Spruce *Picea* species (autumn/winter)
Yew *Taxus* species

Bark – ornamental
Amur cork tree *Phellodendron amurense*
Birch *Betula* (most species)
Cherry *Prunus serrula*
Cork oak *Quercus suber*
David's maple *Acer davidii*
English walnut *Juglans regia*
Eucalypt *Eucalyptus* (most species)
European beech *Fagus sylvatica*
Japanese maple *Acer palmatum* 'Sango-kaku'
London plane *Platanus* × *hispanica*
Manchurian cherry *Prunus maackii*
Paperbark maple *Acer griseum*
Persian parrotia *Parrotia persica*
Red snakebark maple *Acer capillipes*
Rhododendron *Rhododendron barbatum*
Shagbark hickory *Carya ovata*
Snakebark maple *Acer grosseri*
Snowgum *Eucalyptus pauciflora* subsp. *niphophila*
Sorbus *Sorbus alnifolia*
Strawberry tree *Arbutus* species
Striped maple *Acer pensylvanicum*
Violet willow *Salix daphnoides*
White willow *Salix alba*
Zelkova *Zelkova carpinifolia*

Coast redwood *Sequoia sempervirens*
Lace-bark pine *Pinus bungeana*
Scots pine *Pinus sylvestris*

Soil – able to grow in clay
Alder *Alnus* species
Ash *Fraxinus* species
Birch *Betula* species
Cherry *Prunus* species
Cotoneaster *Cotoneaster* species
Crab-apple *Malus* species
Dogwood *Cornus* species
Eucalypt *Eucalyptus* species

Hawthorn *Crataegus* species
Hazel *Corylus* species
Hornbeam *Carpinus* species
Horse chestnut *Aesculus* species
Laburnum *Laburnum* species
Lime *Tilia* species
Maple *Acer* species
Poplar *Populus* species
Rose *Rosa* species
Smoke tree *Cotinus* species
Sorbus *Sorbus* species
Willow *Salix* species
Witch hazel *Hamamelis* species

Arborvitae *Thuja* species
Fir *Abies* species
Larch *Larix* species
Swamp cypress *Taxodium distichum*

Soil – can be acid
Aspen *Populus tremula*
Birch *Betula* species
Chilean fire bush *Embothrium* species
Cotoneaster *Cotoneaster* species
Disanthus *Disanthus cercidifolius*
English oak *Quercus robur*
Franklinia *Franklinia* species
Grey poplar *Populus* × *canescens*
Heath *Erica* species
Holly *Ilex* species
Rhododendron *Rhododendron* species
Sourwood *Oxydendrum arboreum*
White poplar *Populus alba*

Juniper *Juniperus* species
Pine *Pinus* species
Yew *Taxus* species

Soil – can be alkaline
Ash *Fraxinus* species
Beech *Fagus* species
Birch *Betula* species
Cotoneaster *Cotoneaster* species
Crab-apple *Malus* species
Dogwood *Cornus* species
English holly *Ilex aquifolium*
Hawthorn *Crataegus* species
Hedge maple *Acer campestre*
Highclere holly *Ilex* × *altaclerensis*
Hornbeam *Carpinus* species
Horse chestnut *Aesculus* species
Laurel *Laurus nobilis*
Phillyrea *Phillyrea* species
Whitebeam *Sorbus* species

American arborvitae *Thuja occidentalis*
Austrian pine *Pinus nigra*
English yew *Taxus baccata*
Hiba false arborvitae *Thujopsis dolabrata*
Juniper *Juniperus* species
Western red cedar *Thuja plicata*

Soil – must be lime-free (acid)
Bigleaf magnolia *Magnolia macrophylla*
Black tupelo *Nyssa sylvatica*
Camellia *Camellia* species
Campbell's magnolia *Magnolia campbellii, M.c.* subsp. *mollicomata*
Chilean fire bush *Embothrium* species
Clethra *Clethra* species
Gordonia *Gordonia* species
Japanese whiteleaf magnolia *Magnolia obovata*
Japanese willowleaf magnolia *Magnolia salicifolia*

Lily-flowered magnolia *Magnolia liliiflora, M.l.* 'Nigra'
Pin oak *Quercus palustris*
Red oak *Quercus rubra*
Rhododendron *Rhododendron* species
Scarlet oak *Quercus coccinea*
Shumard oak *Quercus shumardii*
Silverbell *Halesia* species
Snowbell *Styrax* species
Sourwood *Oxydendrum arboreum*
Southern beech *Nothofagus* species
Stewartia *Stewartia* species
Sweetgum *Liquidambar styraciflua*
Veitch magnolia *Magnolia* × *veitchii*
Willow oak *Quercus phellos*
Yulan magnolia *Magnolia denudata*

Soil – very dry, either acid or alkaline
Aspen *Populus tremula*
Chestnut *Castanea* species
Grey poplar *Populus* × *canescens*
Robinia *Robinia* species
Siberian elm *Ulmus pumila*
White poplar *Populus alba*

Cedar *Cedrus* species
Cypress *Cupressus* (some species)
Juniper *Juniperus* species
Yew *Taxus* species

Soil – can be badly drained
Alder *Alnus* species
Black tupelo *Nyssa sylvatica*
Caucasian wingnut *Pterocarya fraxinifolia*
Elder *Sambucus* species
European birch *Betula pendula*
Hawthorn *Crataegus laevigata*
Medlar *Mespilus germanica*
Mountain ash *Sorbus aucuparia*
Pear *Pyrus communis*
Pin oak *Quercus palustris*
Poplar *Populus* species
Red maple *Acer rubrum*
River birch *Betula nigra*
Swamp bay magnolia *Magnolia virginiana*
Swamp white oak *Quercus bicolor*
Sweetgum *Liquidambar styraciflua*
White birch *Betula pubescens*
Willow *Salix* species

American arborvitae *Thuja occidentalis*
Dawn redwood *Metasequoia glyptostroboides*
Pond cypress *Taxodium distichum* var. *imbricarium*
Sitka spruce *Picea sitchensis*
Swamp cypress *Taxodium distichum*
Tamarack *Larix laricina*

SPECIAL SITES
Seaside/maritime areas
Aspen *Populus tremula*
Dwarf fan palm *Chamaerops humilis*
English oak *Quercus robur*
Eucalypt *Eucalyptus* species
European ash *Fraxinus excelsior*
Grey poplar *Populus* × *canescens*
Hawthorn *Crataegus* (many species)
Holm oak *Quercus ilex*
Laurel *Laurus nobilis*
Mountain ash *Sorbus aucuparia*
Pittosporum *Pittosporum* species
Sea buckthorn *Hippophae rhamnoides*
Sessile oak *Quercus petraea*

Strawberry tree *Arbutus unedo*
Whitebeam *Sorbus aria*
White poplar *Populus alba*
Willow *Salix* (most species)

Aleppo pine *Pinus halepensis*
Bishop pine *Pinus muricata*
Black pine *Pinus thunbergii*
Juniper *Juniperus* (many species)
Maritime pine *Pinus pinaster*
Monterey cypress *Cupressus macrocarpa*
Monterey pine *Pinus radiata*
Shore pine *Pinus contorta*
Sitka spruce *Picea sitchensis*

Trees which will tolerate industrial or urban atmosphere
Allegheny serviceberry *Amelanchier laevis*
Almond *Prunus dulcis*
Ash *Fraxinus* (most species)
Birch *Betula platyphylla*
Bird cherry *Prunus padus*
Black mulberry *Morus nigra*
Box elder *Acer negundo*
Bull bay magnolia *Magnolia grandiflora*
Camellia *Camellia japonica*
Catalpa *Catalpa* species
Caucasian wingnut *Pterocarya fraxinifolia*
Cherry plum *Prunus cerasifera*
Chinaberry *Melia azedarach*
Chinese toon *Cedrela sinensis*
Clethra *Clethra arborea*
Cotoneaster *Cotoneaster × watereri*
Crab-apple *Malus* species
Cucumber tree *Magnolia acuminata*
English elm *Ulmus procera*
English holly *Ilex aquifolium*
European birch *Betula pendula*
European hornbeam *Carpinus betulus*
European lime *Tilia × europaea*
False acacia *Robinia pseudoacacia*
Gean *Prunus avium*
Hackberry *Celtis* species
Hawthorn *Crataegus* species
Hedge maple *Acer campestre*
Highclere holly *Ilex × altaclerensis*
Holm oak *Quercus ilex*
Honey locust *Gleditsia triacanthos*
Horse chestnut *Aesculus* species
Japanese cherries *Prunus* species
Japanese pagoda tree *Styphnolobium japonicum*
Jersey elm *Ulmus minor* var. *sarniensis*
Laburnum *Laburnum* species
Lime *Tilia × euchlora*
Lucombe oak *Quercus × hispanica* 'Lucombeana'
Magnolia *Magnolia kobus*, *M. × soulangeana*
Medlar *Mespilus germanica*
Norway maple *Acer platanoides*
Paper birch *Betula papyrifera*
Poplar *Populus* (most species)
Red-twigged lime *Tilia platyphyllos* 'Rubra'
Snowy mespilus *Amelanchier canadensis*, *A. lamarckii*, *A. ovalis*
Strawberry tree *Arbutus unedo*
Sycamore *Acer pseudoplatanus*
Tree of heaven *Ailanthus altissima*
Tulip tree *Liriodendron tulipifera*
White birch *Betula pubescens*
White elm *Ulmus americana*
Wych elm *Ulmus glabra*
Yulan magnolia *Magnolia denudata*
Colorado spruce *Picea pungens*

Maidenhair tree *Ginkgo biloba*
White fir *Abies concolor*

For small gardens
Carolina silverbell *Halesia carolina*
Cherry *Prunus* (most species)
Cotoneaster *Cotoneaster × watereri*
Crab-apple *Malus* species
Dawyck beech *Fagus sylvatica* 'Dawyck'
Dogwood *Cornus* species
Eastern redbud *Cercis canadensis*
Flowering ash *Fraxinus ornus*
Fragrant snowbell *Styrax obassia*
Franklinia *Franklinia alatamaha*
Genista *Genista* species
Golden-rain tree *Koelreuteria paniculata*
Hawthorn *Crataegus* (most species)
Hazel *Corylus* species
Heath *Erica* species
Holly *Ilex* species
Japanese maple *Acer japonicum*, *A. palmatum*
Japanese snowbell *Styrax japonicus*
Japanese willowleaf magnolia *Magnolia salicifolia*
Judas tree *Cercis siliquastrum*
Katsura tree *Cercidiphyllum japonicum*
Laburnum *Laburnum* species
Medlar *Mespilus germanica*
Ohio buckeye *Aesculus glabra*
Paperbark maple *Acer griseum*
Photinia *Photinia* species
Quince *Cydonia oblonga*
Rhododendron *Rhododendron* species
Russian olive *Elaeagnus angustifolia*
Sea buckthorn *Hippophae rhamnoides*
Silk oak *Grevillea* species
Snowy mespilus *Amelanchier* species
Sorbus *Sorbus* (most species)
Sourwood *Oxydendrum arboreum*
Stewartia *Stewartia* species
Willow *Salix* (some species)
Willow-leaf pear *Pyrus salicifolia* 'Pendula'
Witch hazel *Hamamelis* species

Dwarf conifers
Eastern hemlock *Tsuga canadensis*
Lace-bark pine *Pinus bungeana*

Street trees – big
Buttonwood *Platanus occidentalis*
Castor-aralia *Kalopanax septemlobus*
False acacia *Robinia pseudoacacia*
Green ash *Fraxinus pennsylvanica*
Honey locust *Gleditsia triacanthos*
London plane *Platanus × hispanica*
Norway maple *Acer platanoides*
Pin oak *Quercus palustris*
Red maple *Acer rubrum*
Red oak *Quercus rubra*
Scarlet oak *Quercus coccinea*
Velvet ash *Fraxinus velutina*
Willow oak *Quercus phellos*

Maidenhair tree *Ginkgo biloba*
Pine *Pinus* (many species)

For narrower streets
Any upright tree (see tables above)
Cherry *Prunus* (most species)
Chinaberry *Melia azedarach*
Chinese toon *Cedrela sinensis*
Euodia *Tetradium daniellii*
European hornbeam *Carpinus betulus*

'Columnaris'
Flowering ash *Fraxinus ornus*
Flowering dogwood *Cornus florida*
Hawthorn *Crataegus* (most species)
Hedge maple *Acer campestre*
Japanese dogwood *Cornus kousa*
Japanese hornbeam *Carpinus japonica*
Lebanon oak *Quercus libani*
Montpelier maple *Acer monspessulanum*
Paperbark maple *Acer griseum*
Red snakebark maple *Acer capillipes*
Tatarian maple *Acer tataricum*

Screening – (very) fast
Bat willow *Salix alba* var. *caerulea*
Black cottonwood *Populus trichocarpa*
Crack willow *Salix fragilis*
Eucalypt *Eucalyptus* species
Lombardy poplar *Populus nigra* 'Italica'
"Robusta" poplar *Populus × canadensis* 'Robusta'

Leyland cypress × *Cuprocyparis leylandii*
Monterey cypress *Cupressus macrocarpa*
Monterey pine *Pinus radiata*
Western hemlock *Tsuga heterophylla*

Tolerant of heavy shade
Beech *Fagus* (some species)
Camellia *Camellia japonica*
Cherry laurel *Prunus laurocerasus*
Common elder *Sambucus nigra*
Holly *Ilex* (some species)
Holm oak *Quercus ilex*
Portugal laurel *Prunus lusitanica*
Sycamore *Acer pseudoplatanus*

Cephalotaxus *Cephalotaxus* species
Coast redwood *Sequoia sempervirens*
English yew *Taxus baccata*
Juniper *Juniperus* (most species)
Western hemlock *Tsuga heterophylla*
Western red cedar *Thuja plicata*

Suitable for tubs
Box *Buxus sempervirens*
Camellia *Camellia* species
Citrus *Citrus* (most species)
Holly *Ilex* (some species)
Laurel *Laurus nobilis*

Chinese arborvitae *Platycladus orientalis* 'Elegantissima'
English yew *Taxus baccata* 'Standishii'
False cypress *Chamaecyparis thyoides* 'Andelyensis'
Lawson cypress *Chamaecyparis lawsoniana* 'Ellwoodii'

Resent moving – except when very small
Birch *Betula* species
Black tupelo *Nyssa sylvatica*
Castor-aralia *Kalopanax septemlobus*
Hickory *Carya* species
Holly *Ilex* species
Kentucky coffee tree *Gymnocladus dioica*
Magnolia *Magnolia* species
Scarlet oak *Quercus coccinea*
Strawberry tree *Arbutus* species
Sweetgum *Liquidambar styraciflua*
Tulip tree *Liriodendron tulipifera*
Walnut *Juglans* species
White oak *Quercus alba*
Cedar *Cedrus* species

Fir *Abies* species
Monterey cypress *Cupressus macrocarpa*
Pine *Pinus* species (except five-needled pines)
Spruce *Picea* species

FEW PESTS OR DISEASES
Alder *Alnus* species
Amur cork tree *Phellodendron amurense*
Castor-aralia *Kalopanax septemlobus*
Dove tree *Davidia* species
English holly *Ilex aquifolium*
Eucalypt *Eucalyptus* species
Glossy privet *Ligustrum lucidum*
Golden-rain tree *Koelreuteria paniculata*
Highclere holly *Ilex × altaclerensis*
Honey locust *Gleditsia triacanthos*
Hornbeam *Carpinus* species
Japanese pagoda tree *Styphnolobium japonicum*
Katsura tree *Cercidiphyllum* species
Kentucky coffee tree *Gymnocladus dioica*
Laburnum *Laburnum* species
Magnolia *Magnolia* species
Persian parrotia *Parrotia persica*
Russian olive *Elaeagnus angustifolia*
Snowbell *Styrax* species
Sweetgum *Liquidambar styraciflua*
Tree of heaven *Ailanthus altissima*
White poplar *Populus alba*

Araucaria *Araucaria* species
Coast redwood *Sequoia sempervirens*
Dawn redwood *Metasequoia glyptostroboides*
False cypress *Chamaecyparis* (most species)
Incense cedar *Calocedrus decurrens*
Japanese umbrella pine *Sciadopitys verticillata*
Juniper *Juniperus* species
Maidenhair tree *Ginkgo biloba*
Podocarpus *Podocarpus* species
Swamp cypress *Taxodium* species
Yew *Taxus* species

The Meanings of Botanical Names

Medieval botanical works used descriptive phrases in Latin to identify plants. Carl Linnaeus (1707–78) did the same, but to each plant (or animal) he also gave a shorter, two-word Latin name: its genus followed by its species. So thorough and indispensable were his books that this convenient two-name system gradually took over completely. Today it is in universal use by all who deal professionally with living organisms.

Every known plant now has a Latin name which identifies it in every country. The advantages of the system are obvious. Most people, however, naturally stick to vernacular names for the few plants they know. For that matter only a minority of plants have vernacular names to stick to. In the United States strenuous efforts are made to give every exotic introduction an English name – but such productions as the Chinese fragrant epaulette tree can hardly be called true vernacular names. And the trouble with the old ones is that they mean different things to different people in different places. Sooner or later one is faced with the need to use, or at least with the value of understanding, the botanical Latin name. The list below demonstrates that for many trees it is a useful description, for others an interesting indication of where it comes from, for others a well-deserved memorial to the explorer who found it. The accent marks the syllable (if any) which is normally stressed. In this list "ii" is normally pronounced as a long "i" and "ae" as a long "e".

A

Abies Old Latin name for fir. From *abire*, to rise, alluding to the great height of some species.
abramsiana In honour of LeRoy Abrams (1874–1956), Professor of Botany at Stamford University, USA.
Acacia Greek name for an African species of *Acacia*; from *akis*, a sharp point.
Acer Old Latin name for maple; also means sharp; the wood was used for spears.
"Aconitifolium" Has leaves like an aconite.
acuminata Slender-pointed.
acuta Sharply pointed.
acutifolia With sharply pointed leaves or leaflets.
acutissima Very acutely pointed.
adamii In honour of M Adam, a nurseryman near Paris, France.
adpressa Closely pressed together (usually of leaves against shoots).
Aesculus Latin name for a kind of oak, but applied by Linnaeus to the horse chestnut.
aetnensis From Mount Etna, Sicily.
agrifolia With rough leaves.
Ailanthus Latinized version of the native Moluccan name *ailanto*, sky tree.
alatamaha From the Native American name Altamaha for the river where it was first found.
alba White.
albertiana Native of Alberta, Canada.
albicaulis White-stemmed (*albus*, white; *caulis*, stem).
albidum Whitish.
Albizia In honour of F del Albizzi, a Florentine nobleman who in 1749 introduced this (the "silk tree") into cultivation.
albolimbatum Edged with white.
albosinensis *Albus*, white; *sinensis*, China.
Aleurites From the Greek *aleuron*, floury; in some species the young growth has a flour-dusted appearance.
alexandrae In honour of Queen Alexandra (1844–1925), wife of Edward VII.
alnifolia With leaves like those of alder.
Alnus Old Latin name for the alder.

alpinum From alpine regions.
altaclerensis Raised at Highclere, a mansion in Hampshire, England.
altissima Very tall.
amabilis Pleasant, but often used in the sense of lovely.
ambigua Uncertain or doubtful.
Amelanchier From the French Provençal name *amelancier* for *A. ovalis*.
americanus, -a From the Americas.
amurénsis, -e From the region of the Amur river in Manchuria.
amygdalo-persica A combination of *Prunus amygdalus* (*dulcis*), the almond, and *Prunus persica*, the peach.
amygdaloides Resembling almond.
Amygdalus The Greek name for the almond.
anagyroides Resembling *Anagyris*.
"Andelyensis" From les Andelys, near Paris.
andinus From the Andes of South America.
andrachne Ancient Greek name.
andrachnoides Resembling *Arbutus andrachne* (ancient Greek name for the strawberry tree, used by Linnaeus).
angustifolia Narrow-leaved.
annularis Ring-shaped.
antarctica Literally from the South Polar region, but used in biology for plants and animals native to south of 45°S.
apiculata Leaves abruptly tipped with a short point.
aquatica Growing in or near water.
aquifolium Latin vernacular name for holly, meaning having leaves with points.
Aralia From the Latinization of the French Canadian name Aralie.
araucana From the Araucani Indians of central Chile, in whose territory the first discovered species is native.
Araucaria See above.
arborea, -escens, -um Tree-like.
Arbutus Latin name for strawberry tree.
Archonthophoenix From the Greek *archon*, a chieftain, and *phoenix*, the date palm; refers to the stately appearance of these palms.
Arecastrum "-astrum" Indicates a resemblance to *Areca*, the Indian betelnut palm.
argentea, -um Silvery.

"Argenteomarginata" With silver edges.
aria Old Latin name for whitebeam.
aristata From the Latin *aristatus*, bearded; so, bearded with awns like ears of barley.
arizonica From the state of Arizona, USA.
armandii The Christian name of l'Abbé Armand David, the French missionary and naturalist who worked in China.
armeniaca From Armenia, on the Black Sea.
articulata Articulated or jointed.
ascendens From the Latin *ascendens*, rising upward.
Asimina The Latinized version of the French form of the Indian vernacular *assimin*.
asperata From the Latin *asperatus*, roughened.
asplenifolia With leaves like the spleenwort i.e. fine, feathery, and fern-like.
Athrotaxis From the Greek *athroos*, crowded together, and *taxis*, arrangement; in reference to the crowded nature of the cone scales.
atlantica From the Atlas Mountains of North Africa (also from the shores of the Atlantic).
atropurpurea, -um Dark purple.
attenuata Narrowing to a point.
aucuparia Old Latin name for rowan tree; from *aucupor*, bird catching, it being used as bait and as an ingredient of bird lime by bird catchers.
aurea, -um, -us From *aurum*, the Latin for gold; literally, golden.
"Aurea Nana" Golden-yellow, dwarf.
"Aureomarginatum" With golden edges.
"Aureospica" With golden spikes.
"Aureovariegata" Gold-variegated.
"Aurora" A fanciful resemblance to the electrical phenomenon of the same name.
australis Southern, in the sense of southern hemisphere or sometimes of southern Europe.
Austrocedrus Literally, southern cedar.
autumnalis Flowering in the autumn.
avellana From Avella Vecchia in southern Italy.
avium Of the birds (eaten by).
ayacahuite The Mexican name for *Pinus ayacahuite*.
azedarach A contracted version of the Persian (Iranian) name azaddhirakt for *Melia azedarach*.
azorica From the Azores Islands.

B

babylonica From Babylon.
baccata From the Latin *baccatus*, berry-like.
"Bacciflava" Yellow-berried.
baileyana In honour of Frederick Manson Bailey (1827–1915), an Australian botanist.
bakeri In honour of John Gilbert Baker (1834–1920), the British botanist who was Keeper of Kew Herbarium 1890–99.
balsamea Balsam-like.
balsamifera Balsam-bearing.
banksiana In honour of Sir Joseph Banks (1743–1820), British naturalist who accompanied Captain Cook's round-the-world expedition of 1768–71 in the *Endeavour*.
barbatum Bearded.
battandieri In honour of Jules Aimé Battandier (1848–1922), a French botanist and an authority on the flora of Algeria.
"Baumannii" In honour of E N Baumann of Bolivia.
"Belgica" From Belgium or the Netherlands.
Betula The old Latin name for the birch tree.
betuloides Resembling the birch tree, *Betula*.
betulus Resembling the birch tree, *Betula*.
bicolor Two colours.
bidwillii In honour of J C Bidwill (1815–53). Director of the Botanic Garden, Sydney, who collected plants in New Zealand and Australia.
"Biflora" Two-flowered.
bignonioides Resembling the trumpet creeper (*Bignonia*).
biloba Two-lobed.

blanda Mild.
boaria The Chilean vernacular name meaning loved by cattle (referring to the leaves of *Maytenus boaria*).
borealis Northern.
bracteata Having conspicuous bracts.
brevifolia With short leaves.
breweriana In honour of William Henry Brewer (1828–1910), American pioneer botanist in California and professor of agriculture at Yale University.
brilliantissimum The most brilliant.
"Briotii" In honour of Pierre Louis Briot (1804–88), a French horticulturist.
Broussonetia In honour of Pierre Marie Auguste Broussonet (1761–1807), professor of botany at Montpellier, France.
brunonii In honour of the botanist Robert Bruon, F R S (1773–1858).
buergerianum In honour of Heinrich Bürger (1804[?]–58), German plant collector.
bungeana In honour of Alexander von Bunge (1803–90), a Russian botanist and professor of botany at Dorpat in Estonia.
Buxus The classical Latin name for the box tree.

C

caerulea Dark blue.
caesia Lavender-blue.
californica From California, USA.
calleryana In honour of J M Callery (1810–62), a Roman Catholic missionary and botanist in China and Korea.
Callitris From the Greek *kali*, beautiful, and *tris* indicating the arrangement in three of leaves, cone scabs, etc.
Calocedrus From the Greek *kalos*, beautiful, and the Latin *cedrus*, cedar.
calophytum From the Greek *kalos*, beautiful, and *phytum*, a plant.
camaldulensis Named after Camaldoli in Italy.
Camellia In honour of George Joseph Kamel (1661–1706), a Jesuit pharmacist and botanist from Moravia who studied the plants of the Philippines and wrote an account of them.
camelliifolia With leaves like a camellia.
campanulata Bell-shaped.
campbellii In honour of Dr Archibald Campbell, Superintendent of Darjeeling and companion of Sir Joseph Hooker on his journey through the Sikkim Himalaya in 1849.
"Camperdownii" Castle in Angus, Scotland, where the elm *Ulmus glabra* 'Camperdownii' arose.
campestre Of the fields.
camphora Like camphor.
canadensis From Canada.
"Canaertii" In honour of the late 19th-century Belgian nurseryman.
canariensis From the Canary Islands.
candicans Shining.
canescens Off-white or ash-grey.
capillipes With slender feet.
capitata With a dense knob-like head, meaning flowers, fruits, or whole plant.
cappadocicum From Cappadocia, an ancient province of Asia Minor.
caprea Favoured by goats.
caramanica From Caramania (Karamania) in southern Asia Minor.
caribaea From the Caribbean area.
carica From Caria, a district in Asia Minor.
carnea Flesh or deep pink.
carolina, caroliniana From North or South Carolina, USA.
carpinifolia, -um With leaves like those of *Carpinus* (hornbeam).
Carpinus The classical Latin name for the hornbeam.
Carya from the Greek *karya*, walnut tree. Its fruit was known as *karyon*, which was also applied to other nuts.

cashmiriana From Kashmir.
cassine The North American Indian name for dahoon holly.
Castanea The Latin name for the sweet chestnut, from Kasanaia, a city in Asia Minor.
castaneifolia With leaves like those of sweet chestnut (*Castanea*).
Castanopsis Related to *Castanea*, the chestnut.
Casuarina The branches have been said to resemble cassowary (*Casuarinus*) feathers.
Catalpa The North American Indian name for *Catalpa bignonioides*, the first species in this genus to be cultivated in Europe.
Cedrela Diminutive of the Latin *cedrus*; from a similarity in the wood and its fragrance.
Cedrus The Latin name for the cedar, though originally the Latin *cedrus* and Greek *kedros* may have applied to species of juniper.
Celtis The Greek name of an unrelated tree, taken up for the hackberry by Linnaeus.
cembra The Italian name for the Swiss arolla or stone pine.
cembroides Resembling *Pinus cembra*, the Swiss arolla pine.
cephalonica From the island of Cephalonia in the Ionian Sea.
Cephalotaxus From the Greek *kephale*, a head, and *taxus*, the yew tree; referring to the relationship to the yew, and resemblance of some species.
cerasifera Cherry-like, referring to the fruits.
cerasus The Latin name for cherry.
Ceratonia The Greek name *keratonia* for the carob.
Cercidiphyllum With leaves like those of the Judas tree (*Cercis*).
Cercis From the Greek name *kerkis* used for the Judas tree.
cerifera Wax-bearing.
cerris A kind of oak, from the old Latin name.
Chamaecyparis From the Greek *chamai*, dwarf, and *kuparissos*, cypress. This now inapt name seems to have been derived from the dwarf juvenile forms formerly known as retinospora.
Chamaerops From the Greek *chamai*, dwarf, and *rhops*, a bush.
"Charles Raffill" In honour of C P Raffill (1876–1951), Curator of Kew Gardens.
"Chermesina" *Salix alba* 'Chermesina' was named after the Chermes (*Adelges*) insect because of the similar reddish colour of the young shoots.
chilensis From Chile.
chinensis From China.
Chionanthus Having snow-white flowers; from the Greek *chion*, snow, and *anthos*, a flower.
chrysocoma With golden hairs.
chrysolepis With golden scales.
chrysophylla Having golden leaves.
cinerea Ash-coloured.
cinnamomum From the classical Greek name for cinnamon.
circinatum Rolled in a circular fashion.
citriodora *Citrus* (lemon) scented.
Citrus The Latin name for the citron (*Citrus medica*).
Cladrastis From the Greek *klados*, branch, and *thraustos*, fragile; alluding to the brittle twigs.
Clethra From the Greek *klethra*, the white alder tree, which has similar leaves.
cliffortioides Like *Cliffortia*, a South African shrub of the rose family.
coccifera Bearing red berries.
coccinea, -um Scarlet.
Cocos From the Portuguese for monkey (*Cocos*), referring to the nut which suggests a monkey's face.
coerulea Blue.
coggygria An inaccurate rendering of the Greek *kokkugia*, the vernacular name for the smoke tree (*Cotinus*).
colorata Coloured.
"Columnaris", -e In the shape of a column.
colurna Old Latin name for hazel nut or wood.
"Commelin" In honour of the Dutch botanists Johan (1629–92) and Caspar (1667–1731) Commelin.
commixta Mixed together.
communis Growing in company, or common.
commutata Changed or changing.
compressa Compressed or flattened.
concolor The same colour throughout.
conferta Crowded.
conica Cone-shaped.
conradinae In honour of Conraddine, wife of the German botanist B A E Koehne (1846–1918).
conspicua From the Latin *conspicuous*, conspicuous, distinctive, remarkable.
contorta From the Latin *contortus*, twisted or irregularly bent.
controversa Controversial, doubtful, or questionable.
cordata Heart-shaped.
cordifolia With heart-shaped leaves.
cordiformis Heart-shaped.
Cordyline From the Greek *kordyle*, a club.
cornubia Of Cornwall, England.
cornubiensis Of Cornwall, England.
Cornus The old Latin name for the cornelian cherry (*Cornus mas*).
coronaria Used for garlands.
Corylus The Greek name for the hazel bush (*Corylus avellana*).
costata Ribbed.
Cotinus From the Greek *kotinos*, a name for the wild olive and therefore of obscure meaning, though possibly used as a term for several bushes of no economic value.
Cotoneaster From the Latin *cotoneum*, a quince, and *aster*, superficial or incomplete resemblance to; used to denote a wild or inferior species. Some species of *Cotoneaster* have leaves like those of quince.
coulteri In honour of Thomas Coulter (1793–1843), an Irish botanist who travelled in the western States.
coxii In honour of E H M Cox, plant collector and author.
crataegifolium With leaves that resemble those of hawthorn (*Crataegus*).
x Crataegomespilus The combination of two generic names *Crataegus* and *Mespilus*, the fusion of their tissues forming this graft hybrid.
Crataegus The Greek name for hawthorn.
crenata Cut in rounded scallops.
"Crippsii" In honour of the nurseryman Cripps of Tunbridge Wells, Kent, England.
crispa Closely curled.
cristata Crested.
crus-galli Cock's spur.
Cryptomeria From the Greek *krypto*, to hide, and *meris*, a part: all parts of the flowers (strobili) are concealed.
Cunninghamia In honour of James Cunningham, East India Company surgeon and plant collector.
"Cuprea" Copper-coloured.
cupressinum Resembling *Cupressus* (cypress).
x Cuprocyparis Compounded from *Cupressus* and *Xanthocyparis*, the parents of this bigeneric hybrid.
cupressoidas Resembling *Cupressus* (cypress).
Cupressus The Latin name for *Cupressus sempervirens*, the Italian cypress tree.
cuspidata From the Latin *cuspidatus*, bearing a stiff point.
cyathea From the Greek *kyatheion*, a little cup; the membrane covering the spore cases on the underside of the leaves breaks at the top at maturity, forming a cup.
Cydonia The Latin name for this fruit tree, derived from Cydon, a town in Crete.
Cytisus Broom, from the Greek *kytisos*, a name used for several shrubby members of the pea family (Leguminosae).

D

dacrydioides Resembling *Dacrydium*.
Dacrydium From the Greek *dakrydion*, a small tear; some exude small drops of clear resin.
dactylifera Finger-like.
dactylis From the Greek *daktylos*, a finger; the inflorescence appears finger-like.
"Dalecarlica" From the Swedish province of Dalecarlia or Dalarna.
daniellii After the surgeon-collector who toured Tientsin in 1860–62.
daphnoides Resembling *Daphne*, from the Greek name for the laurel.
dardarii In honour of M Dardar of Bronvaux, Metz, in whose garden the graft hybrid + *Crataegomespilus dardarii* arose.
Davidia In honour of l'Abbé Armand David (1826–1900), a French missionary and naturalist who worked and collected in China.
davidii In honour of l'Abbé Armand David (see Davidia).
"Dawyck" From the famous Scottish garden on the banks of the river Tweed.
dealbata Whitened, as with a powder.
"Decaisneana" In honour of Joseph Decaisne (1807–82), a director of the Jardin des Plantes, Paris, and a famous botanist and horticulturist.
decidua Deciduous (leafless in winter).
decora, -um Decorative.
decurrens Running down; used when leaf bases are extended down the stem in wing-like tapering ridges.
delavayi In honour of l'Abbé Jean Marie Delavay (1834–95), a missionary who collected plants in western China.
delonix From the Greek *delos*, evident, and *onux*, a claw; referring to the long-clawed (stalked) petals.
deltoides Triangular, like the Greek letter *delta*.
densiflora, -us Densely flowered.
dentata Toothed.
denudata Bare or naked.
deodara From the native Indian name for the deodar cedar.
deppeana Named after Ferdinand Deppe (1795–1861), German botanist.
diabolicum Devilish, applied to a plant with horned fruit.
Dicksonia In honour of James Dickson (1738–1822), a British nurseryman and botanist.
dicksonii In honour of James Dickson (see Dicksonia).
dioicus Dioecious; the separate-sexed flowers are borne on individual trees.
Diospyros From *dios*, divine, and *pyros*, wheat, a Greek name transferred to the persimmon (*Diospyros kaki*).
Disanthus Paired flowers; from *dis*, twice, and *anthos*, flower.
discolor Of two colours.
dissectum Deeply cut, divided into segments.
distichum From the Latin *distichus*, in two parallel ranks.
diversifolia From the Latin *diversus*, diverse, and *folium*, leaf; with diverse leaves.
dolabrata Hatchet-shaped.
dombeyi In honour of Joseph Dombey (1742–94), a French botanist.
domestica Domesticated; much used in gardens.
'Donard Gold' From the Slieve Donard Nursery Co Ltd, Ireland.
'Dovastonii', -iana *Cephalotaxus* × *media* 'Dovastonii' was "found" by Mr John Dovaston of Westfelton, Shrewsbury, England.
'Drummondii' In honour of Thomas Drummond (d. 1835), nurseryman.
drupacea Bearing a fleshy fruit with a single seed within, like a cherry, an olive, or a plum.
dulcis Sweet.

E

ebenum Ebony-black.
echinata Prickly.
edulis Edible.
Ehretia In honour of Georg Dionysius Ehret (1708–70), a German botanical artist.
Elaeagnus Said to be derived from the Greek *elaia*, the olive tree, and *agnos*, a name for the chaste tree (*Vitex*), but more likely from *heleagnos–helodes*, marshy, and *hagnos*, pure (or white), in allusion to the white fruits.
elata Tall.
elegans Elegant.
elegantissima Italian word for "very elegant".
'Ellwoodii' In honour of Mr Ellwood, who discovered *Chamaecyparis pisifera* 'Ellwoodii' *c.*1920 as a chance seedling at Swanmore Park, Bishop's Waltham, England, where he was gardener.
embothrium From the Greek *en*, in, and *bothrion*, a little pit; the stamens are borne in recesses or pits in the petals.
engelmannii In honour of Georg Engelmann (1809–84), a German-born American doctor, greatly interested in plants. He did some collecting, but did more by encouraging others.
erecta, -um Erect or upright.
erecta viridis Erect or upright, and green.
Erica From the Latin *erice*, heath.
ericoides Resembling *Erica*.
ermanii In honour of G A Erman (1806–77) of Berlin, who collected in eastern Asia and elsewhere.
erubescens Reddening.
erythrocladum Red-branched; from the Greek *erythro-*, red, and *klados*, branch.
erythroflexuosa Red, bent alternately, zigzag.
Eucalyptus From the Greek *eu*, well, and *kalypto*, to cover (as with a lid); alluding to the petals and sepals which are fused together to form a cap, shed when the flower opens.
euchlora From the Greek *eu*, good, and *chloros*, green.
Eucommia From the Greek *eu*, good, and *kommi*, gum; the name for the only hardy rubber-producing tree.
Eucryphia From the Greek *eu*, well, and *kryphios*, covered; referring to the sepals which are joined at the tips to form a protective cap to the bud.
Euodia From the Greek word meaning a sweet scent; the leaves are fragrant when bruised.
europaea From Europe.
excelsa Tall or high.
excelsior Taller.

F

Fagus The Latin name for the beech tree (*F. sylvatica*).
falcata The Latin word for sickle-shaped.
falconeri In honour of Hugh Falconer (1808–65), a Scottish doctor, geologist, and botanist in India (1830–55).
fargesii In honour of Paul Guillaume Farges (1844–1912), a French missionary and naturalist who worked in central China.
fastigiata, -um Having erect growth and a columnar habit from the Latin for a gable (which it does not resemble).
ferox Ferocious, very thorny.
femina A female.
ficifolia With leaves like those of a fig (but not the edible fig, *Ficus carica*).
fictolacteum From the Latin *fictus*, false, and *Rhododendron lacteum*, a species with which *Rhododendron fictolacteum* was confused.
Ficus The Latin name for the fig.
filicifolia Fern-leaved.
filicoides Resembling fern, Latin *filix*, a fern.
filifera Thread-bearing.
filiformis Thread-like.
filipes With thread-like stalks.
Fitzroya In honour of Vice-Admiral Robert Fitzroy (1805–65), who commanded the five-year surveying expedition in HMS *Beagle* (Charles Darwin was the naturalist aboard).
flava Yellow.
"Fletcheri" Named after Fletcher's Nursery, Kent. Fletcher first distributed the branch mutant *Chamaecyparis lawsoniana* 'Fletcheri', which originally arose in 1913 in the Ottershaw Nursery, Chertsey, Surrey, England.
floribunda Profusely flowering.
florida Flowery.
fordii In honour of Charles Ford (1844–1927), Superintendent of the Hong Kong Botanical Garden.
formosana From the island of Formosa (Taiwan).

forrestii In honour of George Forrest (1873–1932), from Scotland, who collected large numbers of seeds and specimens in western China.
fortunei In honour of Robert Fortune (1812–80), Scottish horticulturist who collected plants in China and Japan (and introduced tea to India).
Fortunella The kumquat, an evergreen shrub, named after Robert Fortune (*see* fortunei).
fragilis Brittle or fragile and easily broken.
frainetto The native name of the Hungarian oak.
Franklinia In honour of Benjamin Franklin (1706–90), American scholar and statesman.
franklinii In honour of Sir John Franklin (1786–1847), naval captain and arctic explorer, also governor of Tasmania 1836–43.
fraseri In honour of John Fraser (1750–1811).
fraxinifolia With leaves like *Fraxinus* (ash).
Fraxinus The old Latin name for the ash tree (*Fraxinus excelsior*).
frigidus Coming from cold regions.
'Frisia' From Friesland, west Germany and east Netherlands.
fructu-luteo Yellow-fruited.
fulvum Tawny-orange.
funebris Funereal, grown in graveyards.
fusca Dusky brown.

G

garryana In honour of Nicholas Garry, Secretary of the Hudson's Bay Company, who in 1820–30 helped David Douglas in western North America.
generosa Of noble appearance.
Genista The Latin name for dyer's greenweed, but said to have come from the Celtic *gen*, a bush.
germanica From Germany.
giganteum From the Latin *giganteus*, gigantic.
Ginkgo A misrendering of the ancient (and now obsolete) Japanese name *gin-kyo* (silver apricot).
ginnala The native vernacular name for this Asiatic maple.
giraldii In honour of Giuseppe Giraldi, an Italian missionary in China, who in 1890–95 collected plants in Shensi province.
glabra, -ata Smooth, without hairs.
glauca Blue-grey or blue-white.
Gleditsia In honour of Johann Gottlieb Gleditsch (1714–86), Director of the Berlin Botanic Garden.
globosa, -um Round, or spherical.
globulus Rounded or globular.
glutinosa Gluey or sticky (the buds and young leaves).
glyptostroboides Resembling *Glyptostrobus*.
Glyptostrobus From the Greek *glypto*, to carve, and *strobilos*, a cone; referring to the depressions on the cone scales of *Glyptostrobus*, the deciduous Chinese cypress.
gmelinii In honour of Gottlieb Gmelin (1709–55), a German naturalist who travelled extensively in Siberia and Kamchatka.
'Goldsworth Purple' From the Goldsworth Old Nursery, Surrey.
Gordonia Named in honour of James Gordon (d. 1781), a correspondent of Linnaeus and a nurseryman of Mile End, London.
goveniana In honour of J R Gowen, of Highclere, Secretary of the Royal Horticultural Society 1845–50.
gracilistyla With a slender style.
grandidentata With large teeth.
grandiflora Large-flowered.
grandifolia Big-leaved.
grandis, -e Big or showy.
Grevillea In honour of Charles Francis Greville (1749–1809), a founder of the Horticultural Society of London and a Vice-President of the Royal Society.
griffithiana, -um In honour of William Griffith (1810–45), a British doctor and botanist who collected plants in India and Afghanistan.
griffithii In honour of William Griffith (*see* griffithiana).
grignonensis Grignon is a town in France where *Crataegus grignonensis* was first discovered.
griseum Grey.
grosseri Named after WCH Grosser (1869–1942), German botanist. *Acer grosseri* was discovered by Giraldi (*see* giraldii) in Shaanxi, China, and brought to England in 1927.
gunnii In honour of R C Gunn (1808–81) of Tasmania.
Gymnocladus From the Greek *gymnos*, naked, and *klados*, a branch; the tree is leafless in winter

H

Hakea In honour of Baron Christian von Hake (1745–1818), German patron of botany.
halapensis From Aleppo (Halep) in Syria.
Halesia In honour of the Rev. Stephen Hales (1677–1761), curate of Teddington, England, and a physiologist, chemist, and inventor.
halliana In honour of George Rogers Hall (1820–99), an American doctor who introduced many plants from Japan to the USA.
Hamamelis A Greek name, possibly for the medlar.
harrowiana In honour of George Harrow, manager of Veitch's Coombe Wood Nursery where many of E H Wilson's Chinese plants were raised from seeds.
hastata Spear-shaped.
Hebe Evergreen shrubs from South America and Australasia, named after the Greek goddess of youth.
henryae *Chaemaecyparis henryae* is named after Mary Gibson Henry (1884–1967), American botanist and gardener.
henryi, -ana In honour of Augustine Henry (1857–1930), Irish doctor and plant collector who travelled in China and Formosa (Taiwan), later professor of forestry.
"Heptalobum" Seven-lobed.
hersii In honour of Joseph Hers, a Belgian railway engineer who, while in China during the 1920s, sent back plants to the Arnold Arboretum.
heterophylla From the Greek *heter* or *hetero*, various or diverse, and *phyllum*, leaf.
"Hibernica" From Ireland.
highdownensis After Sir Frederick Stern's garden at Highdown Towers, Sussex.
hillieri After the tree and shrub specialist nurserymen, Hillier & Sons, Winchester, Hampshire.
hippocastanum The Latin name for the horse chestnut; there are horseshoe-shaped leaf scars on the twigs.
Hippophae An old Greek name for a spiny spurge, later transferred to this shrub.
hispanica From Spain.
hispida Bristly.
Hoheria Latinized version of the New Zealand Maori name for these evergreen trees and shrubs, *houhera*.
hollandica From the Netherlands (Holland).
holophylla From the Greek *holo*, entire, and *phyllum*, leaf.
homolepis From the Greek *homo*, same, one kind of, and *lepis*, scale.
horizontalis Horizontal.
"Hudsonia" In honour of William Hudson (1730–93), a London apothecary.
humilis Low-growing, or relatively so.
hupehensis From Hupeh, China.
hybrida Of hybrid origin, or appearing so.

I

Idesia In honour of Eberhard (or Evert) Ides (d. *c.*1720), a Dutch (or German) explorer of northern Asia.
Ilex From the Latin name for the evergreen holm oak, the more toothed-leaved forms of which it resembles.
illinoensis From the state of Illinois, USA.
imbricaria Overlapping in regular order, like the tiles on a roof.
imperialis Imperial or showy.
incana Hoary or grey.
incisa Deeply cut.
indica From India.
inermis Unarmed, i.e. having no prickles.
insignis Remarkable.
intermedia Intermediate.
intertexta Inertwined.
involucrata Surrounded or enclosed by an involucre (leaf, scale, or petal-like bracts).
ioensis From the state of Iowa, USA.
irroratum From the Latin *irroro*, to wet with dew or to besprinkle.
"Italica" From Italy.

J

Jacaranda The Latinized form of the Brazilian name for these flowering trees.
jacquemontii In honour of Victor Jacquemont (1801–32), a French naturalist who visited Asia.
japonica, -um From Japan.
"Jaspidea" Like jasper.
jeffreyi In honour of J Jeffrey, a Scottish gardener who collected in Oregon, USA, between 1850 and 1853.
jessoensis, jezoensis From Jezo, Japan.
"John Downie" In honour of the Scottish nurseryman and florist.
Jubaea Named for King Juba of Numidia who killed himself when his kingdom was taken over by the Romans in 46 BC.
Juglans The Latin name, from *jovis*, of Jupiter, and *glans*, acorn.
julibrissin The Persian (Iranian) name for the silk tree.
Juniperus Old Latin name for juniper tree.

K

kaempferi In honour of Engelbert Kaempfer (1651–1716), a German physician who travelled widely in the East and lived for two years in Japan.
kaki An abbreviation of the Japanese name *kaki-no-ki* for this fruit.
Kalopanax From the Greek *kalos*, beautiful, and *Panax*, a related genus.
kelloggii In honour of Albert Kellogg (1813–87), American doctor and botanist.
"Kiftsgate" After the garden at Kiftsgate Court, Gloucestershire.
"Kilmacurragh" From the Irish garden of this name.
kobus Latinization of the Japanese name *kobushi*.
Koelreuteria In honour of Joseph Gottlieb Koelreuter (1733–1806), natural history professor and pioneer in plant hybridization.
Koraiensis, koreana From Korea.
kousa The Japanese name for this *Cornus*.

L

x Laburnocytisus A combination of *Laburnum* and *Cytisus*, parents of this unique graft hybrid.
Laburnum Old Roman name for laburnum trees.
laciniata, -um Deeply cut or slashed.
laeta Bright or vivid or pleasing.
laevigata Smooth.
laevis Smooth or hairless.
lamarckii In honour of Jean Baptiste Antoine Pierre Monet, Chevalier de la Marck (1744–1829), a distinguished French biologist: his work on evolution foreshadowed that of Darwin (but was wrong).
lambertiana In honour of Aylmer Bourke Lambert (1761–1842), English botanist and author of *The Genus Pinus*.
lanata Woolly.
lanceolata, -um Lance-shaped.
"Lanei" Named after Lane's Nursery, Berkhamsted, Hertfordshire.
lantana The Latin name for *Viburnum*.
laricina Resembling (common or European) larch.
Larix The old Latin name for the common larch (*L. decidua*).
lasianthus With woolly anthers.
lasiocarpa Woolly fruited.
latifolius, -a Broad-leaved; from the Latin *latus*, broad, and *folium*, leaf.
Laurelia From the Spanish name for *Laurus*, the laurel or bay, referring to fragrant leaves.
laurifolia With leaves like those of *Laurus nobilis* (bay laurel).
laurina Resembling *Laurus*.
laurocerasus A combination of the Latin for laurel and cherry, hence "cherry laurel".
Laurus Latin name for the bay laurel.
lavallei In honour of M Lavalle, a French nurseryman, *c.*1870.
lawsoniana In honour of Charles Lawson (1794–1873), nurseryman and author of *Pinetum Britannicum*, whose Edinburgh nursery received the first *Chamaecyparis lawsoniana* seeds from Oregon in 1854.
laxifolia Loose or open; from the Latin *laxus*, loose, and *folium*, leaf.
'Lemoinei' In honour of Victor Lemoine (1823–1911) and his son Émile (1862–1942), French nurserymen and plant breeders who introduced many new garden plants.
'Lennei' In honour of M Lenné, Director of Royal Gardens of Prussia.
lenta Tough but flexible.
leucodermis White-skinned; from the Greek *leukos*, white, and *dermis*, skin.
leucoxyla, -on White-wooded.
leylandii In honour of C J Leyland, of Haggerston Hall, Northumberland (see page 143).
libani From Mount Lebanon.
Libocedrus From the Greek *libos*, a tear, and *cedrus*, cedar: referring to drops of resin exuded by the incense cedar.
Ligustrum The Latin name for privet.
likiangensis From the Lichiang range, Yunnan western China.
liliiflora Lily-flowered.
"Linearilobum" From the Latin *linear* and *lobum*, lobed.
linearis Narrow, with nearly parallel sides.
Liquidambar from the Latin *liquidus*, fluid or liquid, and the Arabic *ambar*, amber, alluding to the flagrant gum which exudes from this tree.
Liriodendron Derived from the Greek *leirion*, a lily, and *dendron*, a tree.
Lithocarpus From the Greek *lithos*, stone, and *karpos*, fruit; a reference to the hard nuts (acorns).
Litsea Latinization of an old Japanese name.
Livistona In honour of Baron of Livingstone, founder of Botanic Garden, Edinburgh.
lobbii In honour of Thomas Lobb (d. 1894) who collected plants for the nursery firm of James Veitch (from 1843 to 1860) in Java, India, Malaya, Borneo, and Singapore.
loebneri After Max Loebner, Prussian nurseryman.
longaeva Of great age.
longifolia With long leaves.
lowiana In honour of Henry Stuart Lowe (1826–90), a nurseryman of Clapton, London.
lucidum Shining.
"Lucombeana" Originating in the nursery of Lucombe & Pince, Exeter, Devon.
lusitanica From Lusitania – better known as Portugal.
lutea Yellow.
lutescens Yellowish.
lyallii In honour of Dr David Lyall (1817–95), botanist on HMS *Terror* and other ships, and who collected plants in New Zealand.
lycopodioides Resembling *Lycopodium* (ground pine).
lyrata Lyre-shaped.

M

Maackia In honour of Richard Maack (1825–86), a Russian naturalist who explored the Ussuri river region in eastern Asia.
maackii (as above).
Maclura In honour of William Maclure (1763–1840), an American geologist.
macranthera Having large anthers.
macrocarpa With large fruits.
macrophyllus, -a, -um Large-leaved.
magnifica Splendid, magnificent, or great.
Magnolia In honour of Pierre Magnol (1638–1751), Professor of Botany and Director of the Montpellier Botanic Garden, France.
"Majestica" Majestic.
major Bigger or larger.
Malus The Latin name for the apple.
mandshurica from Manchuria.
margarita Pearl-like.
mariana Of Maryland, USA.
mariesii In honour of Charles Maries (d. 1902), a plant collector who worked for the nursery firm of Veitch in China and Japan.
'Marilandica' From Maryland, USA.
maritima From the Latin *maritimus*, maritime, by or near the sea.
mas Male or masculine, formerly used to distinguish delicate species from robust ones.
matsudana In honour of Sadahísa Matsúdo (1857–1921), a Japanese botanist who wrote a flora of China.
maxima, -um Largest.
maximowicziana, -ii In honour of Carl Ivanovitch Maximowicz (1827–91), a Russian botanist who specialized in the plants of eastern Asia.
Maytenus From the Chilean name *maitén* for *Maytenus boaria*.
media Intermediate between two types.
medullaris Pithy.
Melia The Greek name for an ash tree, in allusion to the similarity of the leaves.
menziesii In honour of Archibald Menzies (1754–1842), a Scottish naval doctor who sailed with Vancouver on his 1790–95 expedition to the Northwest Pacific.
mertensiana In honour of Franz Carl Mertens (1764–1831), Professor of Botany at Bremen, Germany.
Mespilus The Latin name for medlar, the fruit.
Metasequoia From the Greek *meta*, meaning 'with', 'after', 'sharing', or 'changed in nature': alluding to an ancient relationship with *Sequoia*.
'Meyen' In honour of Franz Julius Meyen (1804–40), German physician and plant collector.
meyeri In honour of Frank N Meyer (d. 1918), a Dutch-American plant collector.
michauxii In honour of André Michaux (1746–1803), French explorer and plant collector who visited North America, Persia, and Madagascar.
microphylla With small leaves.
mimosifolia With leaves resembling those of *Mimosa*.
miniata Vermilion-coloured.
minima Smallest.
mitchellii In honour of Dr John Mitchell (1711–68), American physician.
'Moerheim', -ii Named after a Dutch nursery.
molle As *Rhus molle*, the Peruvian vernacular name for this tree. (*Molle* in Latin means soft, usually in the sense of hairy.)
mollicomata With soft hairs.
mollis Softly hairy.
mollissima Very soft, or velvety hairy.
monogyna Having one pistil.
monophylla One-leaved.
monosperma Having one seed.
monspessulanum From Montpellier in France.
montezumae To honour Montezuma, emperor of Mexico in the 16th century.
monticola Growing on mountains.
Mórus The Latin name for mulberry.
moupinensis From Mupin (now Baoxing), western China.
mucronatum Mucronate, ending with a spiny tip.
mugo Or *mugho*, the old Tyrolese name for *Pinus mugo*.
muhlenbergii In honour of Gotthilf Henry Ernest Muhlenberg (1753–1815), Lutheran minister in Pennsylvania, USA, and a noted amateur botanist.
mume From the Japanese name *ume* for the species *Prunus mume*.
muricata (The cone) roughened with hard points, like the seashell *Murex*.
Myrtus The Latin name for myrtle (*Myrtus communis*), a plant sacred to Aphrodite and later associated with the Virgin Mary.

N

nana From the Latin *nanus*, dwarf.
nana gracilis Dwarf (see above), and *gracilis*, slender; very dwarf and graceful.
negundo From the Sanskrit and Bengali *nirgundi*, used for a tree with a similar leaf to *Acer negundo*.
nidiformis Nest-shaped.
niedzwetzkyana In honour of the Russian judge Niedzwetzky.
nigra, -um Black.
nikoense From the mountains near Nikko in Honshu, Japan.
niphophila Snow-loving, from the Greek *niphas*, snow, and *phileo*, to love.
nitida Shining.
nivalis Snow-white, or growing near snow.
nobilis Notable, renowned, famous, stately, or noble.
nootkatensis From the Nootka Sound, Vancouver Island, British Columbia.
nordmanniana In honour of Alexander von Nordmann, Professor of Zoology at Odessa and Helsinki.
Nothofagus From the Greek *nothos*, false, and *fagus*, beech.
nucifera Nut-bearing.
numidica From Algeria.
nuttallii In honour of Thomas Nuttall (1786–1859), an English botanist who collected in North America in 1811–34.
nymansensis From Nymans, a famous garden in Sussex.
Nyssa Named from Nysa or Nyssa, a water nymph.

O

obassia Said to be from the Japanese name for *Styrax obassia*, but not mentioned in the standard *Flora of Japan*, by Chwi.
obliqua Oblique or unequal.
oblonga Oblong.
obovatus, -a An inverted egg-shape.
obtusa Blunted.
occidentalis From the Occident or Western world.
odorata Fragrant.
Olea The Latin name for the olive.
oleracea Of the vegetable garden.
oliveri In honour of David Oliver (1830–1916), who was Keeper of Kew Herbarium from 1864 to 1890.
omorika The local name of the Serbian spruce.
opaca Dark or dull.
opalus Probably from *opulus*, an old Latin name for maple, also given to the guelder rose (*Viburnum opulus*) with its maple-like leaves.
opulus *See* **opalus**.
orientalis From the Orient, or eastern world (eastward from Europe and Africa).
ornus The old Latin name for the manna ash.
osteosperma With hard seeds; from the Greek *osteon*, a bone, and *sperma*, a seed.
ostrya From the Greek name *ostrys* for the hop-hornbeam tree.
ovalis Oval.
ovata Egg-shaped.
oxyacantha Having sharp thorns.
oxycarpa Bearing pointed fruits.
oxycedrus Ancient Greek name for a juniper.
Oxydendrum From the Greek *oxys*, sharp (tasting), and *dendron*, a tree; the leaves are bitter.

P

padus The Greek name for a wild cherry.
palmatum Having hand-like leaves with outspread finger-like lobes.
palmetto The Spanish name for the cabbage palm.
palustris Growing in marshes or liking wet places.
paniculata Having flowers in panicles.
papyrifera Paper-bearing.
paradisi From Paradise.
'Parkmanii' In honour of Francis Parkman, American horticulturist.
Parrotia In honour of F W Parrot (1792–1841), a Russian naturalist who climbed Mount Ararat in 1834.
parviflora Bearing small flowers.
parvifolia Small-leaved.
patula Spreading.
Paulownia In honour of the Princess Anna Paulowna (1795–1865), the daughter of Czar Paul I of Russia.
pavia From an old generic name in honour of Peter Paaw (d. 1617).
pedunculosa With well-developed flower stalks (peduncles).
'Pembury Blue' From the estate of this name in Kent, England.
pendula, -um Hanging down.
pensylvanica, -um From Pennsylvania, USA.
pentandra Each flower having five stamens.
perado The native (Azores) name for *Ilex perado*.
pernyi In honour of Paul Hubert Perny (1818–1907), a French missionary in China.
persica From Persia (now Iran).
petiolaris With a leafstalk.
petraea Rock-loving, growing in stony places.
peuce The Greek name for *Pinus peuce*.
pfitzeriana In honour of Ernst Hugo Heinrich Pfitzer (1846–1906), a Stuttgart nurseryman.
phaenopyrum Having the appearance of a pear.
Phellodendron From Greek *phellos*, cork, and *dendron*, a tree, these trees have a corky bark.
phellos The Greek name for the cork oak (*Quercus suber*), which Linnaeus used for *Quercus phellos*.
Phillyrea The classical Greek name.
Phoenix The Greek name for the date palm.
Photinia Said to be from the Greek *phos* or *photos*, light, in allusion to the glossy shining leaves of some species.
Picea From the old Latin name for pitch pine, now used for spruce.
Picrasma From the Greek *picrazein*, bitter.
pictus Painted or coloured.
pinaster The old Latin name for a wild pine – as distinct from one (*Pinus pinea*) that was cultivated.
pindrow The native Himalayan name for *Abies pindrow*.
pinea Of the pines or growing on pines (as a parasite).
pinnatifida With leaves pinnately lobed.
pinsapo The Spanish name for *Abies pinsapo*.
Pinus The old Latin name for a pine tree, in particular the stone pine (*Pinus pinea*).
pisifera Pea-bearing.
'Pissardii' In honour of M Pissard, the Shah of Persia's French gardener 100 years ago.
Pistacia From the Greek *pistake* (in turn probably derived from a Persian name) for the pistachio nut.
Pittosporum From the Greek *pitta*, pitch, and *spora*, seed; the seeds of this genus are embedded in a sticky, resinous substance.
plantierensis Named after the Plantier nursery near Metz, France.
platanoides Resembling the plane tree, *Platanus*.
Platanus The Greek name for the Oriental plane tree.
platyphylla, -os With broad leaves.
'Plena' Full or double flowered.
plicata Pleated or folded.
plumosa Feathery.
Podocarpus From the Greek *podos*, a foot, and *karpos*, a fruit; the fruits of these trees (some species, not all) are borne at the summit of a broad, fleshy stalk or receptacle.
pohuashanensis From Po hua shan, China.
polita Elegant or neat.
polycarpa Many fruited.
pomifera Bearing pomes or apples.
Poncirus from the French *poncire*, a kind of citrus.
ponderosa Heavy or weighty.
populifolia Having leaves like those of poplar (*Populus*).
Populus The Latin name for the poplar tree.
postelense Named after town in Silesia.
potaninii In honour of Grigori Potanin (1835–1920), a Russian who explored and collected in Asia and China.
praecox Developing or flowering earlier than most of its genus.
prattii In honour of Antwerp E Pratt, an English zoologist who travelled in China in the 1880s.
prinus The Greek name for the common oak (*Quercus robur*), which Linnaeus used for *Quercus prinus*.
procera Tall or slender.
procumbens Prostrate.
prunifolia, -um With leaves like those of the cherry (*Prunus*).
Prunus The latin name for the cherry tree.
pseudoacacia False acacia.
pseudocamellia False camellia.
Pseudolarix Literally, false larch, the tree; resembles true larch (*Larix*).
pseudoplatanus The false plane tree.
Pseudotsuga False *Tsuga* (the hemlock); Douglas fir is not related to *Tsuga* (nor does it look like one).
Ptelea The Greek name for an elm tree; possibly so named because of the somewhat similar flattened, circular, winged fruits.
Pterocarya From the Greek *pteron*, a wing, and *karyon*, a nut.
Pterostyrax From the Greek *pteron*, a wing, and the genus *Styrax*: this genus is distinguished by its winged fruits.
pubescens Covered with soft, downy hairs.
pumila Dwarf.
punctata Marked with dots.
pungens From the Latin *pungens*, piercing, so sharp-pointed.
purpurascens Purplish.
purpurea, -um Purple.
pygmaea Pygmy.
pyramidalis Pyramidal.
pyraster The wild pear; from the Latin *pyrus*, pear, and *aster*, superficial or incomplete resemblance to; used to denote a wild or inferior species.
Pyrus the Latin name for the pear.

Q

quadrangulata With four angles.
Quassia In honour of Graman Quasi, a Negro slave who used this tree bark as a remedy for fever.
quercifolia With leaves like *Quercus* (oak).
Quercus The Latin name for the oak.

R

racemosa The flowers are borne in a raceme.
radiata Emitting or having rays.
reclinata Bent backward.
recurva Curved backward.
regia Royal in appearance.
rehderiana In honour of Alfred Rehder (1863–1949), a German–American dendrologist, Curator of the Herbarium at the Arnold Arboretum, and author of the classic *Manual of Cultivated Trees and Shrubs*.

resinosa Bearing resin.

reticulata Having a net-like pattern.

retusus With a rounded or slightly notched tip.

rhamnoides Resembling *Rhamnus*, the buckthorn.

Rhododendron From the Greek *rhodon*, a rose, and *dendron*, a tree.

rhombifolium Rhomboid-leaved.

Rhus The Greek vernacular name for a species of sumach (probably *Rhus coriaria*) and now used as the scientific generic name.

'Richardii' In honour of IC Richard (1754–1821), a French botanist.

rigida Stiff or rigid.

'Riversii' Raised at England's oldest nursery firm of Thomas Rivers in Hertfordshire, England.

'Rivers Purple' A beech from the Rivers nursery (see 'Riversii').

Robinia In honour of Jean Robin (1550–1629), herbalist to Henri IV and Louis XIII of France, and his son Vespasian Robin, who first grew this tree in Europe.

robur Latin name for any hard wood, especially oak.

robusta Robust, strong in growth.

romanzoffianum In honour of the Russian, MP Romanzoff (1754–1826).

Rosa The Latin name for the rose.

rosacea Rose-like.

rosea, -um Rose-coloured.

'Roseoplena' Rosy in colour and double.

Roystonea In honour of General Roy Stone (1836–1905), American army engineer in Puerto Rico.

rubens From the Latin *rubens*, red or reddish.

rubra, -um Red.

rubrifolia Red-leaved.

rufinerve Red-veined when leaves first emerge.

rugosa Wrinkled.

rupestris Liking a rocky habitat.

'Rustica rubra' From the Latin *rusticus*, relating to the country, and *ruber*, red.

S

Sabal Possibly derived from the South American name for these palms.

saccharinum Sugary.

saccharum Sugarcane (from the Greek *sakcharon*, in turn probably derived from the Malay word for the sweet juice of the sugarcane, *singkara*).

salicifolia With leaves like those of *Salix* (willow).

salignus Resembling *Salix* (willow).

Sambucus The old Latin name for these shrubs and small trees, perhaps connected with *sambuca*, the Biblical sackbut.

Sapium The Latin name for a resinous pine; presumably applied here because of the sticky sap which is exuded.

sargentiana, -ii In honour of Charles Sprague Sargent (1841–1927), American botanist and dendrologist, first Director of Arnold Arboretum.

sarniensis From Guernsey (once Sarnia) in the Channel Islands.

sasanqua From the Japanese name *sasankwa* (more correctly *sazanka*) for the species *Camellia sasanqua*.

Sassafras Probably from a North American Indian name used in Florida.

sativa Cultivated.

Saxegothaea In honour of Prince Albert of Saxe-Coburg-Gotha (1819–61), Consort of Queen Victoria.

Schinus From the Greek *schinos*, a name for the mastic tree, which this genus resembles in its production of resin and a mastic-like juice.

schrenkiana After the Russian botanist, Schrenk (1816–76).

Sciadopitys From the Greek *skiados*, an umbel, and *pitys*, a fir tree; the whorls of spreading, straight, needle-like leaves resemble the spokes of an umbrella.

Scindapsus Ivy arum; from the Greek for an ivy-like plant.

scopulina Bearing small brushes.

scopulorum Growing on cliffs, crags, or rocky outcrops.

selaginoides Resembling club-moss.

semiplena Semi- or partly double flowered.

sempervirens Evergreen.

septentrionalis Northern.

sepulcralis Growing in burial places.

Sequoia Sequoia or Sequoiah was the Cherokee Indian name (meaning opossum) of the mixed race George Gist (1770–1843), who had a German–American father. He invented the Cherokee alphabet.

Sequoiadendron From *Sequoia* and the Greek *dendron*, a tree.

sericea Silky.

serotina Late flowering or ripening.

serrata Saw-toothed.

serrula, -ata With small saw-like teeth.

shastensis From Shasta in California, USA.

shumardii In honour of Benjamin Franklin Shumard (1820–69), State Geologist of Texas in 1860.

sibirica From Siberia.

sieboldii In honour of Philipp Franz von Siebold (1796–1866), a German doctor who collected in Japan.

siliqua, -astrum Bearing pods with partitions like the fruits (siliquae) of certain members of the cabbage family (Cruciferae).

simonii In honour of Gabriel Eugene Simon (b. 1829), French consul and plant collector.

sinensis From China.

sinogrande A species allied to *Rhododendron grande*, from China.

sitchensis Of Sitka, Alaska.

smithiana In honour of James Edward Smith (1759–1828), Founder and first President of the Linnean Society of London.

smithii In honour of John Smith (1798–1888), Scottish gardener and fern specialist at the Royal Botanical Gardens, Kew.

solanderi In honour of Daniel Carl Solander (1736–82), a pupil of Linnaeus and botanist with Sir Joseph Banks on Captain Cook's first voyage to the Pacific.

Sophora From the Arabic name for these shrubs and trees.

Sorbus From the Latin *sorbum*, for the fruit of the service tree (*Sorbus domestica*).

soulangeana In honour of the Chevalier Étienne Soulange-Bodin (1774–1846), a French horticulturist.

'Spaethii' After the Späth Nursery, Berlin.

speciosa Showy.

spectabilis, -e Showy or spectacular.

'Spek' After the Dutch nurseryman.

spinosa Rearing spines.

spiralis Spiral.

splendens Splendid.

spontanea Spontaneous, in the sense of not being planted; growing wild.

squamata Scaly, with small scale-like leaves.

squarrosa Spreading or curved.

standishii In honour of John Standish (1809–75), a Surrey nurseryman who raised the plants introduced from China and Japan by Robert Fortune.

stellata Starry.

stenoptera With narrow wings.

Stewartia (*Stuartia* of some botanists) In honour of John Stuart (1713–92), 3rd Earl of Bute (and Prime Minister 1762–3), a keen patron of botany and horticulture.

stricta Upright.

strobus The old Latin name for an incense-bearing tree.

styraciflua Flowing with gum.

Styrax A Greek name derived from the original Semitic name for these plants.

suaveolens Fragrant.

suber Cork.

subhirtella Somewhat hairy.

'Suecica' Swedish.

sylvatica, sylvestris Forest-loving, growing in woods.

T

taeda The old Latin name for resinous trees, the wood of which was used for torches – *taeda*, a torch.

Taiwania From Taiwan, formerly Formosa.

tataricum From Tartary, a region in central Asia.

Taxodium The name for the swamp cypress; from the Latin *Taxus*, the yew, and the Greek *eidios*, resemblance, the leaves of the swamp cypress and yew are somewhat similar in shape.

Taxus The old Latin name for the yew tree.

Tetraclinis From the Greek *tetra*, four, and *kline*, bed; referring to leaves in fours.

tetragona aurea From the Greek *tetra*, four, *gonus*, angled, and the Latin *aurum*, gold.

tetraptera From the Greek *tetra*, four, and *terus*, winged. Four wing-like appendages or projections.

thompsonii In honour of Thomas Thompson (1817–78), Scottish physician and Superintendent of Calcutta Botanic Garden.

Thuja From the Greek name *thuia* (*thya* or *thyia*) for a resin-bearing tree, almost certainly a kind of juniper.

Thujopsis Resembling *Thuja*.

thunbergii In honour of Carl Peter Thunberg (1743–1828), a student of Linnaeus and later Professor of Botany at Uppsala, Sweden, who collected in Japan and Batavia.

thuringiaca From Thuringia, central Germany.

thyoides Resembling *Thuja* (in turn derived from *thya*, *thyia*, or *thuia*, Greek names for a resin-bearing tree, probably a juniper).

tibetica Of Tibet.

Tilia The Latin name for the lime or linden tree.

tobira Local name.

tomentosa Covered with short woolly or matted hairs.

toringoides Like toringo.

torminalis Used to alleviate (or capable of giving) a stomach ache.

Torreya In honour of Dr John Torrey (1796–1873), one of the most famous of America's botanists who described many thousands of plants brought back by explorers and collectors. Also co-author with Asa Gray of *The Flora of North America*.

torreyana After Dr John Torrey (see Torreya).

tortuosa Twisted as contorted.

totara The native Maori name (New Zealand).

Trachycarpus From the Greek *trachys*, rough, and *karpos*, a fruit.

tremula Trembling.

tremuloides Like *Populus tremula* (aspen) in general appearance.

triacanthos Three-thorned.

triandra Each flower having three anthers.

trichocarpa With hairy fruits.

trichotomum Having all divisions in threes.

triflorum *Tri* means three, and *flora*, a flower; the flowers are borne in groups of three.

trifoliata Having leaves composed of three leaflets.

triloba Three lobes.

'Triomf van Boskoop' From the nursery area of this name in Holland.

tristis Dull or sad.

tschonoskii In honour of Tschonoski (Chonosuke Sukawa) (1841–1925), who collected for Maximowicz in Japan.

Tsuga The Japanese name for hemlock.

tulipifera Tulip-bearing.

turbinata Shaped like a spinning top.

Typhina Resembling reed-mace (*Typha*).

U

'Ukon' From the Japanese for yellow or yellowish.

ulmoides Like *Ulmus* (elm).

Ulmus The Latin name for the elm.

Umbellularia Bearing umbels; the floral clusters are of this form.

umbraculifera Shade-giving.

undulatum Undulating.

Unedo The Latin name for the strawberry tree and its fruit.

ussuriensis From the vicinity of the Ussuri river in eastern Asia.

V

variabilis Variable.

variegata, -um Variegated.

vegeta Vigorous.

veitchii In honour of the Veitch family, nurserymen of Exeter, Devon, and Chelsea London. The original nursery was founded in 1808 at Exeter, dividing in 1854, the Chelsea firm finishing on Sir Harry James Veitch's retirement in 1914.

velutina Velvety.

vera True to type.

verniciflua Yielding varnish.

verticillata Whorled.

vestita Covered.

Viburnum The Latin name of one species of the wayfaring tree.

vilmoriniana, -ii In honour of the French nurseryman Vilmorin-Andrieux.

viminalis With long slender shoots like the willow or osier.

violacea Blue-red colour.

virginiana From the states of East or West Virginia, USA.

virginicus From Virginia, USA.

viridis Green.

viscosa Sticky.

vitellina The colour of an egg yolk.

'Vitifolium' With leaves like *Vitis* (grapevine).

vomitoria Emetic.

vossii In honour of Andreas Voss (1857–1924), German nurseryman.

'Vranja' Named after Vranje in Yugoslavia.

W

wallichiana In honour of Nathaniel Wallich (1786–1854), a Danish doctor and botanist who became Superintendent of the Calcutta Botanic Garden in India from 1814 to 1841.

Washingtonia In honour of George Washington (1732–99), first President of the United States.

watereri For the nursery firm of John Waterer Sons & Crisp Ltd, Twyford, Berkshire (in the case of *Laburnum × watereri*). In honour of the nurseryman Waterer of Surrey (in the case of *Cotoneaster × watereri*).

wilsonii In honour of Ernest Henry Wilson (1876–1930), the famous plant collector who travelled widely in China and later became Curator of the Arnold Arboretum.

'Wisselii' In honour of the Dutch nurseryman Wissel.

X

xanthocarpa Yellow-fruited.

Y

yedoensis From Tokyo (formerly Yedo), Japan.

'Youngii' In honour of the Young family, nurserymen at Epsom, Surrey, during the first half of the 19th century.

Yucca From the Carib name for cassava, a member of the *Euphorbia* family.

yunnanensis From Yunnan, western China.

Z

zebrina Zebra-striped.

Zelkova From the Caucasian native name.

Index of Tree Species

Descriptions of trees in the index follow their botanical (Latin) names. To find botanical or horticultural details of (e.g.) western cedar look up the English names to find the Latin; then look up the Latin (*Thuja plicata*). General notes on the genus *Thuja* are at the head of the entry. Specific notes on the species (*T. plicata*) are listed alphabetically below.

The hardiness zones (e.g. "Zone 9") referred are those mapped on pages 36–37.

The sign × before a name denotes a natural hybrid, the sign + a graft hybrid.

The words "big", "medium", "small" in relation to leaf size are based on the following measurements: "big" – 7 inches upward; "medium" – 2–6 inches; "small" – to 2 inches.

The height given is approximately that of a mature tree in its native land.

The speed of growth given is that of a known tree in good but not exceptional conditions. Statistics are not available for all species. ~For space reasons, certain geographical abbreviations have been used including C. – central, E. – eastern etc.

Page numbers in *italic* refer to illustrations/captions.

A

ABIES
Evergreen trees, big, mostly conical. Leaves either spreading round shoot or directed forward, sometimes parted on upper or lower side of shoot, often notched at tip, leaving disc-like leaf scars, aromatic when bruised. Sexes separate on same tree: male cones in clusters on underside of branches, often brightly coloured; female flowers catkin-like. Cones have spreading scales, cylindrical, upright on branches, breaking up and leaving a central spike on the tree. Some years many, others few or none. Seeds winged. Bark usually resinous. Most dislike shallow chalk. Susceptible to adelgids in Europe and America. 92–9

A. alba Common silver fir, European silver fir. Europe. Zone 4. To 150 ft (45m), sometimes with huge branches. Leaves in 2 ranks, upper shorter than lower, dark shiny green. Cones clustered near top of tree, pale green becoming brown-red. Young foliage susceptible to late frosts. 35ft (10.5m) in 20 years. 96, *98*

A. amabilis Pacific silver fir, Red fir. W North America. Zone 5. To 250ft (76m). Leaves flattened, crowded on upper surface of twig, brilliant white bands beneath. Cones dark purple-brown. Most soil types, except dry or chalky. Requires deep soil or high rainfall. 30ft (9m) in 20 years. 94–5, *94*

A. balsamea Balsam fir, Balm of Gilead. NE North America. Zone 2. To 75ft (23m), dense, narrow. Leaves flat, dark green above whitish near tip, silvery-white beneath, very sweet-scented. Cones purple. Not suitable for chalk. 15ft (4.5m) in 20 years. 94

A.b. 'Hudsonia' Dwarf, to 2ft (60cm) in 25 yrs *159*

A. bracteata Bristlecone fir, Santa Lucia fir, Fringed spruce. Mts of S California. Zone 7. To 150ft (45m), broadly conical shape with hanging branchlets. Buds pale, slender, elongated. Leaves large, spread horizontally into 2 opposite sets, curved and flattened with sharp tips, icy-white underneath. Will grow on deep soil over chalk. 30ft (9m) in 20 years. 95, *98*

A. cephalonica Greek fir. Mts of Greece. Zone 5. To 100ft (30m), broadly conical, becoming more open. Leaves spread round shoots, mostly above; dark shiny green above, 2 narrow white bands beneath, prickly. Male flowers in dense clusters beneath shoots. Cones often at top of crown, green-brown. Early foliage often frost-damaged. Disease-free. Likes chalk. 28ft (8.5m) in 20 years. 96, *98*

A. concolor Colorado white fir, White fir. SW North America. Zone 4. To 160ft (48m), narrow. Leaves 2in (5cm) long (longer than most other species), blue-grey, dull green with age. Cones oblong, olive-green becoming purple. 30ft (9m) in 20 years. 95, *95*

A.c. 'Candicans' Handsome form. Leaves silver-grey to white. 95

A.c. subsp. lowiana Leaves arranged in 2 distinct ranks. 95

A.c. 'Violacea' Leaves waxy bluish-white. 95

A. delavayi W China, N India. Zone 7. Medium-size. Densely set leaves, bright shiny green above, gleaming white below. Cones dark bluish-violet. Buds very resinous. 99

A. durangensis Durango fir. Mexico. Zone 7. To 130 ft (40m). Leaves spirally arranged, in 2 ranks. Cones short. 95

A. fargesii W and C China. Zone 6. To 100ft (30m), with massive branches. Leaves long, spreading horizontally in 2 or more ranks, upper ranks half length of lower. Cones purple to red-brown. 35ft (10.5m) in 20 years. 99, *99*

A. firma Japanese fir: big tree, leaves bright green, thick and broad; bark pinkish. Zone 6. 150

A. forrestii Forrest's fir. W China. Zone 7. To 60ft (18m). Cones deep blue. 40ft (12m) in 20 years. *98*, *99*

A. grandis Grand fir, Giant fir. W North America. Zone 6. To 250ft (76m), columnar, becoming rounded, sometimes multiple-topped. Leaves double-ranked horizontally, upper shorter than lower, bright green above, 2 bands of silver-white beneath. Cones green-purple in summer. Moderately lime tolerant. Vigorous on many soils. 55ft (17m) in 20 years. 92, 94, *94*

A. holophylla Manchurian fir. Manchuria, Korea. Zone 5. 100–150ft (30–45m), narrow conical or spreading. Leaves medium length, bright green, 2 grey-green bands beneath, sharp pointed. Cones green ripening to light brown. 99

A. homolepis Nikko fir. Japan. Zone 4. 80–90ft (24.5–27.5m). Leaves lower at right angles to shoot, upper directed outwards and upward, dark green shiny above, 2 conspicuous white bands on lower leaves. Cones purple, brown when mature. Tough, adaptable, tolerates some pollution. 25ft (7.5m) in 20 years. 98–9

A. koreana Korean fir. Korea. Zone 5. To 40ft (12m), conical, or low, bushy. Leaves nearly covering upper side of shoot, curve upward or near vertical; black-green above, 2 broad white bands below. Male flowers on side-shoots, ovoid, dark red-brown to purple or pink to bright yellow. Good forms cone freely from early age, dark purple. 10ft (3m) in 20 years. 98, *98*, *99*

A. lasiocarpa Subalpine fir. W North America. Zone 5. Sometimes to 130ft (40m), beautifully narrow spire (Washington). Leaves irregularly double-ranked, medium length, shorter on higher branches, waxy, blue-green. Cones dark purple. Seeds with a shining purple wing. Does not grow well in lime as *Adelges* make gouty swellings on shoots. 95

A. l. var. arizonica Arizona cork fir, Corkbark fir. SW America. Zone 5. 100–130ft (30–40m), neat, conical, blue tree. Leaves 1in (2.5cm) long, striped blue-grey each side; flat, upward-curving, those in midline covering the stem. Cones dark purple, seeds with a purple wing. Tolerant of many soils. 15ft (4.5m) in 20 years. 95

A. magnifica California red fir, Red fir. W United States. Zone 5. To 200ft (60m), narrow. Leaves blue to blue-green, almost rounded in cross-section, upper leaves curve upward. Cones purple-brown. Dislikes chalk, 30ft (9m) in 20 years. 95

A. nebrodensis Small silver fir from Sicily. Zone 7. 98

A. nordmanniana Caucasian fir. W Caucasus and N Turkey. Zone 4. To 200ft (60m), conical, becoming columnar, pointed or flat-topped. Bright green leaves, 2 white bands below, forward-pointing, 2 sets on lower sides of shoots, upper set shorter. Cones pale green becoming brown, resinous. Generally disease-free. 35ft (10.5m) in 20 years. 98, *98*, *99*

A. numidica Algerian fir. E Algeria. Zone 6. 70–100ft (21.5–30m), conical. Leaves short, flattened, stiff, broad and thick, dark shining green, broadly banded grey, all round shoot, curving upward. Cones brown. Best *Abies* near towns. 96

A. pindrow West Himalayan fir. Afghanistan to Nepal. Zone 8. To 200ft (60m), narrow, conical with short branches becoming open with age. Leaves ascending above shoot, spreading below. 2–3in (5–7.5cm) long, dark shiny green. Cones deep purple, becoming brown. Requires high rainfall and cool summers. 99

A. pinsapo Spanish fir, hedgehog fir. Mts of S Spain. Zone 8. To 100ft (30m), narrow cone, becoming more irregular. Leaves all round shoot, very short, blunt, broad, straight or slightly curved, noticeable bands on both sides. Male flowers abundant, round, beneath shoot, bright red, opening late spring. Cones tapered, purple-brown. Tolerates lime, chalk, and any dry soils. 25ft (7.5m) in 20 years. 96, *96–7*, *98*

A. procera Noble fir. W United States. Zone 5. To 250ft (76m), dome-headed. Leaves slight 4-angled in section, flatter than *A. magnifica*, blunt, crowded on upper side of twig, blue-green. Cones olive-green to purple. Dislikes chalk. 30–45ft (9–13.5m) in 20 years. 92, 95, *95*

A. sibirica Siberian fir. NE and E Russia. Zone 1. To 100ft (30m). Narrowly conical crown with short slight ascending branches. Leaves forward-pointing and parted above and below shoot. Extremely hardy in Continental climates but liable to frost damage elsewhere.

A. smithii A form of *A. forrestii* from NW Yunnan. 98

A. spectabilis East Himalayan fir. Nepal, Sikkim, Bhutan. Zone 7. To 150ft (45m), broad, columnar, eventually flat-topped. Leaves on a large scale, dense, in 2 ranks, lying slightly forward. Cones pale grey-blue, becoming dull, dark purple by winter. Susceptible to late frosts. 35ft (10.5m) in 20 years. 99

A. veitchii Veitch's silver fir. C Japan. Zone 3. 60 or 70ft (18 or 21.5m). Leaves as *A. nordmanniana* but softer to touch. Male flowers orange-red, minute, becoming round, red-brown; female "flowers" red cylinder. Cone purple-blue to brown. Trunk deeply fluted. Dislikes chalk; thrives in semi-urban conditions. 40ft (12m) in 20 years. 99

ACACIA
Mostly evergreen trees and shrubs. Leaves usually doubly pinnate, but some species are replaced in adult life by flattened structures developed from leafstalks. Flowers bisexual, usu. yellow, winter or spring. Full sun, acid or neutral, dry soils. Tender. 189, 202

A. baileyana Cootamundra wattle, Bailey's mimosa. Australia. Zones 9–10. To 20ft (6m) or more, often weeping. Leaves evergreen, ultimately divided into numerous long, narrow silver-grey leaflets. Yellow flowers in clusters of small, roundish heads. Fruits in pods 2–3in (5–7.5cm) long. 203

A.b. 'Purpurea' Foliage tinged purple. 203

A. dealbata Silver wattle, Mimosa. Australia. Zones 8–9. To 50ft (15m), spreading. Leaves fern-like, downy, silver-green. Flowers yellow, in large clusters of small, round heads, fragrant. Seed pods flat, 2–3in (5–7.5cm) long, blue-white. Sun-loving, can be killed by severe or prolonged frost. 50ft (15m) in 20 years. 202, *202*, 203

A. longifolia Sydney golden wattle. Australia. Zones 8–9. To 20ft (6m), spreading. Leaves evergreen, simple, leathery, dark green. Flowers small, round heads, bright yellow, in spikes 2–3in (5–7.5cm) long. Seed pods 3–4in (7.5–10cm) long. Fairly lime tolerant. 203

A. mearnsii Black wattle. Australia. Zones 9–10. To 40ft (12m). Leaves feathery, grey-green. Flowers bright yellow. 203

A. melanoxylon Blackwood. Australia. Zone 9. To 50 ft (15m), upright. Leaves simple, leathery. Flowers pale yellow. Invasive in tropics. 203

A. pravissima Oven's wattle. Australia. Zone 9. To 26ft (8m). Leaves, neat, triangular, densely placed on stems, greyish-green. Flowers bright yellow. 203

A. podalyriifolia Queensland silver wattle. Australia. Zone 9. To 16ft (5m). Leaves narrow, blue-green with white hairs. Flowers bright yellow. 203

ACER
Maples. Deciduous, rarely evergreen trees. Opposite, usually lobed leaves often with brilliant autumn colours. Clustered uni- or bisexual flowers. Fruits are paired nuts with membranous wings (keys). Susceptible to *Acer* gall mite, sycamore suffers from tar spot and horse chestnut scale. Tolerant of most soil types. 19, *23*, *25*, 31, 105, 294, 298–311, *299*

A. buergerianum Buerger's-type maple, Trident maple. E China, Japan, Korea. Zone 6. To 20ft (6m) high, rounded. Leaves 3-lobed, simple, bright green above, paler, waxy beneath, toothed, stalked. Flowers in a flattish cluster. 307

A. campestre Field maple. Europe (incl. British isles), W Asia. Zone 4. To 25ft (7.5m), rounded, dense. 5-lobed leaves, 4in (10cm) across, lobes blunt, usually butter-yellow in autumn.Small green flowers in small clusters, early May. Keys 1in (2.5cm) long. Grows on chalk. A good hedge or screen plant. *310*, 311

A.c. 'Postelense' Zones 5–6. To 25ft (7.5m), rounded. Leaves golden yellow in spring, turning clear yellow, sometimes flushed red, in autumn. 311

A. capillipes Small Japanese snakebark maple. Zone 5. 307

A. cappadocicum Caucasian

maple. Caucasus and W Asia to Himalaya. Zone 6. To 65ft (20m), spreading. Leaves 5- to 7-lobed, 3–6in (7.5–15cm) across, glossy green, yellow in autumn. Small yellow flowers. Keys in clusters. Shade tolerant. 30ft (9m) in 20 years. 69, 311, *311*

A.c. **'Rubrum'** New shoots blood-red. 311

A. carpinifolium Hornbeam maple. Japan. Zone 5. To 35ft (10.5m), vase-shaped with a number of stems from the base. Leaves opposite, otherwise hornbeam-like, unlobed, 3–4in (7.5–10cm) long, golden brown in autumn. Green flowers in spiky clusters. Keys have curved wings. 15ft (4.5m) in 20 years. 306

A. caudatifolium Kawakami maple. Taiwan. Zones 8–9. To 40ft (12m). Bark green, faintly striated with white. Leaves pointed, unlobed, or with 3 weak lobes, dark green. Keys red. 307

A. circinatum Vine maple. W United States. Zone 5. Small tree, to 40ft (12m), widespreading with several branches from the base. Leaves almost circular, 7–9 lobes, orange to red in autumn. Flowers white to purple in drooping clusters, late April. Red keys. Shade tolerant. 15ft (4.5m) in 20 years. *301*, 302, *302*

A. x conspicuum Hybrid between *A. davidii* and *A. pensylvanicum* giving rise to vigorous snakebark cultivars with prominent white-striped bark. Leaves unlobed or 3-lobed, usually large, to 12in (30cm) long in some cultivars. 302, 307

A. x c. **'Phoenix'** Red winter shoots prominently striped white. 302

A. x c. **'Silver Vein'** Branches green with white stripes. Leaves with red stalks. 302

A. davidii David's maple. China. Zone 6. To 50ft (15m), rounded head. Bark green striped with white. Oval, pointed leaves up to 8in (20cm) long, dark, shining green, yellow and purple in autumn. Keys green flushed with red, hanging along branches in autumn. 30ft (9m) in 20 years. 307

A. forrestii Zone 6. Lovely little Chinese snakebark maple with long graceful branches, red young shoots and leafstalks. The centre 1 of 3 lobes is much the biggest. Likes acid soil. 12ft (3.5m) in 20 years. 307

A. x freemanii Zone 3. Hybrid between *A. rubrum* and *A. saccharinum*. Most widely grown as the cultivar Autumn Blaze ('Jeffersred') which has deeply lobed leaves turning orange-red in autumn. 300

A. griseum Paperbark maple. C China. Zone 5. To 30ft (9m). Leaves compound, 3 leaflets, white downy beneath, scarlet in autumn. Downy winged fruit. Orange-red bark exfoliates. Most striking 3-leafed maple. 15ft (4.5m) in 20 years. 306, *307*

A. grosseri Grosser's maple. C China. Zone 6. Small, to 20ft (6m). Bark green, striped white. Leaves sometimes lobed, toothed, good autumn colour. Keys in long hanging clusters. 307

A.g. **var. *hersii*** Hers's maple. C China. Zone 6. To 30ft (9m). Leaves usually 3-lobed, dark dull green above, paler beneath. Fruits in hanging clusters. Young bark striped with white. Slightly hardier than *A. grosseri*. 30ft (9m) in 20 years. 307

A. japonicum Full-moon maple. Japan. Zone 5. Small bushy tree, 30ft (9m), rounded. Leaves 2–5in (5–12.5cm) long, roundish, 7–11 lobes, bright red in autumn. Purplish-red flowers, April. Keys in clusters. Best in woodland shelter. 10ft (3m) in 20 years. 306

A.j. **'Aconitifolium'** Japan. Small tree or large bush. Leaves deeply cut and divided, soft green, turning rich crimson in autumn. Flowers red in drooping clusters. Grows well in moist, well-drained positions sheltered from cold winds. 306

A.j. **'Vitifolium'** Wide, fan-shaped leaves, 10–12 lobes. Beautiful autumn colour. 306

A. macrophyllum Oregon or Big-leaf maple. W North America. Zone 6. To 100ft (30m), round-headed. Leaves, 6–12in (15–30cm) across, dark shining green, orange and yellow in autumn. Flowers small, yellow, hanging in clusters, fragrant, May. Keys bristly, in hanging clusters. 35ft (10.5m) in 20 years. *303*

A. maximowiczii Maximowicz's maple. C and W China. Zone 6. To 50ft (15m). Snakebark maple with 3-lobed leaves to 3in (8cm) long, the central lobe being the longest, turning yellow in autumn. Flowers red, keys in hanging clusters. 306–7

A. monspessulanum Montpelier maple. S Europe, W Asia. Zone 5. To 30ft (9m), rounded. Leaves 3-lobed, simple, virtually smooth, dark green above, paler beneath. Flowers in drooping, loose clusters, greenish. Fruit red, profuse, winged. *310*, 311

A. negundo Box elder, Ash-leaved maple. North America. Zone 2. To 65ft (20m), spreading and rather open. Leaves compound, 3–5 leaflets, bright green above, paler beneath. Sexes on separate trees. Male flowers in string tassles, more elegant than the female flowers which are in slender drooping clusters. Keys in drooping clusters. 20ft (6m) in 20 years. 300–2, 306

A.n. **'Elegans'** Leaves with bright yellow margins. Young shoots with white bloom. 300–2

A.n. **'Flamingo'** Leaves patterned like 'Variegatum' but pink-flushed when young. 302

A.n. **'Variegatum'** Leaves with white margins. 302, *302*

A.n. **var. *violaceum*** Young shoots purple with white bloom. Pinkish-red male flower tassles. 302

A. opalus Italian maple. S Europe. Zone 5. To 50ft (15m), rounded leaves 2½–4½ (6.5–11.5cm), 5 shallow lobes, toothed, dark green and hairless above, paler and downy beneath. Clusters of small yellow flowers, March. 20ft (6m) in 20 years. 311

A. palmatum Japanese maple. Japan, C China, Korea. Zone 5. To 25ft (7.5m), low, rounded head. Bright green leaves, 5–7 toothed lobes, bronze or purplish in autumn. Small purple flowers on erect stalks. Fairly chalk tolerant; prefers shelter from cold winds. Moist well-drained loam preferred. 15ft (4.5m) in 20 years. *68*, 304–6, *304–5*

A.p. **'Atropurpureum'** Bloodleaf Japanese maple. Leaves with 5–7 lobes, crimson-purple in summer. 20ft (6m) in 20 years. 304

A.p. **Dissectum Group** To 10ft (3m), dome-shaped. Fern-like, green leaves cut to the base into 7, 9, or 11 lobes. Slow-growing. 304

A.p. **'Dissectum Atropurpureum'** As above, but leaves dark red. *307*

A. p. **var. *dissectum* 'Seiryu'** Orange leaves in spring and autumn. *306*

A.p. **'Linearilobum'** Leaves very finely dissected. 306

A.p. **'Osakazuki'** Leaves green, turning bright scarlet in autumn. Possibly the most brilliantly coloured Japanese maple. 306, *307*

A.p. **'Sango-kaku'** Coral bark maple. Shrub or small tree. Young branches coral-red, effective in winter. Leaves yellow in autumn. 304–6

A. pensylvanicum Striped or Snakebark maple, Moosewood. E North America. Zone 3. To 40ft (12m), open and irregular. Young stems green, becoming striped with pale green and white. Leaves 3-lobed, up to 7in (17.5cm) long, bright yellow in autumn. Flowers yellow, in hanging clusters, May. Unhappy on chalk. 20ft (6m) in 20 years. 302

A.p. **'Erythrocladum'** Young winter shoots pink-red, distinctive. 302, *302*

A. pictum Japan, China, Korea, Zone 6. Lobed leaves turning yellow in autumn. *68*

A. platanoides Norway maple. Europe, Caucasus. Zone 3. To 100ft (30m), rounded head. Bright green leaves, 5-lobed, 4–7in (10–17.5cm) wide, yellow in autumn. Small yellow flowers in clusters, late April. 35ft (10.5m) in 20 years. 308, 310, *311*

A.p. **'Crimson King'** Dark reddish-purple leaves and red-tinged yellow flowers. 310

A.p. **'Drummondii'** Leaves with broad white margin but with awful tendency to revert to green as branches rise above head height. 310–11

A.p. **'Goldsworth Purple'** Red leaves. 310

A.p. **'Reitenbachii'** Leaves emerge red, change to green, good autumn colour. 310

A.p. **'Schwedleri'** Leaves and young shoots red-purple, bronzy green later. 310

A. pseudoplatanus Sycamore or Sycamore maple. Europe and W Asia. Zone 5. To 100ft (30m), widespreading. Leaves usually 5-lobed, dark green above, paler below. Yellowish-green flowers in large drooping clusters. Keys in clusters. Succeeds in exposed positions, any soil. 35ft (10.5m) in 20 years. 308, *309*

A.p. **'Atropurpureum'** Leaves purple-tinged. 308

A.p. **'Brilliantissimum'** Small dense tree, young leaves pink, later yellow-green, finally green. 12ft (3.5m) in 20 years. 308–10

A.p. **'Prinz Handjéry'** Leaves bronzy-yellow in spring. 308, *310*

A.p. **'Worley'** Golden sycamore. Medium-sized tree. Leaves yellow-green when young, leafstalks reddish. 20ft (6m) in 20 years. 308

A.pseudosieboldianum* subsp. *takesimense Ullung-do maple. Ullung-do island, Korea (formerly Takesima island). Zones 6–7. Small tree to 16ft (5m). Neat, rounded leaves with 9–11 short lobes, dark green, good autumn colours. 306

A. pubinerve Chocolate maple. China. Zone 6. Small tree 16–23ft (5–7m). Leaves emerge chocolate-brown, turning green, 5-lobed. Keys pale green. 307

A. rubrum Red maple, Swamp maple, Canadian maple. E North America. Zone 3. To 120ft (36m), round head, dense foliage. Leaves medium length, with 3–5 toothed lobes, dark shiny green above, waxy bloom underneath. Small red flowers on slender stalks, early spring. Bright red autumn colour. 35ft (10.5m) in 20 years. 298–300, *301*

A.r. **'Autumn Spire'** Broadly columnar, good early colouring. 300

A.r. **'Columnare'** Columnar red maple. Big, upright, wide. 300

A. saccharinum Silver maple. E North America. Zone 3. To 120ft (36m), spreading head. Leaves medium length, 5 deep toothed lobes, bright green above, silvery white underneath. Clear yellow in autumn. Small greenish-yellow flowers. Keys in hanging clusters. 50ft (15m) in 20 years. 300

A.s. **'Laciniatum'** Leaves deeply lobed. 300

A.s. **'Lutescens'** Pale yellow-leafed form. 300

A. saccharum Sugar maple. C and E North America. Zone 3. To 120ft (36m), fluted trunk, rounded head. Leaves medium length, 3–5 toothed lobes, bright green, yellow, orange, or scarlet in autumn. Small greenish-yellow flowers hanging in clusters. Grey furrowed bark. 35ft (10.5m) in 20 years. 298, *300–1*

A.s. **subsp. *nigrum*** Black maple. Yellow in autumn. 298

A. sempervirens Cretan maple, similar to *A. monspessulanum*; keeps shiny leaves very late. 311

***A. shirasawanum* 'Aureum'** Japan. Soft yellow leaves, liable to scorch in full sun. Slow-growing. 306, *306*

A. sieboldianum Siebold's maple. Japan. Zones 4–5. Like *A. palmatum*, but hardier. Excellent red and gold autumn colour. 306

A.tataricum* subsp. *ginnala Amur maple. China, Japan, Manchuria. Zone 4. To 20ft (6m), upright, rounded, dense branching, vigorous, 3-lobed, toothed margins, bright dark green above, scarlet in autumn. Keys red and conspicuous. Flowers yellowish-white, very fragrant, May. 15ft (4.5m) in 20 years. 307

A. velutinum* var. *vanvolxemii Van Volxem's maple. Caucasus. Zone 5. Large tree to 100ft (30m). Noted for its exceptionally large leaves, to 1ft (30cm) across, bluish-green underneath. Very vigorous, like a bigger-leaved sycamore. 310

A. **'White Tigress'** Hybrid snakebark maple, with exceptionally crisp white markings on trunk. 307

AESCULUS

Deciduous trees or shrubs, opposite, palmate leaves usually with 5–9 leaflets. Flowers in large conical heads, late spring to summer. 1 or 2 large seeds in a smooth or prickly husk. Hardy. All soils, if not too dry. Prefers open, sunny position. Leaf blotch in Europe and America, horse chestnut scale. 31, 294–6

A. californica Californian buckeye. California. Zone 7. To 35ft (10.5m), spreading tree or shrub. Leaves medium length, early metallic-green. Flowers white to rose-coloured, 1in (2.5cm) long, fragrant. April to Aug. Egg-shaped fruit, Aug. and Sept. 296, *297*

A. x carnea Red horse chestnut. Zone 4. Hybrid between *A. hippocastanum* and *A. pavia*, to 70ft (21.5m), pyramidal when young, becoming round-headed with age. Leaves smaller and darker than common horse chestnut. No autumn colour. Flowers rose-pink in upright clusters 10in (26cm) high, mid-May. 25ft (7.5m) in 20 years. 296

A. flava Sweet buckeye. SE United States. Zone 5. To 90ft (27.5m), round-headed. Leaflets narrow, medium length, finely toothed, downy beneath, usually good autumn tints. Flower elongated, pale yellow, May to June. Fruits smooth. 20ft (6m) in 20 years. 296, *297*

A. glabra Ohio buckeye. SE and C United States. Zone 5. To 30ft (9m), round-headed. Leaflets oval to wedge-shaped, smooth with age. Flowers green-yellow. Fruits ovallish, prickly. 20ft (6m) in 20 years. 296

A. hippocastanum Common horse chestnut. Greece. Albania. Zone 3. To 120ft (36m), with a rounded, spreading head. Leaves medium to long, scattered hairs below, hairless above. Flowers white, May. Fruit spiny. 40ft (12m)

Index of Tree Species 375

in 20 years. 12, *12*, *13*, *16*, 20, 274, 294–6, *294–5*, *296*
A.h. **'Baumannii'** Baumann's horse chestnut. Zone 3. To 85ft (27m), rounded at maturity. Flowers double, white, mid-May, lasting longer than the type. No fruit. 296
A. indica Indian horse chestnut. NW Himalaya. Zone 7. To 100ft (30m), often with a short, thick trunk. Leaves large, toothed, hairless, dark shiny green above. Flowers 1in (2.5cm) long, white flushed with pink, July. Fruit tough. 30ft (9m) in 20 years. 296
A.i. **'Sydney Pearce'** Free-flowering selection from tree at Kew. 296
A. **x *neglecta* 'Erythroblastos'** Zone 5. Slow-growing small American buckeye comparable with *A. pseudoplatanus* 'Brilliantissimum' for the pale creamy-pink of its spring foliage, yellow-green later. Subject to spring frost damage in parts of Britain.
A. parviflora Bottle brush buckeye. SE United States. Zone 5. To 15ft (4.5m), a spreading, suckering shrub with several slender stems. Leaves medium to long, leaflets shallow toothed. Flowers white with protruding stamens bearing red anthers. July, Aug. Fruit hairless, nearly egg-shaped. Tolerant of shade. 10ft (3m) in 20 years. 296
A. pavia Red buckeye. S United States. Zone 5. To 20ft (6m), shrub or small tree. Leaves medium length. Flowers crimson, June. Fruit nearly egg-shaped, August. 296
A. turbinata. Japanese horse chestnut. Japan. Zone 6. To 100ft (30m), with a thick trunk. Giant leaves with leaflets up to 16in (41cm) long, toothed, on a long stalk. Attractive autumn tints. Flowers yellowish-white, June. Fruit pear-shaped, rough. 35ft (10.5m) in 20 years. 296

AFROCARPUS gracilior see *PODOCARPUS gracilior*

AGATHIS
A. australis Kauri pine. New Zealand. Zone 9. Leaves to 3¼in (8cm) long, evergreen, leathery, stalked, in opposite pairs, 1in (2.5cm) apart, tapering to a blunt point, dull grey-green. Male and female cones on same tree, but separate. Cones round to oval. Seeds winged. Tender in Britain. 12ft (3.5m) in 20 years. 125
A. robusta Queensland kauri. Australia. To 160ft (50m). Zones 10–11. Large tropical conifer with broad leathery leaves and large round cones. *125*

AILANTHUS 47
A. altissima Tree of heaven. N China. Zone 4. To 90ft (27.5m) usually less. Leaves compound, deciduous, leaflets ovallish, with unpleasant smell, 1 or 2 teeth near base. Flowers in small clusters, greenish-yellow, sexes on separate tree, male flowers with unpleasant smell. Fruits dry, winged, reddish-brown, profuse, on female trees only. Tolerant of dryness, shade, acid soil, but beware suckers; sun-loving. 45ft (13.5m) in 20 years. 314, *314–15*

Alaska Cedar *see CHAMAECYPARIS nootkatensis*
Aleppo Pine *see PINUS halepensis*

ALBIZIA
A. julibrissin Silk tree, Pink siris. Iran to China. Zone 7. To 40ft (12m), broad, spreading, flat-topped. Leaves doubly pinnate, 9–18in (23–46cm) long. Flowers light pink, brush-like, in rounded heads, June–August. Pods 5in (13cm) long, narrow between seeds, flat, Sept.–Nov. 15ft (4.5m) in 20 years. 202, *203*
A.j. f. ***rosea*** Smaller form, pinker flowers, hardy to Zone 5. 202–3
A.j. **'Summer Chocolate'** Dark brown foliage, deep pink flowers. 203

Alder *see ALNUS*
Algerian Oak *see QUERCUS canariensis*
Allegheny Serviceberry *see AMELANCHIER laevis*
Alligator Juniper *see JUNIPERUS deppeana*
Almond *see PRUNUS dulcis*

ALNUS
Very hardy deciduous trees or shrubs. Toothed leaves. Long male catkins, short female catkins on the same tree form woody cones, ripe in autumn. Seeds are small flat nuts. Flowers produced before the leaves or in autumn. Mostly moisture-loving, some lime tolerant. Susceptible to die-back, leaf spot, tent caterpillar. 20, 280
A. cordata Italian alder. Corsica and S Italy. Zone 6. To 80ft (24.5m). Leaves roundish with abrupt point, medium length, bright green, glossy, finely toothed. Male catkins 2–3in (5–7.5cm) long, groups of 3–6. Fruit erect egg-shaped, about 1in (2.5cm) long, in 3s. Suited to all types of soil. A tall, narrow formal tree. 50ft (15m) in 20 years. 280, *280*, *281*
A. formosana Taiwan alder. Zone 8. Handsome tree to 65ft (20m), pyramidal in shape. Leaves narrow, glossy green. Male catkins to 3in (8cm), fruits in clusters. Hardy in Britain. 280
A. glutinosa Common alder, European alder. Europe (incl. British Isles), W Asia, N Africa. Zone 3. To 80ft (24.5m), with sticky young growth. Broad, pear-shaped leaves, medium length, dark green, pale green below. Male catkins in groups, 2–4in (5–10cm) long. Egg-shaped fruits in clusters. Suited to boggy ground. 40ft (12m) in 20 years. 280, *281*
A.g. **'Imperialis'** Leaves deeply and finely lobed, the lobes slender and pointed, not toothed. 25ft (7.5m) in 20 years. 280
A.g. **'Laciniata'** Sturdy, rather stiff form; leaves not so finely divided as 'Imperialis'. 280
A. incana Grey alder, American speckled alder. Europe, Caucasus, E North America. Zone 2. Large shrub to medium-sized tree, to 65ft (20m). Leaves dull green above, grey beneath, downy, oval. Male catkins 2–4in (5–10cm) long in groups. Oval fruits clustered. Good for cold or wet situations. 50ft (15m) in 20 years. 280, *281*
A.i. **'Aurea'** Young shoots and foliage yellow; catkins red-tinted. 280
A. maximowiczii Maximowicz alder. Japan, Korea, Kamschatka. Zone 4. Shrub or small tree, 16–33ft (5–10m). Leaves broad, dark green. Male catkins fat, to 2in (5cm), yellowish green to maroon. Fruits in clusters of 4–5, large. 280, *281*
A. pendula Asian green alder. Japan, Korea. Zones 5–6. Shrub with arching branches, 10–25ft (3–8m). Leaves narrow, pointed, strongly veined, very elegant. Male catkins to 5in (12cm). Fruits in small clusters. 280
A. rubra Red alder. W North America. Zone 6. Medium-sized tree, to 65ft (20m), narrow pyramidal head, pendulous branches. Oval leaves medium length, upper surface dark green, lower surface pale or greyish. Male catkins 4–6in (10–15cm) long in groups. Fruits barrel-shaped: clusters. 40ft (12m) in 20 years. 280
A. rugosa Speckled alder. E North America. Zone 2. Very hardy shrub to 20ft (6m). 280

AMELANCHIER
Small trees or shrubs. Leaves deciduous, good autumn colour. Flowers in small clusters, white, 5-petalled, spring, before leaves fully developed. Fruit berry-like, small. Some species lime tolerant. Subject to rusts and fireblight. 47, 62, 204, 206
A. arborea Downy serviceberry. E North America. Zone 4. Large shrub or tree to 50ft (15m). Leaves ovate, emerging white hairy at same time as clusters of white flowers. Fruits ripening to purplish-black, edible. Needs lime-free soil but tolerant of moist conditions. 206
A. canadensis Serviceberry, Juneberry, Shadblow. North America. Zone 4. Suckering shrub to 20ft (6m). Leaves oval to heart-shaped, simple, saw-toothed, stalked. Flowers in upright or drooping clusters, snow-white. Fruit round, fleshy, berry-like, green turning red and maroon-purple, early summer. Prefers lime-free soils. Sun or semi-shade. 206, *206–7*
A. **x *grandiflora*** Apple serviceberry. Zone 4. Hybrid between *A. arborea* and *A. laevis* to 27ft (8m). Leaves emerging purplish and downy. 206
A x *g.* **'Ballerina'** A profusely flowering selection with red to purple autumn colour. 206
A. laevis Allegheny serviceberry. North America. Zone 4. To 27ft (8m), spreading. Leaves ovate, simple, smooth; young leaves purple-bronze, turning rich orange-red. Flowers small, in drooping clusters, fragrant. Fruit berry-like, purple-black changing to red, summer to autumn. Prefers lime-free soils. Sun or semi-shade. Susceptible to same pests as *A. canadensis*. 206, *207*
A. lamarckii Juneberry. Accounts for the majority of amelanchiers cultivated. Europe. Zone 4. To 20–30ft (6–9m), spreading. Leaves oval to oblong, simple, silky at first, red-copper. Flowers in large, lax bunches. Fruit small, purplish-black, containing several seeds. Prefers lime-free soils. Susceptible to pests as *A. canadensis*. 206, *206*

American Aspen *see POPULUS tremuloides*
American Beech *see FAGUS grandifolia*
American Elm *see ULMUS americana*
American Holly *see ILEX opaca*
American Hornbeam *see CARPINUS caroliniana*
American Mountain Ash *see SORBUS americana*
American Sycamore *see PLATANUS occidentalis*
American Walnut *see JUGLANS nigra*
American Yellowwood *see CLADRASTIS lutea*
Amur Cork Tree *see PHELLODENDRON amurense*
Amur Maple *see ACER tataricum* subsp. *ginnala*
Apricot *see PRUNUS armeniaca*

ARALIA
American and Asiatic deciduous shrubs or trees. Bisexual flowers in clusters. Small berry-like fruits. Mostly hardy. All aralias do well under city conditions. 349
A. elata Japanese angelica tree. Japan. Zone 4. To 50ft (15m). Most often a shrub renewing by suckers. Leaves compound, 3–4ft (1–1.2m) long, 2ft (60cm) wide, many oval, pointed leaflets, dark bright green. Flowers small, white, in large panicles. Black fruits. Sun or semi-shade. Fast-growing.
A.e. **'Aureovariegata'** Leaflets blotched and margined with yellow, fading to cream. 349
A.e. **'Variegata'** Leaves green and white. Both variegated clones are slow-growing and apt to produce reverted suckers. 349

ARAUCARIA
Evergreen trees, big, conical or rounded. Leaves oval, spirally arranged, leathery, overlapped dark green. Male and female cones on separate trees usually; male cones cylindrical, possibly clustered. Cones round to oval. Seeds winged. 22, 122–5, 127
A. angustifolia Parana pine, Candelabra tree. S Brazil. Zone 9. To 110ft (33m), few branches, flat crown with long hanging branchlets. Leaves scale-like with curved spiny tips. 122
A. araucana Chile pine, Monkey puzzle. Chile, Argentina. Zone 8. To 80ft (24.5m). Leaves triangular, long, yellow-green, hard, pointed. Male cones drooping at end of shoots, 6in (15cm) long, cylindrical, pollen in summer, female round, on upper side of branches, ripening after 3 years dark green with long, yellowish spines. Disease-free. Tolerant of all but driest and most boggy soils. 20ft (6m) in 20 years. 122, *122–3*, *124*, *125*
A. bidwillii Bunya bunya. Queensland. Zone 9. To 150ft (45m), young trees broadly conical. Leaves overlapped, stiff. Male cones to 7in (17.5cm) long. Cones upright, round, 12in (30cm) long. Fairly resistant to insect pests. 122, 124, *124*
A. excelsa (syn. *A. heterophylla*) Norfolk Island pine. Norfolk Island. Zones 10–11. To 200ft (60m), with symmetrical branches when young, ragged in age. Leaves small, narrow. Cones rounded, 4in (10cm) across. 122–4, *124*

Arborvitae *see THUJA*

ARBUTUS
Small to large evergreen trees. Leaves simple, dark, shiny green, leathery. Flowers white or pink, pitcher-shaped. Fruits orange-red, globular, fleshy, edible but tasteless. Subject to crown die-back. Leaf canker in America. 332–5
A. andrachne Grecian strawberry tree. E Mediterranean region. Zone 8. To 35ft (10.5m) with cinnamon-brown stems. Leaves simple, ovallish. Flowers dull white, in broad clusters, March and April. Fruit ½in (1cm). 335
A. **x *andrachnoides*** Hybrid between *A. andrachne* and *A. unedo*. 20–30ft (6–9m), with cinnamon-red branches. Leaves dark green, toothed. Flowers white, pitcher-shaped, ¼in (6mm) long in clusters, late autumn or spring. Fruit smaller and smoother than *A. unedo*. Tolerant of lime. Zone 8. *334*, 335
A. menziesii Madrone. W North America. Zone 7. To 100ft (30m). Leaves oval, medium length, usually with entire panicles 3–9in (7.5–23cm) long, whitish, small. Fruits pea-sized berries. Transplants badly. Soils should be lime-free but can be poor and dry. Sun or semi-shade. 35ft (10.5m) in 20 years. *332–3*, 335, *335*
A. unedo Killarney strawberry tree. Mediterranean region, SW Ireland. Zone 8 (Zone 7 with shelter). To 40ft (12m). Leaves medium length, narrow, oval. Flowers late autumn, white, in drooping clusters, small. Fruit strawberry-like, late autumn. Sun or semi-shade. Tolerant of dry and limey soils. 15ft (4.5m) in 20 years. 332–5, *332*, *334*

ARCHONTOPHOENIX
alexandrae King palm. Australia. Zone 10. To 70ft (21.5m), trunk 6ft (1.8m) thick, surface ringed with leaf scars. Short-stalked

often black-dotted beneath, leathery, dark green, short-stalked. Flowers single, red, waxy, possibly slightly fragrant, late winter. 15ft (4.5m) in 20 years. *52, 326, 329*
C. reticulata Yunnan, China. Zone 8. To 35ft (10.5m), evergreen tree or compact shrub. Rigid leathery leaves, dark dull green. Flowers rose–pink. March. Prefers slightly acid soil. *326*
C. sasanqua Japan. Zone 7. Shrub or small tree. To 20ft (6m). Leaves narrowly oval, dark green, leathery. Flowers white, fragrant, early winter. Needs shelter. 12ft (3.5m) in 10 years. *326*
C. sinensis Tea tree. Assam to China. Zone 7. To 45ft (13.5m), compact shrub or tree. Leaves variable, broad to lance-shaped. Flowers white, nodding, spring. Slow-growing. *326, 326–7*

Camphor Laurel see *CINNAMOMUM camphora*
Canary Island Date Palm see *PHOENIX canariensis*
Candelabra Tree see *ARAUCARIA angustifolia*
Canoe Birch see *BETULA papyrifera*
Canyon Live Oak see *QUERCUS chrysolepis*
Caribbean Cabbage Palm see *SABAL palmetto*

CARICA papaya Papaya (also pawpaw). Tropical America. Zone 10+. Often unbranched tree with softy trunk showing large leaf scars. Leaves large, palmate and deeply lobed. Flowers of different sexes on same tree, creamy-white, the females borne on the trunk much larger than males on long sprays. Fruits the familiar papaya of the tropics, greenish yellow ioutside, orange within, with many black seeds: the perfect breakfast fruit. *177*

Carob see *CERATONIA siliqua*
Carolina Hemlock see *TSUGA caroliniana*

CARPINUS
Hardy deciduous trees or shrubs. Leaves alternate, sharply toothed, often doubly so. Hanging male catkins, spring. Narrow female catkins. Fruit in small ribbed nut, ripe in autumn. Bark smooth or scaly grey. Easily grown. Clay- and chalk-tolerant. *282–3*
C. betulus Common hornbeam, European hornbeam. Europe, Asia Minor. Zone 5. To 75ft (23m), broadly conical, becoming rounder. Leaves oval, tapering; dark green, downy beneath, turning yellow and smooth by autumn. Unisexual flowers. Male catkins 1½in (4cm); fruiting catkins 1½–3in (4–7.5cm), with large, 3-lobed bracts. 35ft (10.5m) in 20 years. *75, 276, 282, 282*
C.b. 'Fastigiata' Medium-sized tree, erect, conical, broadening with age, 30ft (9m) in 20 years. *282*
C. caroliniana American hornbeam, Blue beech, Ironwood.

E United States. Zone 5. To 40ft (12m), branches spreading, arching at tips. Leaves oval, bright green, downy when young, sparsely hairy later; scarlet and orange in autumn. Male catkins 1–1½in (2.5–4cm) long. Fruiting clusters 3in (7.5cm) long. 15ft (4.5m) in 20 years. *282*
C. cordata Japan, NE Asia, N and W China. Zone 5. To 40–50ft (12–15m). Leaves heart-shaped, pointed, simple, deeply veined, hairy below, slightly so above. Male catkins with long hairs, female catkins 3in (7.5cm) long. Fruits small, green, clustered. *283*
C. fangiana Monkey-tail Hornbeam. China. Zone 7–8. Tree to 65ft (20m). Leaves large, strongly ribbed, soft green. Female catkins develop into long fruiting bodies to 20in (50 cm) long. *283, 283*
C. japonica Japanese hornbeam, Japan. Zone 8. To 50ft (15m), widespreading. Leaves oval, tapering; medium length, upper surface dark green, lower surface downy on veins, corrugated. Male catkins 1–2in (2.5–5cm) long with conspicuous scales. Fruit clusters to 2½in (6.5cm) long, persist on tree. 15ft (4.5m) in 20 years. *283*

CARYA
Big deciduous trees related to walnuts. Pinnate leaves. Sexes on same tree. Male flowers in branched, slender catkins; female flowers in small clusters. Fruit, nut surrounded by a husk. Deep loamy soil. Subject to canker, leaf blotch and scab in America; pest-free in Europe. Resent transplanting. *46, 47, 284, 286–9*
C. cordiformis Bitternut. E United States. Zone 4. To 90ft (27.5m), broad, rounded. Winter bud scales yellow. Leaves long, 5–10 oval, toothed leaflets, yellow in autumn. Male flowers downy. Fruit in 2s or 3s. 25ft (7.5m) in 20 years. *288, 289*
C. glabra Pignut. E United States. Zone 4. To 120ft (36m), narrow to round. Leaves long, 5–7 oval, toothed leaflets, yellow in autumn. Nuts vary in size and shape. *288, 288*
C. illinoinensis Pecan. E United States. Zone 5. To 150ft (45m), round head, huge branches. 11–17 toothed, pointed leaflets, yellow in autumn. Fruits clustered, oblong, edible. *286, 286–7, 289*
C. ovata Shagbark hickory. E United States. Zone 4. To 120ft (36m), narrow and upright, irregular, grey bark shedding in strips. Leaves long, 5–9 leaflets, hairy beneath, rich yellow to golden brown in autumn. Male catkins hairy, in 3s. Fruits roundish. 20ft (6m) in 20 years. *286–8, 289*
C. tomentosa Mockernut, Bigbud hickory. SE Canada, E United States. Zone 4. To 90ft (27.5m), upright with a round head. Leaves long, 5–9 toothed, pointed leaflets, dark green above,

yellowish, downy below. Male catkins downy, 3–5in (7.5–13cm) long. Fruit roundish. 20ft (6m) in 20 years. *288, 289*

Cashews *312–13*

CASTANEA
Leaves toothed, deciduous. Flowers, sexes on separate catkins; males long, hanging, females shorter; both pale yellow, July. Fruit, nuts surrounded by prickly husks. Tolerant of shade. Not suited to shallow, chalky, or sandy soils. Subject to chestnut blight in America, die-back, leaf spot. *256, 274–5, 274*
C. crenata Japanese chestnut. Japan. Zone 4. To 30ft (9m). Leaves narrow oval, often hairy beneath. Nuts in 1in (2.5cm), 2 or 3 in each husk. *274–5*
C. dentata American sweet chestnut. E United States. Zone 4. To 100ft (30m), upright, broad. Leaves hairless, dull green; narrow oblong, tapering. Flowers in catkins, unisexual. Fruit 1–3 nuts. 15ft (4.5m) in 20 years. *274*
C. mollissima Chinese chestnut. China, Korea. Zone 4. To 65ft (20m), rounded head. Leaves oval, tapering, sometimes hairy, with short, hairy stalk. Usually 2 or 3 nuts in husk. 10ft (3m) in 20 years. Resists chestnut blight. *274–5*
C. sativa Spanish chestnut, Sweet chestnut. S Europe, N Africa, Asia Minor. Zone 5. To 100ft (30m), spreading. Leaves oblong, tapering hairy beneath at first. Flowers yellowish-green, unpleasant-smelling. Ornamental, especially in flowers. Nuts edible. 35ft (10.5m) in 20 years. *254, 274, 275*
C.s. 'Albomarginata' Leaves with white margin, fruits with white spines. *275*

CASTANOPSIS cuspidata Japan, Korea. Zone 7. To 50ft (15 m). Evergreen tree between oak and chestnut. *52, 274*

CASUARINA She-oaks. Australia, Pacific Region. Often coastal trees with pendulous branches bearing long green twigs with tiny leaves, resembling the shoots of an Equisetum. *189*

CATALPA
Deciduous. Big, opposite leaves, long-stalked, sometimes lobed. Flowers in erect, branched clusters, bisexual. Long, narrow, bean-like seed capsule. All well-drained soils. Sun-loving. Subject to leaf spot in America. *346, 349*
C. bignonioides Indian bean tree, Southern catalpa. E United States. Zone 5. To 65ft (20m), widespreading. Leaves broadly oval, tapering, light green above, pale downy beneath, disagreeable odour when crushed. Flowers, white with yellow and purple throat. 20ft (6m) in 20 years. *346–8, 348*
C.b. 'Aurea' Golden Indian bean tree. Zone 4. Leaves rich yellow when young. *348, 348–9*

C. x erubescens To 30ft (9m). Leaves ovallish, 3-lobed, and whole on same tree, purple when unfolding. Flowers similar to *C. bignonioides* but smaller and more numerous, late July. *348*
C. x e. 'Purpurea' Shoots and leaves black-purple, becoming greener. *348*
C. fargesii Zone 6. Smaller-leaved catalpa with excellent summer flowers.
C.f. f. duclouxii A form with wonderfully soft and smooth hairless leaves with longer points. *348, 349*
C. ovata Zone 5. Leaves usually 3-lobed. Small white flowers with yellow and red throat markings. *348*
C. speciosa Western catalpa. C United States. Zone 5. To 100ft (30m). Leaves heart-shaped, tapering, lower side downy. Flowers slightly larger than *C. bignonioides*, less spotted with purple. 20ft (6m) in 20 years. *346–8, 347*

Caucasian Fir see *ABIES nordmanniana*
Caucasian Maple see *ACER cappadocicum*
Caucasian Oak see *QUERCUS macranthera*
Caucasian Wingnut see *PTEROCARYA fraxinifolia*
Cedar see *CEDRUS*
Cedar of Goa see *CUPRESSUS lusitanica*

CEDRELA odorata West Indian cedar. Tropical/sub-tropical tree with aromatic wood, from Caribbean, now naturalized elsewhere. *315*

CEDRUS
Cedars. Hardy evergreen, initially conical, spreading later. Leaves needle-like, in rosettes on old shoots, single on new. Male and female cones usually on same trees, males yellow, profuse. Cones barrel-shaped, break up on tree after 2 years. Seeds winged. Few fungal diseases, susceptible to drought.
C. atlantica Atlas cedar. Atlas Mts. Zone 6. To 120ft (36m), pyramidal when young. Leaves slightly shorter than *C. libani*, sometimes more waxy and more numerous in the rosettes. Male cones erect sugarloaf-shape, pale green in summer, purplish Sept; females green, tinged pink in centre of rosette. 40ft (12m) in 20 years. *114, 114*
C.a. 'Glauca' Blue cedar. To 120ft (36m), more pointed, less liable to have multiple stems. Leaves light blue or waxy. Bark pale grey. 40ft (12m) in 20 years. *114*
C. brevifolia Cyprus cedar. Mts of Cyprus. Zone 7. Narrowly conical. Leaves shorter than the other species, rich green to blue- or yellow-green. Cones smooth pale green. 20ft (6m) in 20 years. *115, 115*
C. deodora Deodar. W Himalaya. Zone 5. To 200ft

(60m), pendulous. Leaves longer than other species, dark green, waxy or silvery, sharply pointed. Cones less common than in other species. 45ft (13.5m) in 20 years. *68, 114, 115*
C. libani Cedar of Lebanon. Asia Minor and Syria. Zone 5. To 70–100ft (21.5–30m), often erect or ascending, flat-topped with age. Leaves short, green, or waxy. 30ft (9m) in 20 years. *112–14, 113, 114*
C.l. subsp. *stenocoma* Turkey. Zone 5. Narrow-shaped tree from Taurus mts, hardier than subsp. libani, so more suitable for colder gardens. *112*

CELTIS
Medium to large, related to elms. Leaves deciduous, simple. Flowers small, greenish. Separate sexes on same tree, male flowers clustered, female flowers in groups of 1–3. Fruits fleshy, single-stoned, sweet, ripe in autumn. Subject to leaf spot where native, witches' broom in America. Any soil. *248–9*
C. australis European nettle tree. S Europe, N Africa, Asia Minor. Zone 6. To 80ft (24.5m), round-headed. Leaves toothed, narrowly ovate, soft down beneath, harsher above, short-stalked. Sun-loving. 10ft (3m) in 20 years. *249, 249*
C. laevigata Sugar hackberry, Mississippi hackberry. SE United States. Zone 5. To 100ft (30m), upright and spreading. Leaves oval to lance-shaped, smooth, dark green above, paler with some hairs beneath, few teeth. Male flowers separate from solitary females. Fruits small, orange to black. Sun-loving. Fairly resistant to witches' broom. 12ft (3.5m) in 20 years. *249*
C. occidentalis Hackberry, Sugarberry. N United States. Zone 2. Usually to 40ft (12m), occasionally much more. Leaves oval to oblong, possibly downy, short-toothed near base. Flowers early spring. Fruit ripens purple, small, profuse. Bark warty, corky, rough. 12ft (3.5m) in 20 years. *249*
C. sinensis Chinese hackberry. E China, Korea, Japan. Zone 9. To 65ft (20m), spreading. Leaves shiny, dark green, broad oval. Flowers as *C. laevigata*. Fruit small, red or yellow, edible. Sun-loving. Resistant to witches' broom. 20ft (6m) in 20 years. *249*

Cembran Pine see *PINUS cembra*

CEPHALOTAXUS
Hardy evergreen shrubs or small trees. Leaves narrow, pointed, dark green above, obvious midrib. Sexes usually on separate trees; male cones round, females scarce, cup-shaped. Fruit oblong, fleshy and brown. Shade tolerant. *47, 126*
C. fortunei Chinese plum yew. C and SW China. Zone 7. To 30ft (9m), slender or bushy. Medium length leaves. Male cones on short stalks below shoot, females on separate trees. Small fruit, oval,

green-white vertically striped green. lime tolerant. 12ft (3.5m) in 20 years. 126, *129*

C. harringtonia* var. *drupacea Cow's tail pine, Japanese plum yew. Japan, C China. Zone 7. To 20–40ft (6–12m), widespreading, bushy crown, rounded. Leaves short, in double ranks, forming narrow V-shaped channel on upper surface of branchlets. Male cones small, short scaly stalks. Fruit egg-shaped, oily, brown, or olive-green. Thrives on chalk soils. 8ft (2.5m) in 20 years. 126, *127*

CERATONIA siliqua Carob. St John's bread. Mediterranean. Zone 10. To 50ft (15m), rounded shrub or small tree. Leaves evergreen, pinnate but lacking terminal leaflet, glossy, dark green. Male and female flowers on separate trees, small, red, clustered, spring. Fruits, pods 1ft (30cm) oblong, edible, early autumn 200

CERCIDIPHYLLUM

C. japonicum Katsura tree. Japan, China. Zone 5. To 50ft (15m). Pendulous branches, often with several stems. Leaves simple, deciduous, heart-shaped, similar to Judas tree (*Cercis*), but opposite, smaller and toothed, superb varying autumn colours, fallen leaves smell of burnt sugar. Flowers, sexes on separate trees, females small with green sepals. Small clustered fruit pods. Chalk-tolerant, but not reliably coloured on these soils; tolerates dry conditions. Sun or semi-shade. 40ft (12m) in 20 years. 56, 182, 184, *184*

***C.j.* 'Heronswood Globe'** 184

***C.j.* 'Pendulum'** Elegant weeping shape. 184

***C.j.* 'Rotfuchs'** Stiffly-erect branches, red foliage. 184

CERCIS

Small group of trees and shrubs. Leaves simple, deciduous, heart-shaped, medium length, untoothed, bisexual. Flowers usually in dense clusters on stems, before or with leaves. Fruits narrow, flat pods holding several flattened seeds. Prefers well-drained, loamy soils. Sun-loving. Hardy. Subject to coral spot fungus, leaf spot, and canker in America. 202

C. canadensis Redbud. SE Canada, E United States, NE Mexico. Zone 4. To 40ft (12m) broad, rounded head. Leaves bright green. Flowers pale rose, spring. Acid and alkaline soils. Less free-flowering than *C. siliquastrum* in Britain. 15ft (4.5m) in 20 years. 202, *203*

***C.c.* 'Forest Pansy'** More shrubby and less vigorous than the species, to 10ft (3m) with spreading habit. Leaves dark reddish-purple, retaining colour well in summer. 202

C. racemosa China. Small tree to 30 ft (9m), similar to other Cercis, but has hanging clusters of pale pink flowers below the leaves. Not easy to grow. 202

C. siliquastrum Judas tree. E Mediterranean. Zone 6. To 30ft (9m), round-topped, bushy. Leaves roundish, waxy green. Flowers rose-purple, April–May. Pods red, Oct. through winter. 25ft (7.5m) in 20 years. 202, *203*

CHAMAECYPARIS

False cypresses. Evergreen, conical, becoming widespread. Leaves scale-like, small, awl-shaped in early stages. Male and female cones separate on same tree; males oval to oblong, tiny, red or brown; cones round. Seeds winged. Dislike dry, chalky soil. Subject to root rot and rusts in America. 138–41

C. henryae Mrs Henry's false cypress. A form of C. thyoides from the southeastern USA, with slight differences such as larger leaves, now included in C. thyoides by botanists. 139

C. lawsoniana Lawson cypress, Port Orford cedar. SW Oregon, NW California. Zone 6. To 175–200ft (52–60m). Leaves in 4s, 2 unequal pairs, leaf sprays flattened. Male cones pink or red. Cones ⅓in (7mm) wide, red-brown in autumn. Tolerates sunshine or shade. 35–45ft (10.5–12m) in 20 years. 104, 133, 134, 138–9, *139*, 140, 148

***C.l.* 'Alumii'** A spire-like blue-grey variety. To 60ft (18m). *140*

***C.l.* 'Columnaris'** Upright, narrow. To 30ft (9m). Leaves a good blue-grey. 25ft (7.5m) in 20 years. 139, *140*

***C.l.* 'Ellwoodii'** To 25ft (7.5m), upright. Leaves juvenile, feathery, turning grey-green. 15ft (4.5m) in 20 years. 139

***C.l.* 'Erecta'** Branches short, upright, deep green and erect; branching from base of tree. Cones narrow. 30ft (9m) in 20 years. 134, *140*

***C.l.* 'Fletcheri'** Columnar, often many tops. Foliage blue-grey and green, feathery, juvenile. 20ft (6m) in 20 years. 139, *140*

***C.l.* 'Imbricata Pendula'** *139*

***C.l.* 'Intertexta'** Large open tree, with b road sprays of foliage. 139, *140*

***C.l.* 'Kilmacurragh'** Narrow-columnar, like a rich green Italian cypress. To 40ft (12m).

***C.l.* 'Lanei'** Medium–small, columnar, dense. Leaves bright gold in ferny sprays. Male "flowers" red. 30ft (9m) in 20 years. 139

***C.l.* 'Lutea'** To 40ft (12m), columnar, level branches, drooping shoots. Winter colour good. 25ft (7.5m) in 20 years. 139, *140*

***C.l.* 'Pembury Blue'** To 30ft (9m), conical, very attractive, loosely branched bushy tree. Silver-blue leaves. 12ft (3.5m) in 20 years. 139

***C.l.* 'Pottenii'** An upright form with very soft apple-green foliage, partly juvenile. To 30ft (9m). 139

***C.l.* 'Spek'** Upright with good grey-blue coloration. 139

***C.l.* 'Stewartii'** Commonly planted conical golden variety turning yellow-green in winter. To 50ft (15m). *140*

***C.l.* 'Triomf van Boskoop'** To 75ft (23m), coarse, open sprays of grey-blue, broadly columnar. Good bole. 35ft (10.5m) in 20 years. 139

***C.l.* 'Wissellii'** To 60–85ft (18–27m), columnar, branches upward-twisting. Leaves deep blue-green, foliage bunched. Male cones prominent crimson. Very vigorous after slow start. 35ft (10.5m) in 20 years. *138, 140*

C. nootkatensis Nootka cypress, Yellow cypress. W United States. Zone 5. To 120ft (36m), weeping foliage. Leaves hard, rough, dull green, sprays in drooping clusters. Male cones small, yellow, spring. Cones green-brown, late spring or early summer. Withstands any soil, considerable exposure. 25–50ft (7.5–15m) in 20 years. The correct generic name for this tree is in dispute; it is not a true *Chamaecyparis* 54, 140–1

***C.n.* 'Pendula'** To 60ft (18m), raised branches, strongly weeping shoots. Leaves dull green. Cones navy-blue first year. 30ft (9m) in 20 years 140

C. obtusa Hinoki cypress. Japan. Zone 6. To 50–75ft (15–23m), broadly conical. Fine branching system bears blunt-tipped leaves in unequal pairs, 1 opposite pair larger than other, shiny bright green, leaf margin white below. Male cones tiny, terminal. Cones green, orange-brown when ripening. Prefers moist soils and air. 25ft (7.5m) in 20 years. 139–40

***C.o.* 'Crippsii'** Broadly conical. Leaves in fronds, drooping slightly at tip, bright golden yellow. 15ft (4.5m) in 20 years. 139

***C.o.* 'Filicoides'** 140

***C.o.* 'Lycopodioides'** Small, sparsely branched, open-crowned. Leaves closely pressed to shoot, dark green, blunt-tipped, moss-like. 12ft (3.5m) in 20 years. 140

***C.o.* 'Nana Gracilis'** To 15ft (4.5m). Shell-shaped sprays of dark green foliage. 6ft (1.8m) in 20 years.139

***C.o.* 'Tetragona Aurea'** To 35ft (10.5m). Young trees gaunt, level branches upswept at tips with yellow to bronze foliage in dense 4-angled clusters, widespreading. 15ft (4.5m) in 20 years. 140

C. pisifera Sawara cypress. Japan. Zone 6. To 50–75ft (15–23m), conical. Leaves in upturned spreading sprays, narrow-pointed, deep shiny green, becoming darker, white markings underneath. Male cones minute, pollen shed in spring. Cones abundant, small, green turning brown. Prefers lime-free soils. Fairly slow. 139

***C.p.* 'Boulevard'** United States. 12ft (3.5m). Leaves mossy, densely bunched silver-blue; coloration best in summer and in slight shade. 6ft (1.8m) in 10 years. 139, *140*

***C.p.* 'Cumulus'** Dwarf, forming soft rounded green cushion. *158*

***C.p.* 'Filifera Aurea'** To 30–40ft (9–10.5m), pyramidal or conical, spreading branches, hanging sprays of golden yellow foliage. *140*

C. thyoides White cypress. E United States. Zone 5. To 70–90ft (21.5–27.6m), conical. Tiny leaves laterally paired, sharp-pointed, with resin gland on reverse of leaves, waxy green, white edged. Male flowers tiny, dark brown. Cones small, waxy, blue-purple becoming red-brown. No serious pests or diseases. 25ft (7.5m) in 20 years. 139

***C.t.* 'Andelyensis'** To 20ft (6m), narrow, columnar sometimes with multiple spires. Branches crowded with short, blue-green aromatic leaves, turning bronze in winter. Bears a mass of pink male cones and tiny cones. Prefers lime-free soils. Slow. 139

***C.t.* 'Ericoides'** Low, compact, pyramidal, usu. reddish-brown in winter. 139

***C.t.* 'Rubicon'** Compact dwarf juvenile form, reddish in winter. *140*

CHAMAEROPS humilis European fan palm. Mediterrranean. Zone 9. A clump-forming palm to 6ft (2m), rarely more. Palmate leaves 2–3ft (60cm–1m) across on long, slender stalk. Sexes on separate trees, small yellow flowers. Brown or yellow rounded fruit ½–1½in (1–4cm). 162, *163*

CHAENOMELES japonica Japanese quince. Japan. Shrub with orange-red flowers in early spring. Most plants under this name are really C. speciosa, from China. 210

Cherry *see PRUNUS*
Cherry Birch *see BETULA lenta*
Cherry Laurel *see PRUNUS laurocerasus*
Cherry Plum *see PRUNUS cerasifera*
Chestnut *see CASTANEA*
Chestnut-leafed Oak *see QUERCUS castaneifolia*
Chile Pine *see ARAUCARIA araucana*
Chilean Cedar *see AUSTROCEDRUS chilensis*
Chilean Firebush *see EMBOTHRIUM*
Chilean Wine Palm *see JUBAEA chilensis*
Chinaberry *see MELIA azedarach*
Chinese Ash *see FRAXINUS chinensis*
Chinese Chestnut *see CASTANEA mollissima*
Chinese Cork Oak *see QUERCUS variabilis*
Chinese Elm *see ULMUS parvifolia*
Chinese Hackberry *see CELTIS sinensis*
Chinese Juniper *see JUNIPERUS chinensis*
Chinese Plum Yew *see CEPHALOTAXUS fortunei*
Chinese Silver Lime *see TILIA oliveri*
Chinese Thuja *see PLATYCLADUS orientalis*
Chinkapin, Golden *see CHRYSOLEPIS chrysophylla*
Chinkapin Oak *see QUERCUS muehlenbergii*

CHIONANTHUS

Deciduous trees or shrubs. Leaves simple, opposite. Flowers, sexes on separate trees. Pure white, in loose, branched clusters. Fruit egg-shaped or oblong, fleshy dark blue with single seed. Moist loam and sun. 345

C. retusus Chinese fringe tree. China. Zone 6. To 20ft (6m), spreading. Leaves oval, entire. Flowers in broad heads, petals strap-shaped, June–July. Fruits ½in (1cm) long, blue-mauve, Sept.–Oct. 8ft (2.5m) in 20 years. 345, *345*

C. virginicus Fringe tree. E United States. Zone 4. To 30ft (9m), shrub or tree, spreading. Leaves narrow, oblong, bright green above, paler, slightly downy beneath, yellow in autumn. Flowers in May–June. Fruit ½–¾in (1–2cm) long, Sept. 8ft (2.5m) in 20 years. 345

Choke Berry *see PRUNUS virginiana*

CHRYSOLEPIS chrysophylla Golden chinkapin. W United States. Zone 8. To 100ft (30m), pyramidal. Leaves leathery, pointed, dark green above, yellow beneath, evergreen. Flowers creamy-white unisexual on same tree, fluffy spikes, June and July. Fruit, green prickly husk 1–3 nuts. Prefers well-drained acid or neutral soils. Tolerates drought. 274, 275, *275*

Chusan Palm *see TRACHYCARPUS fortunei*
Cider Gum *see EUCALYPTUS gunnii*

CINNAMOMUM

C. camphora Camphor laurel. Tropical Asia, China, Japan. Zone 9. To 100ft (30m), upright. Leaves slender, oval, simple, evergreen, shiny, fragrant, stalked. Flowers greenish-white in clusters, spring. Sun-loving, needs shelter. 174, *176, 177*

C.verum Cinnamon. Sri Lanka, southern India. Zone 10+. The aromatic bark is the spice cinnamon. Tree is similar to the above. *176*

CITRUS

Evergreen or semi-evergreen trees or shrubs. Leaves simple and aromatic, leafstalk sometimes winged. Flowers usually fragrant, Fruits are oranges, lemons, etc. Sun-loving. Medium growth. 316–19

C. aurantiifolia Lime. India, SE Asia. Zone 9. Small, spreading. Leaves ovate, evergreen, shiny green, edges slightly crinkly. Flowers in small clusters, white. Fruit round-oblong, pale green, acid-tasting, ripens irregularly. Pests as *C. sinensis*. 316–18, *317*

C. aurantiifolia* x *Fortunella marginata Limequat. Zone 10. Branches become weighed down by fruits. Leaves lance-shaped, evergreen, thick, dark green above, paler beneath. Fruits oval, light yellow, shiny peel, acid-tasting. Almost hardy, but needs a long, warm summer. 319

C. aurantium Seville orange. S Europe. Zone 9. To 20–30ft (6–9m), spreading. Leaves oval, pointed, evergreen, winged stalk. Flowers small, white, in flat heads.

Index of Tree Species 379

Fruit round, orange-red, aromatic, bitter, acid. Hardier than *C. sinensis*. 316, *318*
C. limon Lemon. India. Zone 9. To 20ft (6m), spreading. Leaves ovate, evergreen, smooth, dark green, crinkly-edged. Flowers white, slightly flushed, single or paired. Fruit ripens irregularly. Pests as *C. sinensis*. 316, *318*
C.l. 'Meyer' Hardiest lemon, from China. 319
C. x *paradisi* Grapefruit. S E China. Zone 9. To 30–50ft (9–15m), pyramidal. Leaves ovate, evergreen, light green becoming darker, smooth, edges crinkly, stalked. Flowers in flat heads. Fruits large, pale yellow to orange, flavour pleasant, acid, bitter. Pests as *C. sinensis*. *318*, 319
C. reticulata Mandarin orange, tangerine. China. Zone 9. Small, spreading. Leaves narrow to oval, pointed, evergreen, dark green, stalked. Fruit flattened at both ends, smaller than *C. sinensis*, yellow to orange-red, sweet-tasting. Hardier than *C. sinensis*. Pests as *C. sinensis* 316
C.sinensis Sweet orange. Cochin, China. Zone 9. To 25–40ft (7.5–12m), pyramidal. Leaves ovate, evergreen, smooth, shiny, dark green above, lighter beneath. Flowers white, solitary or in small clusters. Fruit roundish, orange or orange-red, smooth, large. Pests included scales, aphids, mites, and also fungi and bacterial diseases 316, *316–17*, *318*
C. x *tangelo* 'Ugli' Ugli tree. Zone 9. Upright and straggly tree. Leaves ovate, dark green. Flowers white. Fruit usually light lemon-yellow flushed with apple-green, roundish with thick, nobbly peel, sweetish, flavour between tangerine and grapefruit. 319
C. trifoliata (formerly Poncirus trifoliata). Hardy orange. Zone 5. Small tree to about 15–20ft (5–7m), with gren stems bearing vicious spines. Leaves trifoliolate, dark gren, turning gold in autumn before falling. Flowers white in spring, not scented, followed by hard, inedible but fragrant fruits. 319
C.t. 'Flying Dragon' A form with contorted stems, striking in winter. 319

CLADRASTIS
Small group of deciduous trees. Leaves pinnate, leaflets big. Flowers bisexual, white, rarely pinkish, in flat-topped clusters, buds enclosed in swollen base of leafstalk. Fruit narrow, oblong pods, flattened. Wood brittle. Sun-loving.
C. lentukea see *C. lutea*
C. lutea (syn. *C. kentukea*) Yellowwood. S E United States. Zone 3. To 65ft (20m), rounded, spreading. Leaves long, wide oval leaflets, hairless, turning yellow before falling, stalked. Flowers slightly fragrant, June. Seed pods, autumn. Loamy soil; not long-lived on chalk. 199

Claret Ash *see FRAXINUS angustifolia* var. *oxycarpa* 'Raywood'

CLETHRA
Small ericaceous trees or shrubs; deciduous and hardy or evergreen and tender. Leaves alternate, simple. Flowers fragrant, white, usually in clusters. Fruit is a capsule holding many seeds. Lime-free soil. 335
C. alnifolia Sweet pepper bush. Eastern N. America. Zone 3. Deciduous shrub to about 8ft (2.5 m). Flowers white in late summer, sweetly scented. *335*
C. arborea Lily-of-the-valley clethra. Maderia. Zone 9. To 30ft (9m), large shrub or small tree. Leaves lance-shaped, evergreen, finely toothed, dark green above, pale with scattered hairs below. Flowers ½in (1cm) wide. Requires mild area, sun or semi-shade sheltered position. 30ft (9m) in 20 years. 335, *335*

Coast Live Oak *see QUERCUS agrifolia*
Coast Redwood *see SEQUOIA sempervirens*
Coconut Palm *see COCOS nucifera*

COCOS nucifera Coconut palm. Tropics. To 80ft (24.5m), always leaning; trunk thickened at base, surface ringed. Pinnate leaves 10–20ft (3–6m) long, yellowish green. Unisexual flowers on same tree, on 4–6ft (1.2–1.8m) long stalk. Fruit is a big nut surrounded by tough fibrous husk. *161*, 162, *162–3*
Coigue *see NOTHOFAGUS dombeyi*
Colorado Spruce *see PICEA pungens*
Colorado White Fir *see ABIES concolor*
Cork Oak *see QUERCUS suber*
Cornelian Cherry *see CORNUS mas*

CORNUS
Deciduous, seldom evergreen trees or shrubs. Simple leaves, usually opposite. Small, bisexual flowers mostly white, in clusters. Oval, fleshy fruits containing a stone. Normally easy to cultivate. Subject to mildew and anthracnose in Britain, leaf spot, crown canker, and anthracnose in America. 105, 320–3
C. alternifolia 'Argentea' A white-variegated form, extremely graceful and eye-catching. To 15ft (4.5m). 323
C. capitata Bentham's cornel, Evergreen dogwood. Himalaya. Zone 8. Evergreen small tree, bushy. Young shoots downy. Leaves leathery, oval, lance-shaped, dull grey-green, hairy. Minute flowers surrounded by bracts, sulphur-yellow. Fruit strawberry-shaped. 20ft (6m) in 20 years. 323, *323*
C. chinensis Chinese cornelian cherry. China. Zone 8–9. Resembles a large C. mas with elongated leaves and larger flowers. *323*
C. controversa Giant dogwood. Japan, China. Zone 5. To 65ft (20m), horizontal branches in tiers. Leaves alternate, oval, pointed, dark glossy green above, waxy beneath. Flowers white, in flattish heads, profuse. Fruit blue-black, roundish. Chalk-tolerant. 25ft (7.5m) in 20 years. 320, 322, *323*
C.c. 'Variegata' Zone 5. Small tree, form similar to type. Leaves with silver-white to pale yellow border. 12ft (3.5m) in 20 years. 323, *323*
C. 'Eddie's White Wonder' Zone 6. A showy American hybrid (*C. florida* x *C. nuttallii*) with giant white flowers in spring. 322
C. florida Flowering dogwood. E United States. Zone 5. To 40ft (12m), widespreading. Leaves oval, dark green with scattered down above, pale beneath. Flowers insignificant, spring, but 4 white bracts, 2in (5cm) long, expand in May after enclosing bud in winter. Red berries in winter. Sun or semi-shade. Dislikes poor, shallow chalk soils. 10ft (3m) in 20 years. 320, *320–1*, *325*
C.f. f. *rubra* Variable, flower bracts rosy-pink. Less hardy than the type. *322*
C. hongkongensis Hong Kong dogwood, though found through much of southern and western China. Zone 8–9. Small evergreen tree resembling C. capitata. Likes hot summers. 323
C. 'Kn30–8' Venus Hybrid dogwood with enormous 'flowers' to 8in (20 cm) across. 322
C. kousa Japanese dogwood. Japan, Korea. Zone 5. To 25ft (7.5m), shrub or small tree. Leaves oval, pointed, crimson-bronze in autumn. Flowers small, inconspicuous, attractive large white bracts. Fruits strawberry-like. Dislikes poor, shallow, chalky soils. Moderately sunny location. 12ft (3.5m) in 20 years. 56, 320, *323*
C.k. var. *chinensis* Chinese dogwood. China. Taller, more open than *C. kousa*. Leaves slightly bigger, flowers bracts slightly longer. 320–2
C. mas Cornelian cherry. C and S Europe. Zone 5. To 25ft (7.5m), shrub or small tree, dense branching. Leaves oval, pointed, dark dull green, reddish-purple in autumn. Flowers small, yellow, late winter, in short-stalked, spherical clusters on leafless twigs. Fruit bright red, edible. Resists insect or disease pests in America. Sun-loving; tolerates dryness and exposure. 20ft (6m) in 20 years. 323, *323*
C. 'Norman Hadden' Semi-evergreen hybrid between *C. capitata* and *C. kousa*. Creamy-white bracts age to pink. 322
C. nuttallii Pacific dogwood. W North America. Zone 7. To 80ft (24.5m). Leaves oval, downy, good autumn colour. Flowers small, purple and green, surrounded by 4–8 large white bracts, flushed with pink; attractive. Dislikes shallow chalk soils, extreme cold. Sun or semi-shade. 25ft (7.5m) in 20 years. 320, 322, *323*
C. 'Ormonde' Hybrid between *C. florida* and *C. nuttallii*, free-flowering and flourishing in British gardens. 322
C. 'Porlock' 322
C. Venus SEE Cornus 'K30-8'. 322

Corsican Pine *see PINUS nigra* subsp. *laricio*

CORYLUS
Deciduous shrubs, rarely trees. Leaves oval, rounded, toothed. Flowers, male and female on same tree, male catkins hanging in winter, female flowers in tiny red clusters, opening slightly later. Fruit, ovoid nut held in toothed cup, mostly edible, ripen in autumn. lime tolerant, suited to chalk soil. Occasional bacterial blight in America, fungal fruit rot, bark fungi in Europe. Nut weevil and squirrels attack young nuts. 283
C. avellana Hazel, Cobnut. Europe, W Asia, N Africa. Zone 4. To 20ft (6m), rarely 40ft (12m), shrub or small tree. Leaves roundish, double-toothed. Nuts in shallow lobed husk. Mainly attractive for male catkins, although leaves turn yellow in autumn. Sun or shade. 15ft (4.5m) in 20 years. *282*
C. californica California hazel. Western N America. Zone 7. Multistemmed large shrub to 25 ft (8m). Nuts enclosed in long tubular calyx. Regarded by most botanists as a variety of *C. cornuta*. *283*
C. colurna Turkish hazel, Turkish filbert. S E Europe, W Asia. Zone 4. To 80ft (24.5m), symmetrical, pyramidal form. Leaves pointed at apex; upper side dark green, lower downy along midrib. Nuts held in husk fringed with lobes, fine down, in groups of 3 or more. Thrives in hot summers and cold winters, as in central Europe. Sun-loving. 35ft (10.5m) in 20 years. *282*, 283
C. fargesii Farges's hazel. China. Zone 5. Large straight-trunked tree to 80ft (25m), with handsome peeling bark at least when young: resembles *C. colurna*. 283
C. maxima 'Purpurea' Purple-leaf filbert. Leaves and catkins coppery purple. 283

CORYMBIA Australia. Genus of trees very similar to Eucalyptus, but with compound terminal inflorescences holding the flowers beyond the leaves. 186, 188

COTINUS
Deciduous shrubs or trees. Simple alternate leaves with slender stalks. Sexes sometimes on separate trees. Small, egg-shaped, fleshy fruit. 12ft (3.5m) in 20 years.
C. coggyria Smoke tree, Venetian sumach. C Europe to China. Zone 5. Shrub or small tree, to 15ft (4.5m). Smooth rounded green leaves, good autumn colours. Flowers fawn-coloured plumes, turning smoky-grey. 313, *313*
C. 'Flame' Hybrid between *C. coggyria* and *C. obovatus* with excellent autumn colour. 313
C. 'Grace' Parentage as 'Flame', with purple leaves all summer. 313
C. obovatus American smoke tree. S United States. Zone 5. To 40ft (12m), shrub or small tree. Leaves wedge-shaped ovals, medium length, often reddish-purple; brilliant autumn colouring. Flowers in greenish feathery masses; sexes separate. Fruit sparse. Richer soil gives poorer colour. Tolerant of dry soils, Sun-loving. 313

COTONEASTER
Evergreen and deciduous trees and shrubs, thornless. Leaves simple, without teeth and lobes, unlike the closely allied *Crataegus*. Flowers all similar, ⅓–½in (7mm–1cm) wide, white or rose-tinted, usually in profuse clusters, attractive to bees. Fruits round to oval; best types brilliant red. All soils. Sun-loving; some species tolerant of shade. Hardy. Susceptible to fireblight and some rusts. *205*, *230*
C. franchettii var. *sternianus* Stern's Cotoneaster. Tibet. Zone 6. Evergreen shrub to 10ft (3m). Leaves dark green, white underneath. Fruit red. Now usually regarded as a full species, *C. sternianus*. *68*
C. x *watereri* Zone 6. To 15ft (4.5m), spreading, hybrid. Leaves tapered at both ends, semi-evergreen, smooth when fully developed. Fruits small, in clusters. Fast-growing. 207
Cottonwood *see POPULUS*
Cockspur Thorn *see CRATAEGUS crus-galli*
Crab-apple *see MALUS*
Crack Willow *see SALIX fragilis*

CRATAEGUS
Hawthorns. Deciduous trees or shrubs. Usually thorny. Leaves simple, toothed or lobed. Flowers usually white and clustered, late spring/early summer. Fruit apple-like, small, variously coloured. Susceptible to fireblight, juniper rust, hawthorn blight, lacebugs, mites, leaf miners, various borers, and others. Very hardy. All soils. *34*, 208–9, 230
C. crus-galli Cockspur thorn. E United States. Zone 4. To 40ft (12m), widespreading, long thorns to 3¼in (8cm). Leaves wedge-shaped, shiny, smooth, dark green, stalked. Flowers small. Fruits, crimson, winter. Thorns 1½–3in (4–7.5cm) long, branched. 209, *209*
C. laciniata (syn. *C. orientalis*) S E Europe, W Asia. Zone 5. To 20ft (6m), rounded spreading head. Leaves with 5–9 toothed lobes, hairy above, woolly below. Flowers June. Fruit roundish, hairy, orange-red, October. 209
C. laevigata (syn. *C. oxyacantha*) Midland hawthorn, May. N W and C Europe. Zone 4. To 25ft (7.5m) spreading. Leaves 3- to 5-lobed, smooth except at first, glossy, dark green, toothed, stalked. Flowers small, in flat heads. Fruit scarlet, autumn, several-seeded. 208
C.l. 'Punicea' Flowers single, red with white eye. *209*
C. x *lavallei* Zone 5. To 20–25ft (6–7.5m), spreading to upright, hybrid. Leaves long, shiny dark green, often remain until mid-winter. Flowers small, numerous.

Ryukyu Islands. Zone 7. To 35ft (10.5m), large shrub or small tree, deciduous. Young shoots slightly downy. Leaves oblong, downy hairs. Flowers clustered, small, white, beautifully fragrant, June. Fruit greenish-yellow. Young shoots slightly susceptible to frost. 20ft (6m) in 20 years. 315

ELAEAGNUS
E. angustifolia Russian olive, Oleaster. S Europe, W Asia. Zone 2. To 25ft (7.5m), deciduous shrub or small tree, often spiny branches. Leaves medium length, alternate, silver-grey, scaly, simple, lance-shaped. Flowers clustered, bisexual, small, silvery outside, yellow inside, fragrant, bell-shaped. Fruit silvery-yellow, oval, scaly berries, fleshy, sweet, edible. Dislikes shallow chalk soils. Sun-loving, tolerates dryness. Normally disease- and pest-free. 10ft (3m) in 20 years. 189
E. umbellata var. *parvifolia* Himalaya. Zone 3. Shrub or small tree, 12-33ft (4-10m) with silvery growth. Flowers fragrant, creamy white, inconspicuous. Fruits orange. 189

Elder see *SAMBUCUS*
Elm see *ULMUS*
Elm Zelkova see *ZELKOVA carpinifolia*

EMBOTHRIUM
E. lanceolatum Chilean firebush. Norquinco valley, Chile. Leaves long, narrow, strap-shaped. Flowers red, profuse, clusters. Dislikes alkaline soils. Hardy. 25ft (7.5m) in 20 years. 189, 189
E.l. 'Norquinco' Selection from Norquinco valley, Chile. Semi-deciduous, has particularly abundant flowers. 189

Empress Tree see *PAULOWNIA*
Engelmann Spruce see *PICEA engelmannii*
English Elm see *ULMUS procera*
English Oak see *QUERCUS robur*
Epaulette Tree see *PTEROSTYRAX hispida*

ERICA arborea Tree heath. Mediterranean. Zone 7. To 15-20ft (4.5-6m), rounded shrub or tree. Leaves minute, evergreen. Flowers small, round, whitish, fragrant, in big pyramidal clusters, spring. Fruit, small roundish capsules holding minute seeds. Dislikes lime, thrives on sandy soil. 335, 335

ERIOBOTRYA japonica Loquat. China, Japan. Zone 7. The largest-leaved evergreen hardy in most of Britain. Noble pointed leathery leaves with fawn felt beneath. Fragrant whitish flowers in mild winters. Edible yellow fruit. Best in shelter. 15ft (4.5m) in 10 years. 211

EUCALYPTUS
Evergreen trees, rarely shrubs. Adult leaves alternate and hanging, juvenile leaves opposite, often waxy, greyish-white. Flowers white, yellow, or red, usually in small clusters; numerous stamens

are main feature. Fruit funnel-shaped capsule holding many small seeds. Deep, moist loam. Fast-growing. 53, 141, 154, 186-8
E. camaldulensis River red gum. Australia, along rivers. Zone 10. Massive tree to 150ft (45m) with thick trunk. Leaves greenish-grey. Flowers white. Widely planted around the world for forestry, but causing environmental degradation through its thirstiness. 186
E. citriodora Lemon-scented gum. Queensland. Zone 10. To 80-130ft (24.5-40m), narrow. Juvenile leaves medium length, oblong, stalked, bristly; adult leaves narrow, lance-shaped. Flowers clustered in leaf axils. Fruit on long stalks. Bark smooth, white to pinkish. 188
E. dalrympleana Mountain gum. SE Australia. Zone 8. To 130ft (40m). Juvenile leaves rounded to ovate, unstalked, pale green; adult leaves lance-shaped, green. Flowers in 3s or 7s. Bark smooth, blotched white, grey, and green. 68, 188
E. diversicolor Karri. Western Australia. Zone 10+. Giant tree, to nearly 300ft (95m), forming spectacular forests. Not hardy in temperate gardens. 187
E. globulus Blue gum. SE Australia. Zone 9. To 227ft (70m). Juvenile leaves medium length, oval, unstalked; adult leaves longer, narrow, deep lustrous blue-green, stalked. Flowers white, in clusters of 1-7. Bark blue-grey, smooth. 70-90ft (21.5-27.5m) in 20 years. 187-8, 189
E. gunnii Cider gum. Tasmania. Zone 7. To 81ft (25m). Juvenile leaves short, ovate, green or bloomed, stalked. Flowers in 3s on stalk in leaf axil. Bark smooth, green, and white. 75ft (23m) in 20 years. 188
E. neglecta Omeo gum. SE Australia. Zone 7. Small tree with thickened base, to 20ft (6m). Leaves large, green. Flowers white. One of the hardiest eucalypts. 188
E. nitens Shining gum. SE Australia. Zone 8. Incredibly fast-growing large tree; can reach over 50ft (15m) in 4 years in England; in Australia achieves over 230ft (70m). Bark peels to reveal patterns on smooth trunk. Leaves glossy green in adulthood, grey in juvenile stages. Flowers white. 188
E. papuana Ghost gum 188
E. parvifolia Kybean gum. SE Australia. Zone 8. To 30ft (9m). Juvenile leaves elliptic, unstalked, green; adult leaves lance-shaped, green. Flowers in clusters of 7. Bark smooth, grey-green. 188
E. pauciflora Snow gum. SE Australia, Tasmania. Zone 8. To 65ft (20m). Juvenile leaves lance-shaped or ovate depending on subspecies, blue-green; adult leaves lance-shaped, green or blue-green. Flowers in clusters of 7-15, white. Bark smooth, white, grey or green in patches.

E.p. subsp. *debeuzevillei* Jounama snow gum. SE Australia. Zone 8. To 60ft (18m). Similar to *E. pauciflora* but juvenile leaves lance-shaped. The hardiest of the genus. 188
E.p. subsp. *niphophila* Snow gum. SE Australia. Zone 7. Crooked tree to 65ft (20m). Leaves large, leathery, grey-green. Trunk is green, grey, and cream. 50ft (15m) in 20 years. 187, 188, 188
E. regnans Victoria, E Tasmania. Zone 9. To 244ft (75m), exceptionally to 325ft (100m). Bark fibrous below, white, smooth. Before logging was tallest tree in the world, now exceeded by redwoods. 186-7, 187
E. subcrenulata Tasmanian alpine yellow gum. Tasmania. Zone 9. To 60ft (18m). Juvenile leaves elliptic to orbicular, green; adult leaves lance-shaped with slightly wavy margin. Flowers in 3s, white. Bark smooth, grey, white, or yellowish-green. 188

EUCOMMIA ulmoides Hardy rubber tree. China. Zone 5. To 65ft (20m). Leaves simple, deciduous, oval, slightly hairy, becoming smooth above, leathery, glossy containing rubbery sap. Flowers, sexes on separate trees, insignificant. Fruit compressed, winged. 30ft (9m) in 20 years. 249

EUCRYPHIA
Evergreen or occasionally deciduous small trees or shrubs from Australia and Chile. Leaves opposite, simple or pinnate, entire or toothed. Flowers white, with numerous pink anthers on long filaments, fragrant, bisexual, solitary in leaf axils, summer. Fruit a woody capsule. Prefer acidic soil.
E. cordifolia Ulmo. Chile. Zone 9. Columnar, multi-stemmed tree to 65ft (20m), or a large shrub. Leaves simple with wavy margins, softly hairy beneath. Flowers to 2in (5cm) diameter, appearing in late summer. Can tolerate some lime.
E. glutinosa Nirrhe. Chile. Zone 8. Large shrub or small tree, up to 33ft (10m). Leaves evergreen, sometimes deciduous in cultivation, pinnate with five leaflets, glossy dark green but turning red before falling. Flowers extremely freely produced, up to 2.5in (6cm) diameter, making a marvellous display in high summer.
E. × *intermedia* (*E. glutinosa* and the Tasmanian *E. lucida*). Zone 8. Fast-growing shrub or small tree. Leaves may be simple or with 3 leaflets. Extremely floriferous in late summer, though flowers smaller than *E. glutinosa*. 'Rostrevor' is a recommended cultivar.
E. × *nymansensis* (*E. cordifolia* and *E. glutinosa*). Zone 7. First raised at Nymans, West Sussex: the clone 'Nymansay' remains the best. Leaves simple or with three leaflets. Flowers can be up to 3in (7.5cm) across. There is a very pretty white-variegated version.

Euodia, Korean see *TETRADIUM daniellii*
European Walnut see *JUGLANS regia*

EXBUCKLANDIA populnea Malayan aspen. Eastern Asia. Zone 8-9. Large evergreen tree potentially to 65ft (20m) or more. Leaves leathery, rounded with pointed tip, emerging bronze. Flowers greenish-yellow, inconspicuous. 184

F

FAGUS
Big, hardy, deciduous. Leaves entire or shallowly toothed. Male flowers in round clusters, females in 2s or 3s on same tree. Fruits are paired triangular nuts in spiny husk. Moist soils. Surface-rooting and competitive. 19, 19, 22, 46, 252, 254-9, 254-5, 256-7, 276
F. engleriana Bluish-foliaged beech from central China. 258
F. grandifolia American beech. E United States. Zone 4. To 100ft (30m), pyramidal, spreads by means of root suckers. Leaves medium length, oval, tapering, toothed, hairy at first; green above, paler below. 256
F. sylvatica Common beech, European beech. Zone 5. To 120ft (36m), densely pyramidal or rounded. Leaves medium length, oval, pointed, glossy, veins and stalk silky-hairy; red-brown in autumn. Suitable for extremes of alkalinity or acidity. 35ft (10.5m) in 20 years. 256
F.s. 'Aspleniifolia' Fern-leaved beech. Leaf-shape varies, being narrow and cut or lobed, with feathery texture. 30ft (9m) in 20 years. 257-8, 258
F.s. 'Cuprea' 257
F.s. 'Dawyck' Medium to large, narrow-columnar form, broadening with age. 35ft (10.5m) in 20 years. 259, 259
F.s. 'Riversii' Rivers' purple beech. Big tree with purple leaves through summer. 257
F.s. 'Rotundifolia' A bizarre beech, short-branched with round cockleshell leaves. Rather pretty. 258
F.s. 'Zlatia' Leaves soft yellow at first; green in late summer. 257
False Acacia see *ROBINIA*
False Cypress see *CHAMAECYPARIS*; x *CUPROCYPARIS*
Fan Palm see *TRACHYCARPUS wagnerianus*
Farges Fir see *ABIES fargesii*

FICUS
Evergreen or rarely deciduous trees or shrubs. Leaves alternate, sometimes toothed or lobed. Flowers usually unisexual on same plant, uniquely borne inside hollow fruit, pollinated by tiny wasps. Fruit fleshy. 250
F. carica Common fig. Zone 7-8. To 30ft (9m), low, rounded,

spreading head. Leaves big, rounded lobes, rough with short, stiff hairs, deciduous. Fruit greenish-brown and violet. 250, 250
F.c. 'Black Mission' 250
F.c. 'Brown Turkey' 250, 250
F.c. 'Bursa' Rounded fruits, paler flesh. 250
F.c. 'Excel' White fig, green with pale flesh. 250
F.c. 'Panachée' Figs are striped green and yellow. 250
F.c. 'Violette' Flatter fruits with a wonderful violet bloom. 250

Fig see *FICUS*
Filbert see *CORYLUS*
Fir see *ABIES*

FITZROYA cupressoides Chile, Argentina. Zone 8. To 45ft (13.5m), leaning over at top. Leaves free, in 3s, hard, blunt ends curling out from the shoot; deep blue-green with 2 bright bands on each face. Cones, separate sexes, can be on same tree, Small cones, wide and persistent. Seeds winged. Needs good rainfall. 15ft (4.5m) in 20 years. 137

Flowering Cherry see *PRUNUS*
Flowering Dogwood see *CORNUS florida*
Fontainebleau Service Tree see *SORBUS latifolia*
Forrest's Birch see *BETULA forrestii*

FORTUNELLA margarita Nagami, Oval kumquat, SE China. Zone 8. 10-15ft (3-4.5m), spreading. Leaves lance-shaped, dark green, paler beneath, partly crinkle-edged, evergreen. Flowers solitary or in small clusters, white, spring. Fruit oval-oblong, pale orange, sweet, slightly acid. 319

FRANKLINIA alatamaha Georgia, United States, extinct n wild. Zone 8. To 35ft (10.m), upright. Leaves medium length, narrow, oblong, deciduous, simple, shiny above, dark green, then red in autumn, minutely toothed. Flowers to 3in (7.5cm), white, cupped, waxy, single, fragrant, late summer. Fruit woody, round, capsule. Tender in Britain; requires hot summers. Acid or alkaline soil. 10ft (3m) in 20 years. 326, 329. 329

Fraser's Fir see *ABIES fraseri*

FRAXINUS
Deciduous trees, few shrubs. Leaves mostly pinnate. Flowers mainly insignificant, bisexual or unisexual on same tree. Fruits narrowly oval, winged, often propellor-like. Hardy. All *Fraxinus* are susceptible to emerald ash-borer, now spreading rapidly in North America. 63, 338-41, 338-9
F. americana White ash, American ash. E United States. Zone 3. To 135ft (41m), upright. Leaves pinnate, leaflets oval or near, dark green above, whitish-green and downy beneath;

end of leaflet may be toothed. Flowers petalless. Prefers loamy soil, much moisture. Subject to oyster scale, canker in America. 25ft (7.5m) in 20 years. 338, *341*

F.a. 'Autumn Purple' Form selected good autumn colour. 338
F. angustifolia S Europe to Iran and Turkestan. Zone 6. To 30ft (9m), upright to spreading, but compact. Leaflets narrow, lance-shaped, shiny dark green, downy beneath, sharply toothed. Flowers without petals. Sun-loving. Susceptible to oyster scale. 30–50ft (9–15m) in 20 years. 340
F.a. 'Lentiscifolia' 340
F.a. var. *oxycarpa* Leaves smaller, not very distinct. 340
F.a. var. *oxycarpa* 'Raywood' Bred in Australia, smoky-purple autumn colour. 340, *340*
F. chinensis Chinese ash. China. Zone 6. To 45ft (13.5m). Leaves with 5–9 leaflets, dark dull green above, paler beneath, often purple in autumn. Flowers in big loose clusters, fragrant, May. Fruit narrow, 1½in (4cm) long. 345
F. excelsior Common ash. Europe (incl. British Isles), Caucasus. Zone 4. To 130ft (40m), spreading. Leaves pinnate, oblong, lance-shaped, smooth above, furry brown by lower midrib, dark green, toothed; leaflets stalkless. Flowers greenish-yellow, clustered, spring. Likes chalk. Susceptible to oyster scale. 30ft (9m) in 20 years. 338, *339*, 340
F.e. 'Diversifolia' One-leaved ash. Leaves simple or 3-part. 340
F.e. 'Jaspidea' Branches yellowish, young shoots golden. 340–1
F.e. 'Pendula' Weeping ash. Initial height according to graft; spreading. Vigorous. Branches and branchlets stiffly hanging. 341, *341*
F. latifolia Oregon ash. W United States. Zone 6. To 80ft (24.5m), narrow, upright
to broad. Leaves pinnate, leaflets oval or oblong, pointed, downy, dark green above, paler below; main stalk grooved. Flowers petalless. Fruit to 2in (5cm) long. Susceptible to oyster scale. 20ft (6m) in 20 years. 340
F. nigra Black ash. E United States. Zone 7. To 75ft (23m). 7–11 slender pointed leaflets, dark green above, paler beneath. Sexes on separate trees. Fruits 1½in (4cm) long 340
F. ornus Manna ash. S Europe, Asia Minor. Zone 6. To 50ft (15m), spreading. Leaves pinnate, broad, oblong leaflets, rusty hairs along midrib beneath, dull green, toothed. Flowers clustered, off-white, spring. Fruit notched at tip. Sun-loving. Susceptible to oyster scale. 15ft (4.5m) in 20 years. 341, *341*
F. pennsylvanica Red ash. E United States. Zone 4. To 60ft (18m). Leaves big, 7 or 9 narrow oval leaflets, dull green, downy beneath. Sexes on separate trees.

Fruit to 2in (5cm) long. Fast-growing. 338–40
F. quadrangulata Blue ash. C and E United States. Zone 4. To 80ft (24.5m), with square branchlets. Leaves big, 7 or 11 narrow oval leaflets, yellow-green. Bisexual flowers in clusters. Fruit, 1–2in (2.5–5cm) long. 340
F. velutina Velvet ash. S W United States, N Mexico. Zone 7. To 45ft (13.5m), round-headed, fairly open. Suitable for dry alkaline soils. 340

Fringe Tree see *CHIONANTHUS*
Fulham Oak see *QUERCUS* x *hispanica* 'Fulhamensis'

G

Gean see *PRUNUS avium*

GENISTA aetnensis Mount Etna broom. Sardinia, Sicily. Zone 8. To 20ft (6m). Tiny, slender leaves, green, rush-like, sparse, simple, deciduous. Flowers profuse, pea-shaped, golden yellow, summer. Prefers well-drained, light loam; tolerates chalk. Sun-loving. Subject to rust, fungal die-back. 200, *201*

Giant Dogwood see *CORNUS controversa*
Giant Sequoia see *SEQUOIADENDRON giganteum*

GINKGO biloba Maidenhair tree. S E China. Zone 4. To 80ft (24.5m), varies from narrow, upright to broadly spreading. Leaves larger and more deeply cleft at first, green both sides, deciduous. Male "flowers" rare, thick, yellow; female flowers single or paired, pale yellow becoming orange.
Fruits yellow, plum-shaped; offensive smell if crushed. Seeds edible. Tolerates lime. 25ft (7.5m) in 20 years. 46, 53, 78–9, 78–9, 153

Glastonbury Thorn see *CRATAEGUS monogyna* 'Biflora'

GLEDITSIA
Deciduous trees. Leaves pinnate. Flowers small, insignificant, greenish, regularly petalled, not pea-like as other Leguminosae. Most species have large thorns, some thornless. Sun-loving.
G. caspica Caspian locust. N Persia.
Zone 6. To 35ft (10.5m), very spiny. Leaves pinnate with 12–20 oval, toothed leaflets, or doubly pinnate. Pod thin, curved, 8in (20cm) long. *199*
G. japonica Japanese locust. Japan. Zone 6. To 70ft (21.5m), pyramidal, spiny bole. Leaflets roughly lance-shaped, midrib and stalk slightly hairy. Flowers, male and female on separate spikes, yellow-green, bell-shaped, June. Fruit, pods to 10in (26cm) long, curved, eventually twisted. 15ft (4.5m) in 20 years. 199
G. triacanthos Honey locust. C and E United States. Zone 3.

To 140ft (42m), broad, open, spiny trunk and branches. Leaves pinnate, medium to long, glossy dark green. Flowers unisexual on same tree, clustered; males green, females more sparse, June. Pod brown, to 18in (45cm) long, curved, Oct.–Dec. Tolerates chalk soils and drought. 25ft (7.5m) in 20 years. 199
G.t. 'Moraine' Thornless male clone. 199
G.t. 'Rubylace' Maroon-red young foliage matures to dark green in summer. 199
G.t. 'Sunburst' Medium-sized, thornless stems. Young leaves bright yellow. 199

GLYPTOSTROBUS pensilis Chinese deciduous cypress. S China. Zone 8 at best. Rare small bush or tree. Leaves either short in 3 ranks, or scale-like, over-lapping: both types pale sea-green, rich brown in autumn. Cones pear-shaped, long-stalked. Seeds winged. 152

Golden Chinkapin see *CHRYSOLEPIS chrysophylla*

GORDONIA
Tender evergreen trees or shrubs similar to *Camellia*. Dark glossy green leaves. Conspicuous flowers during autumn and winter. Dislikes lime. 326, 329
G. lasianthus Loblolly bay. S E United States. Zone 9. To 60ft (18m), narrow. Leaves oblong, pointed, evergreen, simple, shiny, smooth, leathery, dark green, shallow-toothed. Flowers 3–6in (7.5–15cm) across, white, single, fragrant, mid-summer. Fruits oblong, hard. 329

x *GORDLINIA* Hybrid between *Gordonia lasianthus* and *Franklinia*. Good foliage, large white flowers. Needs hot summers to thrive. 329

Gowen Cypress see *CUPRESSUS goveniana*
Grand Fir see *ABIES grandis*
Grapefruit see *CITRUS* x *paradisi*
Greek Fir see *ABIES cephalonica*
Green Ash see *FRAXINUS pennsylvanica*
Green Thorn see *CRATAEGUS viridis*

GREVILLEA robusta Silk-oak grevillea. Australia. Zone 10. To 150ft (45m) in Australia. Leaves feathery, 2-pinnate, evergreen, white hairs on underside. Flowers bunched, honeysuckle-like, orange to golden yellow. Fruit is a boat-shaped capsule. Needs well-drained, lime-free soil. Sun-loving. Tender in Britain. 189, *189*

Grey Alder see *ALNUS incana*
Grey Birch see *BETULA populifolia*
Grey Poplar see *POPULUS* x *canescens*
Guelder Rose see *VIBURNUM opulus*
Gum Tree see *EUCALYPTUS*

GYMNOCLADUS dioica

Kentucky coffee tree. E and C United States. Zone 4. To 90ft (27.5m), large branches, open. Leaves 2-pinnate (i.e. leafstalk has secondary stalks bearing leaflets), medium to long, green above, grey-green and hairy beneath, yellow in autumn. Small flowers unisexual, clustered, greenish-white, on same or different trees, June. Pods oblong, flat, Oct. through winter. Tolerates chalk. 18ft (5.5m) in 20 years. 199

H

Hackberry see *CELTIS*

HAKEA laurina Pincushion hakea. W Australia. Zone 10. Large shrub to 20ft (6m). Leaves broad, to 6in (15cm). Flowers white fading red in large rounded heads. 189

HALESIA
Shrubs or small trees. Leaves simple, deciduous. Flowers hanging, snowdrop-like, clustered. Fruits pear-shaped winged pods, pale brown. Tolerates lime if enriched with peat or leaf soil. Sun-loving. 331
H. monticola Mountain silverbell, Pearwood. Mts of S E United States. Zone 5. To 90ft (27.5m), pyramidal. Leaves medium length, ovallish, yellow in autumn. Flowers white, spring. Resistant to pests and disease. Sun or semi-shade. 25ft (7.5m) in 20 years. 331, *331*
H. tetraptera Snowdrop tree. S E United States. Zone 5. Now regarded as a form of the variable *H. carolina* forming a large shrub to 20ft (6m), producing masses of pure white flowers in spring. 331
HAMAMELIS
Deciduous shrub or small trees. Leaves alternate, with waxy or toothed edge. Flowers bisexual, clustered, thin yellow petals, late autumn to spring. Fruit, capsule holding 2 shiny black seeds. 35, 182
H. japonica 'Arborea' Zone 5. Widespreading form sometimes making a small tree. Flowers rich yellow. 10ft (3m) in 20 years. 182, *185*
H. mollis Chinese witch hazel. China. Zone 6. To 25ft (7.5m), bushy. Leaves roundish, simple, hairy. Flowers fragrant. 8ft (2.5m) in 20 years. 182
H. virginiana Virginian witch hazel. E United States. Zone 5. To 30ft (9m), often bushy. Leaves oval to triangular, simple, hairy on veins beneath, toothed, hairy stalk. 6ft (1.8m) in 20 years. 182

Handkerchief Tree see *DAVIDIA involucrata*
Hawthorn see *CRATAEGUS*
Hazel see *CORYLUS*
Heather see *ERICA*
Hedgehog Holly see *ILEX aquifolium* 'Ferox'
Hemlock see *TSUGA*
Hiba see *THUJOPSIS dolabrata*
Hickory see *CARYA*

Himalayan Birch see *BETULA utilis*
Hinoki see *CHAMAECYPARIS obtusa*

HIPPOPHAE rhamnoides Sea buckthorn. Europe, temperate Asia. Zone 3. To 30ft (9m), deciduous shrub or small tree. Leaves short to medium length, narrowly lance-shaped, simple, greyish-green above, silvery-green below. Flowers, sexes on separate trees, small, clustered. Fruit, berry-like, small, orange-yellow, in winter; very acid. Tolerates dry soils, exposure; excellent in coastal areas, sandy soils. Sun-loving. 15ft (4.5m) in 20 years. 189, *189*

HOHERIA
Evergreen or deciduous small trees or
shrubs from New Zealand. Leaves alternate, toothed. Flowers white, bisexual, solitary or clustered, summer. Seed case thin, sometimes winged.
H. glabrata Ribbonwood. Zone 8. To 35ft (10.5m). Leaves alternate, toothed, medium length, lance-shaped. Fragrant white flowers in summer. 292
H.g. 'Glory of Amlwch' (*H. glabrata* and *H. sexstylosa*). Raised in Anglesey, Wales. Very free-flowering. 292
H. lyallii Mountain lacebark. Zone 8. A small tree with leaves becoming grey-green by mid-summer, when it arches with white cherry-like blossom. 292
H. populnea Lacebark. New Zealand. Zone 9. Large shrub or small tree to 33ft (10m). Leaves change shape from small juvenile to larger broader adult phase. Flowers large, white. Less hardy than others. 292
H. sexstylosa Lacebark. Zone 8. Erect-growing evergreen with narrow leaves. Summer-flowering. 25ft (7.5m) in 25 years. 292
H.s. 'Stardust' A particularly free-flowering form. 292

Holly see *ILEX*
Hollywood Juniper see *JUNIPERUS chinensis* 'Kaizuka'
Holm Oak see *QUERCUS ilex*
Honey-locust see *GLEDITSIA*
Hop Tree see *PTELEA trifoliata*
Hop-hornbeam see *OSTRYA*
Hornbeam see *CARPINUS*
Hornbeam Maple see *ACER carpinifolium*
Horse Chestnut see *AESCULUS hippocastanum*
Hungarian Oak see *QUERCUS frainetto*
Huon Pine see *LAGAROSTROBUS franklinii*
Hupeh Crab see *MALUS hupehensis*

HYDRANGEA Genus of shrubs or climbers from Asia and the Americas, noted for their attractive inflorescences, usually with showy sterile florets surrounding small fertile flowers. 354

I

ILEX

Deciduous and evergreen trees and shrubs. Leaves simple, stalked. Flowers small, off-white, males and females on separate trees. Fruit small. Most soils. *I. aquifolium* forms hardier than *I. altaclerensis* forms. Sun or semi-shade; injured by cold in N Europe and America. Pest-free, except holly leaf miner. *260–1, 350–3*

I. x altaclerensis Highclere holly. Zone 6.
A group of excellent vigorous hybrids. Small tree or large shrub. Leaves evergreen, large, less prickly than *I. aquifolium*. Tolerates seaside and industrial areas. 25ft (7.5m) in 20 years. *352–3, 352*

I. x a. 'Belgica Aurea' Female form. Leaves dark green, mottled pale green and grey, margins pale yellow or whitish, flat, sparsely spined. 20ft (6m) in 20 years. *353*

I. x a. 'Camelliifolia' Vigorous, conical. Leaves evergreen, nearly spineless, purplish at first, dark green later. Large fruits. Bark purple. 20ft (6m) in 20 years. *352*

I. x a. 'Golden King' Variegated female form. Leaves green, edges bright yellow, virtually spineless. 20ft (6m) in 20 years. *352*

I. x a. 'Lawsoniana' Female form of *I.a.* 'Hendersonii' with large, sparsely spined leaves, medium length, yellow edge, marbled centre, deep and pale green. May revert to plain green. 20ft (6m) in 20 years. *353*

I. x a. 'Purple Shaft' Very vigorous purple-shooting form. Masses of berries. 20ft (6m) in 20 years. *353*

I. x a. 'W J Bean' Female clone producing abundant red fruits. *352–3*

I. aquifolium English holly, S Europe, N Africa, W Asia to China. Zone 6. To 45–70ft (13.5–21.5m), short spreading branches, dense, pyramidal. Leaves oval, spiny, shiny green, evergreen. Flowers small, white, fragrant, May–June. Fruit, red berries, Sept. through winter. Suited to industrial and coastal areas. 15ft (4.5m) in 20 years. *350–1, 350, 353*

I.a. 'Argentea Marginata Pendula' Perry's weeping silver holly. Silver-margined foliage on drooping branches. Berries profuse. 10–15ft (3–4.5m) in 20 years. *352, 353*

I.a. 'Bacciflava' Profuse yellow berries. 10–15ft (3–4.5m) in 20 years. *352*

I.a. 'Crispa' Leaves thickened, warped and twisted, with only terminal spine. *352*

I.a. 'Ferox' Hedgehog holly. Leaves with clusters of short, sharp spines. Male form, no berries. 10–15ft (3–4.5m) in 20 years. *352*

I.a. 'Golden Milkboy' Male. Leaves big, spiny, green, centre marked with gold. 10–15ft (3–4.5m) in 20 years. *352*

I.a. 'Golden Queen' Male. Young shoots green or reddish. Leaves broad, dark shiny green with pale green shading, yellow margin. 10–15ft (3–4.5m) in 20 years. *352, 352, 353*

I.a. 'Silver Milkboy' To 30ft (9m), pyramidal. Leaves spiny, dark green, with central creamy spot. Flowers negligible, male. Most soils. Holly leaf miner. 10–15ft (3–4.5m) in 20 years. *352*

I.a. 'Silver Queen' Male. Leaves broad, oval, dark green, creamy-white margin. 10–15ft (3–4.5m) in 20 years. *353*

I. cassine Dahoon holly S E United States. Zone 6. Large shrub or tree, occasionally to about 40ft (12m). Leaves evergreen, glossy green, to 6in (15cm). Fruits red. Not for chalky soil. *353*

I. cornuta Chinese holly. Zone 6. Dense, rounded, shrub to about 12ft (3.5m). Leaves curiously rectangular, usually with 5 spines. Fruits large, red, abundant. *351*

I. decidua Possum-haw. S E and C United States. Zone 6. Large shrub, usually about 12 ft (3.5 m), sometimes taller. Leaves deciduous, bright green, toothed but not spined. Fruits orange-red, persist well. Not for chalky soil. *350*

I. x koehneana 'Chestnut Leaf' (*I. aquifolium* and *I. latifolia*). Zone 7. Evergreen to 23ft (7m). Leaves elliptic, to 4¾in (12cm), shiny and yellowish-green with prominent marginal spines. Fast-growing for a holly. *353*

I. latifolia Tarajo, Lustreleaf holly. China, Japan. Zone 7. To 60ft (18m). Biggest leaves of any holly, oblong, evergreen, serrated, dark green above, yellow beneath, stalked. Fruit round, orange-red, profuse clusters. Shade-tolerant; best in shelter. 25–30ft (7.5–9m) in 20 years. *353*

I. x meserveae (*I. aquifolium* and *I. rugosa*). Blue holly. Zone 6. Evergreen to 6ft (2m) with purple shoots. Leaves smaller than *I. aquifolium*, spiny, dark bluish-green. Female cultivars such as 'Blue Angel' produce red fruit, 'Black Prince' is male. *351*

I. x m. 'Blue Princess' Larger leaves, good red fruits. *351*

I. opaca American holly. E and C United States. Zone 5. To 50ft (15m), pyramidal. Leaves evergreen, spiny, dull green above, yellowish beneath, stalk grooved. Male flowers in slender heads, female flowers solitary, white. Fruits red. Dislikes chalky soils. 8ft (2.5m) in 20 years. *351, 353*

I.o. f. *xanthocarpa* Fruits yellow. *351*

I. pedunculosa Longstalk holly. China, Japan. Zone 5. Large shrub or tree to 33ft (10m). Leaves evergreen, untoothed, dark glossy green. Fruits held beyond leaves on long stalks, red. *353*

I. perado Madeira Holly. Madeira, Azores, Canary Islands. Zone 9. Tree to 50ft (15m). Leaves broad, often not very spiny. One parent of the important *I. x altaclerensis* hybrids. *351, 352*

I. pernyi C and W China. Zone 5. To 30ft (9m), narrowly pyramidal. Tiny leaves evergreen, triangular, with 2 big, lateral spines, leathery, dark shiny green. Flowers yellow, tiny clusters. Fruit small, red, clustered berries. 15ft (4.5m) in 20 years. *353*

I. purpurea Kashi holly. China, Japan. Zone 8. Round-crowned tree to 33ft (10m), forming a canopy of long glossy leaves. Flowers are uniquely purplish-pink. Fruits red, used as Chinese New Year decorations. *353*

I. rugosa Siberian holly. Japan, Far-eastern Russia. Zone 3. Low shrub, best known as a parent of *I. x meserveae*. *351*

Incense Cedar see CALOCEDRUS decurrens
Indian Bean Tree see CATALPA
Indian Horse Chestnut see AESCULUS indica
Irish Yew see TAXUS baccata 'Fastigiata'
Ironwood see CARPINUS; PARROTIA persica
Italian Alder see ALNUS cordata
Italian Maple see ACER opalus

J

JACARANDA mimosifolia
Jacaranda, Sharpleaf jacaranda. Bolivia, Argentina. Zone 10. To 50ft (15m). Foliage fine, fern-like. Flowers lilac-purple, numerous. Hot, dry position. *346, 346*

Jack Pine see PINUS banksiania
Japanese Angelica Tree see ARALIA elata
Japanese Apricot see PRUNUS mume
Japanese Bitter Orange see CITRUS trifoliata
Japanese Black Pine see PINUS thunbergii
Japanese Cedar see CRYPTOMERIA japonica
Japanese Chestnut see CASTANEA crenata
Japanese Dogwood see CORNUS kousa
Japanese Fan Palm see TRACHYCARPUS wagnerianus
Japanese Fir see ABIES firma
Japanese Horse Chestnut see AESCULUS turbinata
Japanese Larch see LARIX kaempferi
Japanese Maple see ACER japonicum; A. palmatum
Japanese Plum Yew see CEPHALOTAXUS harringtonia
Japanese Red Pine see PINUS densiflora
Japanese Silver Birch see BETULA mandschurica var. japonica
Japanese Snowbell see STYRAX japonicus
Japanese White Pine see PINUS parviflora
Japanese Yew see TAXUS cuspidata
Japanese Zelkova see ZELKOVA serrata
Japonica see CHAENOMELES japonica
Jeffrey Pine see PINUS jeffreyi
Jelecote Pine see PINUS patula

JUBAEA chilensis
Chilean wine palm. Chile. Zone 8. To 80ft (24.5m), trunk 4–6ft (1.2–1.8m) thick, covered with leaf-base scars. Pinnate leaves to 12ft (3.5m) long on stalks. Flower stalks in axils of lower leaves. Yellow oval fruit. *162*

Judas tree see CERCIS siliquastrum

JUGLANS
Deciduous trees, sometimes shrubs. Big, pinnate leaves, toothed or entire. Male flowers in slender hanging catkins, female flowers sparse, sexes separate on same tree. Fruit is hard-shelled nut, autumn. Subject to honey fungus, crown rot, twig die-back, bacterial blight; leaf blotch in Britain. Any soil. *284–5, 286*

J. ailanthifolia Japanese walnut. Japan. Zone 5. Medium-sized tree to 50ft (15m) producing huge leaves over 3ft (1m) long. Nuts edible. Closely related to Chinese and Manchurian walnuts. *285*

J. cathayensis Chinese walnut. Zone 5. Long leaves with many leaflets. *285*

J. cinerea Butternut. E United States. Zone 4. To 60ft (18m) or more, widespreading head. 7–19 pointed, toothed leaflets, hairy below. Male catkins. Fruit covered with sticky hairs. 25ft (7.5m) in 20 years. *285*

J. hindsii Northern California walnut. W United States. Zone 7. West Coast version of *J. nigra*, not so large, but a handsome tree. *284–5*

J. x intermedia (*J. regia* and *J. nigra*). Potentially becoming large and magnificent. *285*

J. mandshurica Manchurian walnut with very long leaves. Zone 5. *285*

J. nigra Eastern black walnut. E and C United States. Zone 4. To 150ft (45m), rounded, spreading to upright. Leaves to 2ft (60cm) long. 11–23 leaflets, downy beneath. Fruit roundish, solitary or paired. 35ft (10.5m) in 20 years. *284–5, 285*

J.n. 'Laciniata' Cultivar with elegant dissected leaflets. Exquisite tree for a lawn in a warm garden. *285*

J. regia English or Persian walnut. S E Europe, Himalaya, China. Zone 5. To 90ft (27.5m), broad, rounded head. Leaves long. 5–7 oval leaflets. 25ft (7.5m) in 20 years. *284, 285, 285*

J.r. 'Franquette' Commonly grown commercial cultivar. *284*

Juneberry see AMELANCHIER canadensis

JUNIPERUS
Evergreen trees or shrubs. Leaves initially awl-shaped, adult leaves scale-like. Flowers usually with sexes on separate trees. Fruit berry-like. Good on lime. Susceptible to fungal attacks. Generally slow-growing. *22, 134, 142–5*

J. chinensis Chinese Jupiter. Himalaya, China, Japan. Zone 4. To 75ft (23m), conical. Juvenile leaves prickly, dark green, mixed with or at base of shoots of adult scale leaves. Male flowers crowded on crown, spring. Cones ⅓in (7mm) across, waxy green becoming dark purple. Almost any soil. 20ft (6m) in 20 years. *144, 145*

J.c. 'Aurea' Young's golden juniper. To 35ft (10.5m), narrow, flat-topped column. Leaves bright gold . Male. Good in towns. 12ft (3.5m) in 20 years. *144*

J.c. 'Femina' Female clone. *144*

J.c. 'Kaizuka' "Hollywood" juniper. Small, branches widespreading at all angles. Leaves crowded, lustrous green. *144*

J.c. 'Keteleeri' Narrow, conical. Adult leaves dark, shiny grey-green. Cones ½in (1cm) wide, waxy blue-green. 15ft (4.5m) in 20 years. *144*

J.c. 'Pyramidalis' Conical. Leaves mostly juvenile, prickly, waxy blue. *145*

J. communis Common juniper. North America, Europe, Asia, Korea, Japan. Zone 2. Small, rarely to 15ft (4.5m), sprawling or erect, conical. Cones small, blue, waxy. Tolerates high chalk content. 8ft (2.5m) in 20 years. *142, 143, 145*

J.c. 'Compressa' Noah's Ark juniper. Dwarf form, 18in (45cm) high, narrow, columnar, shapely. Leaves small, crowded. Cones small, very slow. *142, 158–9*

J.c. 'Hibernica' Irish juniper. To 20ft (6m), narrow, smooth, conic-topped column, showing much blue from inner surfaces of leaves. 12ft (3.5m) in 20 years. *142, 145*

J.c. 'Suecica' Medium, columnar. Branches ascending, weeping at tips. Leaves blue-green. Fruits oblong. *142*

J. deppeana Alligator juniper. S W United States. Zones 8–9. Columnar to conical tree to 65ft (20m). Bark forming attractive scale-like plates. Foliage blue-green. Cones large, brown with waxy bloom. Not satisfactory in cool wet climates. *144*

J.d. var. *pachyphlaea* Bark thicker, even more scaly. Excellent glaucous foliage. *144*

J. drupacea Syrian juniper. E Mediterranean. Zone 7. Columnar tree to 50ft (15m) or more, with bright green foliage. Cones to 1in (2.5cm) across, blue-black, edible if you like the taste of resin. *145*

J. occidentalis Western juniper. W United States. Zone 5. Bushy, rounded tree to about 40ft (12m). Foliage of tiny scale-leaves, light green. Cones blue-bloomed. *144*

J. phoenicea Phoenician juniper. Mediterranean. Zone 9. Small, rounded or broadly conical tree to 25ft (8m), from dry Mediterranean hillsides. Foliage green. Cones brown. *144*

years. 105
P. rubens Red spruce. NE United States. Zone 2. 60–70ft (18–21.5m), narrowly conical. Leaves dark yellow-green, wiry, curved. Cones to 2in (5cm) long, brown at maturity, falling during first winter or following spring. Likes moisture, not chalk. 30ft (9m) in 20 years. 102, *104*
P. sitchensis Sitka spruce. W United States. Zone 7. To 180–200ft (55–60m), conical. Leaves flattened, bright blue-green above, blue-white beneath, sharp-pointed. Cones to 4in (10cm) long, falling by winter of first year. Pale trunk of big trees splays out at base. Likes cool, damp summers; acid or any soils. 55–65ft (17–20m) in 20 years. 42, 43, 44, 54, *72*, 102–4, *102–3*, 106
P. smithiana West Himalayan spruce. Morinda spruce. W Himalaya. Zone 6. To 200ft (60m), branchlets pendulous. Leaves slender, curving, medium length, dark green. Male "flowers" small, end of shoots. Cones green becoming brown, to 7in (17.5cm) long. 35ft (10.5m) in 20 years. 108, *108*, *109*
P. torano Tiger-tail spruce. Japan. Zone 6. To 150ft (45m), usually less in gardens. Foliage radiating from shoot, extremely sharp, green. Cones reddish-brown, to 5in (12.5cm). 108

Pignut *see CARYA glabra*
Pin Oak *see QUERCUS palustris*
Pinaster *see PINUS pinaster*
Pine *see PINUS*
Piñon Pine *see PINUS cembroides*

PINUS
Evergreen trees, conical, becoming flat-topped or bushy. Leaves in groups of 2–5 usually, needle-like, long, minutely toothed in cross-section, semi-circular or 3-sided. Male and female flowers on same tree; males red or yellow, cylindrical; females scaly, woody. Cones open when ripe or remain closed. Seeds may be wingless. Dislikes shallow chalk soils. Sun-loving. Susceptible to insect attack, dry-rot fungi, honey fungus, white-pine blister rust. *13, 17, 20*, 46, *47, 62*, 80–91
P. aristata Bristle-cone pine. SW United States. Zone 3. 15–40ft (4.5–12m). Leaves short, in 5s, deep green, often white speckled with resin; persist up to 17 years; pressed closely to branchlets. Cones to 3½in (9cm) long, scales spiny. Very slow-growing. 26, *26, 47*, 82
P. armandii Armand's pine. W China, Taiwan, Korea. Zone 7. To 60ft (18m), spreading. Leaves drooping in 5s, soon shed, medium-length, pointed, waxy, green above, inner faces white. Cones in 2s or 3s, to 7in (17.5cm) long, becoming pendulous. 86
P. attenuata Knobcone pine. W United States. Zone 7. Often multi-stemmed, medium-sized pine, to 82ft (25m). Leaves in 3s, long, yellow green. Cones persist for years on branches. 82
P. ayacahuite Mexican white pine. Mexico and Central America. Zone 7. 100–160ft (30–48m), spreading. Leaves medium to long, slender, persistent, waxy-green. Cones to 8in (20cm), cylindrical. 83
P. banksiana Jack pine. N United States, Canada. Zone 2. 25–60ft (7.5–18m), branches crooked, sometimes bushy. Leaves short, rigid, paired, olive-green. Cones small, oval, often closed for several years. 30ft (9m) in 20 years. 81
P. bungeana Lacebark pine. China. Zone 5. 80–100ft (24.5–30m), possibly a small bush, often with several stems, conical. Leaves medium length, smooth, rigid, flattened. Bark pale, flaking, leaving patches of white, green, and purple. Cones solitary or paired, oval. Likes limestone. 25ft (7.5m) in 20 years. 86, *86*
P. canariensis Canary Island pine. Canary Islands. Zone 9. Magnificent but tender tree with slender 9in (23cm) needles in 3s, long branches, and drooping sprays. To 90ft (27.5m) in sub-tropical conditions. Tolerates alkaline soil.
P. cembra Arolla pine. Mts of C Europe and N Asia. Zone 4. 60–80ft (18–24.5m), narrow, densely columnar, becoming more open. Leaves in 5s, crowded, dark shining green, medium length. Male flowers crowded at base of weak shoots. Cones 3in (7.5cm) long, deep blue to purple, erect; scales never open on tree. Susceptible to honey fungus. 25ft (7.5m) in 20 years. 88, *89*, 90
P. cembroides Pinyon pine. SW United States, Mexico. Usually a low-growing, domed tree to about 33ft (10m). Leaves usually in 3s, short, thick. Cones rounded, containing large seeds ("pine nuts") that are sought after by Native Americans and wildlife. 83
P. contorta Shore pine. W United States. Zone 7. To 80ft (24.5m), occasionally bushy. Leaves in 2s, medium length, characteristically twisted, yellow-green. Cones ovallish, small, paired or clustered. Much planted on light, stony or sandy land. 50ft (15m) in 20 years. 82
P.c. var. ***latifolia*** Lodgepole pine. W United States. Zone 7. To 60ft (18m). Leaves slightly broader, medium length. Cones larger than for above. Not lime tolerant. Susceptible to pine-shoot moths in Britain. 43
P.c. **'Spaan's Dwarf'** Low-mounding dwarf form. *158–9*
P. coulteri Big-cone pine, Coulter pine. S California, N Mexico. Zone 8. 40–50ft (12–15m). Leaves in clusters of 3, 10in (26cm) long, blue-grey. Cones ovallish, very heavy, 10in (26m) long (biggest of any pine), scales clawed, light brown. 40ft (12m) in 20 years. 82, *82*
P. densiflora Japanese red pine. Japan. Zone 4. 70–120ft (21.5–36m), often twisted or crooked, broad and irregularly conical. Leaves in 2s, medium length, bright green, sharply pointed. Female "flower" turns purple-brown, becomes a cone 2in (5cm) long; pink-brown or purplish. 30ft (9m) in 20 years. 84
P.d. **'Oculus-draconis'** Dragon's eye. When viewed from above leaves show alternate rings of yellow and green. 84
P. echinata Short-leaf pine. E and SE United States. Zone 6. To 120ft (36m), narrow, possibly drooping. Leaves usually in 2s, persistent, medium length, slender, flexible, grey-green. Cones small, clustered, persist after seeds shed. Needs some lime. 81
P. elliottii Slash pine. SE United States. Zone 9. Small tree. Leaves in 2s or 3s, long. Cones conical, 5½in (14cm). Tender. Needs lime-free soil. 81
P. halepensis Aleppo pine. Mediterranean, W Asia. Zone 7. 50–60ft (15–18m), irregularly branched. Leaves in 2s, medium length, slender, pointed, hard, green. Cones in groups of 1–3, to 4½in (11.5cm), often persist many years. Tolerates chalk. 30ft (9m) in 20 years. *90*, 91, *91*
P. heldreichii Bosnian pine. Italy, Balkan peninsula. Zone 5. To 90ft (27.5), narrow, oval. Leaves in 2s, dense, rigid, medium length, dark green, pointed. Cones oval, blue in summer, dull brown. Found on driest limestone; also likes acid soil. 30ft (9m) in 20 years. 91
P.h. **'Schmidtii'** Desirable dwarf clone. *158–9*
P x holfordiana (*P. ayacahuite x P. wallichiana*). A vigorous hybrid named for the founder of Westonbirt Arboretum, England, where it originated. A fast-growing, long-branched, beautifully glaucous, and soft-needled tree, producing many curving resinous cones. To 60ft (18m) in 40 years. 83
P. x hunnewellii (Japanese *P. parviflora* and American *P. strobus*). Zone 6. Usually has twisted blue-green leaves. 80–1
P. hwangshanensis Huangshan pine. C China. Zone 7. Tree to 82ft (25m) forming rounded masses of growth when mature, when very picturesque. Leaves in 2s, dark green, slender. Cones rather small, to 3in (7.5cm). *84–5*, 86
P. jeffreyi Jeffrey's pine. SW United States. Zone 8. To 200ft (60cm), conical. Leaves long, stout, blue-green to grey; when crushed, leaves smell of pineapple. Cones to 10in (26m) long. 35ft (10.5m) in 20 years. 80, *80–1*, 82, *82*
P. koraiensis Korean pine. E Asia. Zone 3. 100–150ft (30–45m), open, conical. Leaves medium length, rough-textured, waxy, deep blue-green. Cones conical to oblong, to 6in (15cm), yellow-brown. 86
P. lambertiana Sugar pine. Oregon and California. Zone 7. Largest member of genus, 220–250ft (66–76m), with very long branches. Leaves medium length, spiralled, in 5s, blue- to grey-green. Cones to 26in (66cm) long, narrow. Susceptible to white-pine blister rust. 40ft (12m) in 20 years. 82
P. massoniana Chinese red pine. China. Zone 7. Large pine, to 82ft (25m) or more. Trunk straight, bark grey, reddish in upper parts. Leaves in 2s, long and slender, dark green. Cones rounded, small. *52*
P. montezumae Montezuma pine. Mts of Mexico. Zone 9. 70ft (21.5m), rounded and spreading. Leaves long, grey-green. Cones to 10in (26cm) long. Susceptible to late spring frosts. 25ft (7.5m) in 20 years. 83, *83*
P. monticola Western white pine. W United States. Zone 4. 150–180ft (45–55m), slender, columnar crown. Leaves to 4in (10cm) long, blue-green. Cones to 15in (38cm) long, often curved, pale brown. Susceptible to white-pine blister rust. 40ft (12m) in 20 years. 82
P. mugo Dwarf mountain pine. Mts of C Europe. Zone 3. Small, bushy, with numerous crooked, irregularly spreading branches. Leaves in 2s, short, dark green. Cones small. 8ft (2.5m) in 20 years. 91
P.m. **'Mops'** Mounding semi-dwarf clone, widely planted. *158*
P.m. **'Moppet'** A truly dwarf clone derived from 'Mops' as a witch's broom. *159*
P. muricata Bishop pine. California. Zone 8. Usually broadly domed, with end branches upturned, descending with age; can also be tall, narrow, and domed. Leaves in 2s, medium length. Cones small, clustered; persist up to 70 years. Very vigorous on poorest acid, sandy soils and in fierce maritime exposure. 40–60ft (12–18m) in 20 years. 83
P. nigra* subsp. *caramanica see *P.n.* var. *pallasiana*
P.n. **subsp. *laricio*** (syn. *P.n.* subsp. *maritima*) Corsican pine. S.Italy, Corsica, Sicily. 40–140ft (12–42m). Crown more open, branches fewer, shorter. More level than *P.n.* subsp. *nigra*. Leaves grey-green. Most soils. 25ft (7.5m) in 20 years. 43, *89*, 90, *90*
P.n. **subsp. *maritima*** see *P.n.* subsp. *laricio*
P.n. **subsp. *nigra*** Austrian pine. S and E Europe. Zone 5. 120–150ft (36–45m), conical, spreading. Becoming flat-topped. Leaves in 2s, persistent. Medium length, rigid, densely crowned, dark green. Cones small, golden yellow. Grows on clay, chalk, limestone. 30ft (9m) in 20 years after a slow start. 90
P.n. **subsp. *pallasiana*** (syn. *P.n.* var. *caramanica*) Crimean pine. SE Europe, W Asia. To 135ft (41m), branches erect, wide, conical. Leaves stiff, twisted, medium length. Cones usually larger than *P.n.* subsp. *nigra*. 90
P. palustris Longleaf pine, Pitch pine. E United States. Zone 8. 80–120ft (24.5–36m), few stout branches, level, then ascending. Leaves in 3s to 18in (45cm), light grassy-green. Cones to 10in (26cm), narrow, dull-brown; scales at base remain when cone is shed. 81
P. parviflora Japanese white pine. Japan. Zone 5. 20–50ft (6–15m), young trees conical, mature trees with flat heads of stout, spreading branches. Leaves in 5s, curved blunt-tipped, small. Cones solitary or in clusters, upright, ovallish, small with few scales. 20ft (6m) in 20 years. 84, *86*
P.p. **'Bonnie Bergman'** Low-growing, multi-stemmed shrubby form, leaves bluish-grey. *87*
P. patula Jelecote or spreading-leaf pine. Mexico. Zone 8. Broadly conical, spreading, then upturned branches. Leaves in 3s, slender, to 1ft (30cm) long, bright grassy-green. Cones in whorls of 2–5, pale brown, persistent. 40ft (12m) in 20 years. *81*, 83
P. peuce Macedonian pine. Balkan Peninsula. Zone 5. To 100ft (30m), columnar–conical. Leaves in 5s, medium length, sharp pointed, blue-green, white inner surface. Cones cylindrical, 6in (15cm) long, in clusters spreading or drooping. Exceedingly healthy and hardy on severe sites. Disease-free. 91, *91*
P. pinaster Maritime pine. W Mediterranean. Zone 8. 90–120ft (27.5–36m), branches upsweeping; old trees with bare trunk and flat-domed head. Leaves in 2s, stiff, medium length, short-pointed, Male "flowers" abundant; female flowers on tips of expanding shoots, dull red. Cones small, dark olive-green ripening to orange-brown. Grows on sand. 50ft (15m) in 20 years. *90*, 91
P. pinea Stone pine, Umbrella pine. Mediterranean. Zone 8. To 80ft (24.5m), flat-topped or umbrella-shaped. Leaves in 2s, slightly twisted, medium length, sharp-pointed. Cones symmetrical, small, pale brown, smooth, closed for 3 years. 46, *47*, 90–1, *90*
P. ponderosa Ponderosa pine, Western yellow pine. W United States. Zone 4. To 235ft (72m), narrowly conical, variable when older. Leaves in 3s, medium to long, dark grey- to yellow-green. Cones to 6in (15cm) long, solitary or clustered, dark brown. Tolerates many soil types, drought. 35ft (10.5m) in 20 years. 82, *82*
P. pseudostrobus Smooth-bark Mexican pine. Mexico. Zones 8–9. Large tree to 150ft (45m). Leaves usually in 5s, long, pendulous, light or greyish green. Cones to 6in (15cm) long. 83
P. radiata Monterey pine. California. Zone 8. 110–120ft (33–36m), domed. Leaves in 3s, widespreading, medium length, deep green, slender, flexible. Cones asymmetrical at base, tawny-yellow, to 7in (17.5cm) long, persist for many years. Many soils. Immensely vigorous. 45–65ft (13.5–20m) in 20 years. 83, *83*
P. resinosa Red pine, Norway pine. E Canada, NE United States. Zone 3. To 80ft (24.5m), symmetrical oval crown. Leaves medium length, paired, dark yellow-green, snap cleanly when doubled over. Cones small. Susceptible to buff-tip moth. 30ft (9m) in 20 years. 81
P. rigida Pitch pine, Northern pitch pine. E United States. Zone 4. In north, small, where on poorest soils; farther south, 50–60ft (15–18m), shape

390 Index of Tree Species

Port Orford Cedar *see CHAMAECYPARIS lawsoniana*
Portugal Laurel *see PRUNUS lusitanica*
Possum Haw *see ILEX decidua*
Post Oak *see QUERCUS stellata*
Prickly Leaved Paperbark *see MELALEUCA styphelioides*
Pride of India *see KOELREUTERIA paniculata*
Prince Albert's Yew *see SAXEGOTHAEA conspicua*
Privet *see LIGUSTRUM*
PRUMNOPITYS andina Plum-fruited yew. Andes of S Chile. Zone 8. Pointed, upswept bush, often on many stems, rarely a conical tree. Leaves pointing forward, short, with acute tips. Male "flowers" stalked, emerging where branches and leaves join. Fruit yellow-white, plum-shaped. 18ft (5.5m) in 20 years. 128

PRUNUS
Big group of deciduous or evergreen trees. Leaves simple, edges toothed, crushed leaves often fragrant. Flowers 5-petalled, white or pink; in doubled forms number of petals is increased; spring. Fruit has 1 cell and 1 seed, fleshy; reduced and inedible in certain ornamental species. Most thrive on lime or chalk. Sun-loving. Usually very hardy. Susceptible to borers, scale and leaf-eating insects and virus diseases. 22, 218–28
***P.* 'Amanogawa'** To 30ft (9m), upright, narrow, tending to splay in age. Leaves greenish-bronze. Flowers double, fringed, pink, profuse, fragrant. Most soils; especially floriferous on chalky soils. 220, *221*
***P.* x *amygdalo-persica* 'Pollardii'** Similar to almond, but flowers richer pink. Fruit between peach and almond. Zone 4. 226, *228*
P. armeniaca Apricot. C Asia, China. Zone 5. To 35ft (10.5m), spreading. Leaves roundish, pointed, deciduous, smooth, shiny green, stalked. Flowers solitary, small. Fruits 1¼in (3cm) wide, larger in cultivation, yellow-orange, red-tinged, early summer, edible. 226
***P.a.* 'Rouge de Roussillon'** *228*
P. avium Gean, Mazzard, Wild cherry. W Asia. Zone 3. To 70ft (21.5m), pyramidal. Leaves oval, long-pointed, deciduous, red in autumn, stalked. Flowers clustered, white, small. Fruit small, round, red, bitter or sweet. 45ft (13.5m) in 20 years. 222, *222–3*
***P.a.* 'Plena'** Double gean. Europe. Zone 3. To 60ft (18m). Flowers profuse, small, double-petalled, drooping masses. Fruits rare. Hardier than oriental cherries. As large as type and longer-lasting. 222
P. campanulata Formosan cherry, Bell-flowered cherry. Formosa, S Japan. Zone 7. To 30ft (9m), bushy. Leaves roughly oval, deciduous, shiny, stalked. Flowers small, rose-pink, bell-shaped at first, opening later. 224
P. cerasifera Cherry plum, Myrobalan. W Asia. Zone 4. To 25ft (7.5m), sometimes thorny, slender branches. Leaves ovallish, toothed, light green. Flowers solitary white, very early, fragrant, March. Fruit round, 1in (2.5cm) across, red or yellow, on mature trees. 227, 228, *229*
***P.c.* 'Pissardii'** Pissard plum, Purple-leaved plum. Iran. Zone 4. To 35ft (10.5m), upright, spreading. Leaves ovallish, deciduous, ruby at first, becoming claret then purple. Flowers single, small, profuse, white, emerging from pink buds. Fruit purple, rarely produced. 25ft (7.5m) in 20 years. 227
P. cerasus Sour cherry, Wild dwarf cherry. SW Asia. Zone 3. To 35ft (10.5m), spreading. Leaves oval, abruptly pointed, deciduous, smooth, shiny, pale green, stalked. Flowers dense-clustered, white. Fruits roundish, red-black, acid-tasting, summer. 222
***P.c.* 'Rhexii'** As *P. cerasus* but flowers doubled, hardier. Last in flower. 222
***P.* 'Chocolate Ice'** (syn. 'Matsumae-fuki') Coppery-bronze leaves, flowers palest pink from pink buds. 220
P. domestica Plum. SW Asia. Zone 5. Small tree to 39ft (12m). Leaves broadly elliptic, deciduous. Flowers white, fruit the familiar edible plum. 228
P.d.* var. *insititia see *P. insititia*
P.d.* var. *syriaca Mirabelle plum. Small round yellow fruit. 228
P. dulcis Common almond. N Africa to W Asia. Zone 4. 20–30ft (6–9m), spreading. Leaves lance-shaped, deciduous, smooth, stalked. Flowers 2in (5cm) across, single or paired, early spring. Fruit to 2½in (6.5cm), velvety. Tolerates dryness. 25ft (7.5m) in 20 years. 226, *226–7*
***P.d.* 'Roseoplena'** Double almond. Flowers pale pink, double, numerous petals. 226, *228*
***P.* x *hillieri* Spire'** Zone 6. To 25ft (7.5m), narrow, pyramidal. Leaves turning red in autumn. Flowers profuse, soft pink. 224, *225*
P. himalaica Himalayan cherry. Nepal. Zone 7. Small tree, to 16ft (5m) with glossy dark brown peeling bark. Leaves softly hairy. Flowers very insignificant, white. 224, *225*
P. hirtipes C China. Zone 8. Open-branched small tree to 40ft (12m). Flowers early in spring, white or pink. 225
***P.h.* 'Semi-Plena'** Flowers semi-double, pale pink. 224
***P.* 'Hokusai'** Vigorous, widespreading. Young leaves brown. Flowers big, semi-double, pale pink, spring. 219
P. insititia (syn. *P. domestica* var. *insititia*) Bullace, Damson. Primitive forms of cultivated plums with dark, blue-bloomed fruit, usually rather bitter. 228
P. jamasakura see *P. serrulata* var. *spontanea*,
***P.* 'Jo-nioi'** Abundant single-white flowers among coppery foliage. 220
***P.* 'Kanzan'** To 35ft (10.5m), upright at first, later spreading. Leaves at first red-brown. Flowers doubled, purply-pink, large. Most soils; especially floriferous on chalky soils. 219, *220*, *221*
***P.* 'Kursar'** (*P. nipponica* var. *kurilensis* x *P. campanulata*) Pretty March-flowering cherry bred by Captain Collingwood Ingram. Dense deep-pink flowers, bronze new leaves. 219
P. laurocerasus Cherry laurel, Common laurel. Europe, Asia Minor. Zone 7. To 40ft (12m), widespreading. Leaves ovate or oblong, tapering, evergreen, leathery, glossy, dark green, short-stalked. Flowers tiny, dull white, upright clusters. Fruit small, conical, red, turning black. Shade tolerant. 30ft (9m) in 20 years. 229
***P.l.* 'Angustifolia'** Narrow-leaved form, branches upright. 229
***P.l.* 'Cameliifolia'** Shrub or small tree. Leaves dark green, contorted. 229
***P.l.* 'Castlewellan'** (syn. *P.l.* 'Variegata'). Leaves marbled with white and grey-green. 229
***P.l.* 'Latifolia'** (syn. *P.l.* 'Magnoliifolia') The biggest-leaved laurel, pale, very shiny, very vigorous, and imposing. 20ft (6m) in 10 years. 229
***P.l.* 'Magnoliifolia'** see *P.l.* 'Latifolia'
***P.l.* 'Rotundifolia'** A short-leaved form making a good dense bush. 229
***P.l.* 'Variegata'** see *P.l.* 'Castlewellan'
***P.l.* 'Zabeliana'** Spreading bush with narrow leaves, very free-flowering. 229
P. lusitanica Portugal laurel. Spain, Portugal. Zone 7. To 60ft (18m), often shrubby. Leaves ovate, medium length, finely toothed, dark shiny green above, paler beneath, evergreen. Flowers tiny, cup-shaped; in slender clusters to 10in (26cm) long. Fruit purple, oval, ½in (1cm) long. 25ft (7.5m) in 20 years. *228–9*, 229
***P.l.* 'Angustifolia'** Leaves smaller and neater than usual. 229
P.l.* subsp. *azorica Azores. Shrub or small tree. Leaves bigger than the type, bright green. 229
***P.l.* 'Variegata'** Leaves variegated white, often with pink flush in winter. 229
P. maackii Manchurian cherry, Amur chokecherry. Manchuria, Korea. Zone 2. To 50ft (15m), spreading. Leaves oval, pointed, deciduous, hairy on veins and midrib. Flowers small, white, in downy clusters. Fruit small, black. Shiny amber bark. 30ft (9m) in 20 years. 224
***P.* 'Matsumae-fuki'** see *P.* 'Chocolate Ice'
***P.* 'Matsumae-hanugasai'** see *P.* 'Pink Parasol'
P. mume Japanese apricot. China, Korea. Zone 6. To 20ft (6m), spreading. Leaves round to oval, long-pointed, deciduous, becoming smooth, short-stalked. Flowers single or paired, early spring, occasionally winter or late spring, almond-scented. Fruit to 1¼in (3cm) across, round, yellow, hardly edible. Best against a wall in cold, exposed areas. *226–7*
***P.m.* 'Beni-chidori'** Flowers double, deep pink. 227, *228*
***P.* 'Okame'** (*P. campanulata* x *P. incisa*). Small round-headed tree to 26ft (8m). Leaves small, deciduous with orange and red tints in autumn. Flowers small, deep pink, appearing profusely in February. 219
P. padus Bird cherry. Europe, N Asia to Japan. Zone 3. To 50ft (15m), spreading. Leaves oval, pointed, rounded at base, deciduous, smooth, dull green above, greyish beneath, short-stalked. Flowers small, in spreading spikes, almond-scented. Fruit small, round, black, harsh, bitter-tasting, summer. Less susceptible to tent caterpillar than other species. 30ft (9m) in 20 years. 222, *224*
***P.p.* 'Colorata'** Flowers pale pink. Leaves initially purplish-brown, purple coloration retained in veins and on undersides. Shoots dark purple 222
P.p.* var. *commutata Eastern Asian form, coming into growth very early. 222
***P.p.* 'Watereri'** Flowers in extra long clusters, up to 8in (20cm) long standing out with great vigour from all over tree. 222
***P. pendula* 'Pendula Rosea'** 15–30ft (4.5–9m), weeping. Flowers single or double, pink, tiny. *220*
P. persica Peach. China. Zone 5. To 25ft (7.5m), spreading. Leaves lance-shaped, deciduous, smooth, short-stalked. Flowers usually single, to 1½in (4cm) across, pink. Fruits to 3in (7.5cm) wide, slightly furry, rounded, yellow-orange, red-tinged on sunny side, juicy. Susceptible to trunk borers and peach leaf curl. Dry soils. 226
***P.p.* 'Klara Mayer'** Double flowers, dark pink. 226
P.p.* var. *nectarina Nectarine. Fruit smooth-skinned, otherwise identical to the type.
P. pensylvanica Pin cherry. E North America. Zone 2. Small tree to 30ft (9m). Leaves narrow, finely toothed. Flowers white, in small clusters. Fruits red. 224
***P.* 'Pink Parasol'** (syn. *P.* 'Matsumae-hanugasa') Spreading habit. Flowers large, double, pale pink. 220
P. sargentii Sargent's cherry. Japan, Sakhalin, Korea. Zone 4. To 40ft (12m), upright, spreading. Leaves oval, slender-pointed, deciduous, smooth, bronze-red, orange, and red in autumn, stalked. Flowers clear pink, in small clusters; one of first cherries to colour in autumn. Fruits tiny. 30ft (9m) in 20 years. 219
P. serotina Black cherry, Rum cherry. E United States, E and S Mexico, Guatemala. Zone 3. To 100ft (30m), spreading. Leaves oval to lance-shaped, tapered both ends, deciduous, smooth, glossy above, lighter beneath, pale yellow in autumn, stalked. Flowers in hanging cylindrical clusters, white, small, early summer. Fruit small, shining, black, late summer, sparse. 40–50ft (12–15m) in 20 years. 224
P.s.* subsp. *capuli Capulin. Mtns of Central and South America. Zones 6–7. Very similar but with much larger, edible fruits. Surprisingly hardy. 224
P. serrula Birch bark tree. W China. Zone 5. To 35ft (10.5m), widespreading. Leaves lance-shaped, deciduous. Flowers small, white, in small groups. Fruits oval, red, small. Bark glossy, dark red, peeling. Most soils. 25ft (7.5m) in 20 years. 224, *224*
P. serrulata Japanese cherry. Zone 5. To 80ft (24.5m), widespreading, flat-topped. Leaves oval to lance-shaped, smooth, waxy beneath, deciduous, short, possibly double-toothed. Flowers to 2½in (6.5cm) across, single or paired. Fruit small, black.
***P.s.* 'Autumn Glory'** A form of *P. verecunda*, selected for good autumn colour. 219
P.s.* var. *spontanea Now known as *P. jamasakura*, the wild "Hill cherry" of Japan, revered for its spring display. 219
***P.* 'Shimidsu'** (syn. *P.* 'Shôgetsu') Small tree with widespread branches. Leaves green, flowers double, white, with fringed petals. 220, *221*
***P.* 'Shirofugen'** 18–25ft (5.5–7.5m), widespreading with long horizontal branches. Young leaves coppery. Flowers doubled, purplish in bud, white when opening, later pinkish, in long clusters, long-lasting, early summer. Most soils. 220
***P.* 'Shirotae'** 18–25ft (5.5–7.5m), widespreading to hanging. Leaves pale green, fringed. Flowers usually single, very big, fragrant, white. Most soils. 220, *221*
***P.* 'Shôgetsu'** see *P.* 'Shimidsu'
P. sogdiana C Asia. Zone 4. Small wild plum related to *P. cerasifera*. Fruits yellow, red, or purple. 224
P. spinosa Sloe, Blackthorn. Europe, N Africa, W Asia. Zone 4. 10–15ft (3–4.5m), spreading and suckering to form thorny thickets. Leaves oval, deciduous. Flowers small, single or pairs. Fruit blue-black, waxy, later shiny, round. Tolerates exposure. *227*, 228
P.* x *subhirtella Spring cherry. Japan. Zone 5. To 30ft (9m), densely branched. Leaves small, toothed. Flowers small, light pink, March to April. Fruit, black. 219
***P.* x *s.* 'Autumnalis'** Autumn cherry. Zone 5. To 30ft (9m). Leaves ovallish, long, pointed, deciduous, hairy beneath. Tiny flowers doubled, white profuse, autumn and intermittently through winter to spring. Fruits tiny, shiny, black. 219
***P.* x *s.* 'Fukubana'** Small. Flowers numerous, rose-pink with many petals, spring. Fruit unlikely. 219
***P.* x *s.* 'Pendula'** Long pendulous branches covered in pink flowers in spring. 219
***P.* 'Taihaku'** Great white cherry. To 35ft (10.5m), spreading. Leaves at first red-brown, deciduous. Flowers single, very big, bright white. Most

single or in 2s on short stalks, cup with downy scales. Bark corrugated, trunk buttressed at base. Any soil. 30ft (9m) in 20 years. 263, 267
Q. ilex Evergreen oak, Holm oak, Holly oak. Mediterranean, SW Europe. Zone 7. To 80ft (24.5m); broad, spreading, round head. Leaves usually narrowly oval, short to medium, pointed, sometimes entire, leathery, glossy green above, greyish, downy beneath, short stalk, evergreen. Acorns to ¾in (2cm), 1–3 on short, downy stalk. Tolerates shade and any well-drained soil. 15ft (4.5m) in 20 years. 264, *266–7*
Q. imbricaria Shingle oak, Northern laurel oak. E United States. Zone 5. To 100ft (30m), pyramidal, round-topped when older. Leaves narrowly oval, medium length, tapering at both ends, usually unlobed, dark, shiny green above, downy beneath, deciduous, rich autumn colours. Acorns usually solitary, to ⅔in (1.6cm) long, short-stalked cup. Not lime tolerant. 20ft (6m) in 20 years. *272, 273*
Q. kelloggii California black oak. California and Oregon. Zone 7. To 90ft (27.5m) rounded, open; stout, spreading branches. Leaves medium length, bristle-toothed, deeply lobed, shiny green. Tolerates dry sandy soils. *268–9*, 272
Q. x kewensis (*Q. cerris* x *Q. wislizenii*) Kew oak. Zone 7. Small to medium-sized, dense compact head. Leaves small, dull green above, shiny below, evergreen. 263
Q. laurifolia Laurel oak. E United States. Zone 7. To 100ft (30m), dense rounded head. Leaves glossy green, oblong, sometimes lobed, deciduous. Small acorns on short stalk. Not lime tolerant. *272*, 273
Q. libani Lebanon oak. Syria, Asia Minor. Zone 6. To 35ft (10.5m), round head, slender branches. Leaves oblong and tapered, sharply toothed, dark glossy green above, paler green beneath, stalked, deciduous. Acorns 1in (2.5cm) long, single or in 2s on thick stalk. 25ft (7.5m) in 20 years. 265–7
Q. lobata California white oak, Valley oak. California. Zone 7. Large deciduous tree to over 100ft (30m). A notable feature of the grasslands of California, not easy in cooler, wetter climates. *271, 272*
Q. lyrata Overcup oak. C and S United States. Large tree to 100ft (30m) with rounded outline. Leaves deciduous, with 3–4 lobes. Acorns almost enclosed in their cup. *272*, 273
Q. macranthera Caucasus, N Iran. Zone 6. To 65ft (20m). Leaves long, oval, lobed, green above, pale and downy beneath, deciduous. Acorns 1in (2.5cm) long on short stalk, cup with downy scales. May be grown in deep soils over chalk. 35ft (10.5m) in 20 years. 265
Q. macrocarpa Burr oak, Mossy-cup oak. NE and NC North America. Zone 3. To 130ft (40m), spreading. Leaves roughly triangular, medium to long, dark glossy green above, downy beneath, downy stalk, deciduous. Twigs corky, bark like white oak. Acorns to 1½in (4cm) long, usually solitary, with a fringed cup. Not lime tolerant. 15ft (4.5m) in 20 years. 272, *272*, 273
Q. marilandica Blackjack oak. E United States. Zone 5. To 30ft (9m), spreading. Leaves roughly triangular, deciduous, wide and 3-lobed at tip, medium length, shiny, dark green, yellow-brown beneath. Acorns ¾in (2cm) long, solitary or in 2s. Good on poor, dry soil. 273
Q. michauxii Swamp chestnut oak. SE United States. Zone 6. Large tree with pale scaly bark, to 100ft (30m). Leaves with many coarse teeth, becoming golden in autumn. Needs hot summers to thrive. *272*
Q. mongolica Mongolian oak. E Asia. Zone 3. Potentially a large tree, but needing warm summers to flourish. Leaves large, deeply round-lobed. 267
Q. muehlenbergii Chinkapin oak, Yellow chestnut oak. S Canada, E United States, NE Mexico. Zone 4. To 80ft (24.5m). Leaves oblong, coarsely toothed, yellow-green above, pale and downy beneath with yellow midrib and stalk; rich autumn colour, deciduous. Roundish acorns, ¾in (2cm) long. Not lime tolerant. *272, 273*
Q. nigra Water oak. S United States. Zone 6. To 80ft (24.5m), conical or round-topped, fine-textured foliage. Leaves roughly triangular, sometimes lobed, sometimes entire, medium length, smooth shiny green, short stalk. Acorns single, ½in (1cm) long. Not lime tolerant, likes moist ground. *272, 273*
Q. palustris Pin oak. SE Canada, E United States. Zone 4. To 120ft (36m), dense pyramidal head, branch ends drooping. Leaves medium length, lobed, toothed near tip, glossy green, hairless except where veins join beneath, slender stalk, deciduous. Acorn about ½in (1cm) long, shallow, saucer-shaped cup. Not lime tolerant, likes moist soil. 30ft (9m) in 20 years. *269*, 272, *272*, 273
Q. petraea Sessile oak, Durmast oak. W, C, and SE Europe, Asia Minor. Zone 4. To 100ft (30m). Leaves medium length, stalked, oval, deeply lobed, dark glossy green above, greyish, downy beneath, deciduous. Acorns to 1¼in (3cm) long, solitary or clustered, stalkless on twig. 30ft (9m) in 20 years. 262, 266
Q. phellos Willow oak, E United States. Zone 5. To 100ft (30m), conical or round-topped head, slender branches. Leaves medium length, narrow, pointed at both ends, entire, pale green, yellow in autumn, deciduous. Acorns tiny. Not lime tolerant. 30ft (9m) in 20 years. 271, *272*, 273
Q. prinus Chestnut oak. E United States. Zone 5. Probably just a form of *Q. michauxii*, resembling it and *Q. muehlenbergii*; American oaks are often difficult to identify. *172*, 273
Q. robur Common oak, English oak. Europe, Caucasus, Asia Minor, N Africa. Zone 5. To 100ft (30m), broad open head, short trunk. Leaves oblong, medium length, lobed, dark green above, greyish, hairless below, deciduous. Acorns to 1¼in (3cm) long, ovoid, 1 or more on long stalk. 30–45ft (9–13.5m) in 20 years. 32, 262, *262–3, 267*
    Q.r. **'Argenteovariegata'** Variegated clone: not as good as the variegated *Q. cerris*. 263
    Q.r. **'Atropurpurea'** Purple English oak. Small to medium. Leaves and shoots rich purple. Slow-growing. 263
    Q.r. **'Concordia'** Golden oak. Small, rounded tree. Leaves golden yellow, scorched in hot sunshine. Very slow. 263
    Q.r. **'Fastigiata'** Cypress oak. Big columnar head, upright branches. 263
***Q. x rosacea* 'Filicifolia'** Leaves divided pinnately into narrow segments. Fruit stalked. 263
Q. rubra Red oak. E United States. Zone 4. To 80ft (24.5m), becoming broad and round-topped. Leaves oval, medium to long, prominently lobed, smooth, dark green above, greyish beneath, stalk yellow, deciduous, red or red-brown in autumn. Acorns to 1¼in (3cm) long. Fast when young. Not very lime tolerant. 270, *271, 272*, 273
Q. rugosa Net-leaf oak. Mexico. Zone 6. Usually a smallish tree up to about 33ft (10m). Leaves evergreen, thick, feeling rather rough. A white oak needing hot summers to thrive. 263
Q. salicina Japan, Korea. Zone 8. Rounded tree to about 50ft (15m), often less. Densely clad in narrow evergreen leaves, silvery-white below. 267
Q. schottkyana China. Zone 8. Medium-sized tree, potentially to 65ft (20m). Evergreen leaves flush reddish, becoming dark green, pale below. 267
Q. shumardii Shumard oak. S and C United States. Zone 5. To 120ft (36m), open, round head. Leaves medium length, obovate with sharp-pointed lobes, smooth, dark glossy green above, red or golden brown in autumn. Acorns 1in (2.5cm) long. Not lime tolerant. 272, *272*, 273
Q. stellata Post oak. W and C United States. Zone 5. To 60ft (18m). Leaves roughly triangular, lobed at tip, medium to long, deciduous, dark green and rough, paler, hairy beneath. Acorns single or in 2s, downy cups. An important timber tree, producing a tough wood. *272*, 273
Q. suber Cork oak. S Europe, N Africa. Zone 8. To 65ft (20m), widespreading and rounded large branches. Leaves oval to oblong, short to medium, toothed, dark glossy green above, downy beneath, evergreen. Acorns ¾in (2cm) long, single or in 2s on short downy stalk. Bark thick, rugged, corky. Needs full sun, dislikes cold exposure. 12ft (3.5m) in 20 years. 263, 264, *264–5*
Q. texana Nuttall oak. Central S United States. Zone 5. To 80ft (24.5m). Leaves medium size, deeply lobed with sharp points, deciduous. Roundish acorns. *272*, 273
Q. x turneri (*Q. ilex* and *Q. robur*). Turner's oak. Large tree to 81ft (25m) with dense, rounded crown. Leaves elliptic with shallow lobes, semi-evergreen. Acorns egg-shaped, to ¾in (2cm) long. 263
Q. variabilis Oriental cork oak. Japan, China, Korea, Taiwan. Zone 4. To 80ft (24.5m), spreading. Leaves oval to oblong, dull green, bristly margin, hairy beneath, deciduous. Bark corky. 267
Q. velutina Black oak, Yellow-bark oak. E and C United States. Zone 4. To 100ft (30m) or more, dense rounded head. Leaves oval, medium to long, deeply lobed, glossy green above, downy beneath, rich autumn colours, deciduous. Acorns usually solitary on short stalk, to ¾in (2cm) long. Inner bark bright yellow. Not lime tolerant. 30ft (9m) in 20 years. *270, 272*, 273
Q. virginiana Live oak. SE United States, NE Mexico, W Cuba. Zone 7. To 60ft (18m), very widespreading, branches nearly horizontal. Leaves oblong, medium length, leathery, glossy green above, pale downy beneath, evergreen. Acorns oval, 1in (2.5cm) long. *271*, 272–3, *272*, 273
Q. x warburgii Cambridge oak. A beautiful hybrid between *Q. rubur* and *Q. rugosa*. Spectacular in the Cambridge Botanic Garden, England, where new leaves and catkins are both reddish. A broad dome, almost evergreen. 263

Quince *see CYDONIA*

R

Rauli *see NOTHOFAGUS alpina*
Red Alder *see ALNUS rubra*
Red Ash *see FRAXINUS pennsylvanica*
Red Beech *see NOTHOFAGUS fusca*
Red Buckeye *see AESCULUS pavia*
Red Cedar, Eastern *see JUNIPERUS virginiana*
Red Cedar, Western *see THUJA plicata*
Red Fir *see ABIES magnifica*
Red Horse Chestnut *see AESCULUS* x *carnea*
Red Maple *see ACER rubrum*
Red Oak *see QUERCUS rubra*
Red Pine *see PINUS densiflora; P. resinosa*
Red Silver Fir *see ABIES amabilis*
Red Spruce *see PICEA rubens*
Redbud *see CERCIS*

RHAPIDOPHYLLUM hystrix
Needle palm. SE United States. Zone 7. Stemless palm, forming thickets in damp shady places. Very hardy but needs summer heat to thrive. *160*, 162

RHODODENDRON
Evergreen or deciduous, some trees, usually shrubs. Simple, entire leaves. Flowers usually in clusters on shoot ends, often funnel-like. Fruit, capsule, usually oval or oblong, minute seeds. Not lime tolerant. 332, 336–7
R. arboreum Temperate Himalaya, Kashmir to Bhutan, Khasia Hills, Sri Lanka. Zone 7. 30–40ft (9–12m) in cultivation; evergreen, thick trunk, wide head. Narrow, oblong leaves, medium to long, dark green above, scaly beneath. Dark red to pink bell-shaped flowers, spring. 336, *336–7*
R. augustinii Medium-sized shrub, evergreen. Beautiful blue flowers in spring. *336*
    R.a. **'Electra'** Cultivar with violet-blue flowers with greenish-yellow blotches. *336*
R. barbatum Nepal, Sikkim, Bhutan. Zone 7. Shrub or small tree, to 40ft (12m), evergreen, smooth blue-grey branches. Oblong leaves, pointed, medium length, dark dull green above, paler beneath. Dark red bell-shaped flowers 4in (10cm) across. Peeling bark. *337*
R. falconeri Sikkim, Nepal, Bhutan. Zone 7. To 30ft (9m). Thick, sparse branches. Oval to oblong evergreen leaves, long, dark green above, rust-coloured felt beneath. Creamy-white flowers shaded with lilac, dark purple blotch at base. 336, *337*
***R.* 'Hydon Down'** *336*
R. maximum Great laurel, Rose bay. E United States. Zone 3. To 40ft (12m). Narrow, oblong leaves, medium to long, dark green above, paler beneath. Rose-purple to pink flowers spotted with olive-green to orange, small, June/July. 336
R. ponticum Iberian peninsular; Balkans and Turkey. Zone 6. Large shrub with dark evergreen foliage. Flowers mauve to lilac-pink. An aggressive if beautiful invader of British woods and heaths. *336*, 337

RHUS
Deciduous or evergreen shrubs, sometimes trees or climbers. Compound, alternate leaves. Uni- or bisexual flowers on same or different trees. Fruits roundish, fleshy, hard stones. Any fertile soil. Subject to coral spot fungus, fungal wilt in America; few diseases in Britain. 313
R. typhina Staghorn sumach. E North America. Zone 3. Small tree or shrub, to 35ft (10.5m), widespreading; sparse branches thick and pithy, yielding thick white juice when cut. Leaves downy at first, large, turning orange, red, purple in autumn. Male and female flowers clustered on separate trees. Fruits closely packed, hairy, decorative. Good in built-up areas. 313, *313*
R. verniciflua Varnish tree. Japan, China, Himalaya. Zone 9. To 65ft (20m). Large leaves. Flowers inconspicuous, on separate trees. Fruit small, yellowish. 313

Ribbonwood see HOHERIA
Rimu see DACRYDIUM cupressinum
River Birch see BETULA nigra
River Red Gum see EUCALYPTUS camaldulensis

ROBINIA
Deciduous trees or shrubs. Leave pinnate. Flowers white to pink or pale purple, pea-like, hanging clusters, summer. Fruit is a flattened brown pod, several seeds. Many soils. Hardy, Sun-loving. 50, 198–9
R. x *hillieri* (*R. kelseyi* and *R. pseudoacacia*). A very pretty hybrid, medium-sized with apple-blossom pink flowers among the delicate pale leaves. 199
R. pseudoacacia Black locust, Common acacia, False acacia. E United States. Zone 3. To 80ft (24.5m), upright, open, few branches. Leaflets opposite, oval, hairy at first. Flowers white, fragrant, June. Pod to 3½in (9cm) long, upper edge winged. Bark rough, furrowed. Tolerates dryness. Good in industrial areas. Subject to locust borer, locust leaf miner, witches' broom, virus growths. 40ft (12m) in 20 years. 198–9, 198
R.p. 'Frisia' Small–medium. Leaves golden yellow all summer. 199, 199
R.p. 'Umbraculifera' Mop-head acacia. Small, compact, round head, branches spineless. Flowers rare. 199

Roble see NOTHOFAGUS obliqua
Rock Elm see ULMUS thomasii
Rocky Mountain Juniper see JUNIPERUS scopulorum

ROSA
Deciduous shrubs, stems thorny, leaves pinnate. Flowers 5-petalled. Fruit, rose 'hip', a fleshy covering of the true fruit which is bony, seed-like. Succeeds in most except acid soils. Sun-loving. Subject to black spot, mildew, etc. 47, 204
R. filipes 'Kiftsgate' Zone 5. Climbing form, to 60ft (18m). Leaves pale green, brown initially. Flowers white, profuse, large clusters. Fragrant. Fruits small, red, numerous. 204
R. 'Wickwar' Vigorous hybrid with slightly greyish leaves and masses of white flowers. 204–5

Rosebay Rhododendron see RHODODENDRON maximum
Rowan see SORBUS aucuparia
Royal Palm see ROYSTONIA regia

ROYSTONIA regia Royal palm. Cuba. Zone 10. To 70ft (21.5m). Pinnate leaves to 10ft (3m) long, 4 rows of leaflets. Flower stalk develops below smooth glossy crown shaft. Purplish fruit, ½in (1cm) long. 162, 163

Rum cherry see PRUNUS serotina
Russian Olive see ELAEAGNUS angustifolia

S
SABAL
S. palmetto Caribbean cabbage palm. Caribbean, Central America. Zone 8. 20–90ft (6–27.5m), very variable. Trunk bare or covered with leaf bases. Palmate leaves 12ft (3.5m) long, green or blue-green. Unisexual flowers on same tree, flower stalk to 3ft (1m) long. Roundish shiny black fruit ⅓in (8mm) across. 162
S. minor Dwarf palmetto. S E United States. Zone 9. Tuft-forming palm, often found wild in dense shade. Leaves palmate. Hardy in areas with hot summers. 162, 163

SALIX
Mostly deciduous trees and shrubs. Leaves simple, typically alternate, long and narrow, pointed, toothed, but other shapes occur. Flowers, sexes usually on separate trees, without petals, in upright silky or hairy catkins, spring, before or after leaves. Seeds in small capsules. Tolerant of moisture; prefer full sun. Very hardy. Fast-growing. Subject to watermark disease, rusts, twig blights, cankers, aphids. 13, 22, 63, 47, 75, 190–3
S. alba White willow. Europe, N Asia, N Africa. Zone 2. To 80ft (24.5m), rather upright. Leaves lance-shaped, deciduous, silky-hairy, yellow in autumn, short-stalked. Flowers, spring. Dislikes shallow chalky soils. Susceptible to leaf-eating insects and other pests and diseases. 70ft (21.5m) in 20 years. 190–1, 193
S.a. 'Britzensis' Scarlet willow. Bark striking orange-scarlet colour, most obvious in winter. 193
S.a. var. *caerulea* Cricket-bat willow. British Isles. Zone 2. To 100ft (30m), pyramidal. Leaves lance-shaped, deciduous, downy at first, smooth later, blue-green. Flowers, only female catkins known. 85ft (27m) in 20 years. 193
S.a. var. *sericea* Very effective smaller form, silvery-white leaves. 193
S.a. 'Tristis' see *S.* × *sepulcralis* var. *chrysocoma*
S.a. var. *vitellina* Golden-stemmed willow. Similar to *S. alba*; only male trees known. Shoots bright yellow. 50ft (15m) in 20 years. 192, 193
S. babylonica Weeping willow. China. Zone 6. To 40ft (12m), weeping. Leaves lance-shaped, deciduous, smooth, except at first, dark green above, paler beneath, short-stalked. Flowers early spring. Pests as *S. alba*. 63–4, 63, 190–2, 191, 194
S.b. 'Annularis' (syn. *S.b.* 'Crispa') Leaves spirally curled. 193
S.b. 'Crispa' see *S.b.* 'Annularis'
S.b. 'Tortuosa' Corkscrew willow, Contorted willow. N China, Manchuria, Korea. Zone 5. To 40ft (12m), branches twisted in corkscrew fashion. Leaves lance-shaped, deciduous, smooth, waxy

above, bright green. Flowers spring. 55ft (17m) in 20 years. 193
S. caprea Goat willow. Common sallow. Europe, W Asia. Zone 5. To 30ft (9m), spreading. Leaves broad, oblong, deciduous, downy at first, woolly beneath, wrinkled above, grey-green, short-stalked. Flowers, sexes on separate trees, spring; male catkins large, yellow; females silver-grey. 35ft (10.5m) in 20 years. 192, 193
S. daphnoides Violet willow. N Europe, C Asia, Himalaya. Zone 5. To 30ft (9m), upright. Leaves oval to lance-shaped, shiny, dark green, short-stalked. Stems purple with white bloom. Flowers late winter. Pests as *S. alba*. 35ft (10.5m) in 20 years. 192, 193
S. 'Erythroflexuosa' Zone 4. 20–30ft (6–9m), hybrid. Shoots orange-yellow, weeping, twisted in corkscrew fashion, as are narrow leaves. 193
S. fargesii C China. Zone 6. Open shrub with thick reddish-brown twigs, to 10ft (3m). Leaves oblong, to 7in (18cm) long, finely toothed, emerging from large red buds, deciduous. Catkins slender, very long for a willow, to 7in (18cm). 193
S. fragilis Crack willow. Europe and N Asia. Zone 5. To 100ft (30m), widespreading. Leaves lance-shaped to narrowly oblong, deciduous, smooth, dark green above, blue-green beneath, stalked. Flowers, spring. Bark roughly channelled, twigs liable to break off. 40ft (12m) in 20 years. 192, 193
S. gracilistyla Japan, Korea, Manchuria. Zone 6. To 10ft (3m), bushy shrub. Leaves oval to oblong, downy at first, shiny green, short-stalked, persist into late autumn. Red catkins before leaves; males grey becoming yellow. 15ft (4.5m) in 20 years. 192
S.g. 'Melanostachys' Form with black male catkins, producing red anthers. 192, 193
S. moupinensis W China. Zone 5. To 18ft (5.5m), often shrubby. Shoots orange and buds shiny red-brown. Leaves big, smooth bright green. Catkins to 6in (15cm) long.
S. × *pendulina* var. *blanda* (syn. *S.* × *p.* 'Blanda') Broad head of weeping branches. 192
S. × *p.* var. *elegantissima* (syn. *S.* × *p.* 'Elegantissima') More strongly weeping. Both are hybrids between *S. babylonica* and *S. fragilis*. Zone 4. 192
S. pentandra Bay willow, Laurel willow. Europe, N Asia. Zone 5. To 65ft (20m), spreading. Leaves ovallish, deciduous, smooth, shiny, dark green above, paler beneath, midrib yellow, short-stalked, aromatic when crushed. Flowers, late spring; male catkins bright yellow. Susceptible to insect pests. 45ft (13.5m) in 20 years. 193
S. purpurea 'Nancy Saunders' Purple Osier. Europe. Zone 5. Densely branching shrub to 16ft (5m). Leaves small, neat. Male catkins freely produced, purplish-brown. 193
S. × *sepulcralis* var. *chrysocoma*

(syn. *S. alba* 'Tristis'). Weeping willow. Zone 6. To 50ft (15m), hybrid, weeping. Leaves lance-shaped, branchlets very long, rich yellow. Flowers occasionally separate, usually both male and female in same catkin. Susceptible to scab and canker. 45ft (13.5m) in 20 years. 192

SAMBUCUS
Hardy deciduous trees or shrubs. Compound leaves with 3–11 toothed leaflets. White flowers in clusters. Fruits are small round berries. Most soils and situations. Subject to leaf spot. 354
S. caerulea see *S. nigra* subsp. *caerulea*
S. nigra Common elder. Europe, N Africa, W Asia. Zone 5. Large shrub or small tree, to 35ft (10.5m). Flowers yellowish or dull white with heavy odour, June, followed by heavy bunches of shining black fruits. Fissured, rugged bark. Chalk-tolerant. 354
S.n. 'Albovariegata' A brightly variegated, vigorous, and attractive form. 354
S.n. 'Aurea' Golden foliage. 354, 354–5
S.n. 'Black Lace' Finely dissected, dark purple-black foliage. Pinkish flowers 354
S.n. subsp. *caerulea* (syn. *S. caerulea*) Blueberry elder. W North America. Berries with dense blue waxy bloom. 354, 355
S.n 'Laciniata' Very finely dissected, dull green leaves give an unusual parsley-like texture. 354

Santa Cruz Cypress see CUPRESSUS goveniana var. abramsiana
Santa Lucia Fir see ABIES bracteata
Sargent Crab see MALUS sargentii
Sargent Spruce see PICEA brachytyla
Sargent's Cherry see PRUNUS sargentii

SASSAFRAS albidum Sassafras. E United States. Zone 5. To 125ft (38m), aromatic. Very various, ovallish or lobed leaves, medium length, bright green above, bloomed beneath, deciduous. Unisexual flowers on separate trees. April/May; males insignificant, females small, yellow, clustered, before leaves. Oval fruit, fleshy, ⅓in (8mm) long, bluish-black. Sept. Bark deeply furrowed. Warm, loamy, lime-free soil. 25ft (7.5m) in 20 years. 174, 176–7, 176

Sawtooth Oak see QUERCUS acutissima
SAXEGOTHAEA conspicua Prince Albert's Yew. Chile. Zone 8. Often bushy, possibly a slender, conical, slightly bent tree, evergreen with hanging branches and shoots. Leaves sparse, irregular, characteristically curved with white bands beneath. Male flowers paired. Cones on same tree, bright powdery blue-grey, short. Bark purple-brown, smooth, scales falling to leave red-brown

areas. 15ft (4.5m) in 20 years. 128, 128

Scarlet Oak see QUERCUS coccinea

SCHIMA Genus of Asian evergreen trees related to *Camellia*. Flowers white with mass of yellow stamens. 329

x *SCHIMLINIA* Hybrids between *Schima* and *Franklinia*, good foliage, white flowers. For areas with hot summers. 329
SCHINUS molle California pepper tree. South America. Zone 9. Small evergreen tree, to 30ft (9m). Drooping branches. Much-divided compound leaves. Female flowers small, yellowish-white. Fruit, clusters of pea-sized red berries. 312–13, 313

SCIADOPITYS verticillata Japanese umbrella pine. Japan. Zone 6. Many-stemmed, broadly conical, or single-stemmed, irregular, narrowly conical, evergreen. Leaves in characteristic united pairs, medium length, glossy green above, yellow-green beneath, arranged in whorls. Male flowers in terminal clusters on small shoots; cones green, ripening to brown, scaly. Bark dark red-brown, peeling. Prefers lime-free soils. 15–25ft (4.5–7.5m) in 20 years. 86, 87

Scotch Laburnum see LABURNUM alpinum
Scots Pine see PINUS sylvestris
Scrub Pine see PINUS virginiana
Sea Buckthorn see HIPPOPHAE rhamnoides
Senegal Date Palm see PHOENIX reclinata

SEQUOIA
S. sempervirens Coast redwood. California, Oregon. Zone 8. Up to 350ft (107m), narrow, irregularly conical. Leaves small, arranged in 2 ranks, longer in middle of shoot. "Flowers", sexes on same tree, solitary; male "flowers" clustered at shoot tips, with spirally arranged filaments; female "flowers" consist of 15–20 pointed scales. Cones small, red-brown, mature after a season. Bark deeply furrowed, to 1ft (30cm) thick. Tolerant of most soil conditions, but not of exposure. World's tallest tree. 45–60ft (13.5–18m) in 20 years. 46, 47, 54, 150, 153, 154–6, 154–5, 156
S.s. 'Adpressa' Shoot tips cream-coloured. Leaves shorter than the type. Often grown as a dwarf but reverts to normal size.

SEQUOIADENDRON
S. giganteum Giant sequoia, Wellingtonia. California. Zone 7. 250–280ft (76–86m). Shorter than coast redwood or douglas fir, but far wider. Evergreen. Leaves short, blue-green, turning brown after 2–3 seasons. Triangular. Cone to 3½in (9cm) long, maturing in 2 years, usually closed for 20 years. Bark to 2ft (60cm) thick on old trunks, cinnamon-red, fissured. Immensely long-lived. Distinct from

396 Index of Tree Species

red, conspicuous in winter. Leaves unequally heart-shaped, densely downy beneath. Flowers June, early July. 290
T. tomentosa Silver linden, Silver lime. S E and E C Europe. Zone 6. To 100ft (30m), pyramidal. Leaves ovallish, simple, vividly silver-downy beneath. Flowers inconspicuous, narcotic to bees, summer. Fruit oval, pointed, downy, warty. 30ft (9m) in 20 years. 290–2, *293*

TOONA
Deciduous or evergreen trees. Leaves alternate, usually pinnate. Flowers small, whitish or greenish, bisexual, in big clusters. Fruit, a capsule holding many winged seeds.
T. sinensis Chinese toon. N and W China. Zone 5. To 70ft (21.5m), rounded and dense. Leaves pinnate, to 2ft (60cm) long, 5–12 pairs of leaflets, open red, yellow in autumn, deciduous, taste of onion. Flower clusters 2ft (60cm) long, fragrant. June. Any soil 315, *315*

TORMINARIA terminalis see *SORBUS torminalis*

Torrey Pine see *PINUS torreyana*

TORREYA
Trees or shrubs with whorled branches. Leaves narrow, spiny tips, pungent. Sexes on same or different trees. Fruit plum-like, thin fleshy coat. Tolerates shade and chalk.
126
T. californica California nutmeg. California. Zone 7. 50–70ft (15–21.5m). Leaves commonly 2-ranked, tapering, dark yellow-green above, two white bands and bright yellow margin below. Strobili, sexes separate on same tree, males solitary, females paired. Fruits fleshy, single, large seed. Bark red-brown, finely latticed. 18ft (5.5m) in 20 years. 126, *129*
T. nucifera Kaya. Japan. Zone 7. 20–80ft (6–24.5m), slender tree or large bush. Leaves narrow, pointed, stiff, curved, pungent when rubbed, shiny, dark green above. Fruit green, tinted purple, edible. Bark reddish, smooth. Likes chalk. Shade tolerant. *52*, 126

Totara see *PODOCARPUS totara*

TRACHYCARPUS
T. fortunei Chinese windmill palm. Asia. Zone 7. To 40ft (12m), trunk covered by mat of fibres. Palmate leaves 3ft (1m) across, dark green above, waxy bloom below. Unisexual flowers on same tree, yellow, fragrant. Fruit small, ripens blue. 160–2
T. wagnerianus Japanese windmill palm. Japan. Zone 7. Form of *T. fortunei* with shorter, stiffer leaflets, giving a much neater appearance. *161*, 162, *163*

Tree of Heaven see *AILANTHUS altissima*
Trident Maple see *ACER buergerianum*

TROCHODENDRON aralioides
184

TSUGA
Hemlocks, Hemlock spruces, Hemlock firs. Evergreen trees, broadly conical. Leaves needle-like, short, usually in 2 ranks. Male and female strobili separate on same tree. Cones small, pendulous, leathery scaled. Seeds winged. Shade tolerant. Virtually disease-free. 110–11
T. canadensis Eastern hemlock, Canadian hemlock. E North America. Zone 4. 60–100ft (18–30m), often many-stemmed, broadly conical. Bark brown to black, deeply furrowed when old. Any soil. 30ft (9m) in 20 years. 110, 111
T.c. 'Pendula' Irregularly dome-shaped, flat-topped bush with branches drooping to ground. Slow. 110, *111*
T. caroliniana Carolina hemlock. S E United States. Zone 6. To 75ft (23m), irregularly conical. Leaves sparse, slightly larger than *T. canadensis*. Bark dark red-brown with large yellow pores, becoming purple-grey, fissured. Dislikes chalk. 20ft (6m) in 20 years. 111
T. chinensis Chinese hemlock. China, Taiwan. Zone 6. Resembling *T. heterophylla* but not quite so large, though potentially to 160ft (50m). A hardy tree apparently resistant to adelgids in E North America, lime tolerant. 111
T. diversifolia Northern Japanese hemlock. Japan. Zone 5. To 45ft (13.5m), usually with many stems, forming a low-domed crown. Leaves regular, oblong with broad, rounded and notched tip, deep glossy green above, two white bands beneath. Male "flowers" bright rusty red. Cones oval, shiny brown. Bark orange-brown, pink fissures. Dislikes chalk. 15ft (4.5m) in 20 years. 111
T. heterophylla Western hemlock. W North America. Zone 6. 200–270ft (60–82m), largest of the hemlocks; regularly broad-conical with nodding leading shoot. Leaves flattened, dark shiny green. Cones long, light brown, numerous. Bark thin, narrowly fissured, russet-brown. Any soil, but growth soon poor on chalk. Likes moisture, shelter; tolerates shade. 50–65ft (15–20m) in 20 years. 44, 104, 110, 111
T. mertensiana Mountain hemlock, Black hemlock. W North America. Zone 4. 75–100ft (23–30m), columnar to conical, slightly pendulous. Leaves thick, grey-green, all round shoot, pointing forward, distinctive amongst hemlocks. Cone 2in (5cm) long, fawn-pink, becoming dark brown then nearly black and spruce-like. Tolerates drought, lime. 25ft (7.5m) in 20 years. 110–11, *111*
T. sieboldii Southern Japanese hemlock. S Japan. Zone 6. To 100ft (30m), possibly bushy. Leaves grooved above, shiny, dark green, white bands beneath. Cones oval, shiny. 111

Tulip tree see *LIRIODENDRON*
Tupelo see *NYSSA sylvatica*
Turkey Oak see *QUERCUS cerris*
Turkish Hazel see *CORYLUS colurna*
Turner's Oak see *QUERCUS turneri*

U

Ugli see *CITRUS* x *tangelo* 'Ugli'

ULMUS
Big deciduous trees. Leaves simple, edges serrated, sometimes doubly so. Flowers tiny, in dense clusters, petalless, bisexual. Fruit, small nut with disk-like papery wing. Hardy. Best in full sun. Fast-growing. Tolerates exposure. Any fertile, moist soil. Subject to Dutch elm disease. *63*, 242–7
U. 'Accolade' (syn. *U.* 'Morton'). A seedling of *U. davidiana* var. japonica with excellent resistance to Dutch elm disease. The original is a very shapely tree at the Morton Arboretum, Chicago. 247
U. americana American elm, White elm. E and C United States. Zone 3. To 130ft (40m), upright. Forming wide fountain-like crown with pendulous branchlets. Leaves ovallish, hairy, upper surface becoming smoother, short-stalked. Flowers green-white, hairy nuts. Bark ash-grey. 30ft (9m) in 20 years. 242, 244, *244–5*, *247*
U.a. 'Princeton' Clone with good resistance to Dutch elm disease. 247
U.a. 'Valley Forge' Also shows good resistance. 247
U. carpinifolia Smooth-leaved elm. Europe. Zone 5. One of the native elms of Europe, also known as *U. minor*, forming large handsome trees before Dutch elm disease. *247*
U. davidiana var. *japonica* Japanese elm. N E Asia, Japan. Zone 5. Large tree to 115ft (35m) with handsome shape. Leaves large, coarsely toothed, rather rough to touch. Valuable as a parent of disease resistant hybrids. *See also U.* 'Accolade', above. 247
U. 'Dodoens' (Himalayan *U. wallichiana* and European *U. glabra*). Has some disease resistance, but superseded. *245*
U. glabra Wych elm, Scotch elm. Europe, N and W Asia. Zone 5. To 130ft (40m), widespreading. Leaves ovallish, undersurface downy, upper surface rough, dull green, short-stalked. Flowers greenish, early spring. Fruits downy at tip. Susceptible to elm leaf miner. 35ft (10.5m) in 20 years. 242, 244, 247, *247*
U.g. 'Camperdownii' Camperdown elm. Weeping form of *U. glabra* with mushroom-shaped head. 244, *247*
U.g. 'Pendula' Tabletop elm, Weeping wych elm. To 30ft (9m), flat-topped, weeping. Otherwise as *U. glabra*. 244
U. x *hollandica* Dutch elm. Zone 5. To 120ft (36m), spreading. Leaves ovallish, upper surface smooth, dark green, lower surface bright green, downy. Flowers as *U. glabra*. Fruit ovallish. 40ft (12m) in 20 years. 244, 247
U. x *h.* 'Belgica' Belgian elm. Netherlands. Broad-crowned tree, straight stems. Leaves oblong to oval. Rough bark. 244
U. x *h.* 'Columella' Hybrid with narrow outline and excellent disease resistance. *247*
U. x *h.* 'Major' Dutch elm. To 120ft (36m), short-trunked, widespreading tree. Leaves oval, dark green above, slightly hairy beneath. 244
U. 'Lutèce' Hybrid incorporating genes from *U. wallichiana* and *U. glabra*, proving the most resistant of hybrids in Europe to date, potentially a magnificent large tree. 247
U. minor Small-leaved elm. Europe, N Africa. Zone 5. To 110ft (33m), spreading, branchlets drooping. Leaves narrow, ovallish, smooth, underside downy, shiny green above, short-stalked. Fruits ovallish, smooth. 244
U.m. 'Cornubiensis' Cornish elm. Narrowly conical, to 100ft (30m), from SW England. *247*
U.m. subsp. *sarniensis* Jersey elm, Wheatley elm. Zone 5. To 100ft (30m), conical. Leaves oval, small. Fruits smooth. 30ft (9m) in 20 years. *246*
U. 'Morton' see *U.* 'Accolade'
U. parvifolia Chinese elm. N and C China, Korea, Formosa, Japan. Zone 5. To 80ft 24.5m), spreading. Leaves very small, oval, pointed, upper surface bright green, lower surface paler, short-stalked. Flowers green-white, Aug. and Sept. Fruit smooth. Bark flaking, mottled. Resistant to disease. 18ft (5.5m) in 20 years. 244, 247, *247*
U. procera English elm. British Isles. Zone 6. To 150ft (45m). Narrow-waisted above broad skirts and below stiff fan-shaped crown. Leaves rounded, rough textured, dark green, paler, downy beneath. Flowers greenish, spring. Seeds infertile, spreads extensively by suckers. Fruit round, smooth. 35ft (10.5m) in 20 years. 242, 243, 244, *245*
U. pumila Dwarf elm, Siberian elm. N Asia. Zone 3. To 40ft (12m), small tree or bush. Leaves broadly lance-shaped, thin, smooth. Flowers, spring. Fruit smooth. 25–35ft (7.5–10.5m) in 20 years. 244, 247
U. rubra Slippery elm, Red elm. C and E United States. Zone 3. To 70ft (21.5m), spreading. Leaves ovallish, lower side downy, upper surface rough, short-stalked. Fruit red-brown, ovallish, hairy at top. Bark slippery. Relatively disease-resistant. 15ft (4.5m) in 20 years. 244
U. 'Sapporo Autumn Gold' (*U. pumila* and *U. davidiana* var. *japonica*). Zone 4. Makes a large rounded tree with good disease resistance. Seldom lives up to its name. 247
U. thomasii Rock elm. E North America. Zone 2. Tall tree with narrowly rounded crown, to 100ft (30m). Seldom cultivated and not disease resistant. 244
U. wallichiana Himalayan elm. Himalaya. Zone 5. Important parent of new hybrids as it is disease resistant, a large tree with leaves up to 8in (20cm) long. 247

UMBELLULARIA californica
Oregon myrtle, California laurel. California, Oregon. Zone 8. To 80ft (24.5m), spreading. Leaves alternate, narrowly oval or oblong, tapered both ends, simple, evergreen, leathery, glossy above, paler beneath, dark green, pungently fragrant. Bisexual flowers, yellowish-green, spring. Fruit roundish, pear-shaped, green changing to purple. Sept.–Oct. Scaly bark. Occasionally injured by severe frost. 20ft (6m) in 20 years. 174–6, *177*

Umbrella Pine see *PINUS pinea*; *SCIADOPITYS*
Ussurian Pear see *PYRUS ussuriensis*

V

Valley Oak see *QUERCUS lobata*
Van Volxem's Maple see *ACER velutinum* var. *vanvolxemii*
Veitch Fir see *ABIES veitchii*
Velvet Ash see *FRAXINUS velutina*

VIBURNUM
Deciduous or evergreen trees and shrubs. Most deciduous species colour well in autumn. Flowers usually white, fragrant, clustered. Fruits brightly coloured, fleshy, with single flat stone. Easily cultivated. 354
V. awabuki Eastern Asia, Japan. Zone 7. Large shrub or small tree, to at least 16ft (5m). Large glossy evergreen leaves support flat heads of small fragrant flowers. 354
V. lantana Wayfaring tree. C and S Europe, W Asia, N Africa. Zone 3. Upright shrub or small tree to 15ft (4.5m). Leaves broad oval, toothed, with a whitish down underneath; turn crimson in autumn. Flowers creamy-white. May and June. Fruit oblong, red turning black, July to Sept. Thrives on chalk. 354, *355*
V. lentago Sheepberry, nannyberry. E North America. Zone 2. Deciduous large shrub or small tree, occasionally to 30ft (9m). Glossy green leaves turn good red in autumn. Flowers in flat heads followed by waxy black berries. 354, *355*
V. opulus Guelder rose. Europe, N Africa, N Asia. Zone 3. Shrub to 12ft (3.5m) , spreading. Leaves maple-like. Flowers white, in flat heads. June and July. Fruits small, round, red, Aug. and Sept., persist through winter. Thrives in wet areas. *355*
V.o. 'Roseum' The snowball bush, round heads of sterile florets turn from green to white. 354
V. prunifolium Black haw. E North America. Zone 3. Resembles *V. lentago* but leaves smaller, less sharply toothed. 354

Vine Maple see *ACER circinatum*

W

Wafer Ash *see PTELEA trifoliata*
Walnut *see JUGLANS*
Washington Thorn *see CRATAEGUS phaenopyrum*

WASHINGTONIA
W. filifera Petticoat palm. California. Zones 8–9. To 50ft (15m); trunk to 3ft (1m) across, usually covered with dead hanging leaves. Palmate leaves 6ft (1.8m) across on 6ft (1.8m) long toothed stalk. White flowers on 9–12ft (2.7–3.5m) long stalk. Oval fruit ½in (1cm) long. 162, *163*
W. robusta Mexican washingtonia. Zone 10. The feather-duster palm of S California: very tall with comparatively slender trunks bearing leaves over 3ft (1m) across. Needs heat with ample water to thrive. 163

Water Larch *see METASEQUOIA glyptostroboides*
Water Oak *see QUERCUS nigra*
Wattle *see ACACIA*
Wayfaring Tree *see VIBURNUM lantana*
Weeping Ash *see FRAXINUS excelsior* 'Pendula'
Weeping Spruce *see PICEA breweriana*
Wellingtonia *see SEQUOIADENDRON giganteum*
West Indian Cedar *see CEDRELA odorata*
Western Hemlock *see TSUGA heterophylla*
Western Juniper *see JUNIPERUS occidentalis*
Western Larch *see LARIX occidentalis*
Western Red Cedar *see THUJA plicata*
Western White Pine *see PINUS monticola*
Westfelton Yew *see TAXUS baccata* 'Dovastoniana'
Weymouth Pine *see PINUS strobus*
White Ash *see FRAXINUS americana*
White Cedar *see THUJA occidentalis*
White Cypress *see CHAMAECYPARIS thyoides*
White Oak *see QUERCUS alba*
White Pine *see DACRYCARPUS dacrydioides*
White Poplar *see POPULUS alba*
White Spruce *see PICEA glauca*
Whitebeam *see SORBUS*
Whiteleaf Magnolia *see MAGNOLIA obovata*
Willowleaf Pear *see PYRUS salicifolia*
Wild Cherry *see PRUNUS avium*
Wild Red Cherry *see PRUNUS pensylvanica*
Willow *see SALIX*
Willow Oak *see QUERCUS phellos*
Willow-leaf Podocarp *see PODOCARPUS salignus*
Windmill Palm *see TRACHYCARPUS fortunei*
Wingnut *see PTEROCARYA*
Witch Hazel *see HAMAMELIS*

WOLLEMIA nobilis Wollemi pine. S E Australia. Evergreen conifer to 130ft (40m) with brown pimply bark. Leaves in 4 ranks, lance-shaped, to 1½in (4cm) long, mid-green. Male and female cones borne at tips of branches. First discovered in 1994, the only representative of its genus, a member of the monkey puzzle family. 124–5, *124, 125*

Wych Elm *see ULMUS glabra*

Y

Yellow Birch *see BETULA lutea*
Yellow Chestnut Oak *see QUERCUS muehlenbergii*
Yellowwood *see CLADRASTIS lutea; PODOCARPUS latifolius*
Yew *see TAXUS*
Yoshino Cherry *see PRUNUS x yedoensis*
Young's Weeping Birch *see BETULA pendula* 'Youngii'
Yulan *see MAGNOLIA denudata*

Z

ZELKOVA
Deciduous trees or shrubs with smooth trunks, related to elms. Leaves like those of the elm but singly toothed; alternate, coarsely toothed. Sexes separate on same tree, small, green; males clustered, females solitary, spring. Fruits small, roundish, autumn. May catch Dutch elm disease 62, 248
Z. carpinifolia Elm zelkova. Caucasus. Zone 5. To 80ft (24.5m), often with many upright branches from short trunk. Leaves oval, simple, hairy beneath, slightly hairy above, dark green, short-stalked. Flowers green; males at base of twigs, females on tip. Fruits ridged above. Bark smooth at first, flaking later. Long-lived. Shade tolerant. 30ft (9m) in 20 years. 248, *248–9*
Z. serrata Japanese zelkova. Japan, Korea, China. Zone 5. To 100ft (30m), upright. Trunk short, branches long and spreading. Leaves ovallish, simple, dark green, slightly hairy above, paler beneath, yellow-brown in autumn, short-stalked. Flowers similar to *Z. carpinifolia*, spring. Fruits woody, pea-like. Bark smooth, grey, becoming flaky. Shade-tolerant; young trees susceptible to frost. 25ft (7.5m) in 20 years. 248, *248*

Index

Photographic acknowledgements

akg-images 56 l

Alamy Adrian Davies 336 br; Alex Ramsay 64, 137 b; Andrea Jones 214, 337 b; Arco Images GmbH 79 a, 104 br, 107 a, /H Reinhard 251 lac, /C Huetter 169 br, /K Irlmeier 252 r; Bill Bachman 194–5; BL Images Ltd 200–1; blickwinkel 78 b, 104 al, 110 l, 125 br, 129 a, /Baesemann 264–5, /Carrasco 226–7, /Hecker 297 al, /Jagel 215 bc, 247 b, 252 l, 275 ar, 288 r, 319 l; Bloom Pictures 141; Bob Gibbons 94 a, 275 rac; botanikfoto/Steffen Hauser 209 ar, 314–5; Classic Image 48 bcl; Clint Farlinger 150–1; Corey Hilz/Danita Delimont 181 br; CuboImages srl 106–7, 111 l, /ParoliGalperti 185 br, 335 ar; D Hurst 87 bc; Dan Perren 108 l; Dave Bevan 25 c; David Bartlett 209 al; David Robertson 279 bcr; Design Pics Inc – RM Content 337 a; Edward Parker 129 l, 132 al; Erin Paul Donovan 271 br; Ern Mainka 186–7; Fernando Quevedo de Oliveira 35 a; Fernando Zabala 122–3; Flowerphotos/Carol Sharp 220 b; Frank Blackburn 25 a; Garden Picture Library/Howard Rice 318 bcr; Hemis 42 l; Ian Grant 134; Ian Vdovin 114 a; imagebroker 6, 116–7; Interfoto 344 l; Jack Sullivan 73 a; Jeremy Pardoe 296 ar; John Glover 188 r, 307 bl; John Kershaw 100 br; Joseph Becker 101; Julie Pigula 256 br; katewarn images 205 c; Krystyna Szulecka 279 al & cl; LusoStock 176 l; Martin Hughes-Jones 189 ar, 224 r, 331 b; Mary Evans Picture Library 48 bl; Mary Liz Austin 203 ar; National Geographic Image Collection/Taylor S Kennedy 102–3; Oliver Smart 162–3; Organica 95 b, 109 l, 140 l; Paul Bradforth 256 bl; Penny Tweedie 283 ar; Peter Arnold Inc 25 al & cl; Peter Titmuss 211 bc; Pictorial Press Ltd 61 bcl; Plantography 307 ar; Richard Becker 24–5; Robert Harding Picture Library Ltd/Adam Woolfitt 19 br, /David Poole 336–7; Roger Eritja 310 br; Rolf Hicker Photography 148; Ross Frid 81 a; RWP 320–1; shapencolour 251 al, 289; Stephen Peter Street 238–9; Susan A Roth 325 br; Suzanne Long 125 al; Terry Whittaker 253 l; The Print Collector 48 br; Tim Scrivener 75 b; Urbanmyth 86 a; V&A Images 242 a; Vario Images GmbH & Co 121 a; Wildlife GmbH 235 bl; William Leaman 25 ac; Wiskerke 216

Andrew Lawson 9, 81 b, 82, 86 b, 87 a, 101 r, 105 b, 107 b, 108 r, 113, 121 b, 124 br & cb, 127, 132 b & c, 139 a & b, 143, 145 l & r, 149, 152 bl, 153 al & ar, 156 al, 161 l, 163 br, 168 l, 170, 178, 180 r, 181 bl, 193 l, c & r, 197 ac, 198, 199 l & r, 203 al, 205 b, 206, 209 cl, 210–11, 211 ar, 232 a, bc & bl, 240 l, 241 ar, 250 c, 254–5, 267, 276–7, 279 ar, bcl & ccr, 281 br, 285 al & cr, 288 l, 290, 291, 292–3, 296 b, 297 ar, 303, 306 r, 308–9, 310 al & bl, 312, 313 ar, 316–7, 323 b, 324, 326–7, 330 al & bl, 339 b, 341 a & bl, 345 r, 346, 348 l, 350–1, 353 bl & br; Arley Hall, Cheshire 266–7, Chatsworth House 257 al, Rofford Manor, Oxford 250 l, Rousham House, Oxon 280, Sezincote, Glos 323 a, The chestnut avenue at Windsor Great Park, by gracious permission of Her Majesty the Queen 294-5, Vann, Surrey 233 c

Arnold Aboretum © President and Fellows of Harvard College. Arnold Arboretum

Archives 57 r, 58 a

Courtesy of **Bleddyn & Sue Wynn-Jones** 61 br

Bob Gibbons 88 l, 89 a, 93, 98, 100 bl, 111 r, 115 a, 118–9, 128 r, 257 bl, 281 a

BotanikFoto Steffen Hauser 125 cr

Bridgeman Art Library Ashmolean Museum, University of Oxford 49 l; Christie's Images 233 a; Philip Mould Ltd, London 62; Ray Miles Fine Paintings 49 bcr; The Stapleton Collection 49 br, 50 al, 52–3 a, 62

Clive Nichols 172 br, 212–3

Colin Roberts 2–3

Corbis David Muench 271 al; Kennan Ward 318 bl; amanaimages/Tomonari Tsuji 218; Charles Mauzy 103 b; Clive Nichols/Arcaid 268; Craig Lovell 304–5; Darrell Gulin 268–9, 271 ar; Douglas Peebles 124 ca; epa 124 a; Gary Braasch 275 b; imagesource 76–7; Kathy Coatney/AgStock Images 286–7; Ken Wilson/Papilio 246 ar; Michael Nicholson 31 c; Owen Franken 191 br; Photolibrary 188 l; Steve Austin/Papilio 120; Steve Terrill 110 r

Courtesy of **Dan Hinkley**, photo Lynne Harrison 61 bcr

FLPA Bob Gibbons 313 al; David Hosking 232 br, 246 al; ImageBroker 251 bcl; Keith Rushforth 99 r, 275 bcr; Martin B Withers 253 br; Maurice Nimmo 245 a; Roger Tidman 251 br; Wil Meinderts/Minden Pictures 197 bl

Fotolia pages 32–33, left to right: Noah Strycker; donkey IA; lubashi; Crimson; lamax; Michael Palis; Olga Vasik; Adrian Hillman (a) and Dmitriy Melnikov (b); DrVIB; robynmac; nyasha (l) and Piotr Marcinski (r); thepoeticimage (a) and tfazevedo (b); Edyta Pawlowska; Marek Kosmal; kmit (a) and Joanna Redesiuk (b); Marina Lohrbach; 50–51, left to right: Liaurinko; Hadi Yuswanto Djunaedi; jStock; RRF; Can Balcioglu; Graham Lumsden; Starjuper; Sascha Burkard; Jonny A; Karen Hadley; Benjamin Christie; Syvana Rega; dabjola 17 bcl; Farmer 17 bl; Jean-marc Richard 17 bc; Joseph Helfenberger 16–17; Leon Forado 177 b; Nataly Korosteleva 23 ar; Patrick Hardy 23 bl; RRF 167; Serjej Razbodowski 17 bcr

Foxhollow Garden Stephen Grubb 159 b

GAP Photos Adrian Bloom 159 a; Charles Hawes 279 acr; Christina Bollen 219, 349 bl; Clive Nichols/Pettifers Garden, Oxfordshire 237; Dianna Jazwinksi 248 a; FhF Greenmedia 169 bl; Fiona Lea 335 ac; Fiona McLeod 229 b; Friedrich Strauss 328 br; Graham Strong 99 al; Howard Rice 293 br; J S Sira 228 bl, 256 bc, 349 al; Jo Whitworth 172–3, 172 ar; John Glover 279 acl; Jonathan Buckley 215 bl; Jonathan Need 279 bl; Lee Avison 19 c; Maddie Thornhill 228 al; Mark Bolton 253 ar; Martin Hughes-Jones 332 l; Maxine Adcock 239 b; Paul Debois 296 al, 313 br; Richard Bloom 87 bl, 99 b, 185 cr; Rob Whitworth 233 bl; Ron Evans 259 br;

S & O Mathews 129 cr, 183, 285 ar; Tim Gainey 225 bl

Garden Collection Jonathan Buckley 185 bl; Torie Chugg 136 r, 225 br

Garden World Images Dave Bevan 72 b; Flowerphotos/Gillian Plummer 318 br; Gilles Delacroix 353 bl, 355 br; Jenny Lilly 185 al, 329 a; John Martin 212; John Swithinbank 136 l, 240–1; Leonie Lambert 129 br; MAP/Arnaud Descat 137 al, 168 r; MAP/Frédéric Didillon 25 br, 269 ar; MAP/Jean-Yves Grospas 24–25 b; MAP/Nicole et Patrick Mioulane 228 bc; MAP/Noun 222–3; Martin Hughes-Jones 79 ccl, 281 bl; Nicholas Appleby 265 r; Paul Lane 173; Philip Smith 166–7; Trevor Sims 140 r, 319 c

Getty Images 33 al, 57 l, 49 bcl; Adam Jones 271 cr; Amy White & Al Petteway 241 b; De Agostini 248; Don Farrall 285 br; FhF Greenmedia 236 a; GAP Photos/Richard Bloom 109 right, /S & O Mathews 235 br; Harald Sund 100 a; Imagemore Co, Ltd 191 bl; Jack Dykinga 142; James Randklev 154–5; Jeff Foott 80–1, 227, 270 br; Ned Therrien 105 al; Panoramic Images 104–5; Richard Bloom 269 br; Ron and Patty Thomas 298–9; Science and Society 33 ar; Tom Mareschal 318 bcl; Vincenzo Lombardo 313 bl

Harpur Garden Images Jerry Harpur 206

Iseli Nursery, Inc Randall C Smith 158

JC Raulston Arboretum at NC State University 61 a

Dr John Grimshaw 79 b, 125 ar, 152 br, 184, 245 b, 283 br, 311 b, 322, 347

Mary Evans Picture Library 48 bcr

National Geographic Stock Michael Nichols 26 r

Nature Picture Library Daniel Gomez 146 a; George McCarthy 196; Juan Carlos Munoz 197 al; Konstantin Mikhailov 319 br; Niall Benvie 339 a; Nigel Bean 251 bl; Philippe Clement 197 ar

Octopus Publishing Group Adrian Pope 48 a, 59, 61 bl, 63 r, 67 b

By kind permission of **Gravetye Manor Hotel & Restaurant**, West Sussex 60 r

Photolibrary Group age fotostock 90 b, /Juan Carlos Munoz 46, 96–7, /Frank Krahmer 311 a, /Henryk T Kaiser 300–1, /Jose Fuste Raga 66, /Martin Siepmann 247 a, /Pablo Galan Cela 275 al, 355 cl, /Tim Isaak 335 al; All Canada Photos Michael Wheatley 332–3; Animals Animals/Breck P Kent 79 c, Jack Wilburn 104 ar; Botanica/Acevedo Melanie 171 br; Britain on View/David Clapp 260–1; Jon Sparks 34; Cuboimages/ Paroli Galperti 228 ac, 328 a, 354–5; De Agostini/C Sappa 91 l, 115 b, /P Jaccod 144 r; First Light/Irwin Barrett 25 cr; Flowerphotos/Carol Sharp 189 al; Garden Photo World/David C Phillips 215 br, 334 al, /Georgianna Lane 140 c, 171 bl, 270 ar; Garden Picture Library/Andrea Jones 58 b, /Brian Carter

23 ac, /Carole Drake 74–5 a, 209 cr, 297 br, /Chris L Jones 137 r, 23 bc, /Christopher Gallagher 90 a, /Eric Crichton 92 r, /Friedrich Strauss 22, John Carey 124 bl, /John Glover 172 al, 302 c, 329b, /Lucy Barden 353 ar, /Mark Bolton 205 a, 275 bcl, 328 bl, 353 al, 355 a, /Mark Turner 176 r, 334 bl, /Martin Page 23 br, 89 b, /Pernilla Bergdahl 301 b, /Photos Lamontagne 207 a, /Ron Evans 228–9, /Stephen Shepherd 87 br, /Victoria Gardner 171 cl; Imagebroker/Arco Images/Daniel Meissner 174–5, /Gerhard Zwerger-Schoner 157, /Horst Sollinger 28, 35 b, /Rudi Sebastian 178–9, /Siepmann 174–5, 342–3; imagesource 72 a; Imagestate/Peter Thompson 257 ar; Japan Travel Bureau 78–9 a, 84–5; John Warburton-Lee 130–1; Nordic Photos/ Ingemar Aourell 224 l; OSF/Bob Gibbons 147, /Konrad Wothe 230–1, /Martyn Chillmaid 338–9, /Tony Howell 10 l; Pacific Stock/Vaughn Greg 302 b; Peter Arnold Images/Ed Reschke 23 al, 152 c, /Manfred Kage 20 bl, /Fritz Polking 103 a; Photodisc 26 l; Photolibrary/Ted Mead 188c; Phototake Science/Dennis Kunkel 20 br & 21bc; Robert Harding Travel/Rob Rainford 88–9; Ron Chapple Stock 156 r; Tips Italia/Focus Database 20–1; White/ Peter Lilja 13

Photoshot JTB 169 a, NHPA 10–11, Photos Horticultural 344–5

Photo Scala, Florence 190 b

Picture Desk Art Archive 49 a

Courtesy of **Practicality Brown**, the specialist tree mover with 30 years' experience 70

Press Association Images 33 ac

Royal Geographical Society Picture Library 56 c

Saling Hall Press 42–3, 44–5, 68–9, 74b, 83 l & r, 91 r, 92 l, 94 b, 95 a, 105 ar, 112, 114 b, 126–7, 133, 135, 138, 144 l, 146 b, cl & cr, 160 b, 171 al, 177 al & ar, 180 l, 181 a, 185 ar & cl, 187, 189 bl & br, 190–1, 192 l & r, 197 br, 204–5, 208, 215 a, 220 a, 221, 229 a, 234–5, 236 b, 238, 239 a, 240 b, 242 b, 243, 244, 246 br, 249, 250 r, 256a, 258 al & bl, 258–9, 279 br & cr, 282–3, 293 bc & bl, 297 bl, 301 a, 302 a, 306 l, 307 al, 310 ar, 315 b, 318 a, 325 a, 336 bl & cr, 340, bl, 341 br, 349 br, 352

Science Photo Library Adrian Thomas 266 bl; Bjorn Svensson 345 l; Bob Gibbons 246 bl; Dr Jeremy Burgess 29 b, 132 ar; Eye of Science 19 ac; Michael P Gadomski 104 bl; Philippe Psaila 21 bl; Power and Syred 19 ar; Rod Planck 19 al; Steve Gshmeissner 21 br

Courtesy of **STIHL** www.stihl.co.uk 73 b

SuperStock Science and Society 57 c; The Irish Image Collection 350

TopFoto 31 l; English Heritage/HIP 60 l; The Granger Collection 31 r